T0310160

POROUS MEDIA TRANSPORT PHENOMENA

POROUS MEDIA TRANSPORT PHENOMENA

FARUK CIVAN

Mewbourne School of Petroleum and Geological Engineering
The University of Oklahoma
Norman, Oklahoma

A JOHN WILEY & SONS, INC., PUBLICATION

Library of Congress Cataloging-in-Publication Data:
Civan, Faruk.
  Porous media transport phenomena / Faruk Civan.
    p. cm.
  Includes index.
  ISBN 978-0-470-64995-4 (hardback)
  1. Porous materials. I. Title.
  TA418.9.P6C58 2011
  620.1'16--dc22
                                   2011006414

Printed in Singapore

eISBN 978-1-118-08643-8
oISBN 978-1-118-08681-0
ePub ISBN 978-1-118-08680-3

10 9 8 7 6 5 4 3 2 1

Dedicated to my family with love and appreciation

CONTENTS

PREFACE

Many processes in the nature and engineering applications occur in porous media and materials. The mathematical description of such processes in porous materials is usually overwhelmingly complicated. Often, a compromising analysis approach is therefore necessary between complexity of modeling and effort required for the solution of problems of practical interest. This is usually accomplished by emphasizing an averaged representative description and neglecting the details of low-order influence.

This book has been designed to provide an understanding of the fundamentals of the relevant processes. The theory and modeling of the porous material, fluid, and species behavior in porous media are reviewed. The methods for prediction, analysis, and description of the commonly encountered porous media processes are discussed. Emphasis is placed upon practical understanding and implementation with straightforward mathematical treatment.

Transport in porous media is an interesting interdisciplinary subject, which attracted many researchers from various disciplines. This book covers the relevant materials with sufficient detail but without overwhelming the readers. This book can be used for understanding and description of the porous media problems. It may serve as a useful reference and text. This book provides knowledge of the theoretical and practical aspects of porous media transport for various purposes, including conducting laboratory and actual-size tests, model-assisted interpretation of test data, and prediction and simulation of various porous media processes.

Obviously, the abundance of literature available in this area is overwhelming because of the great interdisciplinary nature and the wide variety of potential approaches and applications in porous media. Numerous references have been used in the preparation of this book. However, an encyclopedic presentation of all approaches is not attempted for the purpose of this well-focused book. Rather, I have determined and limited the coverage of the relevant materials in this book to the essential critical bottom-line information required for the analyses and description of porous media processes. I believe this book will form an important basis for readers from which they can begin to explore innovative and creative approaches in dealing with applications in porous media.

This book is intended to provide an effective and comprehensive overview of the fundamentals and the experimental and theoretical approaches present in the area of the transport phenomena in porous media. The mechanisms of processes occurring in porous media are discussed. Various approaches used in the modeling of porous media single- and multiphase transport processes, with and without the thermal, species, and particulate processes, are presented systematically. The techniques available for the analysis and modeling of these processes with well-established approaches are described.

Porous media transport phenomena is an interdisciplinary issue of concern for petroleum, chemical, environmental, geological, geothermal, civil, and mechanical engineers, geologists, physicists, materials scientists, and mathematicians; it considers the simultaneous mass, momentum, and energy transport issues associated with porous matrices, pore fluids, and species; it includes a model-assisted analysis of experimental data and parameter determination issues; and incorporates the recent developments in porous media process analysis and modeling techniques.

This book is written in a concise and practical manner. Emphasis is placed upon the practical understanding, formulation, and application of the relevant processes, mechanisms, and methods without overwhelming with unnecessary details and mathematical complexities. The philosophy and theoretical and technical principles involved in the measurement and experimental techniques are explained. It presents an introductory background at the beginning of the chapters, followed by the relevant theoretical development, application problems for illustration and demonstration purposes, instructive questions and exercise problems for self-practice, and many references for further reading and details.

Most of the topics covered in this book are of common interest and interdisciplinary nature. This book is intended for use by researchers, practicing engineers, professionals, and engineering students involved in porous media applications. This book can be used as a textbook for senior undergraduate and early graduate-level engineering courses. The prerequisites are thermodynamics, fluid mechanics, heat transfer, engineering mathematics, basic computer programming, and basic properties of fluids and porous materials.

This book can be used by many disciplines and professionals for instructional and reference purposes because of its interdisciplinary nature. The material presented in this book has been tested in the courses taught by Dr. Civan at the University of Oklahoma over a period of 30 years. Parts of the materials presented in this book were tested by Dr. Civan in his industry short courses presented at various institutions. The materials presented in this book originate from Dr. Civan's key papers, and relevant literatures listed in the references as well as the relevant parts taken from Dr. Civan's course notes prepared for teaching the following courses:

Graduate courses

- Fluid Flow through Porous Media
- Transport Phenomena in Porous Media
- Waterflooding
- Advanced Reservoir Engineering
- Enhanced Oil Recovery
- Process Analysis and Simulation
- Numerical Simulation by Digital Computation

Undergraduate courses

- Oil Reservoir Engineering
- Natural Gas Engineering
- Applied Reservoir Engineering

This book can be used in the following related courses in petroleum, chemical, environmental, civil, and mechanical engineering:

- Transport Phenomena in Porous Media
- Fluid Flow through Porous Media
- Mass, Momentum, and Energy Transfer in Porous Media

Writing this book was an overwhelming accomplishment, which required the effort and dedication of the author for more than 10 years. Analyzing and condensing the vast amount of information available on this subject to the bottom-line critical presentation in this book was a tedious and laborious task. Some topics were more emphasized than others in the literature. Special undertaking was necessary to tie the loose ends. My dedication and motivation in taking such a responsibility was to make my lifelong learning, teaching, development, and expertise on this topic available for the readers of the book. Like any other information available, the readers should critically examine, test, and apply the information provided in this book in view of the particular conditions and requirements of problems of their interest.

I gratefully acknowledge the following organizations and companies for granting permissions to use various materials included in this book: American Chemical Society, American Geophysical Union, American Institute of Chemical Engineers, American Institute of Physics, American Society of Civil Engineers, American Society of Mechanical Engineers, Begell House Inc. Publishers, Chemical Engineering Education, Elsevier, John Wiley & Sons, Oklahoma Academy of Science, Society of Petroleum Engineers, Society of Petrophysicists and Well Log Analysts, and Springer.

Any comments, corrections, and suggestions by the readers to improve this book are welcome. After all, the overall objective of this book is to serve the readers by providing quality information in this one source.

I am indebted to the team at John Wiley & Sons Publishing Company for their support in the preparation and realization of this book.

FARUK CIVAN, PhD
Norman, Oklahoma

ABOUT THE AUTHOR

Faruk Civan is an M.G. Miller Professor in the Mewbourne School of Petroleum and Geological Engineering at the University of Oklahoma in Norman. Previously, he worked at the Technical University of Istanbul, Turkey. He formerly held the Brian and Sandra O'Brien Presidential and Alumni Professorships. He published a book, *Reservoir Formation Damage—Fundamentals, Modeling, Assessment, and Mitigation*, from Elsevier Publishing Company, and more than 270 technical articles published in journals, edited books, handbooks, encyclopedia, and conference proceedings, and presented more than 100 invited seminars and/or lectures at various technical meetings, companies, and universities. He holds an advanced degree in engineering from the Technical University of Istanbul, Turkey, a master of science degree from the University of Texas at Austin, and a doctor of philosophy degree from the University of Oklahoma. Civan has served on numerous American Institute of Chemical Engineers and Society of Petroleum Engineers technical committees. He is a member of the American Institute of Chemical Engineers, the Society of Petroleum Engineers, and the editorial boards of several journals. Civan has received 20 honors and awards, including five distinguished lectureship awards and the 2003 SPE Distinguished Achievement Award for Petroleum Engineering Faculty. He teaches short courses on topics of practical importance concerning the various aspects of in situ energy resource development and utilization.

Faruk Civan may be contacted at the Mewbourne School of Petroleum and Geological Engineering, The University of Oklahoma, T301 Energy Center, 100 E. Boyd St., Norman, Oklahoma 73019, USA. Telephone: (405) 325-6778; fax: (405) 325-7477; e-mail: fcivan@ou.edu.

CHAPTER *1*

OVERVIEW

1.1 INTRODUCTION

Processes occurring in porous media and materials are encountered frequently in various engineering applications. The mathematical description of these processes is complicated because of the intricate flow paths, proximity of the pore wall, and mutual interactions of the fluids, particulates, and porous media. Often, a compromising analysis approach is essential in order to circumvent the complexity of modeling in view of the effort required for the solution of problems of practical interest. We can realistically accomplish this task by emphasizing an averaged description capturing the dominant features and neglecting the low order of magnitude details.

Porous Media Transport Phenomena is a comprehensive review and treatise of the fundamental concepts, theoretical background, and modeling approaches required for applications involved in mass, momentum, energy, and species transport processes occurring in porous media. This general inter- and multidisciplinary book provides a comprehensive material and background concerning the description of the behavior of fluids in porous materials.

An updated, concise, practical, convenient, and innovative treatment and presentation of the critical relevant bottom-line issues of transport in porous media is presented, from which all disciplines dealing with processes involving porous materials can benefit. Motivation, description, and executive summary of the various topics covered in this book are presented.

Description and characterization of porous media and processes occurring therein have been attempted by many different methods. The following approaches are among the outstanding methods used for this purpose (Chhabra et al., 2001):

- *Bundle of capillary tube models.* The fluid flowing through the connected pore space is assumed to follow a number of preferential capillary tortuous flow paths (Kozeny, 1927; Carman, 1938, 1956). The wall friction effect of the capillary tubes is considered as exerting resistance to flow. The effect of irregularities (interconnectivity and entanglement) involved in and the interactions (cross-flow) occurring between the fluids flowing through the various capillary flow paths are ignored (Chhabra et al., 2001; Civan, 2001). However, this issue can be alleviated by means of the leaky-tube model (Civan, 2001, 2002b,d, 2003, 2005a).

- *Pack of solid grain models.* The drag exerted against fluid flowing through a pack of solid grains of assumed shapes, such as spherical particles, in a prescribed arrangement, such as cubic packing, is considered for prediction of frictional pressure loss during flow (Chhabra et al., 2001). This method has been implemented by three main approaches: (1) correlating the experimental data to express the drag as a function of porosity (pore volume fraction) or solidity, or packing (grain volume fraction) of porous media without considering the solid particle arrangement; (2) predicting the drag by solution of the momentum equations for a prescribed arrangement of solid grains; and (3) adjustment of the drag of a single grain in a cell for intergrain interactions by simulation (Chhabra et al., 2001).

- *Averaging of microscopic field equations.* The microscopic conservation equations governing the flow of fluids in porous media are averaged over a representative elementary volume of porous media. However, averaging over time is also carried out for rapid turbulent flow of fluids.

- *Dimensionless empirical correlation methods.* Empirical models are developed based on the method of dimensional analysis and empirically obtained mathematical relationships between the relevant dimensionless groups.

- *Hybrid models.* These combine the various features of the aforementioned methods.

This book presents a concise review and treatise of the relevant developments and theoretical foundations required for understanding, investigation, and formulation of processes involving porous media. The overall objective is to provide the readers with one source to acquire the bottom-line information in a convenient and practical manner. This book is written to provide engineering students, scientists, professionals, and practicing engineers with an updated and comprehensive review of the knowledge accumulated in the literature on the understanding, mathematical treatment, and modeling of the processes involving porous media, fluid, and species interactions and transport. However, presentation is limited to the most critical information needed for applications of practical importance and further developments without overwhelming the readers with unnecessary encyclopedic details.

Fundamental theories, principles, and methods involved in the analysis and modeling of single- and multiphase fluid and species transport in porous media are covered. Special emphasis is placed on the phenomenological modeling of the processes involved in the transport of fluids, species, and particulates in porous materials, oil and gas recovery, geothermal energy recovery, and groundwater contamination and remediation. This book presents the fundamental knowledge and the recent developments in the analysis, formulation, and applications of flow through porous materials. Formulation of mass, momentum, and energy transport phenomena and rate processes; model-assisted analysis of experimental data; and modeling of processes occurring in porous media are described in a practical manner and are illustrated by various example problems. Experimental and measurement techniques used for the study of the processes in porous media and for the determination of relevant process parameters are described and demonstrated by applications.

Among the topics covered in this book are characterization of porous materials; phenomenological description of commonly encountered processes, formulation of equations for mathematical description of porous media transport, and their associated initial and/or boundary conditions; analyses of porous media processes by various approaches including constitutive relationships, dimensional analysis, control volume conservation analysis, and representative volume and time averaging; development and applications of the multiphase transport models, including the noncompositional, limited compositional, and fully compositional types; treatment of potential flow, phase transition, physical and chemical reactions, porous media deformation, particulates, heterogeneity, and anisotropy; and basic numerical simulation examples. Examples exploring and demonstrating the applications of the various formulations are presented for instructional purposes, and exercise problems are provided at the end of the chapters for further practice on the subject matter. Most problems dealing with porous media transport involve subsurface porous media, and therefore some examples given in this book relate to processes occurring in such media. The state of the knowledge is presented in plain language with equal emphasis on the various topics in a uniform format and nomenclature using the consistent International System of Units (SI). Instructive figures are presented to explain the relevant phenomena, mechanisms, and modeling approaches and results. A comprehensive list of relevant references is provided at the end of the book.

1.2   SYNOPSES OF TOPICS COVERED IN VARIOUS CHAPTERS

This book contains 11 chapters covering the most critical topics and formulation of porous media transport phenomena. Chapter 1 presents an overview and executive summaries of the various topics covered in this book.

Chapter 2 presents the transport properties of porous media. The effects of pore connectivity, the valve action of pore throats, and cementation are considered in a bundle of tortuous, leaky capillary tubes of flow for a macroscopic model of permeability of porous media. Practical straight-line plotting and parameter determination schemes are presented for convenient correlation of the porosity–permeability of porous media using a simplified macroscopic model. The permeability of porous media is correlated by means of a single continuous function over the full range of porosity using a power-law flow unit equation. The parameters of the power-law flow unit equation incorporate the fractal attributes of interconnected pore space into a bundle of tortuous, leaky hydraulic tube model of porous media. These parameters are strong functions of the coordination number of porous media and are significantly different from those of the Kozeny–Carman equation. The mathematical relationships of the power-law parameters to the coordination number are also presented. The associated analysis also lends itself to the physical interpretation of the pore connectivity and cementation factor in terms of the relationship of permeability to porosity. From a practical point of view, the primary advantage of the present macroscopic modeling approach is that it leads to a single, simple, compact, and convenient equation, which can be readily incorporated into the modeling of porous media processes without adding appreciable complexity and computational burden into large-scale field simulations. This macroscopic model is an improvement over the Kozeny–Carman equation, which has a more limited application. It is more beneficial than the microscopic pore-scale network models because it requires significantly less computational burden while providing sufficient accuracy for large field-scale applications.

Chapter 3 presents the macroscopic transport equations. First, the methodology for temporal, spatial, and double averaging of microscopic conservation equations for derivation of the porous media macroscopic conservation equations is presented and illustrated by several examples. The volume and mass-weighted volume averaging of the microscopic equation of conservation results in different macroscopic equations. Properly formulated closure schemes, such as the gradient theory, are required for formulation of terms involving the averages of the products of various quantities representing deviations from their volume-average values. It is emphasized that both time and space averaging are necessary by means of double decomposition for macroscopic description of processes involving transport through coarse porous media, where the pore fluid volume is large enough to undergo some turbulence effects. Second, the methodology for direct derivation of the porous media conservation equations by control volume analysis is presented as an alternative approach, which can provide different insights into the nature of macroscopic equations.

Chapter 4 presents the scaling and correlation of transport in porous media. Applications of dimensional and inspection analysis methods to porous media processes are demonstrated and elaborated. Their outstanding advantages and disadvantages are delineated. The benefits of using normalized variables are emphasized. Important dimensionless groups and mathematical relationships are derived and applied for several cases. Analysis and interpretation of experimental results using dimensionless groups and self-similarity transformations are presented. Strategies for effective scaling and generalization of experimental results are discussed.

Chapter 5 presents the fluid motion in porous media. The equation of motion for single-phase fluid flow through porous media is derived by various approaches, including the analysis of forces acting on fluid based on dimensional analysis and a control volume momentum balance approach. Resistive forces associated with pore surface and pore throat are characterized by the capillary orifice model of porous media. Several issues are emphasized, including the porous media averaging of the pressure and shear stress terms, the effect of porous media heterogeneity and anisotropy, source terms, correlation of parameters, flow demarcation criteria, entropy generation, viscous dissipation, generalization of Darcy's law, non-Newtonian versus Newtonian fluid rheology, and threshold pressure gradient that must be overcome for fluid to flow through porous media.

Chapter 6 presents the gas transport in tight porous media. The flow of gases through tight porous media is treated differently from liquids. Walls of tight pores in porous media interfere with the mean free motion of gas molecules and cause a strong deviation from Darcy's law. Darcy's law cannot describe the flow of gas in tight formations under the Knudsen and slip flow conditions because Darcy's law was designed to represent the viscous flow by analogy to liquid flow. A multiple-mechanism transfer model can provide an accurate description of gas flow in tight porous media. A Darcy-like equation using an apparent permeability is presented by considering the gas transport by several mechanisms, which may occur in tight pore spaces under different Knudsen number criteria. This equation incorporates the Knudsen, transition, slip, viscous, surface diffusion, and condensate flow mechanisms into the description of gas flow, each prevailing under different conditions. The characteristics of porous media are represented by fractal description and by considering the pore size distribution. Various issues involved in the proper derivation of relevant formulas based on the realization of the preferential flow paths in porous media by means of a bundle of tortuous tubes are presented. The mechanisms, characteristic parameters, and modeling of gas transport through tight porous media under various conditions are reviewed. Formulations and methodology are described for accurate and meaningful correlations of data considering the effect of the characteristic parameters of porous media, including intrinsic permeability, porosity, and tortuosity, as well as the apparent gas permeability, rarefaction coefficient, and Klinkenberg gas slippage factor.

Chapter 7 presents the coupling fluid mass and motion in porous media. The flow of fluids through porous media under isothermal conditions requires the simultaneous solution of the mass and momentum conservation equations. Hence, this chapter describes the coupling of these two equations in various applications of practical importance involving single- and multiphase fluid systems. The concept of the leaky-tank model is introduced, allowing for the determination of the essential parameters of flow occurring in porous media around the wells. Special convenient formulations are derived for fractional flow, end-point mobility, and streamline/stream tube flow descriptions. Applications of the method of superposition, images, and front tracking are described for potential flow problems. Numerical solutions of various problems are presented for instructional purposes.

Chapter 8 presents the characterization of parameters for fluid transfer. A review of the methods required for defining and determining the essential parameters affecting fluid transport through porous media is presented. Wettability and wall drag, wettability index, capillary pressure and measurement, relative permeability and measurement, temperature dependence, and interfacial drag are discussed. Application of the Arrhenius equation is reviewed for the correlation of the temperature effect on wettability-related properties of material, including the work of immiscible displacement, unfrozen water content, wettability index, and fluid saturation in porous media. The correlation with the Arrhenius equation provides useful information about the activation energy requirements associated with the imbibition and drainage processes involving the flow of immiscible fluids in porous materials. Determination of the relative permeability and capillary pressure from laboratory core tests by direct and indirect methods based on steady-state and unsteady-state core flow tests is described.

Chapter 9 presents the modeling transport in porous media. Applications of coupled mass, momentum, and energy conservation equations are discussed and presented for various problems. Transport of species through porous media by different mechanisms is described. Dispersivity and dispersion in heterogeneous and anisotropic porous media issues are reviewed. Formulation of compositional multiphase flow through porous media is presented in the following categories: the general multiphase, fully compositional nonisothermal mixture model, the isothermal black oil model of nonvolatile oil systems, the isothermal limited compositional model of volatile oil systems, and shape-averaged models. Formulation of source/sink terms in conservation equations is discussed. Analyses and formulations of problems involving phase change and transport in porous media, such as gas condensation, freezing/thawing of moist soil, and production of natural gas from hydrate-bearing formations, are presented.

Chapter 10 presents the modeling of particulate transport in porous media. Formulation of deep-bed and cake filtration processes involving particle transport and retention, and resulting porosity and permeability variation in porous media, is described. Phenomenological modeling considering temperature variation and particle transport by advection and dispersion is discussed. Temperature dependence is accounted for through the filtration rate coefficient and porous matrix thermal deformation. Other factors affecting the filtration coefficient and permeability are considered by means of empirical correlations. Applications are presented concerning the transport of colloids and particles through porous media and compressible cake filtration involving smaller particles packing through the large particles that form the skeleton of the filter cake. The effect of dispersion mechanism and temperature variation on particle transport and retention, and the consequent porosity and permeability impairment, is demonstrated by several examples.

Chapter 11 presents the modeling of transport in heterogeneous porous media. This chapter presents formulations and solutions that apply to heterogeneous systems having various transport units in fractured porous media, where the permeability of the fracture system is relatively greater than that of the porous matrix, and therefore the fractures form the preferential flow paths while the matrix forms a source of fluid for fractures. The objectives of this chapter are to develop an understanding of the mechanism of the matrix-to-fracture fluid transfer by the various processes and the formulation of transport in fractured porous media. The analytical and numerical solutions are presented for relatively simplified cases and systems undergoing an imbibition-drive matrix-to-fracture fluid transfer.

CHAPTER 2

TRANSPORT PROPERTIES OF POROUS MEDIA

2.1 INTRODUCTION

Permeability is an important property of porous formation and a complex function of many variables.* The essential variables include (1) the configuration of the pore structure determined by the arrangement of grains and pores, and described by fabric, texture, and morphology; (2) type, size, shape, and composition of pore deposits, cement, and grains; (3) coordination number defined as either the number of pore space connections or the number of grain contacts, here referred to as the pore coordination number and the grain coordination number, respectively; (4) hydraulic diameter, specific pore surface, areosity, or the area of pores open for flow, nonzero transport threshold, and tortuosity of the flow paths; (5) grain consolidation by cementing, fusing, and other means; and (6) for porous media undergoing an alteration, the evolution of the pore structure and creation of noncontributing porosity by porous matrix deformation, rock–fluid interactions, and other relevant processes (Kozeny, 1927; Carman, 1937a,b, 1956; Nelson, 1994, 2000; Saito et al., 1995; Revil and Cathles, 1999; van der Marck, 1999; Civan, 2000a,b). Cement is any material acting as a grain consolidation agent. Obviously, it is impractical to take into account all details in a simplified model. However, the majority of the simplified modeling approaches have strived at determining permeability from porosity alone, as pointed out by Nelson (1994, 2000). Nevertheless, porosity appears to be common to many of the critical factors affecting permeability, such as the coordination number, pore or grain surface area, and tortuosity. Civan (2002d) considered the porosity, coordination number, and cementation factor of natural

---

* Parts of this chapter have been reproduced with modifications from the following:

Civan, F. 2001. Scale effect on porosity and permeability—Kinetics, model, and correlation. AIChE Journal, 47(2), pp. 271–287, © 2001 AIChE, with permission from the American Institute of Chemical Engineers;

Civan, F. 2002d. Relating permeability to pore connectivity using a power-law flow unit equation. Petrophysics Journal, 43(6), pp. 457–476, and Civan, F. 2008c. Correlation of permeability loss by thermally induced compaction due to grain expansion. Petrophysics Journal, 49(4), pp. 351–361, © 2002 SPWLA, with permission from the Society of Petrophysicists and Well Log Analysts; and

Civan, F. 2003. Leaky-tube permeability model for identification, characterization, and calibration of reservoir flow units. Paper SPE 84603, SPE Annual Technical Conference and Exhibition, Denver, Colorado, October 5–8, 2003, © 2003 SPE, with permission from the Society of Petroleum Engineers.

Porous Media Transport Phenomena, First Edition. Faruk Civan.
© 2011 John Wiley & Sons, Inc. Published 2011 by John Wiley & Sons, Inc.

porous media as the primary independent variables and showed that these are sufficient to define a meaningful relationship for permeability. The secondary factors, such as the interconnectivity parameter and exponent of the power-law flow unit equation of permeability suggested by Civan (1996c, 2001, 2002b,d, 2003, 2005a), are related to permeability through porosity.

Many natural and engineering processes occur in porous media. Among other factors, the transport of fluids through subsurface geological formations is primarily influenced by the permeability distribution in these formations. Accurate and theoretically meaningful description and prediction of the permeability of natural and engineering porous materials are of continuing interest. Improved, compact, and convenient macroscopic permeability models are required for representing the variation of the permeability of porous materials in large-scale geological subsurface reservoirs. In such media, the pore structure and thence permeability may vary by rock, fluid, and particle interactions and by thermal and mechanical stresses during flow of fluids. Various theoretical and laboratory studies have been carried out to determine the functional relationships between the hydrodynamic transport properties of porous materials, especially the permeability, with varying degrees of success (Nelson, 1994; Civan, 2000a,b, 2001, 2007a; Singh and Mohanty, 2000). Modeling at both the microscopic and the macroscopic levels has been attempted. Frequently, conventional approaches for the characterization of flow through porous media have resorted to the estimation of permeability based on the description of flow either around the grains (microscopic description) or through a bundle of capillary tubes (macroscopic description) (Rajani, 1988). Macroscopic models require significantly less computational complexity and are preferred to microscopic, locally detailed, pore-level distributed parameter and network models for large field-scale applications.

Among the various hydrodynamic transport properties, the prediction of permeability has predominated because of its importance in determining the rate of fluid transfer through porous materials. In spite of numerous experimental and theoretical works, there is a scarcity of satisfactory practical and simplified models, which can be incorporated into the mathematical description of porous media fluid flow (Civan, 2002b,d) without adding much complication and computational burden. Yet, these models are in great demand for large-scale flow field simulations of subsurface reservoirs. Complicated modeling that provides information about the details of flow patterns at the microscopic scale, such as the pore-level description using network models, may be an instructive and important research tool. However, it is also computationally demanding and impractical when dealing with processes in intricate porous media, and it requires a significant amount of computational effort. Therefore, simplified models are more advantageous for practical application. In this respect, the majority of attempts at improved, simplified modeling have focused on the extension and modification of the Kozeny–Carman equation as described by Nelson (1994), Singh and Mohanty (2000), and Civan (2000b,d, 2001, 2003, 2005a).

Descriptions of the functional relationship between the permeability and porosity of porous media have been attempted in numerous theoretical and laboratory studies with varying degrees of success. Among other factors, permeability and effective (interconnected) porosity of porous materials strongly depend on pore connectivity. This is measured by the coordination number, which expresses the number of pore throats connecting a pore body to other pore bodies or alternatively by the number of grain contacts in porous media. Kozeny (1927), Carman (1937a,b, 1956), Rajani (1988), Nelson (1994), and many others have considered static porous media, in which the pore structure and therefore the pore connectivity and the coordination number remain fixed. However, when the pore system in porous media undergoes an evolution by various alteration processes due to porous material, fluid, and particle interactions, the coordination number, cementation, and thence the porosity–permeability relationship vary. Therefore, a dynamic relationship is required to describe the permeability of such porous media.

Ladd (1990) and Koch and Sangani (1999) determined the permeability of homogeneous packs of spherical grains by numerical simulation of the hydrodynamic interactions between the particles and fluids. Their numerically predicted permeability values can be used as a substitute for experimental data. For practical application to flow in porous media, they presented the simulation results by means of two separate analytical expressions over the low and high porosity ranges. However, the same result can be achieved by a single and compact relationship in a convenient and theoretically meaningful manner using a power-law flow unit equation (Civan, 1996c, 2000a,b, 2001, 2002a).

Various studies have indicated that the frequently used Kozeny–Carman equation (Carman, 1937a,b, 1956), which is referred to as the linear flow unit equation by Civan (2000b), is inherently oversimplified and does not satisfactorily represent the porosity–permeability relationship for natural porous materials (e.g., Wyllie and Gardner, 1958a,b; Nelson, 1994, 2000; Civan, 1996c, 2001, 2002b,d, 2003, 2005a; Koch and Sangani, 1999). Civan (1996c, 2000a,b) proposed that the porosity–permeability relationship for natural porous media could be more accurately described by expressing the mean hydraulic tube diameter as a power-law function of the volume ratio of pore (or void) to solid (or grain) in porous media. The cementation effects are not included in the solid volume. Subsequently, Civan (2001, 2002d) theoretically derived and verified a power-law flow unit equation by incorporating the fractal attributes of pores in irregular porous media into a bundle of tortuous leaky hydraulic tubes as a model of porous media. Civan (2001) extensively verified the power-law flow unit equation using experimental data relating to scale precipitation and dissolution in porous media. He also derived mathematical expressions for the parameters in this equation using a new and improved theoretical model in terms of the phenomenological characteristics of the hydrodynamic interaction processes between the fluid and grains, and of the fractal parameters of the pore geometry in porous media.

This chapter implements the effects of pore connectivity, the valve action of pore throats, and cementation into the bundle of tortuous, leaky capillary tubes as a macroscopic model of permeability, according to Civan (2001, 2002d, 2003). The parameters of the power-law equation are related to pore connectivity measured by the coordination number using suitable functional relationships. The level of modeling is kept to a compact macroscopic formulation, which can be readily incorporated into a large- or field-scale simulation of flow in geological porous formations without significant computational burden. In this model, the pore structure and permeability can vary by porous solid (rock), fluid, and particle interactions and by thermal and mechanical stresses during fluid flow. Civan's approach allows for the incorporation of various data within a single, compact, and simple power-law equation over the full range of porosity. The power-law parameters are formulated and determined as functions of the coordination number or porosity, because the coordination number correlates well with porosity. The analysis of permeability versus porosity data by Civan (2001, 2002b,d, 2003, 2005a) demonstrates that the power-law flow unit model alleviates the deficiencies of the Kozeny–Carman equation (Carman, 1937a,b). The flow unit parameters are determined by simple regression of permeability versus porosity data.

2.2   PERMEABILITY OF POROUS MEDIA BASED ON THE BUNDLE OF TORTUOUS LEAKY-TUBE MODEL

As discussed in this section, a simplified description of transport processes in natural heterogeneous irregular porous media requires some degree of empiricism. Therefore, the equations presented in this chapter are semiempirical formulae, which need to be calibrated for a specific porous formation by core analysis and/or history matching of flow versus time data in order to determine the values of the various constants that characterize the equations. Many mathematical models, such as network models, sometimes incorrectly referred to as analytical models, are also empirical, because they do not represent the real porous medium. The empiricism in such models is introduced at the start when the porous media are assumed to have a prescribed fabric and texture, as with the Bethe network used by Bhat and Kovscek (1999). Because natural porous media are irregular, these models are only an approximation to real porous media, regardless of the method of analytical determination of permeability. However, they are useful and instructive approaches, because they can provide valuable insights into the microscopic details of the processes based on assumed porous media realizations. Thus, they can be applied at small scale, such as in laboratory core tests. However, they do not necessarily represent the real porous medium and they are not practical for simulation of processes in large-scale subsurface reservoirs containing oil, gas, and brine. Therefore, it is rare to find any practical applications of such detailed and computationally demanding analytical pore-scale network modeling in large-scale porous media. For such applications, simplified and compact equations, such as those presented in this chapter, are preferred because they are convenient and workable. Thus, microscopic mathematical modeling may be instructive and attractive, but it certainly has inherent limitations and it does not always pay off. Simplified macroscopic modeling approaches may still be more advantageous and straightforward for large-scale field applications.

This section demonstrates that permeability and porosity can be related satisfactorily by a single expression over a wide range of values by using a power-law flow unit equation. The interconnectivity parameter and exponent of the power-law flow unit equation are shown to be strong functions of the coordination number of porous media. They are significantly different from those of the Kozeny–Carman equation. The analysis also lends itself to the physical interpretation of the pore

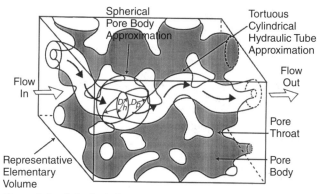

Figure 2.1 Spherical-shaped pore body and cylindrical-shaped tortuous hydraulic tube approximation in a representative elementary volume of a porous medium (Civan, 2001; © 2001 AIChE, reproduced by permission of the American Institute of Chemical Engineers).

connectivity and cementation factor in terms of the relationship of permeability to porosity.

2.2.1 Pore Structure

Consider Figure 2.1, schematically depicting the irregular pore structure in a representative elementary bulk volume of a porous material. The representative elementary volume is an optimum bulk size of porous media over which the microscopic properties of pore fluids and porous matrix can be averaged to obtain a consistent and continuous macroscopic description. Brown et al. (2000) define it as "the range of volumes for which all averaged geometrical characteristics are single-valued functions of the location of that point and time." As pointed out by Nelson (2000), the total porosity of porous formations can be classified into the contributing and noncontributing porosities. The interconnected pore space allowing flow-through constitutes the contributing porosity. The remaining pore space, physically isolated, forms the noncontributing porosity. However, some connected pores may be trapped within dead or stationary fluid regions and therefore may not be able to contribute to flow under prescribed flow conditions. This is another permeability-affecting factor, dependent on the prevailing flow conditions. As pointed by Nelson (2000), the porosity term appearing in the Kozeny–Carman equation refers to the contributing or effective porosity. Frequently, for convenience in modeling, pore structure in porous media is viewed as a collection of pore bodies connected with pore throats as shown in Figure 2.2, which are referred to as the nodes and bonds, respectively, in the network models. The number of the pore throats emanating from a pore body to surrounding pore bodies is a characteristic parameter and denoted by Z, called the coordination number. Its average value over the representative elementary volume for a prescribed pore structure is uniquely defined.

As described by Civan (2000a), alteration in porous media occurs when immobile deposits are formed within the pore space. The flow characteristics are

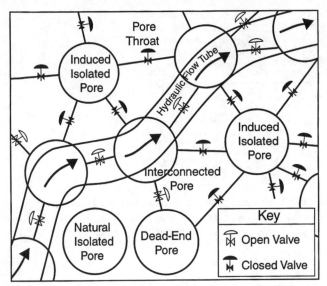

Figure 2.2   Leaky-tube model of flow through porous media considering the valve effects of pore throats in a network of pore bodies connected with pore throats in a porous medium (Civan, 2001; © 2001 AIChE, reproduced by permission of the American Institute of Chemical Engineers).

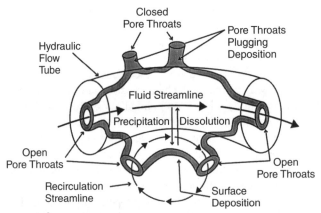

Figure 2.3   Surface and pore-throat plugging deposition processes occurring in a pore body and in associated pore throats (Civan, 2001; © 2001 AIChE, reproduced by permission of the American Institute of Chemical Engineers).

determined by several factors, including coordination number, mean pore diameter, mean hydraulic tube diameter, specific pore surface, and tortuosity. When a suspension of particles flows through porous media, particles may deposit over the pore surface and/or accumulate behind the pore throats, as depicted in Figure 2.3, when the conditions are favorable for deposition. Pore-throat plugging primarily occurs by particulates suspended in the fluid and causes severe permeability reduction, because the plugs formed by the jamming of the pore throats, which act like valves, constrict and/or limit the flow-through. When the pore throats are plugged, the interconnectivity of the pores is reduced and some pores may in fact become dead-end pores or isolated pores as shown in Figure 2.2. Consequently, the permeability of porous media diminishes to zero when a sufficient number of pore throats are closed to interrupt the continuity of the flow channels. Therefore, during the alteration of porous media, the fluid paths continuously vary, adapting to the least resistant paths available under the prevailing conditions. However, dissolution and precipitation processes in porous media most likely occur at the pore surface (Le Gallo et al., 1998), which is in abundant availability in porous materials. As depicted in Figure 2.3, dissolution and precipitation processes result in scale removal and scale formation at the pore surface, respectively. Therefore, for the most part, it is reasonable to assume that the pore-throat plugging phenomenon can be neglected, except at the limit when the pore-throat opening is significantly reduced by surface deposition to an extent that it can exert a sufficient resistance to prevent flow through the pore throat, referred to as the threshold pressure gradient. Thus, for all practical purposes, it is reasonable to assume that the coordination number remains unchanged during the surface dissolution and precipitation processes.

The alteration of the flow paths as described earlier can only be determined by means of the network modeling. Although more instructive, such internally detailed elaborate description of flow in an intricately complicated and varying porous structure may be cumbersome and computationally demanding, and therefore is not warranted for most practical problems. Frequently, simplified models, such as those developed by Civan (2001), based on a lumped-parameter representation of the processes over the representative elementary bulk volume of porous media, are sufficient and in fact preferred. The lumped-parameter model developed here assumes that a hydraulic flow tube or a preferential flow path is formed by interconnecting the pore bodies, like beads on a string as described in Figures 2.1 and 2.2. Consequently, the pore body and pore-throat diameters are averaged as the mean hydraulic tube diameter. The coordination number and tortuosity are assumed to remain constant during dissolution and precipitation because the pore-throat plugging effect is negligible. The rate of variation of the pore volume is assumed as directly proportional to the instantaneous available pore volume and the participating pore surface, over which scale can be formed because of the affinity of the precipitating substance to the substrate present at the pore surface. In addition, fractal relationships are facilitated, with the fractal parameters determined empirically, to describe in a practical manner the various geometrical attributes of the flow paths in random porous media. Ultimately, the fractal coefficients and dimensions are incorporated into the lumped parameters. Civan (2001, 2002b,d, 2003, 2005a) demonstrated that the equations derived based on these assumptions can accurately represent the typical experimental data.

2.2.2 Equation of Permeability

Let L, W, and H denote the length, width, and thickness of the porous medium in the x-, y-, and z-directions, and the representative elemental volume is given by $V = LWH$.

For example, the frictional drag force in the x-direction is given by

$$F_{Dx} = A\Delta p_x. \tag{2.1}$$

Thus, the frictional drag per unit volume in the x-direction is given by

$$f_{Dx} = \frac{F_{Dx}}{V} = \frac{A\Delta p_x}{AL} = \frac{\Delta p_x}{L}. \tag{2.2}$$

In general, the frictional drag per unit volume in vector form is given by

$$\mathbf{f}_D = \frac{\mathbf{F}_D}{V} = -\rho\nabla\psi. \tag{2.3}$$

Consequently, permeability of porous media can be related to frictional drag or coefficient of hydraulic resistance per representative bulk volume element, $\mathbf{f}_D$, to volumetric flux of flowing fluid according to Darcy's law as (Chapter 5)

$$\mathbf{f}_D = -\rho\nabla\psi = \mu\mathbf{K}^{-1}\cdot\mathbf{u}. \tag{2.4}$$

Here, ψ denotes the flow potential given by

$$\psi = \int_{p_o}^{p}\frac{dp}{\rho} + g(z - z_o), \tag{2.5}$$

in which p is the fluid pressure at a depth of z; p_o is the fluid pressure at a reference depth of z_o; ρ and μ denote the density and viscosity of the fluid, respectively; $\mathbf{K}$ is the permeability tensor; and $\mathbf{u}$ is the superficial velocity (Darcy velocity) or the volume flux of the flowing fluid.

The grain packing fraction or solidity, ε, and the pore space volume fraction or porosity, ϕ, of porous media are related according to

$$\phi = 1 - \varepsilon. \tag{2.6}$$

Frequently, permeability K has been expressed in terms of the characteristic parameters of porous media using the Kozeny–Carman equation (Carman, 1937a,b, 1938, 1956), which uses the analogy to a bundle of capillary hydraulic tubes. Amaefule et al. (1993) have rearranged the Kozeny–Carman equation in a more physically usable form as

$$\sqrt{\frac{K}{\phi}} = \Gamma\left(\frac{\phi}{1-\phi}\right), \tag{2.7}$$

where K and ϕ are the permeability and porosity, respectively; $\sqrt{K/\phi}$ is the Leverett pore size (or diameter) factor (Leverett, 1941); and Γ is an intrinsic parameter, which they called the "flow zone indicator," given by

$$\Gamma = \left(\Sigma_g\sqrt{2\tau}\right)^{-1}, \tag{2.8}$$

where Σ_g denotes the grain (or pore) surface area per unit volume of the grains present in a representative element of a bulk porous medium (Civan, 2002b,d) and has a reciprocal length dimension; and τ denotes the tortuosity of the porous medium, defined as the ratio of the apparent length of the effective mean hydraulic tube to the physical length of the bulk porous medium. Here, Γ will be called the interconnectivity parameter because it is a measure of the pore space connectivity. Γ is implicitly dependent on porosity because the specific grain surface and tortuosity depend on porosity. For example, the various inverse power-law correlations of tortuosity versus porosity imply increasing tortuosity trends for decreasing porosity (see Lerman, 1979). Data from Salem and Chilingarian (2000) show similar trends for randomly packed beds of glass spheres $(0.10 < \phi < 0.47)$ and fine sands $(0.34 < \phi < 0.45)$, respectively. The data of Haughey and Beveridge (1969) and of Ertekin and Watson (1991) provide evidence of increasing grain (or pore) surface area per grain volume for increasing porosity. Inherently, the effective pore surface area of the tortuous flow channels vanishes when the effective porosity becomes zero. Hence, the pore size (diameter) is implicit in the grain (or pore) surface area and consequently in the interconnectivity parameter. The interconnectivity parameter shows a decreasing trend for increasing porosity (Civan, 2002b,d).

Eq. (2.7) expresses the mean pore diameter represented by $\sqrt{K/\phi}$ as a linear function of the pore volume-to-solid volume ratio, $\phi/(1-\phi)$. Note that this equation does not consider any cementation effect. Although it has been frequently used in the literature, the Kozeny–Carman equation, Eq. (2.7), suffers from inherent limitations as described in the comprehensive reviews presented by Wyllie and Gardner (1958a,b) and by Civan (2000a,b, 2001, 2002b,d). Civan (1996c, 2000a,b, 2001) showed that the relationship of permeability of real porous materials may deviate significantly from the idealized porous medium model of a bundle of capillary tubes underlying the Kozeny–Carman equation. Civan (2001) theoretically expressed the mean pore diameter as a three-parameter power-law function of the pore volume-to-solid volume ratio based on the fractal attributes of the interconnected pore space in porous media. The model of a bundle of tortuous leaky hydraulic tubes results in a power-law flow unit equation, given by (Civan, (1996c, 2001, 2002b,d, 2003, 2005a)

$$\sqrt{\frac{K}{\phi}} = \Gamma \left(\frac{\phi}{\alpha - \phi} \right)^{\beta}, \tag{2.9}$$

in which β and Γ are the exponent and interconnectivity parameters and α is a cement exclusion factor.

The interconnectivity parameter primarily represents the valve effect of the pore throats (Fig. 2.2) controlling the pore connectivity to other pore spaces in an interconnected network of pore spaces in porous media (Civan, 2001). Thus, the interconnectivity parameter becomes zero when all the pore throats are closed by such mechanisms like obstruction by migrating fines and deposits, sealing with deformable precipitates including gels, wax, and asphaltene, and constriction and collapse of pore throats under thermal and mechanical stresses. The interconnectivity parameter Γ is similar to the generically named pore geometry function G in the permeability function of Rajani (1988), who postulated that permeability could be generally expressed as a product of a pore geometry function G and a porosity function $f(\phi)$ according to

$$K = G f(\phi). \tag{2.10}$$

This idea originates from the method of separation of variables and it is supported by the available models, such as the Kozeny–Carman equation, Eq. (2.7). Note also that Eq. (2.9) can be rearranged as

$$K = \Gamma^2 \phi \left(\frac{\phi}{\alpha - \phi} \right)^{2\beta}. \tag{2.11}$$

Thus, comparing Eqs. (2.10) and (2.11), the pore geometry and porosity functions of the power-law flow unit equation are given by

$$G = \Gamma^2 \quad \text{and} \quad f(\phi) = \phi \left(\frac{\phi}{\alpha - \phi} \right)^{2\beta}. \tag{2.12}$$

The cement exclusion factor α is given by

$$\alpha = 1 - \alpha_c, \tag{2.13}$$

where α_c is the volume fraction of the cementation and grain consolidation in bulk porous media. Ordinarily, the cement exclusion factor $\alpha = 1$ when the cementation effect, which is a measure of grain consolidation by cementing, fusing, and other means, is zero, and thence, $\alpha_c = 0$.

The values of parameters β and Γ may vary significantly if fluid flow through pore throats is restricted. In the following sections, the theoretical expressions are derived, which relate the parameters β and Γ of Eq. (2.9) to the phenomenological parameters of the fluid and rock interaction processes and parameters based upon a fractal view of porous media. There, three aspects of the fractal attributes of the pore space are considered: (1) the total pore surface of all hydraulic tubes, (2) the pore surface of individual hydraulic tubes, and (3) the areosity or effective cross-sectional area open for flow. Therefore, the parameters β and Γ are not just a set of empirically fitted values, but they are expressed in a meaningful phenomenological manner. However, frequently, they are determined empirically by data analysis, because measurements of the fractal parameters are not available and they are tedious to determine by petrographical measurements for frequent applications, as demonstrated by Garrison et al. (1992, 1993). Eq. (2.9) can circumvent the inherent limitations of the previous models because it incorporates the following features of real porous materials:

- Fractal attributes of the irregular interconnected pore space and the prevailing hydraulic flow paths.
- Cementation and grain consolidation effects.
- Gate or valve action of the pore throats expressed by pore connectivity and coordination number.
- Pore evolution by various processes, including the accumulation and removal of various deposits at pore surface and pore throats.

2.2.3 Derivation of the Equation of Permeability

For accurate representation of the flow units in actual porous formations and hence the permeability–porosity data, a formulation is presented based on a bundle of leaky capillary hydraulic tubes for a model of porous media, as depicted in Figure 2.2. This allows for interactions between the hydraulic tubes and characterizing the interconnected pore structure in real porous media with relevant fractal attributes. The leakiness of the hydraulic tubes is related to the pore connectivity measured by the coordination number. For practical applications, the pore connectivity is related to porosity. Because the present formulation is focused on the flow unit characterization and identification, this formulation considers only the contributing porosity or the interconnected pore structure allowing flow through porous media as described by Nelson (1994, 2000). In addition, the fluid flowing through porous media is assumed incompressible, Newtonian, single phase, and isothermal. The flow regime is considered as Darcian. However, the formulation can be readily extended for compressible, non-Newtonian, and multiphase fluid systems, and for non-Darcian flow. After all, permeability is an intrinsic property of porous media and is independent of the type of fluid conditions considered for its determination. The power-law flow unit equation (Civan, 1996c, 2001, 2002b,d, 2003, 2005a) is derived by alternative means with modifications for different descriptions of the functional dependency of its parameters to various fractal attributes of the connected pore space.

The pore volume, V_p, is given by the total volume of the hydraulic tubes in porous media as

$$V_p = nA_h L_h = V_b \phi, \tag{2.14}$$

where n, A_h, P_h, and L_h represent the number, the cross-sectional area open for flow, the perimeter, and the length of the mean hydraulic flow tubes. V_b and ϕ are the volume and porosity of porous medium.

The ratio of the hydraulic tube length L_h to the length L_b of porous media is called the tortuosity τ of the mean hydraulic tube, given by

$$\tau = L_h / L_b. \tag{2.15}$$

The fractal relationship given as follows is used to express the total pore surface of all the hydraulic tubes:

$$\Sigma_p = nP_h L_h = C\left(V_b \Sigma_b\right)^{D/3} = C\left[V_b \Sigma_g (\alpha - \phi)\right]^{D/3}, \tag{2.16}$$

where C and D are the fractal coefficient and dimension, and α is a cement exclusion factor, defined by

$$\alpha = 1 - \phi_c, \tag{2.17}$$

where ϕ_c denotes the volume fraction of the cementing and grain fusing effects present in porous media. Therefore, $(\alpha - \phi)$ represents the volume fraction of the porous matrix excluding such effects.

Combining Eqs. (2.14)–(2.16) yields

$$nP_h \tau L_b = C\left[nA_h \tau L_b \Sigma_g (\alpha - \phi)/\phi\right]^{D/3}. \tag{2.18}$$

The surface area and volume of the mean hydraulic tube can be related by the following fractal relationship:

$$\Sigma_p / n = C\left(V_p / n\right)^{D/3}, \tag{2.19}$$

where C and D are the fractal coefficient and dimension.

Substituting Eqs. (2.14)–(2.16) into Eq. (2.19) yields

$$P_h = C \left(L_b \tau \right)^{-1+D/3} A_h^{D/3}. \tag{2.20}$$

Eliminating P_h between Eqs. (2.18) and (2.20), the cross-sectional area of the hydraulic flow tube can be expressed by

$$A_h = \frac{1}{L_b \tau} \left(\frac{C}{C} \right)^{\frac{3}{D-D}} \Sigma_g^{\frac{-D}{D-D}} n^{\frac{3-D}{D-D}} \left(\frac{\phi}{\alpha - \phi} \right)^{\frac{D}{D-D}}. \tag{2.21}$$

Three approaches can be used to estimate the hydraulic tube diameter. The first is to assume a fractal relationship between the diameter and the cross-sectional area of the hydraulic tube according to

$$D_h = \lambda A_h^\delta, \tag{2.22}$$

where λ and δ are the fractal coefficient and dimension. Then, substituting Eq. (2.21) into Eq. (2.22) yields

$$D_h = \frac{\lambda}{\left(L_b \tau \right)^\delta} \left(\frac{C}{C} \right)^{\frac{3\delta}{D-D}} \Sigma_g^{\left(\frac{-D}{D-D} \right)\delta} n^{\left(\frac{3-D}{D-D} \right)\delta} \left(\frac{\phi}{\alpha - \phi} \right)^{\left(\frac{D}{D-D} \right)\delta}. \tag{2.23}$$

The second approach considers a circular hydraulic tube, for which the cross-sectional area is given by

$$A_h = \pi D_h^2 / 4. \tag{2.24}$$

Thus, comparing Eqs. (2.22) and (2.24) reveals for this special case that

$$\lambda = 2 / \sqrt{\pi}, \delta = 1/2. \tag{2.25}$$

Therefore, substituting Eq. (2.25) into Eq. (2.23) yields

$$D_h = \frac{2}{\sqrt{\pi L_b \tau}} \left(\frac{C}{C} \right)^{\frac{1}{2} \left(\frac{3}{D-D} \right)} \Sigma_g^{\frac{1}{2} \left(\frac{-D}{D-D} \right)} n^{\frac{1}{2} \left(\frac{3-D}{D-D} \right)} \left(\frac{\phi}{\alpha - \phi} \right)^{\frac{1}{2} \left(\frac{D}{D-D} \right)}. \tag{2.26}$$

The third approach uses the hydraulic tube diameter given by (de Nevers, 1970)

$$D_h = 4A_h / P_h. \tag{2.27}$$

Substituting Eqs. (2.20) and (2.21) into Eq. (2.27) yields

$$D_h = \frac{4}{C} \left(\frac{C}{C} \right)^{\frac{3-D}{D-D}} \Sigma_g^{\left(\frac{3-D}{3} \right)\left(\frac{-D}{D-D} \right)} n^{\left(\frac{3-D}{3} \right)\left(\frac{3-D}{D-D} \right)} \left(\frac{\phi}{\alpha - \phi} \right)^{\left(\frac{3-D}{3} \right)\left(\frac{D}{D-D} \right)}. \tag{2.28}$$

On the other hand, an expression for the hydraulic tube diameter can also be derived using Darcy's law and the Hagen–Poiseuille equation given, respectively, by

$$q = \frac{K A_b}{\mu} \frac{\Delta p}{L_b} \tag{2.29}$$

and

$$q = n\frac{\pi D_h^4}{128\mu}\frac{\Delta p}{L_h}, \tag{2.30}$$

in which A_b is the area of the bulk porous media.

The studies with experimental data (Civan, 2001, 2002b,d, 2003, 2005a) indicate that the total pore cross-sectional area permitting flow is given by using a fractal relationship to express the pore connectivity:

$$A_f = A_b c\phi\left(e + \phi^v\right)^2 = nA_h, \tag{2.31}$$

where

$$v = -\frac{1}{2}\left(1 - \frac{d}{3}\right). \tag{2.32}$$

Here, e is an empirical constant concerning the threshold condition of porosity below which fluid cannot flow through porous media, and c and d are the fractal coefficient and dimension. Eq. (2.31) incorporates the "gate or valve effect" of the pore throats (Chang and Civan, 1997; Ochi and Vernoux, 1998). Thus, the fractal dimension d of the pores is given by Eq. (2.32) as

$$d = 3(1 + 2v). \tag{2.33}$$

If $e = 0$, then Eq. (2.24) becomes

$$A_f = A_b c\phi^{d/3}. \tag{2.34}$$

Note that Eq. (2.34) simplifies to the following frequently used expression for a theoretical nonfractal three-dimensional object, for which $e = 0$, $c = 1$, and $d = 3$:

$$A_f = A_b\phi. \tag{2.35}$$

The value of d depends on the degree of constriction of the capillary hydraulic tubes in a fractal porous medium. The value is $d < 3$ for constricted fractal porous medium and close to the value $d = 3$ for nonconstricted fractal porous medium. It is shown in the applications presented in this chapter that $d < 3$ for typical natural porous media.

Consequently, equating Eqs. (2.29) and (2.30) and then substituting Eqs. (2.15), (2.24), and (2.31) leads to the following expression for the mean hydraulic tube diameter:

$$D_h = 4\sqrt{\frac{2\tau}{c}}\sqrt{\frac{K}{\phi}}\left(e + \phi^v\right)^{-1}. \tag{2.36}$$

Using the preceding equations, the following power-law flow unit equation can be derived in alternative ways (Civan, 2002b,d):

$$\sqrt{\frac{K}{\phi}} = \Gamma\left(\frac{\phi}{\alpha - \phi}\right)^\beta. \tag{2.37}$$

The parameters of Eq. (2.37) can be determined by different expressions depending on the various approaches described next.

For example, equating Eqs. (2.23) and (2.36) leads to Eq. (2.37) with the following parameters (Civan, 2002b,d, 2003, 2005a):

$$\Gamma = \gamma\left(\phi^v + e\right); \upsilon < 0; \tag{2.38}$$

$$\beta = \left(\frac{D}{D - D}\right)\delta; \tag{2.39}$$

and

$$\gamma = \frac{\lambda}{4\left(L_b\tau\right)^\delta}\sqrt{\frac{c}{2\tau}}\left(\frac{C}{C}\right)^{\frac{3\delta}{D-D}}\Sigma_g^{\left(\frac{-D}{D-D}\right)\delta}\, n^{\left(\frac{3-D}{D-D}\right)\delta}. \tag{2.40}$$

On the other hand, equating Eqs. (2.26) and (2.36) leads to Eq. (2.38) and the following expressions:

$$\beta = \frac{1}{2}\left(\frac{D}{D - D}\right) \tag{2.41}$$

and

$$\gamma = \frac{n^{\frac{1}{2}\left(\frac{3-D}{D-D}\right)}\left(C/C\right)^{\frac{3}{2(D-D)}}\Sigma_g^{\frac{1}{2}\left(\frac{-D}{D-D}\right)}}{2\tau\sqrt{2\pi L_b/c}}. \tag{2.42}$$

However, equating Eqs. (2.28) and (2.36) yields Eq. (2.38) and the following expressions:

$$\beta = \left(\frac{3-D}{3}\right)\left(\frac{D}{D - D}\right) \tag{2.43}$$

and

$$\gamma = \sqrt{\frac{c}{2\tau}}\frac{1}{C}\left(\frac{C}{C}\right)^{\frac{3-D}{D-D}}\Sigma_g^{\left(\frac{3-D}{3}\right)\left(\frac{-D}{D-D}\right)}\, n^{\left(\frac{3-D}{3}\right)\left(\frac{3-D}{D-D}\right)}. \tag{2.44}$$

2.2.4 Pore Connectivity and Parametric Functions

Two principally different definitions of the coordination number have appeared in the reported studies. Here, they are distinguished as the grain and pore coordination numbers to avoid confusion. The grain coordination number is defined as the number of contact points between a grain and other grains in a representative volume element of a bulk porous medium (Ridgway and Tarbuck, 1967; Makse et al., 2000). The pore coordination number is defined as the number of pore-throat connections between a pore space and other pore spaces in a representative volume element of a bulk porous medium (Bhat and Kovscek, 1999). Usually, the pore coordination number increases when the grain coordination number increases. Therefore, the formulation given in this section is applicable for both definitions of the coordination number. The coordination number for a given pore space or grain is an integer. The

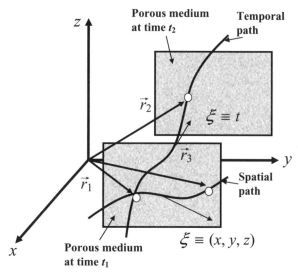

Figure 2.4   Flow units at different locations and time in a porous medium (Civan, 2002d; © 2002 SPWLA, reproduced by permission of the Society of Petrophysicists and Well Log Analysts (SPWLA).).

grain coordination number $Z = 0$ indicates that the grain is buried and isolated inside the cement and cannot contact other grains. The values $Z = 0$ and $Z = 1$ represent the pore coordination numbers for nonconducting isolated pores and dead-end pores, respectively. The lowest value of the coordination number for a conductive pore body is $Z = 2$. The coordination numbers determined analytically for a regular packing of granular materials may also have integer values. However, the coordination numbers measured for an irregular packing of granular materials may not necessarily be an integer because these values have been determined as averages over the representative volume element of a porous medium (Civan, 2002d). This outcome is demonstrated later through an example based on the data of Ridgway and Tarbuck (1967).

As depicted in Figure 2.4, different flow units may be encountered in a heterogeneous porous medium due to pore alteration and/or changing location. Consider that $\mathbf{r} = \mathbf{r}(x, y, z, t)$ denotes a spatial–temporal position vector pointing at a flow unit in the Cartesian coordinates x, y, z and at a time, t, in the porous medium. The flow unit at a given location can vary if the porous medium is undergoing a pore alteration. Thus, for example, the flow unit at a given location (x_1, y_1, z_1) can transform from state 1 at a time, t_1, to state 2 at another time, t_2, as indicated by the position vectors $\mathbf{r}_1(x_1, y_1, z_1, t_1)$ and $\mathbf{r}_2(x_1, y_1, z_1, t_2)$, respectively. On the other hand, the flow units at different points in a heterogeneous porous medium can be different from each other. Consequently, a change in the flow unit from a state 1 to another state 3 can be observed by moving from a position, $\mathbf{r}_1(x_1, y_1, z_1, t_1)$, to another position, $\mathbf{r}_3(x_3, y_3, z_3, t_1)$, at a given time, t_1. The pore connectivity in different state flow units is different and is determined by the prevailing coordination number, Z. Therefore, when pore alteration and/or change of location in a heterogeneous porous medium take place, the interconnectivity parameter Γ and the exponent β can have different values.

The variation of the parameters of the power-law flow unit equation is described here by kinetic equations. Kinetic equations are empirical differential equations and have been applied successfully in many fields, such as population growth, chemical reactions, and radioactive decay. The underlying hypothesis for the formulation of the kinetic equations is that the variation of the parameters with distance and/or time is proportional (or directly related) to the pertinent driving forces causing the variation (Civan, 1998c). The proportionality factor is called the rate constant. Two driving forces are considered: (1) the variation of the coordination number with distance or time and (2) the deviation of the instantaneous parameter values from their maximum or limiting values. Thus, the parameter values remain constant when either or both of the driving forces vanish.

Let the symbol ξ denote a variable referring to either the distance along the flow units in a heterogeneous porous medium or the process time for evolution of a flow unit at a given location (Civan, 2002d) as indicated in Figure 2.4. The interconnectivity parameter Γ and the reciprocal exponent β^{-1} are assumed to vary proportionally with their values relative to the limiting values and the variation of the coordination number by various mechanisms in porous media undergoing an alteration or by moving along different flow units in porous media (Civan, 1998, 2000b, 2002d). Hence, the following empirical kinetic equations can be written:

$$\frac{d\Gamma}{d\xi} = A\left(\Gamma_{max} - \Gamma\right)^p \frac{dZ}{d\xi} \tag{2.45}$$

and

$$\frac{d\beta^{-1}}{d\xi} = C\left[\left(\beta^{-1}\right)_{max} - \beta^{-1}\right]^q \frac{dZ}{d\xi}, \tag{2.46}$$

where the coefficients A and C are the empirical rate constants, the symbols p and q denote the empirical orders of the prevailing process rates, and the subscript max refers to the maximum values of the various quantities. However, the variable ξ disappears when Eqs. (2.45) and (2.46) are rearranged as follows:

$$\frac{d\Gamma}{dZ} = A\left(\Gamma_{max} - \Gamma\right)^p \tag{2.47}$$

and

$$\frac{d\beta^{-1}}{dZ} = C\left[\left(\beta^{-1}\right)_{max} - \beta^{-1}\right]^q. \tag{2.48}$$

Eqs. (2.47) and (2.48) now express the variation of the parameter values Γ and β^{-1} with the coordination number Z by means of the empirical power-law functions of the deviation of the instantaneous parameter values from their maximum limiting values. The analysis of the reported data presented in this section and the fractal-based modeling by Civan (2002d) have indicated a first-order rate dependency, that is, $p = q = 1$, and an interconnectivity parameter expressed by

$$\Gamma = \gamma(\phi^v + e); \, v < 0, \tag{2.49}$$

where the symbols γ, v, and e denote the empirical parameters.

Examination of diverse data reveals a trend of increasing pore interconnectivity Γ and coordination number Z for decreasing porosity ϕ and vice versa, regardless of whether the coordination number is defined in terms of grain contacts or pore space connections. For example, an analysis of the data presented by Bhat and Kovscek (1999) reveals an increasing pore space connectivity parameter and a decreasing power-law exponent with increasing coordination number. The data of Ridgway and Tarbuck (1967), Haughey and Beveridge (1969), and Makse et al. (2000) indicate decreasing coordination number with increasing porosity. This suggests that both Z and Γ increase as ϕ decreases. Further, these data indicate a critical or percolation threshold value of the coordination number, at or below which intergranular permeability vanishes and the interconnectivity parameter Γ and the reciprocal exponent β^{-1} approach very low values, referred to as the minima. Conversely, these parameters attain upper limiting values as the coordination number increases. The physical mechanism for increasing Z and Γ with decreasing porosity is that the number of grain contacts and the pore space connections increase as the porosity ϕ decreases because of the necessity of packing smaller grains in larger numbers to form a prescribed porous medium. This discussion suggests the following trends for the end-point conditions.

The above derivations indicate that the exponent β can deviate from the value of $\beta = 1$ implied by the Kozeny–Carman equation. Eqs. (2.40), (2.41), and (2.43) reveal that the value of the exponent β is determined by the fractal dimensions and can vary in the range of $0 \le \beta < \infty$ (Civan, 2002b,d).

The interconnectivity or flow unit function given by Eq. (2.38) is a mathematical expression of the gate (Chang and Civan, 1997) or valve (Ochi and Vernoux, 1998) effect of the pore throats. Experimental data indicate that interconnectivity and coordination number increase when porosity decreases and vice versa (Civan, 1996c, 2001, 2002b,d, 2003, 2005a). The interconnectivity vanishes as the porosity approaches the unity.

Based on the discussion of trends presented on pore connectivity and parametric functions, Eqs. (2.45)–(2.49) are subject to the following boundary conditions:

$$\Gamma = \Gamma_{min}, \, \beta^{-1} = \left(\beta^{-1}\right)_{min}, \, Z = Z_{min}, \, \phi = \phi_{max} \tag{2.50}$$

and

$$\Gamma = \Gamma_{max}, \, \beta^{-1} = \left(\beta^{-1}\right)_{max}, \, Z = Z_{max}, \, \phi = \phi_{min}, \tag{2.51}$$

where subscripts min and max refer to the minimum and maximum values of the various quantities.

Solving the differential equations given by Eqs. (2.47) and (2.48), subject to Eqs. (2.50) and (2.51), and applying Eq. (2.49) yield, respectively,

$$\ln\left[\frac{\Gamma_{max} - \Gamma}{\Gamma_{max} - \Gamma_{min}}\right] = \ln\left[\frac{\phi^v_{min} - \phi^v}{\phi^v_{min} - \phi^v_{max}}\right] = -A(Z - Z_{min}) \tag{2.52}$$

and

$$\ln\left[\frac{\left(\beta^{-1}\right)_{max} - \beta^{-1}}{\left(\beta^{-1}\right)_{max} - \left(\beta^{-1}\right)_{min}}\right] = -C(Z - Z_{min}) \tag{2.53}$$

or simply

$$\Gamma/\Gamma_{max} = 1 - \exp(-AZ + B) \tag{2.54}$$

and

$$\beta^{-1}/\left(\beta^{-1}\right)_{max} = 1 - \exp(-CZ + D). \tag{2.55}$$

For convenient straight-line plotting, Eqs. (2.54) and (2.55) can be written in linear forms as

$$\ln\left[1 - \frac{\Gamma}{\Gamma_{max}}\right] = \ln\left[1 - \frac{\Gamma\sqrt{\phi_o/K_o}}{\Gamma_{max}\sqrt{\phi_o/K_o}}\right] = -AZ + B \tag{2.56}$$

and

$$\ln\left[1 - \frac{\beta^{-1}}{\left(\beta^{-1}\right)_{max}}\right] = -CZ + D. \tag{2.57}$$

The derived parameters appearing in Eqs. (2.54)–(2.57) are defined as follows:

$$B = AZ_{min} + \ln(1 - \Gamma_{min}/\Gamma_{max}),\ D = CZ_{min} + \ln(1 - \left(\beta^{-1}\right)_{min}/\left(\beta^{-1}\right)_{max}). \tag{2.58}$$

Thus, the parameters Γ and β are related to porosity ϕ through Eqs. (2.49), (2.52), and (2.53) by the following functions:

$$\frac{\left(\beta^{-1}\right)_{max} - \beta^{-1}}{\left(\beta^{-1}\right)_{max} - \left(\beta^{-1}\right)_{min}} = \left[\frac{\Gamma_{max} - \Gamma}{\Gamma_{max} - \Gamma_{min}}\right]^{C/A} = \left[\frac{\phi_{min}^v - \phi^v}{\phi_{min}^v - \phi_{max}^v}\right]^{C/A}; v < 0. \tag{2.59}$$

The fractal dimension d of the pores is given by

$$d = 3(1 + 2v). \tag{2.60}$$

2.2.5   Data Analysis and Correlation Method

The parameters v, A, B, C, and D of the preceding equations are estimated by means of the least squares method to match the measurements of the permeability and porosity of the core samples with Eq. (2.37). This provides a calibration of the power-law flow unit model. Model calibration can be most conveniently accomplished by means of the straight-line plotting of the permeability–porosity data. For this purpose, Eq. (2.37) can be expressed in various linear forms as described with the following three cases (Civan, 2003, 2005a).

Case 1: When Γ, α, and β Have Constant Values

$$\log\left(\frac{K}{\phi}\right) = 2\log\Gamma + 2\beta\log\left(\frac{\phi}{\alpha - \phi}\right) \tag{2.61}$$

$$\log\left(\frac{K}{\alpha - \phi}\right) = 2\log\Gamma + (2\beta + 1)\log\left(\frac{\phi}{\alpha - \phi}\right) \tag{2.62}$$

$$\log\left[\left(\frac{\alpha-\phi}{1-\phi}\right)^{2\beta}\frac{K}{\phi}\right] = 2\log\Gamma + 2\beta\log\left(\frac{\phi}{1-\phi}\right) \tag{2.63}$$

$$\log\left[\left(\frac{\alpha-\phi}{1-\phi}\right)^{2\beta}\frac{K}{1-\phi}\right] = 2\log\Gamma + (2\beta+1)\log\left(\frac{\phi}{1-\phi}\right) \tag{2.64}$$

Case 2: When Γ and β Vary, but α Is Constant

$$(2\beta)^{-1}\log\left(\frac{K}{\phi\Gamma^2}\right) = \log\left(\frac{\phi}{\alpha-\phi}\right) \tag{2.65}$$

$$(2\beta+1)^{-1}\log\left[\frac{K}{(\alpha-\phi)\Gamma^2}\right] = \log\left(\frac{\phi}{\alpha-\phi}\right) \tag{2.66}$$

$$(2\beta)^{-1}\log\left[\left(\frac{\alpha-\phi}{1-\phi}\right)^{2\beta}\frac{K}{\phi\Gamma^2}\right] = \log\left(\frac{\phi}{1-\phi}\right) \tag{2.67}$$

$$(2\beta+1)^{-1}\log\left[\left(\frac{\alpha-\phi}{1-\phi}\right)^{2\beta}\frac{K}{(1-\phi)\Gamma^2}\right] = \log\left(\frac{\phi}{1-\phi}\right) \tag{2.68}$$

Case 3: When Γ, α, and β Are Variable

Eqs. (2.67) or (2.68) can be used.

For convenience in the straight-line plotting of the permeability versus porosity data and improved numerical accuracy in correlation, Eq. (2.65) can also be expressed in a scaled form:

$$\log\left[\frac{K/K_o}{\phi/\phi_o}\right] = 2\log\left[\Gamma\sqrt{\frac{\phi_o}{K_o}}\right] + 2\beta\log\left[\frac{(\phi/\phi_o)}{(\alpha/\phi_o)-(\phi/\phi_o)}\right], \tag{2.69}$$

where K_o and ϕ_o are the base or reference permeability and porosity values, selected appropriately for a specific porous material and process. Frequently, however, K_o and ϕ_o are determined by the initial state of the porous material prior to any alteration.

Alternatively, Eq. (2.9) can be written in a scaled form as follows, by applying at a reference state, denoted by the subscript o, and at any state, without a subscript, and then by dividing the resulting expressions side by side to obtain

$$\frac{K\phi_o}{K_o\phi} = f(\beta,\Gamma)\left[\frac{\phi(1-\phi_o)}{\phi_o(1-\phi)}\right]^{2\beta}, \tag{2.70}$$

in which

$$f(\beta,\Gamma) = \left(\frac{\Gamma}{\Gamma_o}\right)^2\left(\frac{\phi_o}{1-\phi_o}\right)^{2(\beta-\beta_o)}, \tag{2.71}$$

where K_o and ϕ_o denote the reference porosity and permeability values. As demonstrated in the applications section, Eq. (2.70) can be plotted as a straight line on the logarithmic scale once the β and Γ values approach constant terminal values, denoted as β_∞ and Γ_∞, respectively, and therefore Eq. (2.71) assumes a constant value, denoted by f_∞.

The multiparameter estimation is based on a straight-line plotting of data using the least squares linear regression method. The values of the unknown parameters υ, α, ϕ_{min}, ϕ_{max}, $\left(\beta^{-1}\right)_{min}$, $\left(\beta^{-1}\right)_{max}$, Γ_{min}, Γ_{max}, Z_{min}, and (C/A) are adjusted for the best fit of the linear forms of the above-given equations to data. During this process, the values of β and Γ versus ϕ are calculated using Eq. (2.59). The quality of the best straight-line fit is measured by the coefficient of regression, R^2. Hence, the adjustment of the parameter values is guided based on the value of the regression coefficient obtained for a set of trial parameter values. Once the best straight-line fit to data is accomplished, then, the fractal dimension d is calculated by Eq. (2.60).

The statistically best estimate values of the parameters can only be determined when the number of data points exceeds the number of unknown parameters. However, each of these equations individually contains only a few parameters, and not all the parameters mentioned earlier may be present at one time in a given equation. Further, the number of unknown parameters may be reduced when the values of some parameters can be inferred or determined by other means. For example, the cement exclusion factor can be directly measured or assigned a value of $\alpha = 1$ when there are no cementing materials. Again, ϕ_{min} and ϕ_{max} can be measured by an appropriate method, such as the mercury injection technique. Nevertheless, as an approximation, one may consider that the pore connectivity vanishes $\Gamma_{min} = 0$ at the theoretical porosity limit of $\phi_{max} = 1$. The Z_{min} value can be measured at the nonzero transport threshold condition of porous media.

2.2.6 Parametric Relationships of Typical Data

Various reported data are now analyzed and correlated in order to determine the dependency of tortuosity, specific surface, and interconnectivity parameter on porosity. In the following, the data of Haughey and Beveridge (1969), Mowers and Budd (1996), and Ertekin and Watson (1991) are analyzed to investigate the dependency of tortuosity, grain (or pore) surface area, and interconnectivity parameter on porosity. The interconnectivity parameter calculated by Eq. (2.8) is only a first-order approximation, because of the inherent simplifying assumptions of the Carman–Kozeny model. The interconnectivity parameter Γ implicitly depends on porosity because the specific grain surface and tortuosity depend on porosity.

2.2.6.1 Example 1: Synthetic Spheres   Haughey and Beveridge (1969) theoretically calculated their data for regular packing of equal-sized spheres rendering cubic, orthorhombic, tetragonal-sphenoidal, and rhombohedral types. Figure 2.5 shows a plot of their specific grain surface values, calculated from their data and expressed through the reciprocal sphere diameter unit $\left[1/d_g\right]$ versus porosity. The trend indicates an increasing grain (or pore) surface area for increasing porosity. The power-law correlation of these data indicates that the pore surface vanishes at zero porosity.

2.2.6.2 Example 2: Dolomite   Mowers and Budd (1996) studied the dolomite reservoirs of Little Sand Draw Field, Wyoming, and Bindley Field, Kansas, both of which contain late-stage calcite cements. Their data represent the grain (or pore)

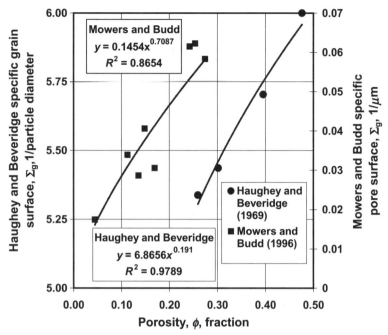

Figure 2.5 Correlation of specific grain or pore surface area versus porosity for the data of Haughey and Beveridge (1969) and Mowers and Budd (1996) (Civan, 2002d; © 2002 SPWLA, reproduced by permission of the Society of Petrophysicists and Well Log Analysts).

surface area per unit pore volume, Σ_p (Nelson, 2002, pers. comm.) in terms of the $[1/\mu m]$ unit, instead of the grain (or pore) surface area per unit grain volume Σ_g that is involved in the flow zone indicator parameter Γ of Eqs. (2.7) and (2.8). However, these two different definitions of the specific grain or pore surface area are related by the expression

$$\Sigma_g = \left(\frac{\phi}{1-\phi}\right)\Sigma_p. \tag{2.72}$$

Therefore, after applying this conversion to the data of Mowers and Budd (1996), we see a trend similar to that of the calculated data of Haughey and Beveridge (1969) (Fig. 2.5).

2.2.6.3 Example 3: Berea Sandstone   Ertekin and Watson (1991) measured the properties of low- and high-permeability Berea sandstones, referred to as range 1 (low) and range 2 (high). However, they did not report the actual permeability ranges. They show the relationship of porosity to pore surface area, expressed in the pore surface area per grain mass unit $[m^2/g]$. This latter parameter was also related to tortuosity expressed as the porous media retention time (s). Figure 2.6 shows that grain (or pore) surface area increases with increasing porosity for both low and high ranges. Figure 2.7 shows that tortuosity broadly increases with decreasing porosity,

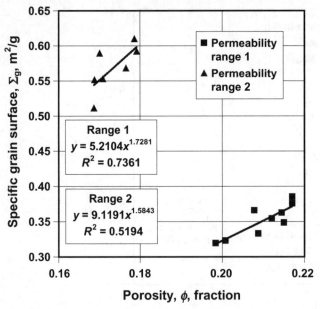

Figure 2.6   Correlation of specific grain surface versus porosity for the data of Ertekin and Watson (1991) (Civan, 2002d; © 2002 SPWLA, reproduced by permission of the Society of Petrophysicists and Well Log Analysts).

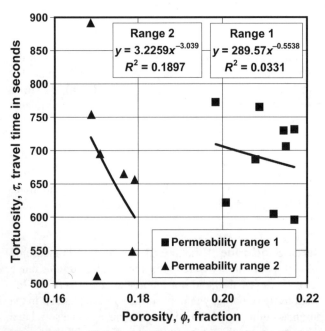

Figure 2.7   Correlation of tortuosity versus porosity for the data of Ertekin and Watson (1991) (Civan, 2002d; © 2002 SPWLA, reproduced by permission of the Society of Petrophysicists and Well Log Analysts).

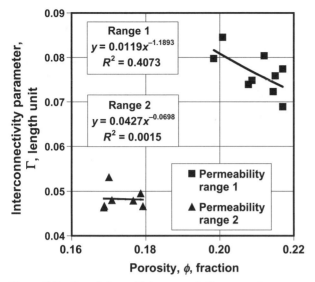

Figure 2.8   Correlation of interconnectivity parameter versus porosity for the data of Ertekin and Watson (1991) (Civan, 2002d; © 2002 SPWLA, reproduced by permission of the Society of Petrophysicists and Well Log Analysts).

but the trend is obscure. Figure 2.8 shows the interconnectivity parameter Γ calculated by Eq. (2.8) using the data of Figures 2.6 and 2.7. The parameter Γ decreases with increasing porosity. The correlation coefficients shown in Figures 2.6–2.8 are low because of significant scatter in the measured data with respect to the power-law correlations. Nevertheless, the correlations are in accord with expected trends and the boundary conditions stated by Eqs. (2.50) and (2.51).

2.2.7   Correlation of Typical Permeability Data

Several reported data are now analyzed and correlated for illustration purposes. Some of these data have been generated numerically by simulation and the others have been measured directly using natural core samples. Table 2.1 presents the best estimates of the parameters in the power-law flow unit equation obtained by least squares regression of these data. These case studies are now presented in a logical sequence of development. It is demonstrated that the interconnectivity parameter can be more rigorously determined and the related data can be effectively correlated using the power-law flow unit equation and its parametric equations. This is done by analyzing the data of Bhat and Kovscek (1999), Ridgway and Tarbuck (1967), Haughey and Beveridge (1969), Koch and Sangani (1999), Couvreur et al. (2001), Hirst et al. (2001), and Smith (1985).

2.2.7.1   Example 4: Synthetic Porous Media   Bhat and Kovscek (1999) theoretically described the variation of permeability with porosity for synthetic porous media described by Bethe network models, which use the following different pore coordination numbers, $Z = 3, 5, 7, 9, 11,$ and 13. The Bethe network gives results

TABLE 2.1   Parameters of the Power-Law Flow Unit Equation for Various Porous Media (Modified after Civan, 2002d; © 2002, Reproduced by Permission of the Society of Petrophysicists and Well Log Analysts)

Parameters	Bhat and Kovscek, (1999)	Ridgway and Tarbuck, (1967) and Haughey and Beveridge, (1969)	Couvreur et al., (2001)	Exercise Problem 1 Koch and Sangani, (1999)
v		−0.05	−0.45	−0.15
d		2.7	0.3	2.1
α	$\dfrac{\phi_o}{0.79}$		1	1
ϕ_{min}		0.03	0.03	0.133
ϕ_{max}		1	1	1
$\left(\beta^{-1}\right)_{min}$			0.05	0.585
$\left(\beta^{-1}\right)_{max}$	1.65		0.349	0.585
Γ_{min}		0	0	0
Γ_{max}	$\dfrac{0.44}{\sqrt{\phi_o/K_o}}$		10.2	6.85
Z_{min}		1.093		
A	0.2567	0.041		
B	0.4883			
C	0.1915			
D	0.1320			
C/A			100	

similar to a cubic lattice. They used this model to predict the permeability versus porosity variation as a result of the redistribution of silica in tight siliceous formations by dissolution and redeposition of silica within the pore space. Figure 2.9 shows the straight-line plots of their numerically simulated data (Kovscek, 1999, pers. comm.) according to Eq. (2.69). Bhat (1998) and Bhat and Kovscek (1999) do not report any ϕ_o and K_o reference or initial porosity and permeability values used in their network model. They only report the ratios of instantaneous to initial permeability and porosity and not the absolute values. Therefore, a best estimate of $\alpha / \phi_o = 1.27$ had to be obtained by least squares regression of their data as shown in Figure 2.9. For this value, all the lines converge to the same abscissa value of 3.7 (i.e., $1.0/((\alpha/\phi_o)-1.0)$) at the reference conditions K_o and ϕ_o. Note that the value of $\phi_o = 0.79$ calculated for $\alpha = 1$ is close to the theoretically calculated fractional porosity value of 0.7766 for the coordination number of $Z = 3$ for an ordered pack of constant diameter spheres (Ridgway and Tarbuck, 1967). Therefore, it seems reasonable for the ordered Bethe network used by Bhat and Kovscek (1999). This value of ϕ_o corresponds to a cement exclusion factor of $\alpha = 1$ in Eq. (2.9) for granular porous media that contain no cement and have no extreme pore-throat constrictions as for the Kozeny–Carman model. Had any cementation effects or a disordered network realization been considered in their model, the ϕ_o value would have been lower than 0.79. Bhat (1998) reports that the typically measured initial porosity

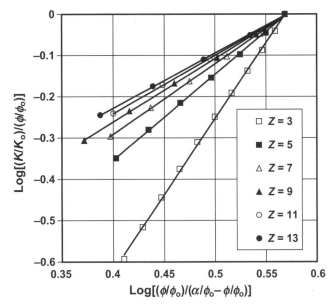

Figure 2.9   Correlation of permeability versus porosity for the data of Bhat and Kovscek (1999) according to Eq. (2.11) (Civan, 2002d; © 2002 SPWLA, reproduced by permission of the Society of Petrophysicists and Well Log Analysts).

fractions ranged from 0.38 to 0.65 and the measured initial permeability ranged from 0.1 to 10 md in South Belridge and Cymric diatomite formations, which should naturally contain some cementation and be disordered. For example, the values of $\alpha = 0.44$ for $\phi_o = 0.35$ and $\alpha = 0.83$ for $\phi_o = 0.65$ are calculated using $\alpha / \phi_o = 1.27$. Figure 2.10 shows the best straight-line fits of Eqs. (2.56) and (2.57) to the intercepts and reciprocal slopes of Eq. (2.69) versus the coordination number, which was obtained from Figure 2.9. The best estimates of the parameter values leading to Figures 2.9 and 2.10 and the coefficient of regression, indicating the quality of regression, are presented in Table 2.1, column 2.

2.2.7.2   Example 5: Glass Bead and Sand Packs   Ridgway and Tarbuck (1967) and Haughey and Beveridge (1969) provide the data of grain coordination number versus porosity determined for packs of glass beads and sands. Figure 2.11 shows that these data can be accurately represented by a regression line according to Eq. (2.52). The regression parameters A, Z_{min}, and R^2 are presented in Table 2.1, column 3. As pointed out earlier, it is not surprising that the abscissa value of $Z_{min} = 1.093$ for $\phi = \phi_{max}$ is not an integer because the volume-averaged coordination number may not necessarily be an integer value. However, the fact that this Z_{min} value is very close to unity indicates a limiting case of dead-end pores and therefore nontransmissive porous media. The fractal dimension $d = 2.7$ reported in Table 2.1 indicates that the porous medium is effectively open for flow. It is a fractal medium that is very close to the theoretical case of $d = 3$ for a nonfractal three-dimensional medium.

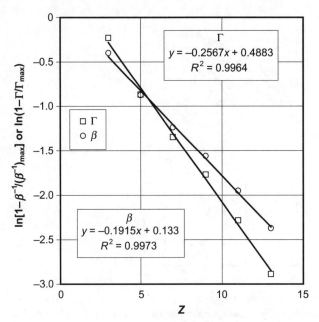

Figure 2.10   Correlation of interconnectivity parameter Γ and exponent β versus coordination number for the data of Bhat and Kovscek (1999) according to Eqs. (2.17) and (2.18) (Civan, 2002d; © 2002 SPWLA, reproduced by permission of the Society of Petrophysicists and Well Log Analysts).

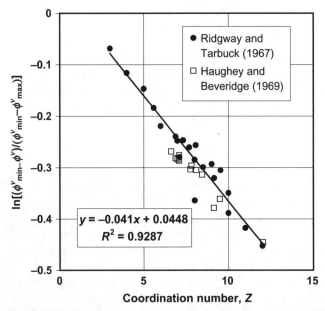

Figure 2.11   Porosity versus coordination number for the data of Ridgway and Tarbuck (1967) and Haughey and Beveridge (1969) according to Eq. (2.16) (Civan, 2002d; © 2002 SPWLA, reproduced by permission of the Society of Petrophysicists and Well Log Analysts).

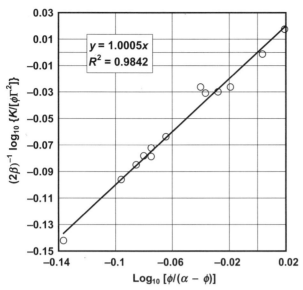

Figure 2.12   Correlation of the data of Couvreur et al. (2001) according to Eq. (2.13) (Civan, 2002d; © 2002 SPWLA, reproduced by permission of the Society of Petrophysicists and Well Log Analysts).

2.2.7.3 *Example 6: Silty Soil*   Couvreur et al. (2001) report measurements of permeability and porosity of dry silty soil samples taken from depths of 0.5–1.0 m in Louvain-la-Neuve, Belgium. Figure 2.12 shows a data fit to a graphical depiction of Eq. (2.65). The best estimates of the parameter values and the coefficient of regression are presented in Table 2.1, column 4. The value of $d = 0.3$ implies that the area open for flow belongs to a constricted fractal porous medium in contrast to the value of $d = 3$ for a theoretical nonfractal three-dimensional medium. Figure 2.13 shows the variation of Γ and β with ϕ according to Eq. (2.59). Caution: the plot of curves constructed in Figure 2.13 covers the range of $0 \leq \phi \leq 1$. In reality, the range utilized in the analysis is limited by the range of actual data. Note that $0 \leq \Gamma \leq 10.2$. The range of the exponent β, that is, $2.87 \leq \beta \leq 20$, is significantly greater than the constant value of $\beta = 1.0$ implied by the Kozeny–Carman equation, that is, Eq. (2.7). This confirms the claim by Couvreur et al. (2001) that the Kozeny–Carman equation does not represent their data.

2.3   PERMEABILITY OF POROUS MEDIA UNDERGOING ALTERATION BY SCALE DEPOSITION

Although complete formulations for the hydraulic tube model incorporating fractal dimensions are presented in this section, the resulting expressions are nevertheless simplified and/or analytically solved considering the conditions of usual laboratory tests using core plugs extracted from porous materials. These simplified expressions are sufficient for purposes of many applications. When applications to

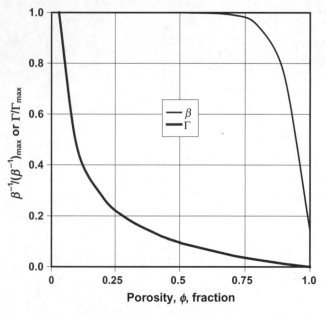

Figure 2.13 Interconnectivity parameter Γ and exponent β versus porosity for the data of Couvreur et al. (2001) according to Eq. (2.19) (Civan, 2002d; © 2002 SPWLA, reproduced by permission of the Society of Petrophysicists and Well Log Analysts).

porous media involve varying fluid and flow conditions, the nonsimplified expressions should be used, thus requiring numerical solutions by appropriate numerical methods.

Following Walsh et al. (1982) and Lichtner (1992), a mineral dissolution (backward) or precipitation (forward) reaction, such as the gypsum and quartz precipitation reactions given by Carnahan (1990),

$$Ca^{2+}_{(aq)} + SO^{2-}_{4(aq)} + 2H_2O \leftrightarrow CaSO_4 \cdot 2H_2O_{(s)} \tag{2.73}$$

and

$$Si(OH)^{\circ}_{4(aq)} \leftrightarrow SiO_{2(s)} + 2H_2O, \tag{2.74}$$

can be represented in general by (Liu et al., 1997)

$$M + \sum_{i=1}^{N} v_i S_i = 0, \tag{2.75}$$

where M denotes a mineral, S_i: $i = 1, 2, \ldots$; N represents the number of aqueous species associated with the mineral reaction; and v_i are some stochiometric coefficients, which are positive for the products and negative for the reactants. The actual ion activity product for Eq. (2.75) is given by

$$K_{ap} = \prod_{i=1}^{N} a^{v_i}_{i,\text{actual}}, \tag{2.76}$$

where a_i is the activity of species i. The equilibrium saturation ion activity product is given by

$$K_{sp} = \prod_{i=1}^{N} a_{i,\text{equilibrium}}^{v_i}.$$ (2.77)

Hence, the saturation ratio can be defined as (Oddo and Tomson, 1994):

$$F_s = K_{ap}/K_{sp}.$$ (2.78)

The rate of pore volume variation by dissolution/precipitation of a solid in porous media is directly related to several driving forces and/or factors (Steefel and Lasaga, 1990; Ortoleva, 1994; Holstad, 1995; and Civan, 1996c): (1) deviation of the saturation ratio, F_s, from unity ($F_s < 1$, $F_s = 1$, and $F_s > 1$ indicate under saturation, equilibrium, and super saturation, respectively); (2) volume of pore fluid (for precipitation, the volume of fluid is equal to the instantaneous pore volume, which is the initial pore volume minus the volume of the precipitating deposits; for dissolution, it is equal to the initial pore volume plus the volume of dissolving deposits); and (3) pore surface available and having affinity for dissolution/precipitation. Hence, a rate equation for porosity variation by dissolution/precipitation can be written as

$$-dV_p/dt = k(F_s - 1)V_p \Sigma_p^m,$$ (2.79)

where k is a dissolution/precipitation rate coefficient, m is an exponent of pore surface participation, and t is time. Note that only the pore surface having affinity to the scale-forming precipitate is considered.

Thus, substituting Eqs. (2.14) and (2.19) into Eq. (2.79) yields the following rate equation for porosity variation by scale:

$$-d\phi/dt = k_\phi \phi^{1+r},$$ (2.80)

where

$$r = mD/3$$ (2.81)

and

$$k_\phi = k(F_s - 1)C^m n^{m-r} V_b^r.$$ (2.82)

The initial porosity is given by

$$\phi = \phi_o, t = 0.$$ (2.83)

When the saturation ratio, F_s, varies at actual pore fluid conditions, Eqs. (2.80)–(2.83) should be numerically integrated along with the reaction–transport equations to consider the effect of the varying saturation ratio of the solution. However, when the saturation ratio is maintained constant, such as the case with the specially designed laboratory tests conducted to generate the experimental data used in the applications section, the solution of Eqs. (2.80)–(2.83) can be obtained analytically as follows:

$$\phi/\phi_o = \exp(-k_\phi t), m = 0$$ (2.84)

and

$$\phi^{-r} - \phi_0^{-r} = k_\phi t, \, m \neq 0, \tag{2.85}$$

which can be rearranged as

$$\phi/\phi_o = \left(1 + k_\phi \phi_o^r t\right)^{-1/r}. \tag{2.86}$$

These analytical solutions are convenient and useful for analyzing and interpreting precipitation/dissolution-induced porosity and permeability variation data obtained under constant saturation conditions. They can also be used to determine the lumped dissolution/precipitation rate coefficient, k_ϕ, and the lumped r parameter under given saturation conditions, that is, when F_s is constant.

Although the above derivation primarily focused on the porosity variation, $(d\phi/dt)$, an analogous equation can be readily derived for the rate of solid dissolution/precipitation, $(d\varepsilon_s/dt)$, by expressing the porosity of the inert porous matrix, ϕ_o, as a sum of the instantaneous porosity, ϕ, and the fractional bulk volume, ε_s, occupied by dissolving/precipitating solid or scales as

$$\phi_o = \phi + \varepsilon_s. \tag{2.87}$$

If σ denotes the scale volume expressed as the occupied fraction of the original pore volume fraction, ϕ_o, then it follows that

$$\varepsilon_s = \phi_o \sigma. \tag{2.88}$$

Thus, invoking Eq. (2.88) into Eq. (2.87), the instantaneous porosity is given by

$$\phi = \phi_o (1 - \sigma). \tag{2.89}$$

2.3.1 Permeability Alteration by Scale Deposition

Substituting Eq. (2.84) for $m = 0$ into Eq. (2.70) yields

$$\frac{K}{K_o} = e^{-\alpha t} \left[\frac{1 - \phi_o}{1 - \phi_o e^{-k_\phi t}} \right]^{2\beta}, \tag{2.90}$$

where

$$\alpha = (1 + 2\beta) k_\phi. \tag{2.91}$$

When k_ϕ is small, Eq. (2.90) can be approximated as

$$K/K_o \cong e^{-\alpha t}. \tag{2.92}$$

Substituting Eq. (2.86) for $m \neq 0$ into Eq. (2.70) yields

$$\frac{K}{K_o} = (1 + \lambda t)^{-\delta} \left[\frac{1 - \phi_o}{1 - \phi_o (1 + \lambda t)^{-1/r}} \right]^{2\beta}, \tag{2.93}$$

where

$$\lambda \equiv k_\phi \phi_o^r \tag{2.94}$$

and

$$\delta \equiv (1+2\beta)/r. \tag{2.95}$$

For short times and/or small k_ϕ, Eq. (2.93) can be approximated by

$$K/K_o \cong (1+\lambda t)^{-\delta}. \tag{2.96}$$

2.3.2 Permeability Alteration in Thin Porous Disk by Scale Deposition

For convenience in laboratory measurements by flowing a solution with constant concentration (hence, F_s = constant also), Schechter and Gidley (1969) used cross-sectional area measurements of single pore sizes during surface dissolution in thin porous disks. Therefore, interpretation of their measurements requires an area formulation for thin slices instead of the bulk volume formulation of porous media given earlier.

The preceding expressions can be applied for a thin slice shown in Figure 2.14, for which $L_h \cong L_b$, and therefore $\tau \cong 1.0$. Consequently, Eq. (2.71) becomes

$$\phi = nA_h/A_b. \tag{2.97}$$

Thus, substituting Eq. (2.97) into Eqs. (2.70), (2.84)–(2.85), (2.90), (2.92), (2.93), and (2.96), a thin-section formulation can be obtained as follows:

$$\frac{KA_{ho}}{K_o A_h} = f(\beta, \gamma)\left[\frac{A_h(A_B - A_{ho})}{A_{ho}(A_B - A_h)}\right]^{2\beta}, \tag{2.98}$$

in which, from Eq. (2.71),

$$f(\beta, \gamma) = \left(\frac{\gamma}{\gamma_o}\right)^2 \left(\frac{A_{ho}}{A_B - A_{ho}}\right)^{2(\beta - \beta_o)}, \tag{2.99}$$

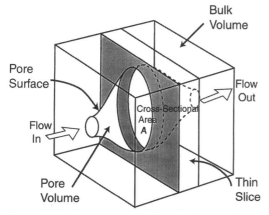

Figure 2.14   Pore space and thin slice in porous media (Civan, 2001; © 2001 AIChE, reproduced by permission of the American Institute of Chemical Engineers).

where

$$A_B = A_b/n. \tag{2.100}$$

For $m = 0$, Eq. (2.84) becomes

$$A_h/A_{ho} = \exp(-k_\phi t). \tag{2.101}$$

For $m \neq 0$, Eq. (2.85) becomes

$$A_h^{-r} - A_{ho}^{-r} = k_A t, \tag{2.102}$$

where

$$k_A = k_\phi/A_B^r. \tag{2.103}$$

Eq. (2.86) becomes

$$A_h/A_{ho} = \left(1 + A_{ho}^r k_A t\right)^{-1/r}. \tag{2.104}$$

For $m = 0$, Eq. (2.90) becomes

$$\frac{K}{K_o} \cong e^{-\alpha t} \left[\frac{A_B - A_{ho}}{A_B - A_{ho} e^{-k_\phi t}} \right]^{2\beta}. \tag{2.105}$$

For small k_ϕ and/or short time, Eq. (2.105) simplifies as

$$\frac{K}{K_o} \cong e^{-\alpha t}. \tag{2.106}$$

For $m \neq 0$, Eq. (2.93) becomes

$$\frac{K}{K_o} \cong (1 + \lambda t)^{-\delta} \left[\frac{A_B - A_{ho}}{A_B - A_{ho}(1 + \lambda t)^{-1/r}} \right]^{2\beta}. \tag{2.107}$$

Eq. (2.96) is used for short time and/or small k_ϕ (or λ) as

$$K/K_o \cong (1 + \lambda t)^{-\delta}. \tag{2.108}$$

2.3.3 Data Analysis and Correlation Method

The model described earlier involves a number of lumped parameters, including β, γ, r, k_ϕ, α, λ, and δ. The basic parameters, such as the dissolution/precipitation rate coefficient and various fractal parameters, have been grouped to derive these parameters. The best estimates of the lumped parameters can be determined by the least squares linear fit of the expressions derived for porosity and permeability variation by scales, after these expressions have been rearranged in forms suitable for straight-line plotting. For given fluid and porous materials, laboratory tests can be conducted, and the data obtained can be utilized in a similar fashion to determine the best estimates of the lumped parameters for prescribed fluid and porous material systems. This approach is referred to as the calibration or tuning of the model and is sufficient for many practical applications. However, if also the specific values of the individual basic parameters, including the dissolution/precipitation rate coefficient and the fractal parameters, are being sought, additional data may be required via measurements by appropriate means. For example, the dissolution/precipitation rate coefficient can be measured by various techniques, such as those described by Civan (2000b,c, 2002c). The participating pore surface can be measured by contact angle measurements, such as those described by Civan (1997). The parameters of the fractal equation can be determined by a procedure similar to Johns and Gladden (2000). The coordination number and pore surface can be determined by the method of Liu and Seaton (1994). The other basic parameters can also be measured by suitable procedures. However, facilitating the defining expressions of the lumped parameters given in this section may reduce the number of basic parameters to be determined by individual measurements. Although it is feasible to determine the values of all the basic parameters in this way, this tedious task is not warranted, and determination of the values of the lumped parameters as described in the applications section, referred to as the model calibration procedure, is sufficient for most practical problems. After all, the sole purpose of developing a lumped-parameter model is to avoid such cumbersome measurements.

2.3.4   Correlation of Scale Effect on Permeability

2.3.4.1   Example 7: Scale Formation   Bertero et al. (1988) measured the permeability reduction caused by scale formation during the simultaneous injection of two incompatible brine solutions into porous rock core samples. They determined the amounts of scales formed in the cores by weighting the cores at various times during the experiments. They measured permeability as percentage of the initial permeability versus the scale volume as percentage of the initial pore volume for three core tests, identified as cores A, B, and C, in the form of smooth curves. However, they do not report the actual values of the measured data points. In order to construct straight-line plots according to Eq. (2.70), these curves were first discretized to generate a set of point readings of the scale quantities at 1% scale volume intervals, and the instantaneous permeability and porosity values were calculated using the reported initial porosity and permeability values of 13.9% and 3.8 md for core A, 12.6% and 45.6 md for core B, and 15.0% and 262.0 md for core C. Note that 1 md (millidarcy) $\cong 10^{-15}\,m^2$. The instantaneous porosity was calculated by Eq. (2.89). Figure 2.15 shows the best straight-line plots of the cores A, B, and C data according to Eq. (2.70). The coefficients of the least squares linear regressions were obtained as $R^2 = 0.9817, 0.9915$, and 0.9958 for cores A, B, and C, respectively, very close to 1.0, indicating that Eq. (2.70) accurately represents the permeability–porosity relationship during scale formation. The best permeability–porosity correlation was obtained as

$$\frac{K}{K_o} = f_\infty \frac{\phi}{\phi_o} \left[\frac{\phi(1-\phi_o)}{\phi_o(1-\phi)} \right]^{2\beta}, \tag{2.109}$$

where

$$K_o = 3.8 \text{ md}, \ \phi_o = 0.139, \ \beta = 8.078, \ f_\infty = 0.59, \text{ core A};$$
$$K_o = 45.6 \text{ md}, \ \phi_o = 0.126, \ \beta = 12.55, \ f_\infty = 0.52, \text{ core B; and}$$
$$K_o = 262 \text{ md}, \ \phi_o = 0.15, \ \beta = 25.65, \ f_\infty = 1.0, \text{ core C}.$$

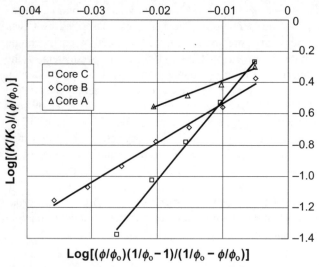

Figure 2.15   Straight-line plots of permeability versus porosity data of Bertero et al. (1998) by Civan's (1996) power-law flow unit equation (Civan, 2001; © 2001 AIChE, reproduced by permission of the American Institute of Chemical Engineers).

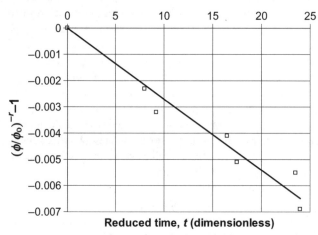

Figure 2.16   Straight-line plot of Glover and Guin (1973) data of porosity variation by acid dissolution (Civan, 2001; © 2001 AIChE, reproduced by permission of the American Institute of Chemical Engineers).

2.3.4.2   *Example 8: Acid Dissolution*   Glover and Guin (1973) measured the permeability and porosity enhancement during acid dissolution of porous media. They injected a dilute aqueous HF acid solution into Pyrex 774 (81% SiO_2) sintered porous glass disks. They monitored the pressure drop across the disk to determine the permeability and measured the porosity of the disks at various time intervals. They reported that determination of the initial porosity may have introduced a small error. Here, the $\phi_o = 0.29$ average value of the initial porosity of the four disks was used. Figure 2.16 presents a straight-line plot of their measured porosity versus

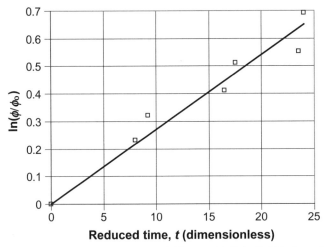

Figure 2.17   Straight-line plot of Glover and Guin (1973) data of porosity variation by acid dissolution (Civan, 2001; © 2001 AIChE, reproduced by permission of the American Institute of Chemical Engineers).

reduced time data for all disks according to Eq. (2.85). The best fit was obtained with a coefficient of the least squares linear regression of $R^2 = 0.9464$. The best porosity versus time correlation was obtained as

$$\phi^{-r} = \phi_o^{-r} + k_\phi t, \phi_o = 0.29, r = 0.01, k_\phi = 3.037 \times 10^{-4}, \tag{2.110}$$

where t is reduced or dimensionless. Because the value of r is small and r appears in the exponent, the accuracy of k_ϕ obtained by Eq. (2.110) is expected to be low. Because $r \cong 0$ and $m \cong 0$ by Eq. (2.81), the best straight-line plot of the same data according to Eq. (2.84) is given in Figure 2.17. The best least squares linear fit of the data was obtained as follows, with a coefficient of regression of $R^2 = 0.9465$:

$$\phi / \phi_o = \exp(k_\phi t), k_\phi = 2.71 \times 10^{-2}, \tag{2.111}$$

where t is reduced or dimensionless. Thus, Eq. (2.84) closely represents their experimental data.

Because their discrete porosity and permeability enhancement data were collected at different time instances, not at the same discrete times, these data were first smoothed by best curve fitting, and then the discrete values of porosity and permeability corresponding to the same time instances were sampled at 10 unit intervals of the reduced time and were plotted according to Eq. (2.70). The best straight-line fit was obtained as the following with a coefficient of the least squares linear regression of $R^2 = 0.992$, indicating that Eq. (2.70) satisfactorily represents the data:

$$\frac{K}{K_o} = f_\infty \frac{\phi}{\phi_o} \left[\frac{\phi(1-\phi_o)}{\phi_o(1-\phi)} \right]^{2\beta}, \phi_o = 0.29, \beta = 0.295, f_\infty = 1.2. \tag{2.112}$$

Note that Glover and Guin (1973) did not report the initial permeability value, K_o, and only reported K/K_o. Figure 2.18 shows a plot of the Glover and Guin (1973) permeability versus reduced time data according to Eq. (2.96). The least squares

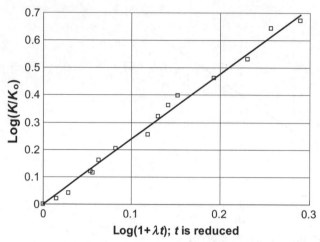

Figure 2.18 Straight-line plot of Glover and Guin (1973) data of permeability variation by acid dissolution (Civan, 2001; © 2001 AIChE, reproduced by permission of the American Institute of Chemical Engineers).

linear fit of the data was obtained as the following with a coefficient of regression of $R^2 = 0.9912$. This indicates that Eq. (2.96) closely represents the data:

$$K/K_o = (1+\lambda t)^\delta, \lambda = 3.5\times10^{-2}, \delta = 2.4, \tag{2.113}$$

where t is reduced or dimensionless.

2.3.4.3 Example 9: Wormhole Development   Schechter and Gidley (1969) measured the pore enlargement and permeability improvement of thin porous disks of an Indiana limestone during wormhole development by exposure to 1% HCl acid at room temperature. They measured the enlargement of a typical pore area, estimated by means of a microscope, and the permeability enhancement versus the exposure time during the test. Because they report the measurements of the cross-sectional area of pores in thin slices of the limestone samples, formed in the shape of wormholes, instead of the porosity of the limestone block, the specifically derived Eqs. (2.102)–(2.104) for cross-sectional area formulation were facilitated in this case. The initial cross-sectional area of the pore has been estimated as $A_{ho} = 0.24$ mm^2 by smoothing their data and extrapolating to the initial time of zero, because their data point indicating the initial pore area somewhat deviates from the general trend of the overall data set and therefore appears to be somewhat inaccurate. Figure 2.19 shows the best fit of Eq. (2.102) to their data, with a coefficient of regression of $R^2 = 0.9944$. Hence, Eq. (2.102) satisfactorily represents their data. The best correlation of the pore area versus time is obtained as

$$A_h^{-r} = A_{ho}^{-r} - k_A t, A_{ho} = 0.24 \text{ mm}^2, r = 2.0\times10^{-3}, k_A = 3.0\times10^{-6} \text{ s}^{-1}. \tag{2.114}$$

The accuracy of k_A as determined by Eq. (2.114) is suspected to be low because r has a small value and appears in the exponent. Thus, for $r \cong 0$ and $m \cong 0$ by Eq. (2.81), the value of k_A can be more accurately determined using Eq. (2.101). The

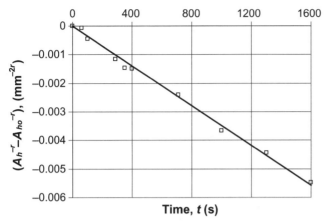

Figure 2.19 Straight-line plot of Schechter and Gidley (1969) data of pore area variation by acid dissolution (Civan, 2001; © 2001 AIChE, reproduced by permission of the American Institute of Chemical Engineers).

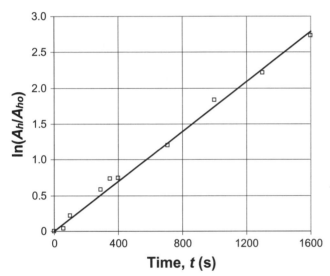

Figure 2.20 Straight-line plot of Schechter and Gidley (1969) data of pore area variation by acid dissolution (Civan, 2001; © 2001 AIChE, reproduced by permission of the American Institute of Chemical Engineers).

best straight-line fit is obtained as follows, with a coefficient of regression of $R^2 = 0.9945$, as shown in Figure 2.20, when Eq. (2.101), which assumes $m = 0$, is used:

$$\ln(A_h / A_{ho}) = k_\phi t, \quad A_{ho} = 0.24 \text{ mm}^2, \quad k_\phi = 1.7 \times 10^{-3} \text{ s}^{-1}. \tag{2.115}$$

Figure 2.21 shows the best straight-line plot of the data of Schechter and Gidley (1969) according to Eq. (2.98). Because they measured the pore cross-sectional area and permeability at different times, their data were first best smooth curve fitted and then the data points were sampled at 200-s time intervals to generate

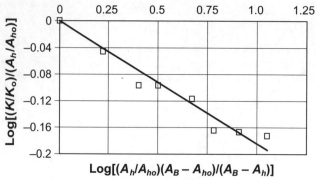

Figure 2.21   Straight-line plot of permeability versus porosity data of Schechter and Gidley (1969) by Civan's (1996) power-law flow unit equation (Civan, 2001; © 2001 AIChE, reproduced by permission of the American Institute of Chemical Engineers).

a data set consisting of the discrete point values of the area and permeability at prescribed time instances. The data points yield the best straight-line fit with a coefficient of regression of $R^2 = 0.9462$, close to 1.0, indicating that Eq. (2.98) accurately represents the data points. The best pore area–permeability correlation is obtained as

$$\frac{K}{K_o} = f_\infty \left(\frac{A_h}{A_{ho}}\right) \left[\frac{A_h (A_B - A_{ho})}{A_{ho} (A_B - A_h)}\right]^{2\beta}, A_B = 30 \text{ mm}^2, A_{ho} = 0.24 \text{ mm}^2, \qquad (2.116)$$
$$K_o = 2 \text{ md}, \beta = -0.092, f_\infty = 1.0.$$

The minus sign in the beta value indicates $K \rangle K_o$ for scale dissolution. Note that the representative bulk area involving the pore area investigated by Schechter and Gidley (1969) was not reported and therefore had to be estimated as $A_B = 30 \text{ mm}^2$ by the best fit of Eq. (2.98) to the data.

The plot of the Schechter and Gidley (1969) permeability versus time data according to Eq. (2.108) similar to Figure 2.10 yields the following permeability versus time correlation. The coefficient of regression, $R^2 = 0.995$, close to 1.0, indicates that Eq. (2.108) accurately represents their data.

$$K/K_o = (1 + \lambda t)^\delta, K_o = 2 \text{ md}, \lambda = 6.0 \times 10^{-4} \text{ s}^{-1}, \delta = 3.1. \qquad (2.117)$$

Figure 2.22 gives the plot of the same data according to Eq. (2.106) as follows, with a coefficient of regression of $R^2 = 0.9824$, lower than that of Eq. (2.108), but still sufficiently high for practical applications.

$$K/K_o = \exp(\alpha t), K_o = 2 \text{ md}, \alpha = 1.4 \times 10^{-3} \text{ s}^{-1}. \qquad (2.118)$$

2.4   TEMPERATURE EFFECT ON PERMEABILITY

Frequently, processes involving in enhanced oil recovery, geothermal energy recovery, flue gas sequestration, and well stimulation require injection of fluids into

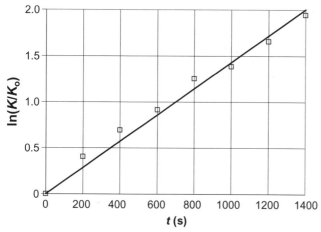

Figure 2.22 Straight-line plot of Schechter and Gidley (1969) data of permeability variation by acid dissolution (Civan, 2001; © 2001 AIChE, reproduced by permission of the American Institute of Chemical Engineers).

subsurface formations, such as petroleum, coal-bed, and geothermal reservoirs, and groundwater aquifers. When temperature of the pore fluid is different from that of the porous matrix, permeability may vary significantly and affect ability of flowing fluids through porous media (Civan, 2006c, 2010c). Permeability may be altered by expansion/contraction of grains and other constituents of porous formation and by creation of microfractures owing to thermal shock caused by suddenly induced thermal stress. Other processes, such as structural alterations by comminuting, fusing, and embedment of grains and various constituents of porous matrix, and particulate migration, may also affect permeability in very complicated ways. Contribution of various factors to permeability alteration has been addressed by nature and mechanism of relevant processes induced by temperature variation (Civan, 2010c). Somerton (1992) points out that thermal expansion of minerals by temperature increase causes pore-throat contraction in porous rocks, which in turn should increase tortuosity and formation resistivity factor, and hence should decrease permeability. Somerton (1992) emphasizes, however, that permeability variation may be reversible or irreversible, and present studies are not conclusive as to the nature of dependence of permeability and formation resistivity factor with temperature through tortuosity.

Literature on the effect of temperature upon permeability is inconsistent, contradicting, and conflicting, indicating decreasing, increasing, or practically no-change trends with an increase of temperature (Gobran et al., 1987). However, this observation should not be surprising when a wide range of factors affecting absolute permeability during measurements is considered. In fact, permeability alteration may occur by many different mechanisms; some of those caused by temperature variation were mentioned above. Other factors may also influence the measured permeability of rocks. For example, results differ between tests conducted with confined and unconfined samples. Further, inadvertently, methods used for measurements of permeability may induce adverse processes, such as rock–fluid incompatibility, particulate migration, and swelling, which might cause irreversible permeability impairment in many complicated ways (Civan, 2007a, 2010c). In another words, permeability measurements reported in various literatures may not necessarily reflect upon the sole effect of temperature on permeability.

A limited number of efforts have dealt with theoretical formulation of permeability of porous formation undergoing temperature variation. Marshall's equation considering pore-throat contraction and porosity reduction by temperature increase as described by Somerton (1992) does not correlate the permeability versus temperature data adequately. Needham (1976), Gupta and Civan (1994b), and Civan (2008c, 2010c) described the thermal effects on permeability of porous formations by taking into account variations of porosity, tortuosity, and pore surface with temperature by the grain expansion effects using the Kozeny–Carman equation. Civan (2008c) described the permeability variation by the mechanism of thermal expansion/contraction of porous media grains. Hence, the other mechanisms mentioned earlier are not considered in this formulation. Civan (2008c) accomplished this by two approaches as described in the following sections.

2.4.1 The Modified Kozeny–Carman Equation

This approach is a mechanistic formulation that incorporates the effect of temperature by modifying the Kozeny–Carman equation (Carman, 1937a,b, 1956) in terms of variation of its parameters by temperature. For this purpose, variations of porosity, tortuosity, and pore surface with temperature are expressed in terms of grain expansion effects. The coefficient of volumetric thermal expansion is facilitated to determine the variation of the petrophysical characteristics of porous media. Hence, the accuracy of this formulation is subject to the validity of the inherent assumptions of the Kozeny–Carman equation. The idealized representation of porous formation grains to be of spherical shapes and flow paths as a bundle of cylindrical capillary tubes may not necessarily be a true representation of complex pore structures and flow patterns encountered in natural porous formations, such as diatomite. Therefore, such simplified description merely serves as a convenient means of correlating the permeability of complex natural formations by a simple model in which parameters are considered as representative apparent lump values at the macroscopic scale. The geometric shape factor included in the formulation is expected to help alleviate some of the problems associated with the deviation of actual core samples from such an idealized realization of porous media.

Consider the Kozeny–Carman equation (Carman, 1937a,b, 1956), written in the following form (Civan, 2002d, 2007a, 2008c):

$$\sqrt{\frac{K}{\phi}} = \frac{1}{\Sigma_g \sqrt{F_s \tau}}\left(\frac{\phi}{1-\phi}\right), \tag{2.119}$$

where K and ϕ are the permeability and porosity, respectively; F_s is a geometric shape factor; Σ_g denotes the grain (or pore) surface area per unit volume of grains present in bulk porous medium; and τ denotes the tortuosity of porous medium, defined as the ratio of the apparent length of effective mean flow path to the physical length of the bulk porous medium. In the formulation presented in the rest of this section, the initial or reference values of all the parameters are indicated by the subscript o.

Assuming spherical grains of diameter d_g, the grain volume is given by

$$V_g = \frac{\pi d_g^3}{6}. \tag{2.120}$$

The grain surface area is given by

$$A_g = \pi d_g^2. \tag{2.121}$$

Then, the specific grain surface is given by

$$\Sigma_g = 6/d_g. \tag{2.122}$$

The bulk volume V_b of porous formation is the sum of the grain volume V_g and the pore volume V_p according to

$$V_g + V_p = V_b. \tag{2.123}$$

The porosity is given by

$$\phi = V_p / V_b. \tag{2.124}$$

Combining Eqs. (2.123) and (2.124) yields the following equation for the grain volume:

$$V_g = (1 - \phi)V_b. \tag{2.125}$$

The volumetric coefficients of thermal variation of pore volume V_p, grain volume V_g, and bulk volume V_b under constant pressure p are defined, respectively, by (Civan, 2008c)

$$\beta_{pV} = -\frac{1}{V_p}\left(\frac{\partial V_p}{\partial T}\right)_p, \tag{2.126}$$

$$\beta_{gV} = \frac{1}{V_g}\left(\frac{\partial V_g}{\partial T}\right)_p, \tag{2.127}$$

and

$$\beta_{bV} = \frac{1}{V_b}\left(\frac{\partial V_b}{\partial T}\right)_p. \tag{2.128}$$

Substituting Eq. (2.120) into Eq. (2.127), the coefficient of volume expansion of grains is expressed by

$$\beta_{gV} = \frac{1}{\frac{\pi d_g^3}{6}}\left[\frac{\partial}{\partial T}\left(\frac{\pi d_g^3}{6}\right)\right]_p = \frac{3}{d_g}\left(\frac{\partial d_g}{\partial T}\right)_p. \tag{2.129}$$

The coefficient of linear expansion of grains is given by

$$\beta_{gL} = \frac{1}{d_g}\left(\frac{\partial d_g}{\partial T}\right)_p. \tag{2.130}$$

Therefore, comparing Eqs. (2.129) and (2.130) renders

$$\beta_{gL} = \frac{1}{3}\beta_{gV}. \tag{2.131}$$

Eqs. (2.126)–(2.131) can be expressed in terms of porosity as the following by means of Eqs. (2.124) and (2.125):

$$\beta_{pV} = -\frac{1}{\phi}\left(\frac{\partial\phi}{\partial T}\right)_p - \beta_{bV}, \tag{2.132}$$

$$\beta_{gV} = -\frac{1}{1-\phi}\left(\frac{\partial\phi}{\partial T}\right)_p + \beta_{bV}, \tag{2.133}$$

and

$$\beta_{gV} = 3\beta_{gL} = \left(\frac{\phi}{1-\phi}\right)\beta_{pV} + \left(\frac{1}{1-\phi}\right)\beta_{bV}. \tag{2.134}$$

The fractional volumetric expansion coefficient α is defined by

$$\alpha = \frac{V_g}{V_{go}} - 1, \tag{2.135}$$

where V_g and V_{go} denote the instantaneous and initial grain volumes, respectively. Thus, substituting Eq. (2.135) into Eq. (2.127) yields

$$\beta_{gV} = \frac{1}{1+\alpha}\left[\frac{\partial(1+\alpha)}{\partial T}\right]_p = \left[\frac{\partial\ln(1+\alpha)}{\partial(T-T_o)}\right]_p, \tag{2.136}$$

where T_o is chosen to be a suitable reference temperature. Inferred by Eq. (2.136), an appropriate empirical equation, such as a power series given by the following expression, can be used to correlate the volumetric coefficient of thermal expansion:

$$\beta_{gV} = a + b(T-T_o) + c(T-T_o)^2 \ldots\ldots, \tag{2.137}$$

where a, b, c, . . . are empirical constants. For example, Civan (2000b) correlated the quartz and calcite thermal expansion data of Prats (1982) very accurately using a linear form of Eq. (2.137). Hence, substituting Eqs. (2.137) into Eq. (2.129) yields

$$\frac{3}{d_g}\left(\frac{\partial d_g}{\partial T}\right)_p = a + b(T-T_o) + c(T-T_o)^2 \ldots\ldots \tag{2.138}$$

Then, integrating Eq. (2.138), the variation of grain diameter by temperature is given by

$$d_g / d_{go} = \exp[f(T)/3], \tag{2.139}$$

where $f(T)$ is a function, given as

$$f(T) = a(T-T_o) + (b/2)(T-T_o)^2 + (c/3)(T-T_o)^3 \ldots\ldots \tag{2.140}$$

Similarly, the porosity variation with temperature can be expressed as the following by substituting Eqs. (2.137) into Eq. (2.133) and then integrating with respect to temperature:

$$\frac{1-\phi}{1-\phi_o} = \frac{V_{bo}}{V_b} \exp[f(T)].$$

(2.141)

The tortuosity of porous formation may be correlated by Archie's empirical power-law expression (Archie, 1942):

$$\tau^m = F_\tau \phi,$$

(2.142)

where F_τ is the formation resistivity factor and m is an exponent representing the cementation factor. The relationship of tortuosity on porosity depends on the flow direction relative to the texture and fabric of porous formation, and the exponent m varies typically in the range of $0.5 \leq m \leq 2.0$ for Eq. (2.142) (Salem and Chilingarian, 2000, and Scholes et al., 2007).

Considering Eqs. (2.122), (2.139), (2.141), and (2.142) in the ratio of Eq. (2.119) applied at the present and initial/reference conditions yields the following relationship:

$$\frac{K}{K_o} = \frac{F_{so}}{F_s} \left(\frac{F_{\tau o}^{\frac{1}{m_o}}}{F_\tau^{\frac{1}{m}}} \right) \left(\frac{V_b}{V_{bo}} \right)^2 \left\{ \frac{\left[1 - (1-\phi_o) \dfrac{V_{bo}}{V_b} \exp[f(T)] \right]^{3-\frac{1}{m}}}{\phi_o^{3-\frac{1}{m_o}}} \right\} \exp\left[-\frac{4}{3} f(T) \right].$$

(2.143)

Recall the quantities indicated by subscript o refer to values taken at a reference temperature T_o that is considered as the initial condition. The above formulation is general up to this point. However, constitutive equations are required to express the effect of temperature on the geometric shape factor F_s, formation resistivity factor F_τ, cementation factor m, and bulk volume of porous formation V_b. For illustration purposes, constant average values are assigned here to these parameters, and deviations of actual parameter values from average values are alleviated by estimating the coefficients of the $f(T)$ polynomial to achieve the best fit of measured permeability versus temperature data. Consequently, Eq. (2.143) may be simplified as the following under these conditions:

$$\frac{K}{K_o} = \left\{ \frac{1 - (1-\phi_o) \exp[f(T)]}{\phi_o} \right\}^{\left(3-\frac{1}{m}\right)} \exp\left[-\frac{4}{3} f(T) \right].$$

(2.144)

The unknown parameters associated with this equation, namely, $a, b, c, \ldots$ and m, are determined to obtain the best regression of experimental data. Eq. (2.144) can be used for the correlation of permeability variation with temperature as demonstrated later in the applications.

2.4.2 The Vogel–Tammann–Fulcher (VTF) Equation

This approach describes the temperature dependency of permeability using the VTF equation (Vogel, 1921; Fulcher, 1925; Tammann and Hesse, 1926). This is a practical equation yielding meaningful information about the energy of activation of the lump effects of the thermal processes caused by temperature variation. Many studies demonstrated that the VTF equation accurately describes the temperature effect on various properties of materials (Civan, 2004, 2005b, 2006a,b, 2007b,c,d). Civan (2008b, 2010c) showed its applicability for variation of permeability of porous rock by temperature. However, assigning physical meaning to its parameters in terms of properties of constituent minerals of porous rocks requires further research. For example, the characteristic limit temperature corresponds to phase transition temperature in the case of polymers and ceramics.

The VTF equation is given by the following general relationship constituting an asymptotic exponential function (Vogel, 1921; Fulcher, 1925; Tammann and Hesse, 1926; Civan, 2005b, 2008b):

$$\ln f = \ln f_c - \frac{E}{R(T-T_c)},\tag{2.145}$$

where f is a temperature-dependent parameter and f_c is a pre-exponential coefficient whose units are determined by the parameter type; T and T_c refer to the actual and characteristic limit temperatures, K, respectively; E is the activation energy of the parameter variation process (joule per kilomole); and R is the universal gas constant (joule per kilomole kelvin). The parameters f_c, E, and T_c can be determined uniquely using the least squares method (Monkos, 2003).

The VTF equation will be written in the following form for the present application:

$$\ln\left(\frac{f}{f_c}\right) = \frac{A}{(T-T_c)},\tag{2.146}$$

where

$$A = -E/R.\tag{2.147}$$

The VTF equation can also be written as the following at a properly selected reference condition, such as the initial condition, denoted by subscript o:

$$\ln\left(\frac{f_o}{f_c}\right) = \frac{A}{(T_o-T_c)}.\tag{2.148}$$

Hence, subtracting Eq. (2.148) from Eq. (2.146) yields the following convenient expression for straight-line plotting of experimental data:

$$\ln\left(\frac{f}{f_o}\right) = A\left[\frac{1}{(T-T_c)} - \frac{1}{(T_o-T_c)}\right].\tag{2.149}$$

In the present case, the property of interest is permeability K. Therefore, replacing the general property symbol f with the permeability symbol K, Eq. (2.149) reads as

$$\ln\left(\frac{K}{K_o}\right) = A\left[\frac{1}{(T-T_c)} - \frac{1}{(T_o-T_c)}\right].\tag{2.150}$$

The two unknown parameters of this equation, namely, A and T_c, are determined to obtain the best straight-line plots of experimental data. Then, the energy of activation E of thermal processes causing permeability variation can be calculated using Eq. (2.147).

2.4.3 Data Analysis and Correlation

As discussed earlier, when permeability variation with temperature is measured using core samples, it is not a priori known if the variation is solely due to temperature. Frequently, the measurement methods used to measure permeability may introduce the influence of other factors. Obviously, the formulations previously given do not consider the effect of such factors. For illustration purposes, Civan (2008c) correlated the permeability loss by the thermally induced diatomite compaction data of Dietrich and Scott (2007) using the Kozeny–Carman and VTF equations. The mineral compositions, and initial porosity, permeability, and temperature conditions of the various cores tested in their studies, referred to as rocks A, B, C, and D, are indicated in Table 2.2 based on the data provided by Dietrich and Scott (2007). The data reveal that these rocks are very different from each other in mineral composition, which significantly affects their properties. In this section, Eqs. (2.144) and (2.150) are applied for correlating the permeability of the diatomite core samples described in Table 2.2 with temperature and under confining stress.

2.4.3.1 Example 10: Correlation Using the Modified Kozeny–Carman Equation   Four parameters, namely a, b, c, and m, are considered in the modified

TABLE 2.2   Initial Temperature, Permeability, and Porosity Conditions and the Estimated Parameter Values of the Modified Kozeny–Carman and Vogel–Tammann–Fulcher Equations for Various Porous Rocks (Civan, 2008c; © 2008 SPWLA, Reproduced by Permission of the Society of Petrophysicists and Well Log Analysts)

Rock Type		A	B	C	D
Opal A (%)		90–95	78–90	35–78	0
Opal CT (%)		–	–	–	100
Detritus (%)		5–10	10–17	17–40	–
Clay (%)		0	0–5	5–25	–
Initial temperature, T_o (°F)		72	72	72	72
Initial porosity, ϕ_o (%)		71.2	71.5	49.9	48.6
Initial permeability, K_o (μd)			0.85–10		
Kozeny–Carman	m	2.0	2.0	0.7	2.0
equation	a	2.55E-03	4.20E-03	4.18E-03	4.96E-03
	b	−1.00E-05	−9.50E-06	−2.25E-05	−1.86E-05
	c	0	0	3.00E-08	0
Vogel–Tammann–	T_c (°F)	−150	−400	−70	−90
Fulcher equation	A (°F)	362.83	3181.4	474.83	1152.4
	E, Btu/lbmol (calculated from $A = -E/R$)	−720.22	−6315	−942.5	−2288

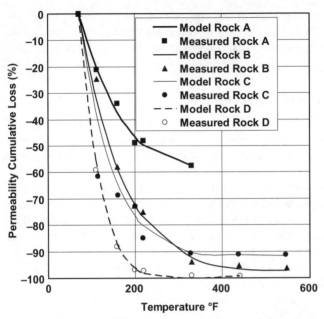

Figure 2.23 Permeability loss versus temperature of diatomite formations with different initial conditions. Correlations by the modified Kozeny–Carman equation are presented using the data of Dietrich and Scott (2007) (Civan, 2008c; © 2008 SPWLA, reproduced by permission of the Society of Petrophysicists and Well Log Analysts).

Kozeny–Carman equation. The variation of permeability with temperature involved in the above-mentioned data has been correlated successfully as shown in Figure 2.23 by adjusting the values of the four parameters of Eq. (2.144). The parameter values have been estimated by a trial-and-error method to maximize the coefficient of regression to be as close to $R^2 = 1.0$ as possible. Table 2.2 presents the estimated parameter values yielding the best correlation for each rock sample, all depicting decreasing trends with increasing temperature. Permeability decreases with increasing temperature because of inward expansion of grains of the core samples confined externally.

The modified Kozeny–Carman equation performs well because parameter variation with temperature has been incorporated into the coefficients of the $f(T)$ polynomial that are determined by fitting permeability versus temperature data. The best estimate values of parameters a and b reported in Table 2.2 are in the same order of magnitudes for all the rocks. However, the value of parameter c is zero for all rocks except for rock C. Further, the value of parameter m is 2.0 for all rocks except for rock C, for which m is equal to 0.7. Hence, this finding signals that rock C is significantly different from the other rocks.

2.4.3.2 Example 11: Correlation Using the VTF Equation
The VTF equation contains two parameters as indicated by Eq. (2.150). The value of parameter A was determined by the slope of the straight-line fit of the above-mentioned data as

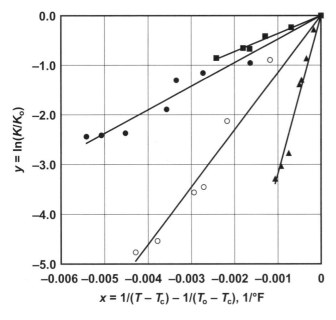

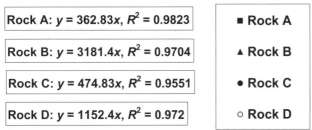

Rock A: $y = 362.83x$, $R^2 = 0.9823$	■ Rock A
Rock B: $y = 3181.4x$, $R^2 = 0.9704$	▲ Rock B
Rock C: $y = 474.83x$, $R^2 = 0.9551$	● Rock C
Rock D: $y = 1152.4x$, $R^2 = 0.972$	○ Rock D

Figure 2.24 Permeability loss versus temperature of diatomite formations with different initial conditions. Correlations by the linearized Vogel–Tammann–Fulcher equation are presented using the data of Dietrich and Scott (2007) (Civan, 2008c; © 2008 SPWLA, reproduced by permission of the Society of Petrophysicists and Well Log Analysts).

shown in Figure 2.24. The value of the activation energy E was calculated using the $A = -E/R$ relationship. The value of the characteristic limit absolute temperature T_c was determined by adjusting to obtain a straight-line fit yielding the highest value for the coefficient of regression, R^2, as close to 1.0 as possible. The values of the parameters of the VTF equation determined in this way have been checked against and found to be close to those calculated uniquely by the least squares method according to Monkos (2003). Table 2.2 summarizes the estimated parameter values yielding the best correlation. Using these values, the variation of permeability of various rock samples with temperature has been correlated successfully as shown in Figure 2.25, all revealing decreasing permeability trends with increasing temperature. The order of magnitude of the best estimate values of parameters T_c and A of the VTF equation varies significantly by different rocks. This may be signaling

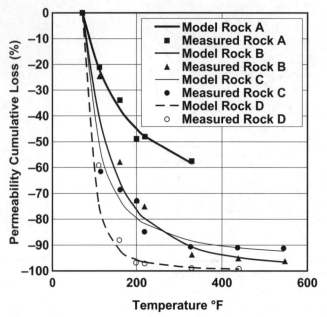

Figure 2.25 Permeability loss versus temperature of diatomite formations with different initial conditions. Correlations by the Vogel–Tammann–Fulcher equation are presented using the data of Dietrich and Scott (2007) (Civan, 2008c; © 2008 SPWLA, reproduced by permission of the Society of Petrophysicists and Well Log Analysts).

significant differences between the microscopic characteristics of these samples which affect the thermally induced deformation of the constituents of the rock matrix.

2.5 EFFECTS OF OTHER FACTORS ON PERMEABILITY

The effect of other factors such as porous media compressibility and particulate deposition/mobilization can be expressed in terms of porosity variation similar to the temperature effect demonstrated earlier and in Chapter 10.

2.6 EXERCISES

1. How much does the permeability vary according to the Kozeny–Carman equation when the porosity is doubled but the other parameters remain unchanged?
2. How much does the permeability change according to the VTF equation when the temperature value is doubled but the other parameters remain unchanged?
3. Does the permeability of porous rock increase or decrease when it is heated?

4. Consider the data of Carman (1937a,b), Ladd (1990), and Koch and Sangani (1999), which encompass a wide range of porosity values. Assume that the cement exclusion factor is taken as $\alpha = 1$.

   (a) Can the Kozeny–Carman equation represent these data?

   (b) Represent the data by a regression line of the form of Eq. (2.65) with a unit slope. Determine the best estimates of the parameter values required for best regression and the coefficient of regression. Compare your answer with the best estimate values reported by Civan (2001) in Table 2.1.

   (c) Prepare the plots of β and Γ versus porosity.

5. Carnahan (1990) simulated the variation of porosity by scale precipitation at constant temperature and also under constant temperature gradient in a 5-m-long porous medium using a reactive chemical transport simulator. For the isothermal precipitation example, Carnahan (1990) simulated the precipitation of gypsum at 25°C according to the reaction given by Eq. (2.73) by injecting

$$10^{-3}\,\mathrm{mole/m^2 s\ Ca^{2+}_{(aq)}} \quad \text{and} \quad 10^{-3}\,\mathrm{mole/m^2 s\ SO^{2-}_{4(aq)}}$$

at the inlet of the porous medium. For the nonisothermal precipitation example, Carnahan (1990) simulated the precipitation of quartz under a −20°C/m temperature gradient, cooling from 150 to 50°C over a 5-m distance along the porous medium according to Eq. (2.74) by injecting

$$10^{-4}\,\mathrm{mole/m^2 s\ Si(OH)^{\circ}_{4(aq)}}$$

at the inlet of the porous medium. Carnahan (1990) calculated the porosity variation at the inlet, where the porosity variation was maximum, and temperature and saturation remained constant at 25°C for gypsum and 150°C for quartz precipitation. Note that, although the temperature and saturation conditions varied along the porous medium for the nonisothermal case, the temperature and saturation conditions nevertheless remained constant at the inlet face of the porous medium and are the same for the injected solution entering the medium. Therefore, only the porosity variation at the inlet side of the porous medium occurs under constant temperature and saturation conditions. Use this data to verify the analytic expressions obtained here for F_s = constant.

   (a) Develop the straight-line correlations of Carnahan's (1990) simulated data of porosity variations by gypsum and quartz precipitation according to Eq. (2.85) with the least squares linear regression.

   (b) Does Eq. (2.85) satisfactorily represent the porosity variation by scale precipitation?

   (c) Confirm that the best correlation of the porosity variation by the gypsum and quartz precipitations, respectively, is obtained as

$$\phi^{-r} = \phi_o^{-r} + k_\phi t, \text{ at } T = 25°C, \phi_o = 0.05, r = 0.81, k_\phi = 5.7\times10^{-6}\,\mathrm{s^{-1}}, \text{gypsum}, \quad (2.151)$$

and

$$\phi^{-r} = \phi_o^{-r} + k_\phi t, \text{ at } 150°C, \phi_o = 0.05, r = 0.81, k_\phi = 1.88\times10^{-8}\,\mathrm{s^{-1}}, \text{quartz}. \quad (2.152)$$

6. Todd and Yuan (1988) only measured the permeability ratio K/K_o in the Clashach sandstone by simultaneous $(Ba, Sr)SO_4$ solid deposition from brines of various concentrations.

(a) Plot the data of the injection port of a test core plug for deposition from brine containing St/Ba in the ratio of 0.1 according to Eq. (2.96) by the least squares linear fit.

(b) Does Eq. (2.95) represent the Todd and Yuan (1988) data accurately?

(c) Show that the permeability versus time correlation is obtained as

$$K/K_o = (1 + \lambda t)^{-\delta}, \lambda = 0.11 \, min^{-1}, \delta = 1.33. \tag{2.153}$$

7. Construct a permeability–porosity cross-plot using Civan's power-law equation of permeability using the parameter values determined by fitting of the data of Couvreur et al. (2001) as reported in Table 2.1.

CHAPTER 3

MACROSCOPIC TRANSPORT EQUATIONS

3.1 INTRODUCTION

Macroscopic transport equations are derived by representative volume averaging, time averaging, and control volume analysis in this chapter.* As stated by Civan (2002e), "Porous media averaging offers potential advantages, but its implementation frequently involves cumbersome procedures and inherent considerations, which limits the applicability of the resultant macroscopic relationships to specific cases." Therefore, control volume analysis is applied as an alternative approach to help resolve such difficulties.

Gray (2000) points out that "the microscale is a continuum length scale that is much smaller than the diameter of a pore" in porous media. In addition, the shape and configuration of the pore structure and the interface boundaries may be irregular and complex, and may vary by various processes. Therefore, microscopic-level (pointwise) description of the behavior of multiphase fluid systems in porous media is not a practical approach in view of the effort required for mathematical formulation and solution of the governing equations and boundary conditions in an intricately detailed geometrical domain, which may vary spatially and temporally.

Frequently, this difficulty can be conveniently alleviated by a macroscopic-level (representative elementary volume [REV]-averaged) description by averaging the microscopic phase properties and governing transport equations over a REV according to a prescribed set of averaging rules and theorems (Wang and Beckermann, 1993). Therefore, Gray (2000) states "Flow in porous media is typically modeled at a length scale, referred to as the macroscale, such that a point of the system encompasses tens to hundreds of pore diameters. To obtain conservation equations at this scale, averaging theorems for the phases have been employed such that quantities appearing in the equations (i.e., density and velocity) are in fact average values from a region surrounding a point of interest." In addition, time–space double averaging based on the double decomposition of the various quantities is required when the

---

* Parts of this chapter have been reproduced with modifications from Civan, F. and Evans, R.D. 1998. Determining the Parameters of the Forchheimer Equation from Pressure-Squared vs. Pseudopressure Formulations. SPE Reservoir Evaluation & Engineering, 1(1), pp. 43–46, © 1998 SPE, with permission from the Society of Petroleum Engineers.

Porous Media Transport Phenomena, First Edition. Faruk Civan.
© 2011 John Wiley & Sons, Inc. Published 2011 by John Wiley & Sons, Inc.

pore fluid volume is large, such as in high-porosity porous media, so that turbulence effects are important (de Lemos, 2008).

Application of volume averaging is a complicated procedure and has varied widely by reported studies. Efforts for the proper implementation and validation of the volume-averaging techniques are continuing. In this chapter, the fundamentals of time, representative volume and mass-weighted volume, and time–space double averaging, and their applications by several examples are presented. The treatise is intended to be instructional and at an introductory level. More elaborate discussion, description, and applications can be found elsewhere, such as by Gray (1999), Whitaker (1999), Civan (2002e, 2010b), and de Lemos (2008). An alternative approach based on the control volume analysis is also presented for purposes of comparison and for gaining useful insights into the modeling of transport in porous media.

3.2 REV

The REV is a prescribed characteristic volumetric size over which microscopic properties and transport equations can be averaged for consistent description of the macroscopic properties and transport equations in multiphase systems. Teng and Zhao (2000) define the REV as "a minimum element at which scale characteristics of a porous flow hold." Brown et al. (2000) define the REV as "the range of volumes for which all averaged geometrical characteristics are single-valued functions of the location of that point and time."

Bear and Braester (1972) suggest that also the REV may be different for different properties of multiphase systems. As stated by Wang and Beckermann (1993), the conventional volume averaging is a single-scale volume averaging. However, they point out that various phases present in heterogeneous multiphase systems may possess different length scales. Therefore, multiscale volume averaging may lead to a better description of heterogeneous systems. For example, they propose a dual-scale volume-averaging procedure for dual-porosity systems involving naturally fractured and aggregated porous media.

Determination of the REV has occupied various researchers for a long time, including Hubbert (1956), Baveye and Sposito (1984), Civan and Evans (1998), and Brown et al. (2000). Figure 3.1 by Brown et al. (2000) describes the conceptualization of the REV. The plateau shown in Figure 3.1 indicates a REV domain of $V_{min} < V < V_{max}$, over which a property remains homogeneous. Figures 3.2–3.4 obtained for dolomite core samples by Brown et al. (2000) confirm that the REV can be determined by proper measurement methods. Brown et al. (2000) show that the REV domain varies with the core material and the considered property (such as porosity and density).

Figure 3.5 by Civan and Evans (1998) shows that the minimum representative elementary core length (REL) for the permeability and inertial flow coefficient of a sandstone core sample is about 8 cm. Figure 3.6 by Civan and Evans (1998) indicate different REL scales, 17 cm for the permeability and 30 cm for the inertial flow coefficient of a simulated tight core.

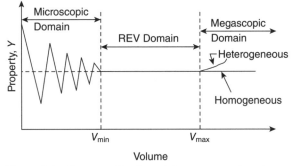

Figure 3.1   Conceptualization of the representative elementary volume (REV) (after Brown et al., 2000, © 2000 American Geophysical Union).

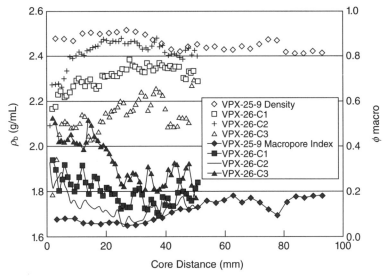

Figure 3.2   Individual slice bulk density and macropore index along each core (after Brown et al., 2000, © 2000 American Geophysical Union).

The REV is a convenient concept used for theoretical macroscopic-scale description, as demonstrated in the next section. The quantification of the REV is also important for the selection of the proper sample size necessary for meaningful tests (Brown et al., 2000).

3.3   VOLUME-AVERAGING RULES

The schematic provided in Figure 3.7 depicts a multiphase system containing N different phases in a representative elementary bulk volume ΔV_b of porous media. The various phases present are denoted by $j = 1, 2, \ldots, N$. This multiphase

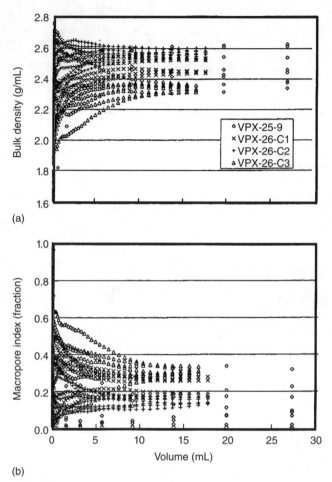

Figure 3.3   Mean volume properties calculated by the expanding prism method for all four cores. (a) Bulk density. (b) Macropore index (after Brown et al., 2000, © 2000 American Geophysical Union).

system may contain both solid and fluid phases. We consider the flow of the j-phase through porous media while interacting with the surrounding solid and fluid phases present in porous media. The volume occupied by the j-phase inside the representative elementary bulk volume is indicated by ΔV_j. Further, we assume that the porous matrix is nondeformable, and therefore ΔV_b does not vary with time.

Thus, the bulk volume of porous media is given by the total volume occupied by various phases present in the multiphase system as

$$\Delta V_b = \sum_j \Delta V_j. \tag{3.1}$$

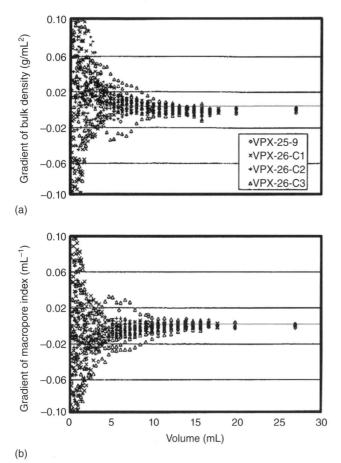

(a)

(b)

Figure 3.4   Gradient of mean volume properties using expanding prism data for all four cores. (a) Bulk density. (b) Macropore index (after Brown et al., 2000, © 2000 American Geophysical Union).

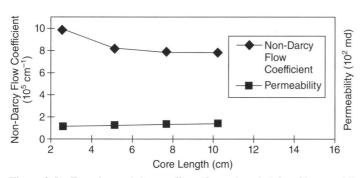

Figure 3.5   Experimental data—effect of core length (after Civan and Evans, 1998, © 1998 SPE; reprinted by permission of the Society of Petroleum Engineers).

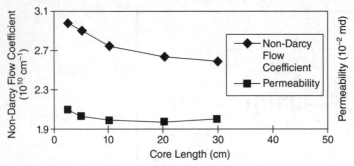

Figure 3.6   Simulated data—effect of core length (after Civan and Evans, 1998, ©1998 SPE; reprinted by permission of the Society of Petroleum Engineers).

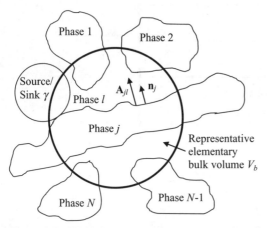

Figure 3.7   Multiphase system in a representative elementary bulk volume (prepared by the author by modifying after Gray, 1975).

The porosity ϕ is defined as the pore volume fraction of the bulk volume, given by

$$\phi = \frac{\sum_{j \neq s} \Delta V_j}{\Delta V_b},$$

(3.2)

where s denotes the sum of the solid phases forming the porous matrix.

The volume fraction of the j-phase in the bulk volume is determined by

$$\varepsilon_j = \Delta V_j / \Delta V_b.$$

(3.3)

Note that these volume fractions add up to unity,

$$\sum_j \varepsilon_j = 1.0,$$

(3.4)

whereas the volume fractions of all the mobile fluid phases present in the pore space add up to the part of porosity occupied by the mobile fluids. Thus,

$$\sum_{j\neq s} \varepsilon_j = \phi. \tag{3.5}$$

For fluid phases, the fractional bulk volume ε_j occupied by the j-phase can also be expressed in terms of the fractional pore volume S_j occupied by this phase, referred to as phase saturation, using

$$\varepsilon_j = \phi S_j. \tag{3.6}$$

The sum of the saturations of all the fluid phases is equal to unity:

$$\sum_{j\neq s} S_j = 1.0. \tag{3.7}$$

Note that a portion of the j-phase may be immobile (Civan, 2010b). Thus, we can write the following expressions for the immobile portion of the j-phase:

$$\varepsilon_{jm} = \phi S_{jm}. \tag{3.8}$$

Therefore, the total of the mobile and immobile portions of the j-phase is given by

$$\varepsilon_{jt} = \varepsilon_j + \varepsilon_{jm} = \phi\left(S_j + S_{jm}\right) = \phi S_{jt}. \tag{3.9}$$

The pore volume occupied by the immobile fluids is given by

$$\sum_{j\neq s} \varepsilon_{jm} = \phi_m. \tag{3.10}$$

Therefore, the total porosity is given by

$$\phi_t = \phi + \phi_m. \tag{3.11}$$

Consider the various properties of the j-phase, such as those denoted by $f, g, h, \ldots$. Such properties of the j-phase are zero everywhere in the bulk volume except for the region occupied by the j-phase. For example (Gray, 1982),

$$f_j \neq 0, x \in \Delta V_j \tag{3.12}$$

and

$$f_j = 0, \text{elsewhere}. \tag{3.13}$$

The same is also true for other properties, $g, h, \ldots$.

Next, we define, for example, the bulk volume average of the property f of the j-phase by (Slattery, 1972; Gray, 1975; Whitaker, 1999)

$$\langle f_j \rangle_b = \frac{1}{\Delta V_b} \int_{\Delta V_b} f_j dV = \frac{1}{\Delta V_b} \sum_j \int_{\Delta V_j} f_j dV = \frac{1}{\Delta V_b} \int_{\Delta V_j} f_j dV. \tag{3.14}$$

The individual phase volume average of the same property is defined in a similar manner as (Slattery, 1972; Gray, 1975; Whitaker, 1999)

$$\langle f_j \rangle_j = \frac{1}{\Delta V_j} \int_{\Delta V_j} f_j dV. \tag{3.15}$$

The point values of the properties f, g, and h of the j-phase can be decomposed into their individual phase average values $\langle f_j \rangle_j$, $\langle g_j \rangle_j$, and $\langle h_j \rangle_j$ and their deviations $\hat{f}_j$, $\hat{g}_j$, and $\hat{h}_j$ from such average values according to Gray, (1975)

$$f_j = \langle f_j \rangle_j + \hat{f}_j, x \in \Delta V_j, \tag{3.16}$$
$$g_j = \langle g_j \rangle_j + \hat{g}_j, x \in \Delta V_j, \tag{3.17}$$

and

$$h_j = \langle h_j \rangle_j + \hat{h}_j, x \in \Delta V_j. \tag{3.18}$$

The averaging of Eqs. (3.16)–(3.18) by means of Eq. (3.15) over the j-phase reveals that

$$\langle \hat{f}_j \rangle_j = \langle \hat{g}_j \rangle_j = \langle \hat{h}_j \rangle_j = 0. \tag{3.19}$$

Further,

$$\langle \hat{f}_j \rangle_b = \langle \hat{g}_j \rangle_b = \langle \hat{h}_j \rangle_b = 0. \tag{3.20}$$

Obviously, we can also write the following expressions for such properties that do not belong to the other phases denoted by l:

$$f_j = \langle f_j \rangle_j = \hat{f}_j = 0, j = l \tag{3.21}$$
$$g_j = \langle g_j \rangle_j = \hat{g}_j = 0, j = l \tag{3.22}$$
$$h_j = \langle h_j \rangle_j = \hat{h}_j = 0, j = l. \tag{3.23}$$

A number of relationships of practical importance can be derived as demonstrated in the following based on the above-given volume-averaging rules.

By means of Eqs. (3.14) to (3.23), the bulk volume and intrinsic phase volume averages of the property f of the j-phase are related by

$$\langle f_j \rangle_b = \varepsilon_j \langle f_j \rangle_j. \tag{3.24}$$

Alternatively, we can obtain the same result by writing the following and applying Eq. (3.3):

$$\Delta V_b \langle f_j \rangle_b = \sum_j \Delta V_j \langle f_j \rangle_j = \Delta V_j \langle f_j \rangle_j, \langle f_j \rangle_b = \frac{\Delta V_j}{\Delta V_b} \langle f_j \rangle_j = \varepsilon_j \langle f_j \rangle_j. \tag{3.25}$$

The following rule is obtained by replacing f_j with $\langle f_j \rangle_b$ in Eq. (3.24) (modified here after Gray, 1975):

$$\langle \langle f_j \rangle_b \rangle_b = \varepsilon_j \langle \langle f_j \rangle_b \rangle_j = \langle f_j \rangle_b. \tag{3.26}$$

The same result can be obtained by starting with $\langle f_j \rangle_b = \langle f_j \rangle_b$ and then taking a bulk volume average and expanding the result. However, the relationship between the bulk- and fluid volume averages given by Eq. (3.24) should be applied with caution because exceptions can apply in some case, such as for pressure (Liu and Masliyah, 2005; Civan, 2010b) as demonstrated in Chapter 5.

Lunardini (1991) provides the following rules:

$$\left\langle \left\langle f_j \right\rangle_j \right\rangle_j = \left\langle f_j \right\rangle_j \tag{3.27}$$

and

$$\left\langle \left\langle f_j \right\rangle_j \right\rangle_b = \left\langle f_j \right\rangle_b = \varepsilon_j \left\langle f_j \right\rangle_j. \tag{3.28}$$

This relationship can be derived as the following:

$$\Delta V_b \left\langle \left\langle f_j \right\rangle_j \right\rangle_b = \sum_j \Delta V_j \left\langle \left\langle f_j \right\rangle_j \right\rangle_j = \Delta V_j \left\langle \left\langle f_j \right\rangle_j \right\rangle_j = \Delta V_j \left\langle f_j \right\rangle_j. \tag{3.29}$$

Rearranging Eq. (3.29) and substituting Eq. (3.3) yields

$$\left\langle \left\langle f_j \right\rangle_j \right\rangle_b = \frac{\Delta V_j}{\Delta V_b} \left\langle \left\langle f_j \right\rangle_j \right\rangle_j = \frac{\Delta V_j}{\Delta V_b} \left\langle f_j \right\rangle_j = \varepsilon_j \left\langle f_j \right\rangle_j = \left\langle f_j \right\rangle_b. \tag{3.30}$$

Alternatively, start with Eq. (3.16), rearranged as

$$\left\langle f_j \right\rangle_j = f_j - \hat{f}_j. \tag{3.31}$$

Then, take the bulk volume average as

$$\left\langle \left\langle f_j \right\rangle_j \right\rangle_b = \left\langle f_j \right\rangle_b - \left\langle \hat{f}_j \right\rangle_b = \left\langle f_j \right\rangle_b = \varepsilon_j \left\langle f_j \right\rangle_j, \left\langle \hat{f}_j \right\rangle_b = 0. \tag{3.32}$$

Gray's (1975) decomposition method as described by Eqs. (3.16)–(3.23) is particularly useful for relating the averages of the products of various phase properties to the products of the averages of these properties. For example, $\left\langle f_j g_j \right\rangle_b$ can be expressed in terms of $\left\langle f_j \right\rangle_j$ and $\left\langle g_j \right\rangle_j$ by a series of operations according to Gray (1975). First, applying Eq. (3.24), obtain

$$\left\langle f_j g_j \right\rangle_b = \varepsilon_j \left\langle f_j g_j \right\rangle_j, \tag{3.33}$$

in which, applying Eqs. (3.16) and (3.17),

$$f_j g_j = \left(\left\langle f_j \right\rangle_j + \hat{f}_j \right) \left(\left\langle g_j \right\rangle_j + \hat{g}_j \right) = \left\langle f_j \right\rangle_j \left\langle g_j \right\rangle_j + \left\langle f_j \right\rangle_j \hat{g}_j + \hat{f}_j \left\langle g_j \right\rangle_j + \hat{f}_j \hat{g}_j. \tag{3.34}$$

Then, the intrinsic phase average over the j-phase is obtained as

$$\left\langle f_j g_j \right\rangle_j = \left\langle \left(\left\langle f_j \right\rangle_j \left\langle g_j \right\rangle_j + \left\langle f_j \right\rangle_j \hat{g}_j + \hat{f}_j \left\langle g_j \right\rangle_j + \hat{f}_j \hat{g}_j \right) \right\rangle_j. \tag{3.35}$$

Expanding Eq. (3.35) yields

$$\left\langle f_j g_j \right\rangle_j = \left\langle \left\langle f_j \right\rangle_j \left\langle g_j \right\rangle_j \right\rangle_j + \left\langle \left\langle f_j \right\rangle_j \hat{g}_j \right\rangle_j + \left\langle \hat{f}_j \left\langle g_j \right\rangle_j \right\rangle_j + \left\langle \hat{f}_j \hat{g}_j \right\rangle_j. \tag{3.36}$$

Because the intrinsic j-phase average value of a property is uniform throughout the j-phase, Eq. (3.36) can be further expanded as

$$\left\langle f_j g_j \right\rangle_j = \left\langle f_j \right\rangle_j \left\langle g_j \right\rangle_j + \left\langle f_j \right\rangle_j \left\langle \hat{g}_j \right\rangle_j + \left\langle \hat{f}_j \right\rangle_j \left\langle g_j \right\rangle_j + \left\langle \hat{f}_j \hat{g}_j \right\rangle_j. \tag{3.37}$$

Thus, invoking Eq. (3.19) into Eq. (3.37) results in (Gray, 1975)

$$\left\langle f_j g_j \right\rangle_j = \left\langle f_j \right\rangle_j \left\langle g_j \right\rangle_j + \left\langle \hat{f}_j \hat{g}_j \right\rangle_j. \tag{3.38}$$

Then, substituting Eq. (3.38) into Eq. (3.33) yields

$$\langle f_j g_j \rangle_b = \varepsilon_j \langle f_j \rangle_j \langle g_j \rangle_j + \varepsilon_j \langle \hat{f}_j \hat{g}_j \rangle_j. \tag{3.39}$$

Alternatively, applying Eq. (3.24) to Eq. (3.39) leads to (Civan, 1996c)

$$\langle f_j g_j \rangle_b = \langle f_j \rangle_b \langle g \rangle_b / \varepsilon_j + \langle \hat{f}_j \hat{g}_j \rangle_b. \tag{3.40}$$

Similarly, the average of the product of three properties can be expressed by (Civan, 1996c)

$$\langle f_j g_j h_j \rangle_j = \langle f_j \rangle_j \langle g_j \rangle_j \langle h_j \rangle_j + \langle f_j \rangle_j \langle \hat{g}_j \hat{h}_j \rangle_j + \langle g_j \rangle_j \langle \hat{f}_j \hat{h}_j \rangle_j + \langle h_j \rangle_j \langle \hat{f}_j \hat{g}_j \rangle_j + \langle \hat{f}_j \hat{g}_j \hat{h}_j \rangle_j \tag{3.41}$$

and

$$\langle f_j g_j h_j \rangle_b = \langle f_j \rangle_b \langle g_j \rangle_b \langle h_j \rangle_b / \varepsilon_j^2 + \langle f_j \rangle_b \langle \hat{g}_j \hat{h}_j \rangle_b / \varepsilon_j$$
$$+ \langle g_j \rangle_b \langle \hat{f}_j \hat{h}_j \rangle_b / \varepsilon_j + \langle h_j \rangle_b \langle \hat{f}_j \hat{g}_j \rangle_b / \varepsilon_j + \langle \hat{f}_j \hat{g}_j \hat{h}_j \rangle_b. \tag{3.42}$$

Lunardini (1991) provides the following additional relationships between various averages:

$$\langle \langle f_j \rangle_j g_j \rangle_j = \langle f_j \rangle_j \langle g_j \rangle_j, \tag{3.43}$$

$$\langle \langle f_j \rangle_j g_j \rangle_b = \langle f_j \rangle_j \langle g_j \rangle_b = \varepsilon_j \langle f_j \rangle_j \langle g_j \rangle_j, \tag{3.44}$$

$$\langle \langle f_j \rangle_b g_j \rangle_b = \langle f_j \rangle_b \langle g_j \rangle_b, \tag{3.45}$$

$$\langle \langle f_j \rangle_b g_j \rangle_j = \langle f_j \rangle_b \langle g_j \rangle_j = \varepsilon_j \langle f_j \rangle_j \langle g_j \rangle_j, \tag{3.46}$$

$$\langle \langle f_j \rangle_b \langle g_j \rangle_b \rangle_j = \langle f_j \rangle_j \langle g_j \rangle_j + \langle \hat{f}_j \hat{g}_j \rangle_j, \tag{3.47}$$

$$\langle \langle f_j \rangle_b \langle g_j \rangle_b \rangle_b = \varepsilon_j \langle f_j \rangle_j \langle g_j \rangle_j + \varepsilon_j \langle \hat{f}_j \hat{g}_j \rangle_j = \varepsilon_j \langle f_j \rangle_j \langle g_j \rangle_j + \langle \hat{f}_j \hat{g}_j \rangle_b, \tag{3.48}$$

and

$$\langle \langle f_j \rangle_b \langle g_j \rangle_b \langle h_j \rangle_b \rangle_j = \langle f_j \rangle_j \langle g_j \rangle_j \langle h_j \rangle_j + \langle h_j \rangle_j \langle \hat{f}_j \hat{g}_j \rangle_j + \langle g_j \rangle_j \langle \hat{f}_j \hat{h}_j \rangle_j$$
$$+ \langle f_j \rangle_j \langle \hat{g}_j \hat{h}_j \rangle_j + \langle \hat{f}_j \hat{g}_j \hat{h}_j \rangle_j. \tag{3.49}$$

We can express the last equation as

$$\langle \langle f_j \rangle_b \langle g_j \rangle_b \langle h_j \rangle_b \rangle_b = \varepsilon_j \langle f_j \rangle_j \langle g_j \rangle_j \langle h_j \rangle_j + \langle h_j \rangle_j \langle \hat{f}_j \hat{g}_j \rangle_b$$
$$+ \langle g_j \rangle_j \langle \hat{f}_j \hat{h}_j \rangle_b + \langle f_j \rangle_j \langle \hat{g}_j \hat{h}_j \rangle_b + \langle \hat{f}_j \hat{g}_j \hat{h}_j \rangle_b. \tag{3.50}$$

The time derivative, gradient, and divergence of a property, f, belonging to the j-phase can be averaged over the REV, respectively, as the following (Whitaker, 1968, 1969, 1999; Slattery, 1969, de Lemos, 2008):

$$\left\langle \frac{\partial f_j}{\partial t} \right\rangle_b = \frac{\partial \langle f_j \rangle_b}{\partial t} - \frac{1}{V_b} \int_{A_j} f_j \mathbf{v}_{Aj} \cdot \mathbf{n}_j dA = \frac{\partial \left(\varepsilon_j \langle f_j \rangle_j \right)}{\partial t} - \frac{1}{V_b} \int_{A_j} f_j \mathbf{v}_{Aj} \cdot \mathbf{n}_j dA, \tag{3.51}$$

$$\langle \nabla f_j \rangle_b = \nabla \langle f_j \rangle_b + \frac{1}{V_b} \int_{A_j} f_j \mathbf{n}_j dA = \nabla \left(\varepsilon_j \langle f_j \rangle_j \right) + \frac{1}{V_b} \int_{A_j} f_j \mathbf{n}_j dA, \tag{3.52}$$

and

$$\left\langle \nabla . \mathbf{f}_j \right\rangle_b = \nabla . \left\langle \mathbf{f}_j \right\rangle_b + \frac{1}{V_b} \int_{A_j} \mathbf{f}_j . \mathbf{n}_j dA = \nabla . \left(\varepsilon_j \left\langle \mathbf{f}_j \right\rangle_j \right) + \frac{1}{V_b} \int_{A_j} \mathbf{f}_j . \mathbf{n}_j dA. \tag{3.53}$$

Here, t denotes time and ∇ denotes the gradient operator. A_j indicates the surface area of the j-phase. $\mathbf{v}_{Aj}$ is the velocity at which the surface area A_j of the j-phase moves. A denotes the variable for the surface area. $\mathbf{n}_j$ is the outward unit normal vector.

Applying Eq. (3.24), the average of a gradient of a scalar property can be expressed as the following (Gray, 1975):

$$\left\langle \nabla f_j \right\rangle_b = \varepsilon_j \left\langle \nabla f_j \right\rangle_j, \tag{3.54}$$

which, upon invoking Eq. (3.16) and expanding yields

$$\left\langle \nabla f_j \right\rangle_b = \varepsilon_j \left\langle \nabla \left(\left\langle f_j \right\rangle_j + \hat{f}_j \right) \right\rangle_j = \varepsilon_j \left\langle \nabla \left\langle f_j \right\rangle_j \right\rangle_j + \varepsilon_j \left\langle \nabla \hat{f}_j \right\rangle_j$$
$$= \left\langle \nabla \left\langle f_j \right\rangle_j \right\rangle_b + \left\langle \nabla \hat{f}_j \right\rangle_b. \tag{3.55}$$

Applying Eq. (3.52) into Eq. (3.55) leads to

$$\left\langle \nabla f_j \right\rangle_b = \varepsilon_j \left[\nabla \left\langle \left\langle f_j \right\rangle_j \right\rangle_j + \frac{1}{V_j} \int_{A_j} \left\langle f_j \right\rangle_j \mathbf{n}_j dA \right] + \nabla \left\langle \hat{f}_j \right\rangle_b + \frac{1}{V_b} \int_{A_j} \hat{f}_j \mathbf{n}_j dA. \tag{3.56}$$

But $\left\langle \hat{f}_j \right\rangle_b = 0$ according to Eq. (3.20) and $\left\langle f_j \right\rangle_j$ does not have jump discontinuities along the interface surface area of the j-phase. Also, $\left\langle \left\langle f_j \right\rangle_j \right\rangle_j = \left\langle f_j \right\rangle_j$ by Eq. (3.27). Then, Eq. (3.56) simplifies as

$$\left\langle \nabla f_j \right\rangle_b = \varepsilon_j \nabla \left\langle f_j \right\rangle_j + \frac{1}{V_b} \int_{A_j} \hat{f}_j \mathbf{n}_j dA. \tag{3.57}$$

Eqs. (3.52) and (3.57) can be used alternatively. Gray (1975) named Eq. (3.57) as the modified averaging theorem. Comparing Eqs. (3.52) and (3.57), Gray (1975) determined the relationship

$$\frac{1}{V_b} \int_{A_j} \left\langle f_j \right\rangle_j \mathbf{n}_j dA = -\left\langle f_j \right\rangle_j \nabla \varepsilon_j. \tag{3.58}$$

3.4   MASS-WEIGHTED VOLUME-AVERAGING RULE

Gray et al. (1993) expressed the mass-weighted volume average of a property f in terms of the volume averages according to

$$\tilde{f}_j = \frac{\dfrac{1}{\Delta V_b} \displaystyle\int_{\Delta V_b} \rho_j f_j dV}{\dfrac{1}{\Delta V_b} \displaystyle\int_{\Delta V_b} \rho_j dV} = \frac{\left\langle \rho_j f_j \right\rangle_b}{\left\langle \rho_j \right\rangle_b} = \frac{\dfrac{1}{\Delta V_j} \displaystyle\int_{\Delta V_j} \rho_j f_j dV}{\dfrac{1}{\Delta V_j} \displaystyle\int_{\Delta V_j} \rho_j dV} = \frac{\left\langle \rho_j f_j \right\rangle_j}{\left\langle \rho_j \right\rangle_j}. \tag{3.59}$$

3.5 SURFACE AREA AVERAGING RULES

Referring to Figure 3.7, consider the area ΔA_j of the surface of the j-phase forming the interface surface areas $\Delta A_{j\ell}$ with the other phases denoted by ℓ. The following relationships can be written (Lunardini, 1991):

$$\Delta A_j = \sum_{\substack{\ell \\ \ell \neq j}} \Delta A_{j\ell}. \tag{3.60}$$

The interface surface area average of a quantity, f, located between the phases j and l is determined by

$$\langle f_j \rangle_{j\ell} = \frac{1}{\Delta A_{j\ell}} \int_{\Delta A_{j\ell}} f_j dA. \tag{3.61}$$

Further, note that

$$f_j = \langle f_j \rangle_{j\ell} + \tilde{f}_j, \tag{3.62}$$

$$\langle \tilde{f}_j \rangle_{j\ell} = 0, \tag{3.63}$$

$$\langle \langle f_j \rangle_{j\ell} \rangle_{j\ell} = \langle f_j \rangle_{j\ell}, \tag{3.64}$$

and

$$\left\langle \langle f_j \rangle_{j\ell} g_j \right\rangle_{j\ell} = \langle f_j \rangle_{j\ell} \langle g_j \rangle_{j\ell}. \tag{3.65}$$

3.6 APPLICATIONS OF VOLUME AND SURFACE AVERAGING RULES

For example, consider the averaging of the mass conservation equation for the property f of a system present in porous media given as (Civan, 1996c)

$$\partial W_j / \partial t + \nabla \cdot \mathbf{w}_j = q_j. \tag{3.66}$$

Let ρ_j and f_j denote the density and an intensive property of the j-phase, respectively; W_j denotes an extensive property of the j-phase per its unit volume; $\mathbf{w}_j$ denotes the total flux of the property f of the system; $\mathbf{v}_j$ is the velocity vector; $\mathbf{D}_j$ is the diffusivity tensor; and q_j denotes the mass added per unit volume of the j-phase per unit of time. Hence, the following relationships can be written:

$$W_j = \rho_j f_j \tag{3.67}$$

and

$$\mathbf{w}_j = \rho_j \left(\mathbf{J}_{Cj} + \mathbf{J}_{Dj} \right). \tag{3.68}$$

Here, the convective flux $\mathbf{J}_{Cj}$ and the diffusive flux $\mathbf{J}_{Dj}$ of the quantity f are described, respectively, by

$$\mathbf{J}_{Cj} = \mathbf{v}_j f_j \tag{3.69}$$

and

$$\mathbf{J}_{Dj} = -\mathbf{D}_j \cdot \nabla f_j. \tag{3.70}$$

Then, the volume average of Eq. (3.66) applying Eq. (3.14) yields

$$\langle \partial W_j / \partial t \rangle_b + \langle \nabla \cdot \mathbf{w}_j \rangle_b = \langle q_j \rangle_b. \tag{3.71}$$

Eq. (3.71) can be processed by means of Eqs. (3.51) and (3.53) to obtain

$$\partial \langle W_j \rangle_b / \partial t - \frac{1}{\Delta V_b} \int_{\Delta A_j} W_j \mathbf{v}_{Aj} \cdot \mathbf{n}_j dA + \nabla \cdot \langle \mathbf{w}_j \rangle_b + \frac{1}{\Delta V_b} \int_{\Delta A_j} \mathbf{w}_j \cdot \mathbf{n}_j dA = \langle q_j \rangle_b. \tag{3.72}$$

Thus, applying the volume, mass-weighted volume, and surface averaging rules presented earlier, Eq. (3.72) yields

$$\frac{\partial}{\partial t} \left(\varepsilon_j \langle \rho_j \rangle_j \tilde{f}_j \right) + \nabla \cdot \left(\varepsilon_j \langle \rho_j \rangle_j \widetilde{\mathbf{v}_j f_j} \right) = \nabla \cdot \left(\varepsilon_j \langle \rho_j \rangle_j \widetilde{\mathbf{D}_j \cdot \nabla f_j} \right)$$

$$+ \sum_l \frac{-1}{\Delta V_b} \int_{\Delta A_{jl}} \rho_j f_j \left(\mathbf{v}_j - \mathbf{v}_{Ajl} \right) \cdot \mathbf{n}_j dA + \sum_l \frac{1}{\Delta V_b} \int_{\Delta A_{jl}} \rho_j \mathbf{D}_j \cdot \nabla f_j \cdot \mathbf{n}_j dA \tag{3.73}$$

$$+ \sum_\gamma \frac{-1}{\Delta V_b} \int_{\Delta A_{j\gamma}} \rho_j f_j \left(\mathbf{v}_j - \mathbf{v}_{Aj\gamma} \right) \cdot \mathbf{n}_j dA + \sum_\gamma \frac{1}{\Delta V_b} \int_{\Delta A_{j\gamma}} \rho_j \mathbf{D}_j \cdot \nabla f_j \cdot \mathbf{n}_j dA + \langle q_j \rangle_b.$$

Here, γ denotes the external sources that can exchange mass with the j-phase, such as a fluid injection into the hole drilled through porous media.

For illustration purposes, invoking $f_j = 1$, $\hat{f}_j = 0$, and $\tilde{f}_j = 1$ into Eq. (3.73) and neglecting all the interface transfer terms, we derive the following equation of continuity of the j-phase (Lunardini, 1991):

$$\frac{\partial}{\partial t} \left(\varepsilon_j \langle \rho_j \rangle_j \right) + \nabla \cdot \left(\varepsilon_j \langle \rho_j \rangle_j \tilde{\mathbf{v}}_j \right) = \langle q_j \rangle_b, \tag{3.74}$$

which contains a mass-weighted volume average velocity, $\tilde{\mathbf{v}}_j$ (Gray, 1975). Applying Eq. (3.59) to Eq. (3.74), we can obtain an alternative form as the following, expressed only in terms of volume averaging:

$$\frac{\partial}{\partial t} \left(\varepsilon_j \langle \rho_j \rangle_j \right) + \nabla \cdot \left(\varepsilon_j \langle \rho_j \mathbf{v}_j \rangle_j \right) = \langle q_j \rangle_b. \tag{3.75}$$

However, it is necessary to apply the following decomposition rule in order to be able to express the average of the product as a product of the averages:

$$\langle \rho_j \mathbf{v}_j \rangle_b = \varepsilon_j \langle \rho_j \mathbf{v}_j \rangle_j = \varepsilon_j \left(\langle \rho_j \rangle_j \langle \mathbf{v}_j \rangle_j + \langle \hat{\rho}_j \hat{\mathbf{v}}_j \rangle_j \right)$$

$$= \varepsilon_j \langle \rho_j \rangle_j \langle \mathbf{v}_j \rangle_j + \langle \hat{\rho}_j \hat{\mathbf{v}}_j \rangle_b, \tag{3.76}$$

where $\langle \hat{\rho}_j \hat{\mathbf{v}}_j \rangle_b$ denotes the bulk volume average of the product of deviations of density and velocity from their average values. This term can be related to the volume average density empirically by applying the gradient theory as a closure method (Bear, 1972; Bear and Bachmat, 1990). Civan (1993, 1994d, 1996c, 1998a, 2002e) proposed the following expression for this purpose:

$$\langle \hat{\rho}_j \hat{\mathbf{v}}_j \rangle_b = \varepsilon_j \langle \hat{\rho}_j \hat{\mathbf{v}}_j \rangle_j = -\mathbf{D}_{jb} \cdot \nabla \langle \rho_j \rangle_b = -\mathbf{D}_{jb} \cdot \nabla \left(\varepsilon_j \langle \rho_j \rangle_j \right), \tag{3.77}$$

where $\mathbf{D}_{jb}$ is a bulk-hydraulic dispersivity tensor. Consequently, combining Eqs. (3.74)–(3.77) yields (Civan, 1993, 1994d, 1996c, 1998a, 2002e)

$$\frac{\partial}{\partial t} \left(\varepsilon_j \langle \rho_j \rangle_j \right) + \nabla \cdot \left(\varepsilon_j \langle \rho_j \rangle_j \langle \mathbf{v}_j \rangle_j \right) = \nabla \cdot \left[\mathbf{D}_{jb} \cdot \nabla \left(\varepsilon_j \langle \rho_j \rangle_j \right) \right] + \langle q_j \rangle_b. \tag{3.78}$$

However, had we considered the following expression instead of Eq. (3.77),

$$\langle \hat{\rho}_j \hat{\mathbf{v}}_j \rangle_b = \varepsilon_j \langle \hat{\rho}_j \hat{\mathbf{v}}_j \rangle_j = -\varepsilon_j \mathbf{D}_{jj} \cdot \nabla \langle \rho_j \rangle_j, \tag{3.79}$$

where $\mathbf{D}_{jj}$ is a phase-hydraulic dispersivity tensor, the result would have been different as the following:

$$\frac{\partial}{\partial t} \left(\varepsilon_j \langle \rho_j \rangle_j \right) + \nabla \cdot \left(\varepsilon_j \langle \rho_j \rangle_j \langle \mathbf{v}_j \rangle_j \right) = \nabla \cdot \left[\varepsilon_j \mathbf{D}_{jj} \cdot \nabla \langle \rho_j \rangle_j \right] + \langle q_j \rangle_b. \tag{3.80}$$

Whether Eqs. (3.77) or (3.79) is the proper choice for closure can be resolved according to Civan (2010b) based on the control volume analysis approach as the following. It is reasonable to assume that the additional mass flux $\left(\langle \hat{\rho}_j \hat{\mathbf{v}}_j \rangle_b \right)_x$ caused by the deviations in density and velocity from their respective mean values, for example, in the x-Cartesian direction, is related proportionally to the difference of the mass of the j-phase $\left(\varepsilon_j \langle \rho_j \rangle_j \right)$ per distance Δx in the flow direction. Thus,

$$\left(\langle \hat{\rho}_j \hat{\mathbf{v}}_j \rangle_b \right)_x = -D_{jbx} \left[\lim_{\Delta x \to 0} \frac{\left(\varepsilon_j \langle \rho_j \rangle_j \right)_{x+\Delta x} - \left(\varepsilon_j \langle \rho_j \rangle_j \right)_x}{\Delta x} \right] = -D_{jbx} \frac{\partial \left(\varepsilon_j \langle \rho_j \rangle_j \right)}{\partial x}. \tag{3.81}$$

In view of this derivation, it is apparent that the expression given by Eq. (3.77) is the proper form of the gradient theory, which can be derived by generalizing Eq. (3.81). A discussion of the "extrapolated limit" concept of Hubbert (1956) for justification of $\Delta x \to 0$ is presented in Chapter 5.

3.7 DOUBLE DECOMPOSITION FOR TURBULENT PROCESSES IN POROUS MEDIA

It is necessary to consider both time and space averaging for the macroscopic description of processes undergoing turbulent fluctuations in time during transport through porous media because the pore fluid volume is sufficiently large to allow turbulence to occur (Antohe and Lage, 1997, Takatsu and Masuoka, 1998, Nakayama and Kuwahara, 1999, Getachew et al., 2000). However, the same results are obtained regardless of whether time averaging or volume averaging is carried out first followed with the other (de Lemos, 2008).

Following de Lemos (2008), a quantity f_j of a phase j, which is undergoing some turbulent fluctuations, f_j', around its time average value $\bar{f}_j$ can be expressed as

$$f_j = \bar{f}_j + f_j', \tag{3.82}$$

where the time averaging is defined by

$$\overline{f}_j = \frac{1}{\Delta t} \int_t^{t+\Delta t} f_j \, dt. \tag{3.83}$$

Here, the time period Δt considered for averaging is selected such that the time average of the fluctuation component vanishes; that is, $\overline{f'_j} = 0$.

Alternatively,

$$\overline{f}_j = \frac{1}{\Delta t} \int_t^{t+\Delta t} f_j \, dt = \frac{1}{\Delta t} \int_t^{t+\Delta t} \left(\langle f_i \rangle_j + \hat{f}_j \right) dt = \overline{\left(\langle f_i \rangle_j \right)} + \overline{\left(\hat{f}_j \right)}. \tag{3.84}$$

We can take the volume average of a quantity involving turbulence using Eq. (3.15) as

$$\langle f_j \rangle_j = \frac{1}{\Delta V_j} \int_{\Delta V_j} f_j \, dV = \frac{1}{\Delta V_j} \int_{\Delta V_j} \left(\overline{f}_j + f'_j \right) dV = \langle \overline{f}_j \rangle_j + \langle f'_j \rangle_j. \tag{3.85}$$

In addition, we can consider turbulent fluctuations of $\hat{f}_j$ around its time average value as

$$\hat{f}_j = \overline{\left(\hat{f}_j \right)} + \left(\hat{f}_j \right)'. \tag{3.86}$$

Similarly, we can consider the spatial deviations of turbulent fluctuations around the volume average value as

$$f'_j = \langle f'_j \rangle_j + \widehat{\left(f'_j \right)}. \tag{3.87}$$

Thus, we can show that

$$\overline{\left(\hat{f}_j \right)} = \left[\overline{\left(\hat{f}_j \right)} \right] + \left(\hat{f}_j \right)' = \left[\overline{\left(\hat{f}_j \right)} \right] \text{ because } \overline{\left(\hat{f}_j \right)'} = 0. \tag{3.88}$$

Similarly,

$$\langle f'_j \rangle_j = \langle \langle f'_j \rangle_j \rangle_j + \langle \widehat{\left(f'_j \right)} \rangle_j = \langle \langle f'_j \rangle_j \rangle_j \text{ because } \langle \widehat{\left(f'_j \right)} \rangle_j = 0. \tag{3.89}$$

We resort to two alternative approaches to develop a double-decomposition rule. First, substitute Eq. (3.82) into Eq. (3.16):

$$f_j = \langle f_j \rangle_j + \hat{f}_j = \langle \left(\overline{f}_j + f'_j \right) \rangle_j + \overline{\left(\overline{f}_j + f'_j \right)} = \langle \overline{f}_j \rangle_j + \langle f'_j \rangle_j + \overline{\left(\overline{f}_j \right)} + \overline{\left(f'_j \right)}. \tag{3.90}$$

Second, we attempt substituting Eq. (3.16) into Eq. (3.82) as

$$\begin{aligned} f_j = \overline{f}_j + f'_j &= \overline{\left(\langle f_j \rangle_j + \hat{f}_j \right)} + \left(\langle f_j \rangle_j + \hat{f}_j \right)' \\ &= \overline{\left(\langle f_j \rangle_j \right)} + \overline{\left(\hat{f}_j \right)} + \left(\langle f_j \rangle_j \right)' + \left(\hat{f}_j \right)'. \end{aligned} \tag{3.91}$$

The time and volume averages should have the following commutative properties for the expressions given by Eqs. (3.90) and (3.91) to be the same (Pedras and de Lemos, 2001):

$$\overline{\left(\langle f_j\rangle_j\right)}=\langle\overline{f_j}\rangle_j,\ \langle f_j'\rangle_j=\left(\langle f_j\rangle_j\right)',\ \overline{\left(\widehat{f_j}\right)}=\widehat{\left(\overline{f_j}\right)},\ \left(\widehat{f_j'}\right)=\left(\widehat{f_j}\right)'. \tag{3.92}$$

As an application of the double-decomposition approach, consider the following instantaneous microscopic equations of continuity and motion, respectively, for the flow of an incompressible fluid through porous media, assuming constant fluid and porous media properties according to de Lemos (2008) with modifications:

$$\nabla.\mathbf{v}=0 \tag{3.93}$$

$$\rho\left[\frac{\partial\mathbf{v}}{\partial t}+\nabla\cdot(\mathbf{vv})\right]=-\nabla p+\mu\nabla^2\mathbf{v}+\rho\mathbf{g}. \tag{3.94}$$

Applying the rules of double averaging yields

$$\nabla.\left(\varepsilon_j\langle\overline{\mathbf{v}}\rangle_j\right)=0 \tag{3.95}$$

and

$$\rho_j\left[\frac{\partial\left(\varepsilon_j\langle\overline{\mathbf{v}}\rangle_j\right)}{\partial t}+\nabla\cdot\left(\varepsilon_j\langle\overline{\mathbf{v}\,\mathbf{v}}\rangle_j\right)\right]=-\nabla\left(\varepsilon_j\langle p\rangle_j\right)+\mu_j\nabla^2\left(\varepsilon_j\langle\overline{\mathbf{v}}\rangle_j\right)$$
$$+\nabla\cdot\left(-\varepsilon_j\rho_j\langle\overline{(\mathbf{v}'\mathbf{v}')}\rangle_j\right)+\varepsilon_j\rho_j\mathbf{g}+\mathbf{R}, \tag{3.96}$$

where $-\varepsilon_j\rho_j\langle\overline{(\mathbf{v}'\mathbf{v}')}\rangle_j$ denotes the macroscopic Reynolds stress tensor, and the residual term is expressed by

$$\mathbf{R}=-\frac{\mu_j\varepsilon_j^2\langle\overline{\mathbf{v}}\rangle_j}{K}-\frac{c_f\rho_j\varepsilon_j^3\left|\langle\overline{\mathbf{v}}\rangle_j\right|\langle\overline{\mathbf{v}}\rangle_j}{\sqrt{K}}, \tag{3.97}$$

where c_f is a pressure coefficient.

Expressing the pore fluid velocity in terms of the Darcy (superficial) velocity by substitution of $\langle\overline{\mathbf{u}}\rangle_j=\varepsilon_j\langle\overline{\mathbf{v}}\rangle_j$, Eqs. (3.95)–(3.97) become

$$\nabla.\langle\overline{\mathbf{u}}\rangle_j=0 \tag{3.98}$$

and

$$\rho_j\left[\frac{\partial\langle\overline{\mathbf{u}}\rangle_j}{\partial t}+\nabla\cdot\left(\frac{\langle\overline{\mathbf{u}\,\mathbf{u}}\rangle_j}{\varepsilon_j}\right)\right]=-\nabla\left(\varepsilon_j\langle p\rangle_j\right)+\mu_j\nabla^2\langle\overline{\mathbf{u}}\rangle_j$$
$$+\nabla\cdot\left(-\rho_j\frac{\langle\overline{(\mathbf{u}'\mathbf{u}')}\rangle_j}{\varepsilon_j}\right)+\varepsilon_j\rho_j\mathbf{g}+\mathbf{R}, \tag{3.99}$$

where the residual term is now expressed by

$$\mathbf{R}=-\frac{\mu_j\varepsilon_j\langle\overline{\mathbf{u}}\rangle_j}{K}-\frac{c_f\rho_j\varepsilon_j\left|\langle\overline{\mathbf{u}}\rangle_j\right|\langle\overline{\mathbf{u}}\rangle_j}{\sqrt{K}}. \tag{3.100}$$

Note the averages of the product of deviations appearing in the above formulations need to be processed by proper closure (Saito and de Lemos, 2010).

3.8 TORTUOSITY EFFECT

Let J denote the fluid volume average, J_b the bulk volume average, and J_I the interstitial value of a transport property. Following Civan (2007b), Consider a REV of a porous medium through which the fluids are flowing.

As stated by Civan (2007b), the length of the actual preferential hydraulic flow paths L_f averaged over the representative elementary porous media volume (REV) is greater than the length of porous media, L (Fig. 3.8). The tortuosity of the actual flow paths, τ, is defined by the ratio of these lengths as

$$\tau = L_I/L. \tag{3.101}$$

Thus, fluid needs to move faster through the preferential flow path compared to flow through a straight path in porous media. Consequently, equating the volumetric flow through the tortuous and straight flow paths yields

$$m = A_b J_b = A_I J, \tag{3.102}$$

where A_b is the bulk surface area normal to flow and A_I is the open flow area, given by (Civan, 2001, 2002d, 2007b)

$$A_I = A_b c \phi^n, \tag{3.103}$$

where n is a fractal dimension and c is the fractal coefficient. The symbol J_b denotes the transport through porous media per unit bulk surface area and per unit of time. This is usually referred to as the superficial transport. The symbol J represents the transport through a straight flow path in porous media (Fig. 3.8). The actual transport J_I through the tortuous preferential flow path L_I is called the interstitial or internal transport. Equating the travel time, t, for transport through the straight and tortuous paths yields

$$t = L/J = L_I/J_I. \tag{3.104}$$

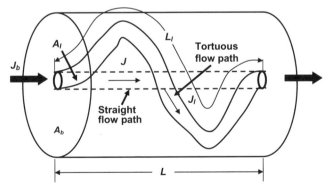

Figure 3.8   Straight and tortuous flow paths in porous media. J_b, J_I, and J represent the superficial, interstitial, and pore volume transports (reprinted by permission from Springer, after Civan, 2007b).

Hence, combining Eqs. (3.101) through (3.104) yields

$$m = A_b J_b = A_b c \phi^n J = A_b c \phi^n J_I / \tau, \tag{3.105}$$

$$J_I = \frac{L_I}{L} J = \tau J, \tag{3.106}$$

and

$$J_I = \frac{\tau J_b}{c \phi^n} \text{ or } J_b = \frac{c \phi^n}{\tau} J_I. \tag{3.107}$$

Therefore, when ϕ or τ vary, they need to be treated as variable properties. For example,

$$\nabla \cdot J_b = \nabla \cdot \left(\frac{c \phi^n}{\tau} J_I \right). \tag{3.108}$$

However, when ϕ and τ are constants, then

$$\nabla \cdot J_b = \frac{c \phi^n}{\tau} \nabla \cdot J_I. \tag{3.109}$$

It is common to assume $n = 1.0$ and $c = 1.0$. For example, Dupuit (1863) expressed the interstitial fluid velocity as

$$v_f = \tau u_f / \phi. \tag{3.110}$$

3.9 MACROSCOPIC TRANSPORT EQUATIONS BY CONTROL VOLUME ANALYSIS

Control volume analysis is a practical approach for the derivation of the macroscopic equations of conservation for various phases present in porous media. However, the control volume surfaces of a j-phase are of the two types, the external surfaces present over the bulk surface of the REV through which j-phase interacts with the surrounding media and the internal control surfaces through which j-phase interacts with other phases, including the solid phase inside the control volume. Derivation of the internal interface interaction terms is a challenging task as demonstrated for the equation of motion in Chapter 5. In this derivation, we will omit the brackets < > indicating volume averages.

The control volume equation for the j-phase contained in the REV can be expressed as (Civan, 2010b)

$$\sum_{CS} f_j \left(\mathbf{n}_j \cdot \mathbf{A}_j \right) + \frac{\partial}{\partial t} \left(f_j V_j \right)_{CV} = \sum_{CV} \dot{r}_j. \tag{3.111}$$

In this equation, the symbols CV and CS denote the control volume and control surface of the j-phase; $\mathbf{A}_j$ is the outward and normal surface area vector for flow; $\mathbf{n}_j$ is the total flux vector of the property transferred across the control surfaces; its components are n_{jx}, n_{jy}, and n_{jz} in the x-, y-, and z-Cartesian directions; V_j denotes the volume of the j-phase; t indicates the time variable; and $\dot{r}_j$ is the amount of property added per unit volume of j-phase. Expanding Eq. (3.111) for the j-phase present in the REV, we obtain

$$
-\left(f_j n_j A_j\right)_x +\left(f_j n_j A_j\right)_{x+\Delta x} -\left(f_j n_j A_j\right)_y +\left(f_j n_j A_j\right)_{y+\Delta y}
$$
$$
-\left(f_j n_j A_j\right)_z +\left(f_j n_j A_j\right)_{z+\Delta z} +\frac{\partial}{\partial t}\left(f_j V_j\right) = \sum_{CV} \dot{r}_j. \tag{3.112}
$$

But we can also write the following:

$$
A_j = A_b c_j \varepsilon_j^{d_j/3}, \quad \alpha_j = A_b c_j \varepsilon_j^{-1+d_j/3}, \quad f_j = f_{jb}/\varepsilon_j, \quad V_j = V_b \varepsilon_j, \tag{3.113}
$$
$$
\text{and} \quad r_j = r_{jb} V_b.
$$

Rearranging Eq. (3.112) in view of Eq. (3.113) yields

$$
-\left(f_{jb} n_j A_b \alpha_j\right)_x +\left(f_{jb} n_j A_b \alpha_j\right)_{x+\Delta x} -\left(f_{jb} n_j A_b \alpha_j\right)_y +\left(f_{jb} n_j A_b \alpha_j\right)_{y+\Delta y}
$$
$$
-\left(f_{jb} n_j A_b \alpha_j\right)_z +\left(f_{jb} n_j A_b \alpha_j\right)_{z+\Delta z} +\frac{\partial}{\partial t}\left(f_{jb} V_b\right) = \sum_{CV} \dot{r}_{jb} V_b, \tag{3.114}
$$
$$
A_{bx} = \Delta y \Delta z, \ A_{by} = \Delta x \Delta z, \ A_{bz} = \Delta x \Delta y, \ V_b = \Delta x \Delta y \Delta z.
$$

Note that we assumed isotropic property for the cross-sectional areas of the j-phase in the above, which may not be true.

Next, we divide Eq. (3.114) by $(\Delta x \Delta y \Delta z)$ and then take a limit as Δx, Δy, and $\Delta z \to 0$. Finally, we generalize the resulting expression in the vector–tensor notation to obtain the following general j-phase macroscopic equation of conservation:

$$
\nabla \cdot \left(\mathbf{n}_j f_{jb}\right) + \frac{\partial f_{jb}}{\partial t} + \left(\mathbf{n}_j f_{jb}\right)\cdot \nabla \alpha_j + \left(\alpha_j -1\right)\nabla \cdot \left(\mathbf{n}_j f_{jb}\right) = \sum_{CV} \dot{r}_{jb}. \tag{3.115}
$$

Note that Eq. (3.115) simplifies as the following when $\alpha_j = 1$ for $c_j = 1$ and $d_j = 3$ (Civan, 2010b):

$$
\nabla \cdot \left(\mathbf{n}_j f_{jb}\right) + \frac{\partial f_{jb}}{\partial t} = \sum_{CV} \dot{r}_{jb}. \tag{3.116}
$$

Note that Liu and Masliyah (2005) use $c_j = 1$ and $d_j = 2$.

The total flux of a property is equal to the sum of transfer by bulk flow (convection/advection) and spontaneous dispersion by diffusion, mixing, and other means:

$$
\mathbf{n}_j f_{jb} = \mathbf{v}_j f_{jb} + \mathbf{j}_{jb}, \ \mathbf{j}_{jb} = -\mathbf{D}_{jb}\cdot \nabla f_{jb}. \tag{3.117}
$$

Substitution of Eq. (3.117) into Eq. (3.115) yields the following generalized transport equation:

$$
\nabla \cdot \left(\mathbf{v}_j f_{jb}\right) + \frac{\partial f_{jb}}{\partial t} + \left(\mathbf{v}_j f_{jb} - \mathbf{D}_{jb}\cdot \nabla f_{jb}\right)\cdot \nabla \alpha_j
$$
$$
+ \left(\alpha_j -1\right)\nabla \cdot \left(\mathbf{v}_j f_{jb} - \mathbf{D}_{jb}\cdot \nabla f_{jb}\right) = \nabla \cdot \left(\mathbf{D}_{jb}\cdot \nabla f_{jb}\right) + \sum_{CV} \dot{r}_{jb}. \tag{3.118}
$$

Note that Eq. (3.118) simplifies as the following when $\alpha_j = 1$ for $c_j = 1$ and $d_j = 3$ (Civan, 2010b):

$$\nabla \cdot \left(\mathbf{v}_j f_{jb}\right) + \frac{\partial f_{jb}}{\partial t} = \nabla \cdot \left(\mathbf{D}_{jb} \cdot \nabla f_{jb}\right) + \sum_{CV} \dot{r}_{jb}. \tag{3.119}$$

3.10 GENERALIZED VOLUME-AVERAGED TRANSPORT EQUATIONS

The following generalized transport equation can be derived for a scalar property f and a vector property $\mathbf{f}$, respectively, as by generalizing Eq. (3.78) (Civan, 2010b):

$$\frac{\partial}{\partial t}\left(\langle f_j\rangle_b\right) + \nabla \cdot \left(\langle f_j\rangle_b \langle \mathbf{v}_j\rangle_j\right) = \nabla \cdot \left[\mathbf{D}_{jb} \cdot \nabla\left(\langle f_j\rangle_b\right)\right] + \langle q_j\rangle_b \tag{3.120}$$

for vector propertie $\mathbf{f}$

$$\frac{\partial}{\partial t}\left(\langle \mathbf{f}_j\rangle_b\right) + \nabla \cdot \left(\langle \mathbf{f}_j\rangle_b \langle \mathbf{v}_j\rangle_j\right) = \nabla \cdot \left[\mathbf{D}_{jb} \cdot \nabla\left(\langle \mathbf{f}_j\rangle_b\right)\right] + \langle \mathbf{q}_j\rangle_b. \tag{3.121}$$

3.11 EXERCISES

1. Derive Eq. (3.58) given by Gray (1975).

2. Some studies implemented a decomposition rule as (Gray, 1975)

$$f_j = \langle f_j\rangle_b + \hat{f}_j \tag{3.122}$$

instead of applying Gray's (1975) decomposition rule,

$$f_j = \langle f_j\rangle_j + \hat{f}_j. \tag{3.123}$$

Show that the former is incorrect because it leads to (Gray, 1975)

$$\langle \hat{f}_j\rangle_b = \left(1 - \varepsilon_j\right)\langle f_j\rangle_b \neq 0. \tag{3.124}$$

3. Prove Eqs. (3.43)–(3.49) given by Lunardini (1991)

4. Simplify Eq. (3.78) for (1) incompressible fluid flowing through heterogeneous porous media and (2) incompressible fluid flowing through homogeneous porous media.

5. Derive Eq. (3.78) by direct volume averaging instead of the mixed use of the volume and mass-weighted volume-averaging rules.

6. Liu and Masliyah (2005) takes exception to the rule given by Eq. (3.24) for applications to pressure and viscosity when the solid matrix is immobile and not moving with the fluid so that the solid phase does not affect the fluid present in the pore space (Chapter 5). Then,

$$\langle p_j\rangle_b = \langle p_j\rangle_j, \langle \mu_j\rangle_b = \langle \mu_j\rangle_j. \tag{3.125}$$

Nevertheless the general rule is still applicable when the solid is suspended in the fluid so that it can be treated like the fluid phases. Investigate the applicability of the exception to the rule according to Liu and Masliyah (2005) by means of theoretical argument.

7. Check the following procedure for averaging the $\rho \nabla \psi$ term. Start with the definition of the flow potential, given by

$$\rho \nabla \psi = \nabla p + \rho g \nabla z. \tag{3.126}$$

Averaging over the REV gives

$$
\begin{aligned}
\langle \rho \nabla \psi \rangle_b &= \langle \nabla p \rangle_b + \langle \rho \rangle_b \, g \nabla z = \phi \left[\langle \nabla p \rangle_f + \langle \rho \rangle_f \, g \nabla z \right] \\
&= \phi \left[\nabla \langle p \rangle_f + \frac{1}{\Delta V_f} \int_{\Delta A} \mathbf{n}_A p_A \mathbf{I} dA + \langle \rho \rangle_f \, g \nabla z \right] \\
&= \phi \left[\nabla \langle p \rangle_f + \langle \rho \rangle_f \, g \nabla z \right] + \frac{\phi}{\Delta V_f} \int_{\Delta A} \mathbf{n}_A p_A \mathbf{I} dA,
\end{aligned}
\tag{3.127}
$$

where $\mathbf{I}$ is a unit tensor.

8. Derive the source terms for the momentum and energy equations listed in Table 3.1.

9. Investigate the effect of potential interactions between pore fluid and solid porous matrix for other properties considered in other macroscopic gradient laws commonly used in phenomenological modeling.

10. Investigate the quantification of internal resistance terms.

TABLE 3.1 Description of Terms for the Generalized Transport Equation (Civan, 2010b)

Quantity	$\langle f_j \rangle_b = \varepsilon_j \langle f_j \rangle_j$	$\mathbf{D}_{jb}$	$\langle q_j \rangle_b$
Mass	$\varepsilon_j \langle \rho_j \rangle_j$	$\mathbf{D}_{jm}$	$\langle q_{jm} \rangle_b$
Species mass	$\varepsilon_j \langle \rho_j \rangle_j \langle w_{ij} \rangle_j$	$\mathbf{D}_{jim}$	$\langle q_{jim} \rangle_b$
Momentum	$\varepsilon_j \langle \rho_j \rangle_j \langle \mathbf{v}_j \rangle_j$	$\mathbf{D}_{vj}$	

$$\langle \mathbf{q}_{jv} \rangle_b = -\nabla \langle p_j \rangle_b + \nabla \langle p_{bj} \rangle_b - \langle \rho_j \rangle_j \nabla \Phi_{bj}$$

$$- \varepsilon_j \langle \mu_j \rangle_j \mathbf{K}_{bj}^{-1} \cdot \langle \mathbf{v}_j \rangle_j - \varepsilon_j^2 \langle \rho_j \rangle_j \beta_{bj} \cdot |\langle \mathbf{v}_j \rangle_j| \langle \mathbf{v}_j \rangle_j$$

$$- \nabla \cdot \mathbf{T}_{bj} + \varepsilon_j \sum_{\substack{j=1 \\ j \neq k}}^{N} f_{jk} \left(\mathbf{T}_{hj} \cdot \langle \mathbf{v}_j \rangle_j - \mathbf{T}_{hk} \cdot \langle \mathbf{v}_k \rangle_k \right)$$

f_{kj} is a shear factor.

$\mathbf{K}_{bk}$ and β_{bk} are the permeability and inertial flow coefficient tensors.

$\mathbf{T}$ is the shear stress tensor.

∇p_{bhk} is the threshold pressure gradient.

Φ_{bk} is the flow potential.

Energy	$\varepsilon_j \langle \rho_j \rangle_j \langle H_j \rangle_j$	$\mathbf{D}_{je} = \dfrac{\kappa}{\rho C}$	$\langle q_{je} \rangle_b = \langle \mathbf{v}_j \rangle_j \cdot \nabla \langle p_j \rangle_b + \dfrac{\partial \langle p_j \rangle_b}{\partial t} + \langle q_{je} \rangle_b$
	$= \varepsilon_j \langle \rho_j \rangle_j C_j \langle T_j \rangle_j$	κ is thermal conductivity tensor	
		C is specific heat capacity at constant pressure	

CHAPTER 4

SCALING AND CORRELATION OF TRANSPORT IN POROUS MEDIA

4.1 INTRODUCTION

This chapter introduces the theoretical foundations and practical applications of scaling and correlation of transport in porous media prior to proceeding with topics where this knowledge is required. Dimensional and inspection analyses are powerful methods resorted for grouping of relevant geometrical and phenomenological variables, quantities, and parameters of processes into dimensionless numbers. These dimensionless groups can provide convenience in formulation and scaling of processes of interest. Combining a number of variables into several meaningful dimensionless groups provides various advantages in developing empirical correlations. This is because the number of dimensionless groups is less than the number of variables by the number of basic dimensions involved in the analysis, namely, mass M, length L, temperature θ, and time T. Excellent treaties are available from many sources, including Ipsen (1960), where details and general discussion can be found about this topic and various approaches. Therefore, the present discussion is limited to the applications concerning porous media processes.

There are various alternative approaches available for the implementation of the dimensional analysis method. Here, a practical simplified approach is taken, involving the three basic steps described by Krantz (2000) as the following: "1. List all quantities on which the phenomena depends, 2. Write the dimensional formula for each quantity, and 3. Demand that these quantities be combined into a functional relation that remains true independently of the size of the units."

However, as pointed out by Krantz (2000), a dimensional analysis of the real problems is not free of difficulties and there are often some glitches that the analyst has to alleviate by intuition and insight into the mechanisms and characteristics of a system of interest. Further, dealing with a dimensional constant, such as universal gas constant $R_g = 8314\,\text{J/kmol-K}$, dimensionless quantities, such as angle, and quantities involving energy and temperature, which are related, requires appropriate treatments, and the type of dimensionless groups to be obtained by dimensional analysis is not unique (Krantz, 2000). In fact, dimensionless groups obtained by one analysis can be combined into other forms of dimensionless groups. Frequently, the types of dimensionless groups are selected for suitability for a given problem and to conform to conventional and preferred standard form, such as the well-known dimensionless numbers, including Reynolds, Prandtl, and Schmidt numbers.

Peters et al. (1993) describe some of the outstanding advantages and disadvantages of inspectional and dimensional analysis methods, which can be summarized as the following:

- The inspectional analysis method requires information about the mathematical description (governing differential equation) and conditions of solution (initial and/or boundary conditions) of the processes of interest. In contrast, dimensional analysis only requires information about the type of variables involved in the processes.

- Often, the physical significance of the dimensionless groups obtained by inspectional analysis is obvious and yields an incomplete set of dimensionless groups. In contrast, dimensional analysis is a general method that can yield many dimensionless groups, some of which may not have any significant relevance to the process of interest.

The application of similarity transformations provides significant convenience and advantage in analyzing and interpreting experimental data, and for the effective modeling of processes and the governing mathematical models (Barenblatt, 1979; Peters et al., 1993). Spatial distribution of properties of self-similar processes follows a similar pattern at different times (Barenblatt, 1979). Frequently, measurements obtained from experimental investigation of processes involve some noise, fluctuations, and various measurement errors, which can make the analyses and interpretation of experimental data difficult. Peters et al. (1993) alleviate the problem by constructing a general dimensionless response function of the process by means of a self-similarity transformation of the measured response of processes.

Churchill (1997, p. 158) explains, "Dimensional analysis is most powerful when it is applied to a complete mathematical model in algebraic form (differential and/or integral), but it is remarkably productive even when a complete model is unknown or unwieldy and the analysis must be applied to a single list of the relevant variables. The utility of dimensional analysis when applied to either a model or a list of variables may often be greatly enhanced by the collateral use of speculative and asymptotic analysis." We describe this approach in the following, according to Churchill (1997).

The speculative analysis starts with a tentative list of potential variables and then investigates their validity by experimental studies. Then, an asymptotic analysis is carried out in an effort to obtain simple limiting relationships by removing some variables of less importance in a systematic manner and then testing the resulting expressions experimentally to determine if they are still applicable and their conditions of applicability if any.

As Churchill (1997) describes, we first start with a speculative list of variables of the dependent, independent, and parametric types, which we consider pertinent to represent a given system. The number of independent dimensionless groups that can be obtained from a list of variables can be predicted by (Van Driest, 1946)

$$i = n - k, \tag{4.1}$$

where i denotes the minimum number of independent dimensionless groups, n denotes the number of variables considered, and k denotes the maximum number of variables noncombinable as a dimensionless group. Buckingham (1914) assumed such noncombinable variables as mass, length, temperature, and time. However, temperature may be treated as a dependent variable because as Hawking (1998) explains, "Temperature is simply a measure of the average energy or speed."

Kline (1965) cautions that an unnecessarily large quantity of dimensionless groups can result when ignoring the fact that some variables may be related by other means. Therefore, the actual minimum number of independent dimensionless groups for a given system may indeed be less than that predicted by Eq. (4.1).

4.2 DIMENSIONAL AND INSPECTIONAL ANALYSIS METHODS

The dimensional and inspectional analysis approaches are demonstrated in the following problem by Greenkorn and Cala (1986) concerning the transient-state transport of species mass by convection/advection and dispersion mechanisms:

$$\rho\left(\phi\frac{\partial w}{\partial t} + u\frac{\partial w}{\partial z}\right) = \frac{\partial}{\partial z}\left(\phi\rho D\frac{\partial w}{\partial z}\right), \tag{4.2}$$

where ϕ, ρ, w, u, D, t, and z denote the porosity, fluid density, species mass fraction, volumetric fluid flux, diffusion coefficient, time, and vertical distance in the gravity direction, respectively.

The equation of motion for fluid flowing through porous media is given by Darcy's law as

$$u = -\frac{K}{\mu}\left(\frac{\partial p}{\partial z} + \rho g\right), \tag{4.3}$$

where K is permeability, μ is fluid viscosity, p is fluid pressure, and g is gravitational acceleration. We consider ϕ and ρ as constants in this exercise.

4.2.1 Dimensional Analysis

Churchill (1997) explains that dimensional analysis can be applied by three different methods:

- Method of inspection
- Method of combination of variables
- Method of algebraic power series of variables

However, for convenience, we apply a step-by-step elimination of the fundamental dimensions one at a time according to Ipsen (1960) as described in the following exercise.

Note that the variables/parameters involved in Eqs. (4.2) and (4.3) are ϕ, w, t, u, z, D, K, μ, ρ, g, and p. There are 11 variables/parameters, but two of them (ϕ and w) are already dimensionless by definition. Hence, we consider the remaining nine variables/parameters for dimensional analysis.

The dimensions of these variables/parameters can be expressed in terms of the basic dimensions as shown in Table 4.1. We expect to find $9 - 3 = 6$ dimensionless groups because three basic dimensions (M, L, and T) are involved. For this purpose, we eliminate the basic dimensions in any arbitrary or convenient order using one variable selected as the scaling variable (Ipsen, 1960) as illustrated in Table 4.1. Hence, the types of dimensionless groups obtained will vary depending on these selections. However, they can be combined to derive other types of dimensionless groups.

Based on Table 4.1, the following dimensionless groups have been obtained:

$$\pi_1 = \frac{K}{z^2}, \pi_2 = \frac{ut}{z}, \pi_3 = \frac{Dt}{z^2}, \pi_4 = \frac{\mu t}{\rho z^2}, \pi_5 = \frac{gt^2}{z}, \pi_6 = \frac{pt^2}{\rho z^2}. \tag{4.4}$$

We can manipulate these groups to obtain the following conventional dimensionless groups:

$$\pi_3' = \frac{\pi_3}{\pi_2} = \frac{D}{uz} \equiv \text{Pe}, \pi_4' = \frac{\pi_2}{\pi_4} = \frac{uz\rho}{\mu} \equiv \text{Re}, \pi_5' = \frac{\pi_1\pi_5}{\pi_2\pi_4} = \frac{K\rho g}{\mu u}, \pi_6' = \frac{\pi_1\pi_6}{\pi_2\pi_4} = \frac{Kp}{\mu uz}. \tag{4.5}$$

Recall that two variables (ϕ and w) are already dimensionless by definition. Thus,

$$\pi_7 = \phi, \pi_8 = w. \tag{4.6}$$

4.2.2 Inspectional Analysis

Here, the approach of Greenkorn and Cala (1986) is presented with modifications. For this purpose, we define the following dimensionless or scaled variables:

$$t^* = \frac{t}{t_c}, z^* = \frac{z}{z_c}, w^* = \frac{w}{w_c}, u^* = \frac{u}{u_c}, p^* = \frac{p}{p_c}, D^* = \frac{D}{D_c}. \tag{4.7}$$

Recall that we consider ϕ and ρ as constants in this exercise and two variables (ϕ and w) are already dimensionless by definition. Substituting these dimensionless variables into Eqs. (4.2) and (4.3) yields

$$\frac{\partial w^*}{\partial t^*} + \frac{1}{\phi}\left(\frac{t_c u_c}{z_c}\right)u^*\frac{\partial w^*}{\partial z^*} = \left(\frac{t_c D_c}{z_c^2}\right)\frac{\partial}{\partial z^*}\left(D^*\frac{\partial w^*}{\partial z^*}\right) \tag{4.8}$$

and

$$u^* = -\left(\frac{Kp_c}{\mu z_c u_c}\right)\frac{\partial p^*}{\partial z^*} - \frac{K\rho g}{\mu u_c}. \tag{4.9}$$

TABLE 4.1 Step-by-Step Elimination of the Fundamental Dimensions

Initial system Variables, Parameters	Dimensions	Step 1 Variables, Parameters	Eliminate M Dimensions	Step 2 Variables, Parameters	Eliminate L Dimensions	Step 3 Variables, Parameters	Eliminate T Dimensions
t	T	t	T	t	T	t: scaling variable	T
u	L/T	u	L/T	u/z	$1/T$	tu/z: dimensionless group	1
z	L	z	L	z: scaling variable	L		
D	L^2/T	D	L^2/T	D/z^2	$1/T$	tD/z^2: dimensionless group	1
K	L^2	K	L^2	K/z^2: dimensionless group	1		
μ	M/LT	μ/ρ	L^2/T	$\mu\rho/z^2$	$1/T$	$t\mu\rho/z^2$: dimensionless group	1
ρ	M/L^3	ρ: scaling variable	M/L^3				
g	L/T^2	g	L/T^2	g/z	$1/T^2$	t^2g/z: dimensionless group	1
p	M/LT^2	p/ρ	L^2/T^2	$p/\rho/z^2$	$1/T^2$	$t^2p/\rho/z^2$: dimensionless group	1

Examination of Eqs. (4.8) and (4.9) reveals the presence of the following four dimensionless groups in addition to $\pi_7 = \phi, \pi_8 = w$:

$$\pi_2 = \frac{t_c u_c}{z_c}, \pi_3' = \frac{t_c D_c}{z_c^2} \text{ or } \pi_3 = \frac{\pi_3'}{\pi_2} = \frac{D_c}{u_c z_c}, \pi_5 = \frac{K\rho g}{\mu u_c}, \pi_6 = \frac{Kp_c}{\mu z_c u_c}. \quad (4.10)$$

It is apparent that the inspectional analysis could not determine the following additional dimensionless groups obtained by the previous dimensional analysis method:

$$\pi_1 = \frac{K}{z^2}, \pi_4' = \frac{uz\rho}{\mu} \equiv \text{Re}. \quad (4.11)$$

Shook et al. (1992) present an application of inspectional analysis for scaling immiscible displacement in porous media.

4.3 SCALING

4.3.1 Scaling as a Tool for Convenient Representation

Note there are five unknown characteristic scaling parameters (t_c, u_c, z_c, D_c, and p_c) but four dimensionless coefficients in Eqs. (4.8) and (4.9). We set $z_c = d$ and determine the other scaling parameters by setting the values of these dimensionless coefficients equal to the unity:

$$\frac{1}{\phi}\left(\frac{t_c u_c}{z_c}\right) = 1, \frac{t_c D_c}{z_c^2} = 1, \frac{Kp_c}{\mu z_c u_c} = 1, \frac{K\rho g}{\mu u_c} = 1. \quad (4.12)$$

Thus, we obtain

$$t_c = \frac{\phi d}{u_c} = \frac{\phi d\mu}{K\rho g}, D_c = \frac{d^2}{t_c} = \frac{dK\rho g}{\phi\mu}, p_c = \frac{\mu d u_c}{K} = d\rho g, u_c = \frac{K\rho g}{\mu}. \quad (4.13)$$

Consequently, Eqs. (4.8) and (4.9) can be simplified in the following convenient forms:

$$\frac{\partial w^*}{\partial t^*} + u^* \frac{\partial w^*}{\partial z^*} = \frac{\partial}{\partial z^*}\left(D^* \frac{\partial w^*}{\partial z^*}\right) \quad (4.14)$$

and

$$u^* = -\frac{\partial p^*}{\partial z^*} - 1. \quad (4.15)$$

4.3.2 Scaling as a Tool for Minimum Parametric Representation

The following exercise according to Krantz and Sczechowski (1994) demonstrates how scaling can be resorted to as a method for minimum parametric representation in porous media.

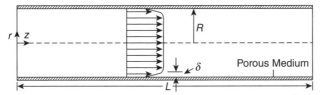

Figure 4.1   Schematic of pressure-driven, steady-state flow of a viscous fluid through a porous medium confined within a horizontal tube of radius R and length L (after Krantz and Sczechowski, 1994; reproduced by permission of Chemical Engineering Education).

Consider the steady-state velocity profile established over a cross-sectional area of a cylindrical porous core sample placed inside a horizontal tube as depicted in Figure 4.1 by Krantz and Sczechowski (1994). The radius, length, and permeability of the core sample are denoted by R, L, and K, respectively. The flow occurs under steady-state conditions because of the pressure difference Δp applied across the core sample.

Under these conditions, the creeping flow through the porous core sample can be described by the following equation of motion, the centerline symmetry condition, and the no-slip wall boundary condition, respectively (Bird et al., 1960):

$$\frac{\Delta p}{L} = -\frac{\mu}{K}u + \mu\frac{1}{r}\frac{d}{dr}\left(r\frac{du}{dr}\right), \tag{4.16}$$

$$\frac{du}{dr} = 0, r = 0, \tag{4.17}$$

and

$$u = 0, r = R. \tag{4.18}$$

The objectives are to determine (1) the condition under which the tube-wall drag can be neglected and (2) the thickness of the near-wall region over which the tube-wall drag cannot be neglected. In an effort to address these issues, consider the following dimensionless variables:

$$r^* = \frac{r}{r_c}, u^* = \frac{u}{u_c}, \tag{4.19}$$

where r_c and u_c denote some characteristic radius and superficial velocity.

Substituting Eq. (4.19) into Eqs. (4.16)–(4.18) and then rearranging yields the following dimensionless equations:

$$1 - \frac{\mu u_c L}{K\Delta p}u^* + \frac{\mu u_c L}{r_c^2 \Delta p}\frac{1}{r^*}\frac{d}{dr^*}\left(r^*\frac{du^*}{dr^*}\right) = 0, \tag{4.20}$$

$$\frac{du^*}{dr^*} = 0, r^* = 0, \tag{4.21}$$

and

$$u^* = 0, r^* = \frac{R}{r_c}. \tag{4.22}$$

Because the porous medium is the dominant source of resistance to fluid flow, the order of magnitude of the Darcy term can be set equal to the unity:

$$\frac{\mu u_c L}{K \Delta p} = 1, \quad \text{and therefore } u_c = \frac{K \Delta p}{\mu L}. \tag{4.23}$$

Consequently, substituting Eq. (4.23) into Eq. (4.20) yields

$$1 - u^* + \frac{K}{r_c^2} \frac{1}{r^*} \frac{d}{dr^*} \left(r^* \frac{du^*}{dr^*} \right) = 0. \tag{4.24}$$

We can now investigate the conditions under which the tube-wall drag effect can or cannot be neglected. If the drag effect is negligible over the full radial distance, then the criterion required for neglecting the tube-wall drag is obtained as

$$\frac{K}{r_c^2} \ll 1, r_c = R, \quad \text{and thus } K \ll R^2. \tag{4.25}$$

On the other hand, if the wall drag cannot be neglected, then the velocity will vary significantly over a boundary layer of thickness, δ. Consequently, the order of magnitude of the Darcy and viscous terms in this region is going to be the same. Then,

$$\frac{K}{r_c^2} = 1, r_c = \delta, \quad \text{and thus } \delta = \sqrt{K}. \tag{4.26}$$

This indicates that the tube-wall effect is observed over the region near the tube wall having a thickness of the order of the square root of permeability, $\sqrt{K}$.

4.3.3   Normalized Variables

Formulation of porous media problems using normalized variables can provide many advantages (Civan, 2011). First of all, the order of magnitude of the normalized variables is the same, and therefore the numerical round-off errors resulting from calculations with different orders of magnitude values are avoided. Second, the formulations expressed in normalized variables are more convenient for solutions of equations, either by analytical or numerical means. Normalized variables can be derived by appropriate scaling methods, such as those described next (Hashemi and Sliepcevich, 1967; Civan and Sliepcevich, 1984; Civan, 1994a). For example, the normalized distance x^* in a finite-length domain of length L can be defined by

$$x^* = x/L, 0 \le x \le L, 0 \le x^* \le 1, \tag{4.27}$$

where x denotes the actual distance.

On the other hand, the normalized distance x^* in a semi-infinite domain can be defined by

$$x^* = \frac{x}{x + x_c}, 0 \le x < \infty, 0 \le x^* \le 1, \tag{4.28}$$

where x_c is a properly selected characteristic length.

The same can be accomplished over a semi-infinite domain according to

$$x^* = 1 - \exp(-x/x_c), 0 \le x < \infty, 0 \le x^* \le 1, \tag{4.29}$$

where x_c is a properly selected characteristic length.

Similarly, the time variable can be expressed in normalized forms using either of the following expressions, where t_c is a properly selected characteristic time:

$$t^* = \frac{t}{t + t_c}, 0 \le t < \infty, 0 \le t^* \le 1 \tag{4.30}$$

$$t^* = 1 - \exp(-t/t_c), 0 \le t < \infty, 0 \le t^* \le 1. \tag{4.31}$$

Frequently, the variation Δf in a quantity f is expressed as a fraction of its maximum variation range $\Delta f_{\max}$, where $f_{\min}$ and $f_{\max}$ are the lowest and highest values of this property:

$$f^* = \frac{\Delta f}{\Delta f_{\max}} = \frac{f - f_{\min}}{f_{\max} - f_{\min}}, f_{\min} \le f \le f_{\max}, 0 \le f^* \le 1. \tag{4.32}$$

An example for the application of Eq. (4.32) is normalizing the water saturation S_w, which varies between the connate water saturation S_{wc} and the saturation $(1 - S_{or})$ when the residual oil saturation S_{or} is attained:

$$S_w^* = \frac{\Delta S_w}{(\Delta S_w)_{\max}} = \frac{S_w - S_{wc}}{(1 - S_{or}) - S_{wc}}, S_{wc} \le S_w \le (1 - S_{or}), 0 \le S_w^* \le 1. \tag{4.33}$$

4.3.4 Scaling Criteria and Options for Porous Media Processes

Scaling criteria are the conditions required for the processes occurring in a model system to be able to represent the processes occurring in a prototype (actual) porous media of interest. Such criteria can be derived by ensuring the equality of the dimensionless groups derived to meet the various relevant similitude conditions (geometric, dynamic, etc.) between the model and prototype systems.

Kimber et al. (1988) explain: "Dimensional analysis requires knowledge of the complete set of relevant variables influencing the process. Inspectional analysis requires the variables in a set of equations which fully describe the process. Dimensional analysis often yields a more complete set of similarity groups, but the physical meaning of the similarity groups themselves is generally more apparent from inspectional analysis. Usually some scaling requirements must be relaxed."

Kimber et al. (1988) and Rojas et al. (1991) elaborate that in inspectional analysis, we derive the scaling groups by first expressing the relevant mathematical model (governing differential equations, conditions of solution, and relevant constraints and constitutive relationships) of the process of interest in dimensionless variables and then making the coefficients of various terms dimensionless. They emphasize, however, that not all requirements of the scaling criteria can be satisfied in practical applications. A compromising approach is thus necessary, emphasizing on scaling of dominant mechanisms of processes satisfactorily while neglecting the effect of minor mechanisms. They demonstrated by several examples that this is an extremely cumbersome and overwhelming task in many cases of practical importance.

A prototype represents a real system of investigation such as the geothermal, groundwater, and petroleum reservoirs. On the other hand, a model represents a scaled (down or up) version of the system considered for analysis of the governing processes that are similar to those occurring in the prototype (Collins, 1961; Bear, 1972). Types of similarities that should be considered depend on the system and problem of interest. For example, Bear (1972) explains that hydraulic systems require the fulfillment of the geometric, kinematic, and dynamic similitude conditions; however, the kinematic similitude condition cannot be satisfied in distorted systems.

For illustration purposes, consider again the example given in the previous section. Following Greenkorn and Cala (1986), consider similar systems of a model and a prototype that are subject to the same initial and boundary conditions. The transport processes prevailing in these systems are considered similar if the following equality conditions can be maintained for various relevant dimensionless groups derived in the above sections, where we assume $z = d$, a characteristic pore diameter:

$$\pi_3' = \left(\frac{D}{uz}\right)_{\text{model}} = \left(\frac{D}{uz}\right)_{\text{prototype}} \tag{4.34}$$

$$\pi_5' = \left(\frac{K\rho g}{\mu u}\right)_{\text{model}} = \left(\frac{K\rho g}{\mu u}\right)_{\text{prototype}} \tag{4.35}$$

$$\pi_4' = \left(\frac{uz\rho}{\mu}\right)_{\text{model}} = \left(\frac{uz\rho}{\mu}\right)_{\text{prototype}} \tag{4.36}$$

$$\pi_1 = \left(\frac{K}{z^2}\right)_{\text{model}} = \left(\frac{K}{z^2}\right)_{\text{prototype}}. \tag{4.37}$$

Note that the inspectional analysis method could not determine the dimensionless groups used in Eqs. (4.36) and (4.37).

As stated by Greenkorn and Cala (1986), when both the model and prototype systems are homogeneous and geometrically similar, then only the conditions dictated by Eqs. (4.34) and (4.35) need to be satisfied. Eq. (4.36) can be omitted if the pore surface friction effect is negligible. However, Eq. (4.37), involving the pore geometry, should also be satisfied for heterogeneous systems. Further, additional variables may have to be considered in heterogeneous systems. For example, Greenkorn and Cala (1986) propose the following variables to describe a field with single heterogeneity according to parameters described in Figure 4.2: the permeability ratio TR $= K_1/K_2$, the penetration ratio PR $= W_2/W_1$, the heterogeneity size SR $= W_1/W_o$, and the length ratio LR $= L_1/L_o$.

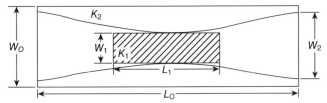

Figure 4.2   Flow field with single heterogeneity (reproduced by permission of the American Chemical Society after Greenkorn and Cala, 1986).

4.3.5   Scaling Immiscible Fluid Displacement in Laboratory Core Floods

The exercise carried out by Peters et al. (1993) is described here with some modifications for illustration of scaling and generalization of typical laboratory core floods involving immiscible displacement of oil and water in different wettability core samples. Peters et al. (1993) observed a higher displacement efficiency in water-wet porous formations compared with those occurring in oil-wet porous formations because of the way instability, wettability, and relative permeability affect the governing mechanisms of the unstable immiscible fluid displacements occurring in such formations. They demonstrated by proper scaling the self-similarity and therefore the predictability of unstable displacements of immiscible fluids occurring in different homogeneous porous formations.

The saturation distribution in geometrically similar core floods can be assumed to be influenced by many variables and parameters, for example, as included in the following unknown function to be characterized here, namely, distance, time, permeability, porosity, density difference between water and oil, water viscosity, oil viscosity, water–oil surface tension, wettability number given as a function of contact angle, gravitational acceleration, volumetric flux, effective water permeability, effective oil permeability, water–oil capillary pressure, core length, mean hydraulic flow tube diameter, core diameter, water mobility, and oil mobility, respectively:

$$S_w = S_w\left(x, t, K, \phi, \rho_w, \rho_o, \mu_w, \mu_o, \sigma_{wo}, f(\theta), g, u, K_w, K_o, p_c, L, D_h, D, M_w, M_o\right).$$
(4.38)

Note that it is customary to combine the water and oil density as the density difference $\Delta \rho \equiv \rho_w - \rho_o$, water–oil surface tension σ_{wo} and wettability number $f(\theta)$ as $\sigma_{wo} f(\theta)$ and effective fluid permeability and viscosity as the mobility $M_w = K_w / \mu_w$, $M_o = K_o / \mu_o$, and finally as the mobility ratio

$$m = \frac{M_w}{M_o} = \frac{K_w / \mu_w}{K_o / \mu_o}$$

Note that $f(\theta) = \cos \theta$ for cylindrical tubes. The mean hydraulic tube diameter is correlated as $D_h \sim \sqrt{K/\phi}$, similar to (Kozeny, 1927; Carman, 1938, 1956)

$$D_h = 4\sqrt{2\tau_h}\sqrt{\frac{K}{\phi}}.$$

TABLE 4.2 Summary of Dimensionless Groups Obtained by Dimensional Analysis

Dimensionless Groups	Expressions
Dimensionless length	$x^* = x/L$
Leverett J-function	$p^* \equiv J = \dfrac{p_c}{\sigma_{wo} f(\theta)} \sqrt{\dfrac{K}{\phi}}$
Capillary number	$N_c = \dfrac{\sigma_{wo} f(\theta)}{Lv\mu_o} \sqrt{\dfrac{K}{\phi}}, \; v = \dfrac{u}{\phi(1 - S_{wc} - S_{or})}$
Gravity number	$N_g = \dfrac{Kg\Delta\rho}{v\mu_o}$
Effective permeability ratio	$K^* = K_o/K_w$
Viscosity ratio	$\mu^* = \mu_o / \mu_w$
Dimensionless time	$t^* = \dfrac{vt}{L}, \; v = \dfrac{u}{\phi(1 - S_{wc} - S_{or})}$
Stability number (derived number; Peters and Flock, 1981)	$N_s = \left(\dfrac{\mu_o}{\mu_w} - 1\right) \dfrac{D^2 v\mu_w}{K\sigma_{wo} f(\theta)}, \; v = \dfrac{u}{\phi(1 - S_{wc} - S_{or})}$
Self-similar dimensionless variable	$\pi^* = \dfrac{x^*}{t^*} = \dfrac{x}{vt}, \; v = \dfrac{u}{\phi(1 - S_{wc} - S_{or})}$

K_w, K_o, and p_c are assumed to be functions of saturation, influenced by pore geometry, wettability, and other relevant parameters. Therefore, we consider these convenient forms in the following dimensional analysis.

A dimensional analysis carried out using the above-mentioned variables yields the additional dimensionless groups, presented in Table 4.2. Note that S_w and ϕ are already dimensionless, but it is convenient to use normalized forms according to Eq. (4.32).

Peters et al. (1993) conducted core flood experiments involving flow of oil and water in unconsolidated sand packs of different properties and under different boundary conditions. They correlated their experimental data using the following self-similar dimensionless variable and normalized saturation, respectively, given by

$$\pi = \frac{\phi x}{ut}, \; S = \frac{S_w - S_{wc}}{1 - S_{wc}}. \tag{4.39}$$

assuming $S_{or} = 0$.

They determined the fluid saturation profiles by means of the X-ray computed tomography (CT) imaging technique. They obtained the typical CT saturation images for displacement of oil (mineral oil) by water (brine) in water-wet and oil-wet sand packs, respectively. In one case, they determined the saturation images at various pore volumes of water injected into an oil-saturated water-wet sand pack for which the conditions were as follows: $K = 9.26\,\text{darcy}$, $\phi = 30.9\%$, $S_{wc} = 12.3\%$, $u = 8.5 \times 10^{-3}\,\text{cm/s}$, $\mu_o/\mu_w = 91.7$, $N_s = 271$, $N_c = 4.62 \times 10^{-4}$, and $N_o = 1.26 \times 10^{-3}$ (Peters et al., 1993). In another case, they determined the saturation images at various pore volumes of water injected into an oil-saturated oil-wet sand pack for

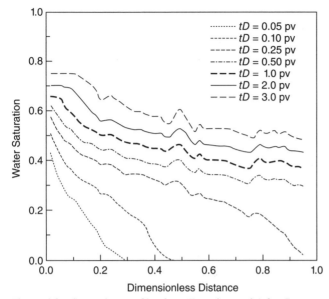

Figure 4.3   Saturation profiles from Experiment 6 (after Peters et al., 1993; reproduced by permission of Elsevier).

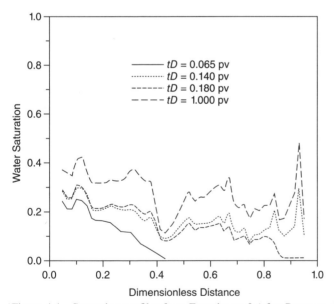

Figure 4.4   Saturation profiles from Experiment 3 (after Peters et al., 1993; reproduced by permission of Elsevier).

which the conditions were as follows: $K = 18.4$ darcy, $\phi = 36.5\%$, $S_{wc} = 0\%$, $u = 1.52 \times 10^{-3}$ cm/s, $\mu_o/\mu_w = 85.5$, $N_s = 1681$, $N_c = 2.87 \times 10^{-3}$, and $N_o = 1.26 \times 10^{-2}$ (Peters et al., 1993). This allowed the construction of Figures 4.3 and 4.4, showing the typical normalized saturation profiles obtained at various dimensionless times for water-wet and oil-wet sand packs, respectively. When tests are repeated with

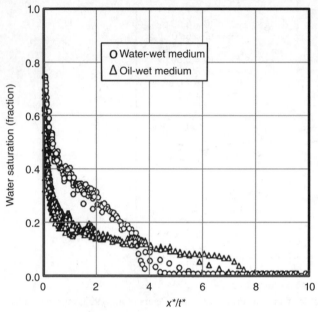

Figure 4.5   A comparison of the dimensionless response curves from water-wet and oil-wet media (after Peters et al., 1993; reproduced by permission of Elsevier).

different sand packs, the dimensionless response curves obtained in different water-wet sand pack tests collapsed into a single curve as shown in Figure 4.5, indicating a well-developed displacement front. In contrast, the displacement front is not well-defined in the oil-wet sand pack tests, although the responses obtained from different tests tend to collapse to a single curve for the most part as shown in Figure 4.5.

4.4   EXERCISES

1. The mean diameter D_h of the hydraulic flow paths in porous media may be assumed as a function of the tortuosity τ_h, porosity ϕ, and permeability K of porous media. Prove that the following dimensionless groups/variables can be used to obtain a functional relationship between these variables/parameters: $\pi_1 \equiv \phi$, $\pi_2 \equiv \tau_h$, $\pi_3 = D_h / \sqrt{K}$. Determine a proper combination of these dimensionless groups/parameters so that the result conforms to (Kozeny, 1927; Carman, 1938, 1956)

$$D_h = 4\sqrt{2\tau_h}\sqrt{\frac{K}{\phi}}.$$

2. Derive the Leverett J-function

$$J(S_w) = \frac{p_c R_h}{\sigma_{wo}}.$$

by carrying out a dimensional analysis with the variables/parameters involved in this function. Note that R_h is the mean radius of hydraulic flow paths.

3. Derive the Leverett J-function

$$J(S_w) = \frac{p_c}{\sigma_{wo} f(\theta)} \sqrt{\frac{K}{\phi}}$$

given in Table 4.2 by carrying out a dimensional analysis with the variables/parameters involved in this function.

4. Hassanizadeh and Gray (1993) related the capillary pressures under dynamic and equilibrium conditions according to the following nonequilibrium expression:

$$p_c^{dyn.} - p_c^{eq.} = -\tau \frac{\partial S}{\partial t}, \tag{4.40}$$

where

$$p_c^{dyn.} \left[ML^{-1}T^{-2}\right], p_c^{eq.} \left[ML^{-1}T^{-2}\right], \quad \text{and} \quad \tau[ML^{-1}T^{-1}]$$

denote the dynamic capillary pressure, equilibrium capillary pressure, and relaxation time or damping coefficient, respectively. S is the average saturation and t is time. Das et al. (2007) propose that the damping coefficient can be correlated using the following variables: porosity ϕ [1], pore size distribution index λ [1], porous medium entry pressure p_e $[ML^{-1}T^{-2}]$, permeability $K[L^2]$, domain volume $V[L^3]$, fluid density ratio (ρ_{nw}/ρ_w) [1], fluids viscosity ratio (μ_{nw}/μ_w) [1], gravitational acceleration $g[L/T^2]$, and saturation S [1]. Show that the following general relationship can be proposed between the various dimensionless groups:

$$\frac{\tau \sqrt{g}}{V^{1/6} p_e} = f\left(\frac{\phi}{\lambda}, \frac{\rho_{nw}\mu_w}{\rho_w \mu_{nw}}, \frac{K}{V^{2/3}}, S_w\right). \tag{4.41}$$

5. The following quantities are considered to be important in the chemical comminution of coal using a solvent: comminution rate R (g/cm^2/s), surface tension σ (dyne per centimeter), density ρ (gram per cubic centimeter), viscosity μ (poise), mean pore diameter (centimeter), and effective mass diffusivity D_e (square centimeter per second). By means of dimensional analysis, determine the dimensionless comminution, surface tension, and Schmidt numbers given, respectively, by (Civan and Knapp, 1991)

$$N_c = Rd/\mu, T_s = \mu^2/(\sigma\rho d), Sc = (\mu/\rho)/D_e.$$

6. The interfacial drag force between a liquid phase 1 and a gas phase 2 can be correlated by considering the following variables: interfacial drag force F_{12}, density difference $\Delta\rho \equiv \rho_1 - \rho_2$, density ρ_1, gravitational acceleration g, liquid viscosity μ_1, neglect gas viscosity and density $\mu_2 \cong 0, \rho_2 \cong 0$, relative interface velocity

$$\left(\frac{u_2}{S_2} - \frac{u_1}{S_1}\right),$$

permeability K, inertial flow coefficient β, interfacial tension σ, and liquid phase saturation S_1. (Note that $S_2 = 1 - S_1$.)

(a) How many dimensionless groups/variables can be found? (Answer: $10 - 3 = 7$)

(b) Determine the independent groups that can be formed with these variables.

7. The interfacial drag force between a liquid phase 1 and a gas phase 2 can be correlated alternatively by considering the buoyant force F_b, viscous force F_μ, inertial force F_i, capillary force F_c, and saturation S_1. A dimensional analysis will be carried out between these variables.

(a) How many dimensionless groups/variables can be found? (Answer: $6 - 1 = 5$ because there is only one combined force unit involved.)

(b) Determine the dimensionless groups that can be formed with these variables.

(Answer: $\pi_1 = \dfrac{F_{12}}{F_c}, \pi_2 = \dfrac{F_b}{F_c}, \pi_3 = \dfrac{F_\mu}{F_c}, \pi_4 = \dfrac{F_i}{F_c}, \pi_5 = S_1$.)

(c) Express these independent groups in terms of the system variables by substituting the following expressions of the various forces: the buoyant force given by

$$F_b = (\rho_1 - \rho_2)g,$$

viscous force $F_\mu = \dfrac{\mu_1}{K}\left(\dfrac{u_2}{S_2} - \dfrac{u_1}{S_1}\right),$

inertial force $F_i = \beta\rho_1\left(\dfrac{u_2}{S_2} - \dfrac{u_1}{S_1}\right)^2,$ and

capillary force $F_c = \dfrac{\sigma}{K}.$

(d) Determine the expressions of the commonly known dimensionless groups that can be formed with these forces. (Answer: the drag coefficient

$$c_D = \frac{F_{12}}{F_i} = \frac{F_{12}}{\beta\rho_1\left(\dfrac{u_2}{S_2} - \dfrac{u_1}{S_1}\right)^2}$$

capillary number $N_c = \dfrac{F_\mu}{F_c} = \dfrac{\mu_1}{\sigma}\left(\dfrac{u_2}{S_2} - \dfrac{u_1}{S_1}\right),$

Reynolds number $Re = \dfrac{F_i}{F_\mu} = \dfrac{K\beta\rho_1}{\mu_1}\left(\dfrac{u_2}{S_2} - \dfrac{u_1}{S_1}\right),$

and Froude number $Fr = \sqrt{\dfrac{F_i}{F_b}} = \left(\dfrac{u_2}{S_2} - \dfrac{u_1}{S_1}\right)\sqrt{\dfrac{\beta}{\left(1 - \dfrac{\rho_2}{\rho_1}\right)g}}.$ Further, note that $\pi_5 = S_1$)

(e) How can you justify the approach taken by Schulenberg and Muller (1987), who considered

• Viscous force and inertia of the gas phase small compared to the above-mentioned forces

- At moderately high saturations, the viscosity effects compensate each other: $F_\mu \cong 0$.

- A correlation in the general form of $\pi_1 = \pi_1(\pi_2, S_1)$, where only the following dimensionless groups were considered:

$$\pi_1 = \frac{F_{12}}{F_b} = \frac{F_{12}}{(\rho_1 - \rho_2)g}, \pi_3 = \frac{F_i}{F_c} = \frac{K\beta\rho_1}{\sigma}\left(\frac{u_2}{S_2} - \frac{u_1}{S_1}\right)^2, \pi_3 = S_1.$$

CHAPTER 5

FLUID MOTION IN POROUS MEDIA

5.1 INTRODUCTION

Description of fluid motion in porous media requires special attention to account for the interactions of fluids with the pore wall and other fluids.* Derivation of a generalized equation of momentum for fluids flowing through porous media is not a trivial task. Since its inception, the fundamental law of flow through porous media given by Darcy (1856) has been scrutinized greatly and been the subject for many modifications to take into account the effect of other factors that are not considered by this law.

Many researchers attempted to improve Darcy's law by various approaches to take into account several conditions of practical importance. The outstanding approaches have been developed primarily on the bases of dimensional analysis (Civan, 1990), control volume balance (Das, 1997, 1999; Civan, 2008d, 2010b), and porous media averaging (Hubbert, 1956; Slattery, 1972; Gray, 1975; Marle, 1982; Whitaker, 1986, 1996; Ruth and Ma, 1992, 1997; Liu and Masliyah, 2005). Civan (2008d) draws attention to the fact that

> "porous media averaging does not only require extremely complicated and intricate algebraic manipulation procedures, but also has been implemented inadequately in many studies. Consequently, debate about proper handling of various issues dealing

* Parts of this chapter have been reproduced with modifications from the following:

Civan, F. and Evans, R.D. 1996. Determination of non-Darcy flow parameters using a differential formulation of the Forchheimer equation. SPE Paper 35621, Proceedings of the Society of Petroleum Engineers Gas Technology Symposium (April 28–May 1, 1996), Calgary, Alberta, Canada, pp. 415–429, © 1996 SPE, with permission from the Society of Petroleum Engineers;

Civan, F. and Evans, R.D. 1998. Determining the parameters of the Forchheimer equation from pressure-squared vs. pseudopressure formulations. SPE Reservoir Evaluation & Engineering, February 1998, pp. 43–46, © 1998 SPE, with permission from the Society of Petroleum Engineers;

Civan, F. 2008d. Generalized Darcy's law by control volume analysis including capillary and orifice effects. Journal of Canadian Petroleum Technology, 47(10), pp. 1–7, © 2008 SPE, with permission from the Society of Petroleum Engineers; and

Prada, A. and Civan, F. 1999. Modification of Darcy's law for the threshold pressure gradient. Journal of Petroleum Science and Engineering, 22(4), pp. 237–240, with permission from Elsevier.

Porous Media Transport Phenomena, First Edition. Faruk Civan.
© 2011 John Wiley & Sons, Inc. Published 2011 by John Wiley & Sons, Inc.

with porous media averaging is still continuing. The results presented by various studies do not agree whether the porous media averaged pressure gradient term should include porosity. The same concern can be raised about shear stress. Liu and Masliyah (2005) emphasize that averaging of pressure and viscosity terms in the microscopic equation of motion should be treated differently than the other terms. Also, the mathematical definition of partial derivatives assumes a limit as control volume approaches zero, i.e., collapses to a point of zero volume. By and large, discussion of differences introduced by the application of such a condition in deriving porous media equations of mass, momentum and energy conservation has been omitted, whereas, representative elementary bulk volume is a finite quantity in definition of volume averaged quantities used for deriving the porous media (macroscopic) equation of motion. Others made concerted efforts to interpret, manipulate and define the various terms resulting from representative elemental volume averaging in a manner to conform the averaged equations to the specific mathematical forms of well-known equations, such as those given by Darcy (1856) for low velocity flow, Forchheimer (1901) for rapid flow, and Brinkman (1949) for considering the boundary effect."

This chapter demonstrates according to Civan (2008d) that the control volume approach is instructive and provides a rigorous equation of motion for porous media flow in a convenient and straightforward manner when interpreted and implemented properly. The equation captures the essential features of porous media flow based on the capillary orifice model of tortuous flow paths formed in porous media, but, in a more general sense, accounting for compressibility, source, and heterogeneity and anisotropy effects. This approach distinguishes and implements the different conventions used to express fluid flow, such as fluid capillary velocity (interstitial velocity), bulk volume average velocity (superficial velocity), and fluid volume average velocity (pore average velocity), defined over representative elementary bulk volume of porous media. Issues pertaining to proper handling of porous media averaging of the pressure and stress terms, effect of porous media heterogeneity and anisotropy, and threshold pressure gradient that must be overcome for fluid to be able to flow through porous media are rendered. Valuable insights are gained into the nature of the sources of such terms on the basis of the capillary orifice model. The generalized Darcy's law is presented here according to Civan (2008d). This represents the conventional Darcy's law and its well-known modifications when their inherent simplifying conditions are imposed.

Several other issues concerning fluid motion in porous media are reviewed, including the porous media averaging of the pressure and shear stress terms, effect of porous media heterogeneity and anisotropy, source terms, correlation of parameters, flow demarcation criteria, entropy generation, viscous dissipation, generalization of Darcy's law, non-Newtonian versus Newtonian fluid rheology, and threshold pressure gradient, which must be overcome for fluid to flow through porous media.

5.2   FLOW POTENTIAL

Flow potential is the available energy of a fluid to do work under favorable conditions. Hubbert's (1940) flow potential, Ψ, referred to as the pseudoflow potential by

Bear (1972), is defined as the total mechanical energy required to move a unit fluid mass from a reference level, z_o, to a different level, z, at a constant velocity. The pseudoflow potential is expressed as (Thomas, 1982)

$$\Psi = -p_o V_o \quad \text{(flow work of collection)}$$
$$- \int_o p\,dV \quad \text{(work of compression)}$$
$$+ g(z - z_o)\,\text{(potential energy variation)} \quad (5.1)$$
$$+ pv \quad \text{(flow work of deliverance)}$$
$$+ \frac{v^2 - v_o^2}{2g} \quad \text{(kinetic energy variation = zero because } v = v_o \text{)}.$$

The subscript o denotes an arbitrarily or conveniently selected reference, datum, or base level. z is the upward vertical distance in the gravity direction. V is the specific fluid volume, which is equal to the reciprocal fluid density, ρ; that is, $V = 1/\rho$. Following Thomas (1982), it can be shown that an integration by parts yields

$$\int_o p\,dV = pV - p_o V_o - \int_o V\,dp. \quad (5.2)$$

Therefore, invoking Eq. (5.2) into Eq. (5.1) leads to Hubbert's pseudoflow potential function as

$$\Psi = \int_{p_o} \frac{dp}{\rho} + g(z - z_o), \quad (5.3)$$

whereas the true-flow potential is defined by (Bear, 1972)

$$\Phi = p - p_o + g\int_{z_o} \rho\,dz. \quad (5.4)$$

Hence, comparing Eqs. (5.3) and (5.4) indicates that these two different definitions of the flow potentials are related by

$$d\Phi = \rho\,d\Psi. \quad (5.5)$$

5.3 MODIFICATION OF DARCY'S LAW FOR BULK- VERSUS FLUID VOLUME AVERAGE PRESSURES

Here, the classical form of Darcy's law is modified according to Civan (2010b).

Pore fluid pressure p is affected by pressure p_e created by the deformation of material of porous matrix and confining pressure p_c applied to porous media. In this case, the pressure (or stress) relationship is given by (Terzaghi, 1923; Biot, 1941)

$$p_e = p_c - \alpha p, \quad (5.6)$$

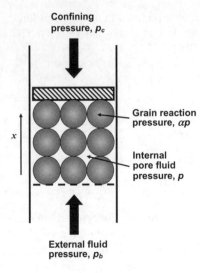

Figure 5.1 Quasi-static analysis of forces involving a fluid-saturated porous core plug subjected to confining pressure and external fluid pressure. External forces acting upon the core plug are balanced with the internal forces (after Civan, 2010b; reprinted by permission of Begell House, Inc.).

where α is Biot's effective stress coefficient. Its value is $\alpha = 0$ when the material of the porous matrix is totally elastic and $\alpha = 1$ when the material of the porous matrix is totally rigid. Its value is determined empirically for other (realistic) cases.

Consider Figure 5.1 (inferred by Bolton, 2002) for a quasi-static analysis of forces involving a fluid-saturated porous core plug subjected to confining pressure and external fluid pressure. Under these conditions, the external forces acting upon the core plug are balanced with the internal forces. A force balance between the various forces involved in porous media in view of Pascal's law of transmission of pressure applied at a point throughout the complete system can be written as the following, for example, in the x-Cartesian direction (Civan, 2010b):

$$\sum F_x = F_b - F_p - F_c + F_e = p_b A_b - pA - A_c p_c + A_e p_e = 0, \tag{5.7}$$

where A_b, A, A_c, and A_e are the surface areas for bulk porous media, pore fluid, material confining stress application, and material deformation (compression/expansion) application, respectively. p_b, p, p_c, and p_e are the corresponding pressures applied on these surfaces, respectively. F_b, F, F_c, and F_e are the corresponding forces applied on these surfaces, respectively.

Assuming the irregular-shaped fluid phase present in porous media is a fractal medium, the following fractal equation can be proposed (Civan, 2001, 2002d, 2010b) (see Chapter 2):

$$A/A_b = c\varepsilon^{d/3}, \tag{5.8}$$

where c is a fractal coefficient and d is a fractal dimension. These are both unitless parameters of fluid-phase distribution (shape). For a two-dimensional nonfractal medium, $c = 1$ and $d = 2$ (Liu and Masliyah, 2005). Further, consider that

$$A_c = A_e = A_b - A. \tag{5.9}$$

Hence, the bulk and fluid volume average pressure values can be related to each other as the following by means of Eqs. (5.6)–(5.9) (Civan, 2010b):

$$p_b = \left[\alpha + (1-\alpha)c\varepsilon^{d/3}\right]p, \ 0 \le \alpha \le 1. \tag{5.10}$$

Consequently, $p_b = p\varepsilon$ is applicable for the pressure variable only if $\alpha = 0$, that is, if the material of the porous matrix is totally elastic, $c = 1$ and $d = 3$ (Das, 1999; Hsu, 2005; Civan, 2008d). A special case occurs where $p_b = p$ if $\alpha = 1$, that is, if the material of the porous matrix is totally rigid (Liu and Masliyah, 2005). Obviously, neither of the special cases is realistic.

In view of the above-mentioned developments, Darcy's law requires modification as the following. For starters, it is important to realize that Darcy's law expresses the flow through a porous medium proportionally to the difference in pressures applied at the outside surfaces of bulk porous medium (fluid plus porous material) divided by the distance between these surfaces (Gray and Miller, 2004). Hence, for example, the rate of fluid flowing in the x-Cartesian direction through a rectangular-shaped porous element of dimensions Δx, Δy, and Δz in the x-, y-, and z-Cartesian directions can be expressed, applying the procedure presented by Civan (2010b), as

$$
\begin{aligned}
q_x &= \frac{K_x}{\mu}\left[\frac{(p_b A_b)_x - (p_b A_b)_{x+\Delta x} - \tau_w A_w \Delta x \Delta y \Delta z}{\Delta x}\right] \\
&= \frac{K_x}{\mu}\left[\frac{(p_b \Delta y \Delta z)_x - (p_b \Delta y \Delta z)_{x+\Delta x} - \tau_w A_w \Delta x \Delta y \Delta z}{\Delta x}\right],
\end{aligned}
\tag{5.11}
$$

where τ_w is the pore wall shear stress and A_w is the pore wall surface area contained per unit bulk volume.

Rearranging, it results in

$$\frac{q_x}{\Delta y \Delta z} = -\frac{K_x}{\mu}\left[\frac{(p_b)_{x+\Delta x} - (p_b)_x}{\Delta x} + \tau_w A_w\right]. \tag{5.12}$$

Then, taking the extrapolated limit according to Hubbert (1956) as $\Delta x, \Delta y, \Delta z \to 0$ and substituting Eq. (5.10) into Eq. (5.12), we obtain a modified Darcy's law as

$$u_x = -\frac{K_x}{\mu}\left(\frac{\partial p_b}{\partial x} - \frac{\partial p_{th}}{\partial x}\right) = -\frac{K_x}{\mu}\left[\frac{\partial\{[\alpha + (1-\alpha)c\varepsilon^{d/3}]p\}}{\partial x} - \frac{\partial p_{th}}{\partial x}\right], \tag{5.13}$$

where

$$-\frac{\partial p_{th}}{\partial x} = \tau_w A_w$$

is the threshold pressure gradient. Therefore, flow can occur only if

$$\left|\frac{\partial p_b}{\partial x}\right| > \left|\frac{\partial p_{th}}{\partial x}\right|.$$

It now becomes clear that the original Darcy's law is valid only if

$$\frac{\partial p_{th}}{\partial x} = 0 \quad \text{and} \quad \alpha = 1,$$

representing a material of porous matrix that is ideally rigid (Liu and Masliyah, 2005; Civan, 2010b). In the rest of the chapter, we assume $\alpha = 1$.

5.4 MACROSCOPIC EQUATION OF MOTION FROM THE CONTROL VOLUME APPROACH AND DIMENSIONAL ANALYSIS

Extending Collins (1961) control volume approach, Civan and Tiab (1989) and Civan (1990) derived the momentum equation for fluid flowing through porous media. For this purpose, they considered a cylindrical core plug as shown in Figure 5.2. The length and cross-sectional area of the core are denoted by L and A, respectively. The mean diameter of the hydraulic flow paths is d. The effective or conductive porosity is ϕ. The angle of inclination with respect to the horizontal is α. The fluid viscosity and density are denoted by μ and ρ, respectively. The fluid pressures at the inlet and outlet are p_o and p_L, respectively. g denotes the gravity of acceleration. The distance in the flow direction in the core plug, measured from the inlet end, is ℓ. The upward vertical distance along the gravity direction is z. q is the fluid volumetric flow rate. u denotes the fluid volumetric flux, given by

$$u = q/A. \tag{5.14}$$

The net force causing the fluid motion is given by

$$F_i = F_p - F_g - F_v, \tag{5.15}$$

in which F_i, F_p, F_g, and F_v represent the inertial, pressure, gravity, and viscous forces, respectively.

The pressure force is given by

$$F_p = A\phi(p_o - p_L) = A\phi\Delta p, \tag{5.16}$$

where the pressure difference across the core plug is given by

$$\Delta p = p_o - p_L. \tag{5.17}$$

The gravity force is given by

$$F_g = mg\sin\alpha, \tag{5.18}$$

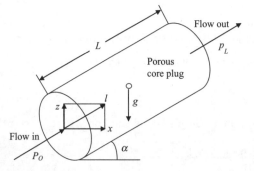

Figure 5.2   Porous core plug considered as a control volume (prepared by the author).

where m is the mass of the fluid contained in the pore volume

$$m = AL\phi\rho, \tag{5.19}$$

where ρ is the density of the fluid.

The viscous force, resisting flow, can be assumed related to a number of variables according to

$$F_v = f_1(\rho, \mu, u, L, d), \tag{5.20}$$

where f_1 is an unknown function. By dimensional analysis, these variables can be combined into three dimensionless groups as

$$\frac{F_v}{\mu u L} = f_2\left(\frac{d}{L}, \frac{\rho u L}{\mu}\right), \tag{5.21}$$

where f_2 represents an unknown function to be determined. Hence, defining an aspect ratio,

$$A_R = d/L, \tag{5.22}$$

and the Reynolds number,

$$\text{Re} = \frac{\rho u L}{\mu}, \tag{5.23}$$

Eq. (5.21) can be written as

$$F_v = \mu u L f_2(A_R, \text{Re}). \tag{5.24}$$

The inertial force is given by Newton's law as

$$F_i = ma, \tag{5.25}$$

where a denotes the fluid acceleration, which is assumed to be a function of a number of variables according to

$$a = f_3(\rho, \mu, u, \Delta p, L, d), \tag{5.26}$$

where f_3 is an unknown function. Applying dimensional analysis, these variables can be combined into four dimensionless groups as

$$\frac{aL}{u^2} = f_4\left(\frac{d}{L}, \frac{\rho u d}{\mu}, \frac{\Delta p}{1/2\,\rho u^2}\right), \tag{5.27}$$

where f_4 is an unknown function to be determined. Define the pressure loss coefficient as

$$C_p = \frac{\Delta p}{\dfrac{1}{2}\rho u^2}. \tag{5.28}$$

Thus, substituting Eqs. (5.22), (5.23), and (5.28) into Eq. (5.27) and then rearranging yield the following expression for the fluid acceleration:

$$a = \frac{u^2}{L} f_4(A_R, \mathrm{Re}, C_p). \tag{5.29}$$

Consequently, invoking Eqs. (5.16)–(5.29) into Eq. (5.15) and rearranging lead to

$$\frac{\Delta p}{L} - \rho g \sin \theta = \mu u \frac{f_2(A_R, \mathrm{Re})}{A\phi} + \rho u^2 \frac{f_4(A_R, \mathrm{Re}, C_p)}{L}. \tag{5.30}$$

Define the core length average permeability and inertial flow coefficient, respectively, by

$$\frac{1}{\tilde{K}} = \frac{f_2(A_R, \mathrm{Re})}{A\phi} \tag{5.31}$$

and

$$\tilde{\beta} = \frac{f_4(A_R, \mathrm{Re}, C_p)}{L}. \tag{5.32}$$

Hence, the problem of determining the unknown f_2 and f_4 is transformed to determining $\tilde{K}$ and $\tilde{\beta}$. Substituting Eqs. (5.31) and (5.32) into Eq. (5.30) and then taking a limit as the core length L approaches the representative elementary core length L_R yields the following differential macroscopic fluid momentum balance equation:

$$-\left(\frac{dp}{d\ell} + \rho g \sin \alpha\right) = \frac{\mu}{K} u + \beta \rho u^2, \tag{5.33}$$

in which

$$K = \lim_{L \to L_R} \tilde{K} \tag{5.34}$$

and

$$\beta = \lim_{L \to L_R} \tilde{\beta}. \tag{5.35}$$

Substituting

$$\sin \alpha = dz/d\ell \tag{5.36}$$

and the fluid flow potential function

$$\Psi = \int_{Po}^{p} \frac{dp}{\rho} + g(z - z_o) \tag{5.37}$$

into Eq. (5.33) yields a convenient and compact form of Eq. (5.33) as

$$-\rho \frac{d\Psi}{d\ell} = \frac{\mu}{K} u + \beta \rho u^2. \tag{5.38}$$

Alternatively, substituting Φ into Eq. (5.38) according to

$$d\Phi \equiv \rho d\Psi \tag{5.39}$$

leads to

$$-\frac{d\Phi}{d\ell} = \frac{\mu}{K}u + \beta\rho u^2. \tag{5.40}$$

When $\beta = 0$, Eqs. (5.33), (5.38), or (5.40) are referred to as Darcy's (1856) law and Forchheimer's (1914) equation for $\beta \neq 0$.

The dimensions of K and β are L^2 and L^{-1}.

For convenience, Eqs. (5.33), (5.38), or (5.40) can be expressed, respectively, as

$$u = -N_{nd}\mu^{-1}K\left(\frac{dp}{d\ell} + \rho g \sin\alpha\right), \tag{5.41}$$

$$u = -N_{nd}v^{-1}K\frac{d\Psi}{d\ell}, \tag{5.42}$$

and

$$u = -N_{nd}\mu^{-1}K\frac{d\Phi}{d\ell}, \tag{5.43}$$

where $v \equiv \mu/\rho$ is the kinematic viscosity and N_{nd} is the non-Darcy number given as the reciprocal Darcy number as

$$N_{nd} = Da^{-1} = (1+Re)^{-1}, \tag{5.44}$$

in which the porous media Reynolds number is given by

$$Re = \frac{\rho u K \beta}{\mu}. \tag{5.45}$$

Thus, comparing Eqs. (5.23) and (5.45), an apparent hydraulic tube diameter can be defined as (Green and Duwez, 1951; Civan, 1990)

$$d = K\beta. \tag{5.46}$$

Note that $0 \leq N_{nd} \leq 1$ for $0 \leq Re \leq \infty$.

5.5 MODIFICATION OF DARCY'S LAW FOR THE THRESHOLD PRESSURE GRADIENT

There is experimental evidence "that fluids can flow through porous media only if the fluid force is sufficient to overcome the threshold pressure gradient and, therefore, Darcy's law should be corrected for this effect" as stated by Prada and Civan (1999). The law of flow, that is, the equation of motion

$$u = \frac{q}{A} = -\frac{K}{\mu}\left(\frac{\partial p}{\partial \ell} + \rho g \sin\alpha\right), \tag{5.47}$$

has been extensively used for description of fluid flow in porous media. However, Prada and Civan (1999) have shown that Darcy's law needs a correction for the effect of the threshold pressure gradient. In Eq. (5.47), q is the flow rate and A is the area normal to flow. An integration of Darcy's law (Eq. 5.47) over a core length L, assuming constant fluid and porous media properties and horizontal flow, results in

$$u = \frac{q}{A} = \frac{K}{\mu} \frac{\Delta p}{L}, \tag{5.48}$$

where Δp denotes the pressure difference across the core and L is the core length.

The plots of q/A versus $\Delta p/L$ data obtained by Prada and Civan (1999) by flowing saturated brine through shaly sandstone, brown sandstone, and sandpacks are shown in Figure 5.2. As can be seen, these data yield straight line plots shifted relative to the origin. Eq. (5.48) can be corrected for these offsets as

$$u = \frac{q}{A} = \frac{K}{\mu}\left[\frac{\Delta p}{L} - \left(\frac{\Delta p}{L}\right)_{cr}\right], \text{ when } \frac{\Delta p}{L} > \left(\frac{\Delta p}{L}\right)_{cr} \tag{5.49}$$

$$u = q/A = 0, \text{ otherwise}, \tag{5.50}$$

where $(\Delta p/L)_{cr}$ is the threshold pressure gradient. When the pressure gradient is below this value, the fluid cannot flow because of the frictional effects. Extrapolating the lines to zero flow, the threshold pressure gradient, $(\Delta p/L)_{cr}$, can be obtained from Figure 5.3. They decrease as the fluid mobility, K/μ, increases. The threshold pressure gradient is negligible for gases but significant for liquids. Figure 5.4 shows a

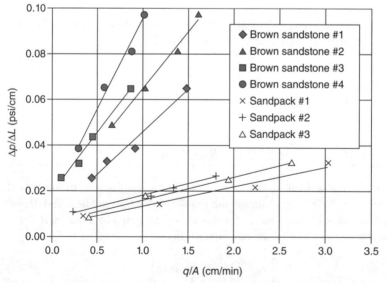

Figure 5.3   Volume flux versus pressure gradient (after Prada and Civan, 1999; reproduced by permission of Elsevier).

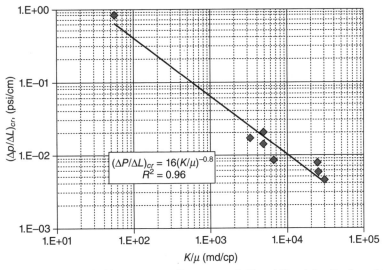

Figure 5.4   Threshold pressure gradient versus fluid mobility (after Prada and Civan, 1999; reproduced by permission of Elsevier).

plot of these data on a log–log scale. The least squares correlation of these data is given by Prada and Civan (1999) as

$$(\Delta p/\Delta L)_{cr} = 16(K/\mu)^{-0.8}.$$ (5.51)

Eq. (5.51) indicates that the threshold pressure gradient is negligible at high fluid mobilities. Hence, for low-mobility fluids, Prada and Civan (1999) corrected Darcy's law for forward flow as

$$u = \frac{q}{A} = -\frac{K}{\mu}\left[\frac{\partial p}{\partial \ell} - \left(\frac{\partial p}{\partial \ell}\right)_{cr} + \rho g \sin \alpha\right], \text{ when } -\left(\frac{\partial p}{\partial \ell} + \rho g \sin \alpha\right) > -\left(\frac{\partial p}{\partial \ell}\right)_{cr}$$ (5.52)

$$u = q/A = 0, \text{ when } -\left(\frac{\partial p}{\partial \ell} + \rho g \sin \alpha\right) \leq -\left(\frac{\partial p}{\partial \ell}\right)_{cr}.$$ (5.53)

Thus, generalizing Eq. (5.51) as

$$-(\partial p/\partial \ell)_{cr} = m(K/\mu)^{-n}$$ (5.54)

and substituting into Eqs. (5.52) and (5.53), Prada and Civan (1999) derived the following modified Darcy's equation:

$$u = \frac{q}{A} = -\frac{K}{\mu}\left[\frac{\partial p}{\partial \ell} + \rho g \sin \alpha\right] - m\left(\frac{K}{\mu}\right)^{1-n}, -\left(\frac{\partial p}{\partial \ell} + \rho g \sin \alpha\right) > m(K/\mu)^{-n}$$ (5.55)

$$u = q/A = 0, -\left(\frac{\partial p}{\partial \ell} + \rho g \sin \alpha\right) \leq m(K/\mu)^{-n}.$$ (5.56)

5.6   CONVENIENT FORMULATIONS OF THE FORCHHEIMER EQUATION

The Forchheimer (1914) equation for rapid, one-dimensional flow of gases through an inclined core plug is given by

$$-(dp/dx + \rho g \sin \alpha) = (\mu/K)u + \beta \rho u^2, \tag{5.57}$$

where p is pressure, x is distance along the flow direction, α is the angle of inclination, ρ is density, μ is viscosity, K is permeability, β is the inertial flow coefficient, and u is the volume flux.

The expressions for the Klinkenberg equation, mass flow rate, and the real gas density are given, respectively, by

$$K = K_\infty(1 + b/p), \tag{5.58}$$

$$w = \rho u A, \tag{5.59}$$

and

$$\rho = MP/(zRT). \tag{5.60}$$

K_∞ is the liquid permeability; b is an empirical parameter; w is the mass flow rate; A is the cross-sectional area of the core plug; M is the molecular weight; z is the real gas deviation factor; R is the ideal gas constant; and T is temperature.

The density is calculated using the real gas equation of state, Eq. (5.60), with the gas deviation factor determined by the equation of state fitted by Gopal (1977) and the viscosity by the Starling and Ellington (1964) viscosity correlation.

Invoking Eqs. (5.58)–(5.60) into Eq. (5.57) yields the following general differential equation for pressure variation along the core plug:

$$\frac{dp}{dx} = -\frac{wRT\mu(p)z(p)}{AMp}\left[\frac{1}{K_\infty(x)(1+b/p)} + \frac{\beta(x)w}{A\mu(p)}\right] - \frac{Mp}{z(p)RT}g\sin\alpha, \tag{5.61}$$

in which the last term involving the gravity effect is negligible for dilute gases.

Firoozabadi et al. (1995) used a simplified integral form of Eq. (5.61) as the basis for the determination of the K and β values. Their equation can be written as follows, and it involves a variety of simplifying assumptions to be able to analytically carry out the integration of Eq. (5.61) over the core length:

$$\frac{MA(p_1^2 - p_2^2)}{2\bar{\mu}\bar{z}RTLw} = \frac{1}{\tilde{K}} + \tilde{\beta}\frac{w}{A\bar{\mu}}, \tag{5.62}$$

where, $\tilde{K}$ and $\tilde{\beta}$ are the approximate values of K and β over the core length obtained by a straight line plot of Eq. (5.62).

Eq. (5.62) assumes the gravity and Klinkenberg effects are negligible and facilitates the use of the average viscosity, $\bar{\mu}$, and real gas deviation factor, $\bar{z}$, defined in an appropriate manner, such as by the mean value theorem as

$$\bar{\mu} = \int_{P_1}^{P_2} \mu dp / (p_2 - p_1) \qquad (5.63)$$

and

$$\bar{z} = \int_{P_1}^{P_2} z dp / (p_2 - p_1). \qquad (5.64)$$

Eq. (5.64) was also proposed by Ikoku (1984).

Inferred by Geertsma (1974), Eq. (5.62) can be reformulated, by multiplying by $\tilde{K}$, as

$$Da = 1 + \mathrm{Re} \qquad (5.65)$$

or, alternatively, by dividing Eq. (5.65) by Re, as

$$f = 1 + 1/\mathrm{Re}. \qquad (5.66)$$

In Eqs. (5.65) and (5.66), Da, Re, and f are the dimensionless numbers known as the Darcy number, Reynolds number, and the friction factor defined, respectively, by

$$Da = \frac{\tilde{K}MA(p_1^2 - p_2^2)}{2\bar{\mu}\bar{z}RTLw} = \frac{\bar{\rho}\tilde{K}A}{\bar{\mu}w}\frac{\Delta p}{L}, \qquad (5.67)$$

$$\mathrm{Re} = \frac{\beta\tilde{K}w}{\bar{\mu}A}, \qquad (5.68)$$

and

$$f = \frac{Da}{\mathrm{Re}} = \frac{MA^2(p_1^2 - p_2^2)}{2\tilde{\beta}\bar{z}RTLw^2} = \frac{\bar{\rho}A^2}{\tilde{\beta}w^2}\frac{\Delta p}{L} = \frac{1}{\tilde{\beta}\bar{\rho}\bar{u}^2}\frac{\Delta p}{L}. \qquad (5.69)$$

The product $\tilde{\beta}\tilde{K}$ in Eq. (5.68) represents the diameter of a representative hydraulic flow tube given by Eq. (5.46) (Civan, 1990). The average pressure, density, and volume flux are defined, respectively, by

$$\bar{p} = (p_1 + p_2)/2, \qquad (5.70)$$
$$\bar{\rho} = M\bar{p}/(\bar{z}RT), \qquad (5.71)$$

and

$$\bar{u} = w/(\bar{\rho}A). \qquad (5.72)$$

The pressure differential is given by

$$\Delta p = p_1 - p_2. \qquad (5.73)$$

A chart of the porous media friction number versus the porous media Reynolds number given by Ahmed and Sunada (1969) indicates that the transition from laminar to rapid flow is smooth in porous media because of the pore size distribution effect of porous media. The laminar regime exists for Re < 1.0. Ahmed and Sunada (1969) provide the following empirical correlation for the porous media friction number versus the porous media Reynolds number:

$$f = 1 + \frac{1}{Re}, 10^{-4} < Re < 10. \tag{5.74}$$

where Re and f are given by Eqs. (5.68) and (5.69), respectively.

Civan and Evans (1995, 1996, 1998) suggested that the accuracy of the preceding analysis could be improved by first expressing the Forchheimer equation in terms of the pseudopressure function (generalized after Al-Hussainy and Ramey, 1966)

$$m(p) = \int_{p_b}^{P} \frac{\rho}{\mu} dp \tag{5.75}$$

and then by integrating to obtain the following alternative expression:

$$\frac{A[m(p_1) - m(p_2)]}{Lw} = \frac{1}{\tilde{K}} + \tilde{\beta} \frac{w}{A\bar{\mu}} \tag{5.76}$$

because Eq. (5.76) involves only $\bar{\mu}$, one average fluid property less than Eq. (5.75) of Firoozabadi et al. (1995) or the present Eq. (5.62), which involves two average fluid properties, $\bar{\mu}$ and $\bar{z}$.

Eq. (5.76) can also be reformulated in the forms of Eqs. (5.65) and (5.66). In this case, while the Reynolds number is still given by Eq. (5.68), the expressions for the Darcy number and the friction factor, respectively, become

$$Da = \frac{\tilde{K}A\Delta m}{Lw} \tag{5.77}$$

and

$$f = \frac{A^2 \bar{\mu}}{\tilde{\beta}w^2} \frac{\Delta m}{L}, \tag{5.78}$$

where Δm is the pseudopressure differential given by

$$\Delta m = m(p_1) - m(p_2). \tag{5.79}$$

In order to account for the variations of density and viscosity along the core, a direct numerical integration of the Forchheimer equation, given by Eq. (5.61), which is an ordinary differential equation, as a function of the distance from the core inlet to outlet, should be carried out by an appropriate numerical method such as the Runge–Kutta method. The unknown parameters $K(x)$ and $\beta(x)$ for heterogeneous cores could then be determined to match the pressure values at the core ends. K and β are uniform values for homogeneous cores and Eq. (5.61) can be written, neglecting the gravity and Klinkenberg effects, as

$$\frac{dp}{dx} = -\frac{wRT\mu(p)z(p)}{AMp} \left[\frac{1}{K} + \beta \frac{w}{A\mu(p)} \right] \tag{5.80}$$

subject to

$$p = p_1, x = 0 \tag{5.81}$$

and

$$p = p_2, x = L. \tag{5.82}$$

Based on Eq. (5.80), the local Darcy number, the Reynolds number, and the friction factor are defined, respectively, by

$$Da = \frac{KMAp}{\mu(p)z(p)RTw}\left(-\frac{dp}{dx}\right) = \frac{KA}{w}\frac{\rho(p)}{\mu(p)}\left(-\frac{dp}{dx}\right) = \frac{KA}{w}\left[-\frac{dm(p)}{dx}\right], \tag{5.83}$$

$$\mathrm{Re} = \frac{\beta Kw}{A\mu(p)}, \tag{5.84}$$

and

$$f = \frac{A^2\rho(p)}{\beta w^2}\left(-\frac{dp}{dx}\right) = \frac{1}{\beta\rho(p)u(p)^2}\left(-\frac{dp}{dx}\right) = \frac{A^2\mu(p)}{\beta w^2}\left[-\frac{dm(p)}{dx}\right]. \tag{5.85}$$

5.7 DETERMINATION OF THE PARAMETERS OF THE FORCHHEIMER EQUATION

This section presents a comparison of the pressure-squared and pseudopressure formulations, a demonstration of the effect of the core length and determination of the representative core length for the simultaneous measurement of permeability and non-Darcy flow coefficient.

The inertial flow coefficient can be predicted by the Liu et al. (1995) correlation given by

$$\beta = 8.91 \times 10^8 \, \tau/(\phi K), \tag{5.86}$$

in which β is in ft^{-1}, K is in millidarcy, and τ is the tortuosity.

Many studies, including that of Firoozabadi et al. (1995), have facilitated the integral forms of the Forchheimer (1914) equation as a convenient means of determining the permeability, K, and the non-Darcy flow coefficient, β, from high-velocity flow data. However, as stated by Civan, (1990) and Civan and Evans (1995, 1996, 1998), the core length averaged $\tilde{K}$ and $\tilde{\beta}$ are functions of length, because the viscosity, μ, and the real gas deviation factor, z, are averaged over the pressure drop across the core ends to obtain $\bar{\mu}$ and $\bar{z}$, respectively. In theory, the pressure-squared function has inherent limitations, because it must satisfy two contradictory conditions in order to obtain an accurate estimation of permeability and inertial flow coefficient from laboratory core tests. The first condition requires very short cores for average viscosity and real gas deviation factor to be close to actual values. The second condition requires sufficiently long cores to approximate the representative core length correctly. Mathematically, these conditions can be expressed by Eqs. (5.87) and (5.88), respectively:

$$\underset{L \to L_R}{\mathrm{Limit}} \ \tilde{K} = K \quad \text{and} \quad \tilde{\beta} = \beta \tag{5.87}$$

and

$$\underset{L \to 0}{\mathrm{Limit}} \ \bar{\mu} = \mu \quad \text{and} \quad \bar{z} = z \tag{5.88}$$

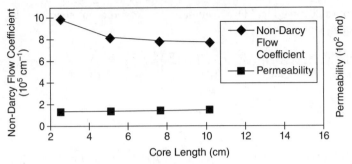

Figure 5.5   Experimental data on the effect of core length on permeability and non-Darcy flow coefficient (after Civan and Evans, 1998; © 1998 SPE, reprinted by permission of the Society of Petroleum Engineers).

where L_R is a representative elemental core length. The pseudopressure formulation by Civan and Evans (1996, 1998) involves only $\bar{\mu}$, which is almost constant for practical purposes, because the effect of pressure on the gas viscosity is negligible. Therefore, the pseudopressure formulation alleviates the need for satisfying both of the above contrasting conditions; that is only the condition stated by Eq. (5.87) needs to be satisfied. Therefore, in the pseudopressure formulation, the limit is taken with respect to the representative length only.

Evans and Civan (1994) and Civan and Evans (1996, 1998) used experimental data to demonstrate the effect of the core length on permeability and non-Darcy flow coefficients. A series of different length Berea cores have been used to generate the pressure differential versus flow rate experimental data at steady-state conditions. These data were then plotted for each different core length, and the permeability and the non-Darcy flow coefficient values were determined. The results were then plotted against core length to determine the sensitivity due to the core length. Figure 5.5 indeed indicated that the core length average permeability and the non-Darcy flow coefficient were dependent on the core length. The representative values of permeability and the non-Darcy flow coefficient were obtained by extrapolation of the core length average values to the representative elemental core length for which these values reach the limiting values given, respectively, as

$$K = \text{Limit}\ \tilde{K} = 148\ \text{md}$$
$$L \to 10\ \text{cm}$$

(5.89)

$$\beta = \text{Limit}\ \tilde{\beta} = 7.77 \times 10^5\ \text{cm}^{-1}$$
$$L \to 10\ \text{cm}.$$

(5.90)

As can be seen, the representative elemental core length necessary for accurate measurements is simultaneously estimated to about 10 cm. However, the extrapolated values obtained in this example represent the average $\tilde{K}$ and $\tilde{\beta}$ values for the whole porous material, not the local values of K and β at a selected location along the core.

Note that the errors due to not using representative core lengths are not negligible. If, for example, had a core of 2.54-cm length instead of the representative length of 10 cm been used, there would have been an error of

$$\frac{9.85-7.77}{9.85}\times100=21\%$$

in the β value and

$$\frac{1.29-1.48}{1.29}\times100=-15\%$$

in the K value according to the data presented in Figure 5.5.

Civan and Evans (1996, 1998) generated the simulated data by solving Eqs. (5.80)–(5.82) using the Runge–Kutta method and substituted for actual experimental data to avoid the affects of measurement errors associated with laboratory core tests. The average viscosity, $\bar{\mu}$, the average real gas deviation factor, $\bar{z}$, and the pseudo-pressure difference, $\Delta m = m(p_1) - m(p_2)$, were calculated according to Eqs. (5.63), (5.64), and (5.75), respectively. For maximum accuracy, Civan and Evans (1996, 1998) calculated the integral terms in these equations by the Runge–Kutta method.

In the first case, the permeability and the inertial flow coefficient values were assumed to be $K = 600$ md and $\beta = 3.0 \times 10^5$ cm^{-1}, respectively, and the temperature was taken as 290°K. The simulated data were obtained for 0.0254, 0.0508, 0.1016, and 0.2032 m (1 in, 2 in, 4 in, and 8 in) long and 0.0254 m (1 ft) diameter core plugs for a range of injection rates of a 0.65-gravity natural gas and for 101.3-, 700.0-, and 3500.0-kPa back pressures. The plots of the data were constructed according to the pressure-squared and pseudopressure formulations given by Eqs. (5.62) and (5.76).

As can be seen, for example, in Figure 5.6 for a 2-in-long core, both data lie on the same straight lines leading to $K = 600$ md and $\beta = 3.0 \times 10^5$ cm^{-1}. These values are the same as those assumed to generate the simulated data used in the calculations. In this case, the core length effect could not be observed, because in the high-permeability cores used in this exercise, only small variations of the gas pressure along the core were observed. Therefore, using high-permeability cores, Firoozabadi et al. (1995) concluded that the core length had no effect on the non-Darcy flow coefficient.

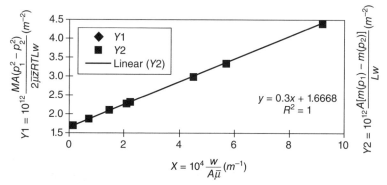

Figure 5.6 Comparison of the pressure-squared ($Y1$) and pseudopressure ($Y2$) approaches for a high-permeability core ($K = 600$ md) (after Civan and Evans, 1998; © 1998 SPE, reprinted by permission of the Society of Petroleum Engineers).

In the second case, a tight formation of porosity $\phi = 0.08$ and permeability $K = 0.02\,\text{md}$ was considered. The inertial flow coefficient was predicted by the Liu et al. (1995) correlation to be $\beta = 2.6 \times 10^{10}\ \text{cm}^{-1}$. The calculations similar to the first case lead to the simulated data indicating significant pressure variations along the core. As seen in Figure 5.7, for example, for a 2-in-long core, the pressure-squared and pseudopressure differences differed significantly in this tight formation than the previous high-permeability formation. The permeabilities and inertial flow coefficients are determined according to Eq. (5.76), which is a better linear representation of the data. The core length average permeabilities and the inertial flow coefficients are dependent on the core length as indicated by Figure 5.8. By extrapolation to the representative core length of about 30 cm, the permeability and the inertial flow coefficient of the tight formation are obtained, respectively, as

$$K = \underset{L \to 30\ \text{cm}}{\text{Limit}}\ \tilde{K} = 0.02\ \text{md} \tag{5.91}$$

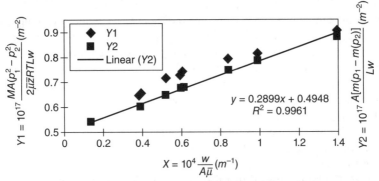

Figure 5.7   Comparison of the pressure-squared ($Y1$) and pseudopressure ($Y2$) approaches for a low-permeability core ($K = 0.02\,\text{md}$) (after Civan and Evans, 1998; © 1998 SPE, reprinted by permission of the Society of Petroleum Engineers).

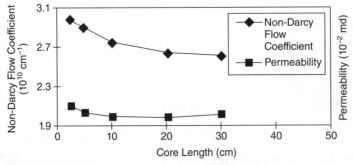

Figure 5.8   Simulation data on the effect of core length on permeability and non-Darcy flow coefficient (after Civan and Evans, 1998; © 1998 SPE, reprinted by permission of the Society of Petroleum Engineers).

and

$$\beta = \text{Limit } \tilde{\beta} = 2.6 \times 10^{10} \text{ cm}^{-1}$$
$$L \to 30 \text{ cm}.$$
(5.92)

These are the same as those values assumed as the core properties in the solution of Eqs. (5.91)–(5.97) to generate the simulated data used in these calculations. Thus, the limit of $\tilde{K}$ and $\tilde{\beta}$ do go to the exact values as the limit length approaches the representative elementary length, which is correctly indicated by Figure 5.8 when the pseudopressure approach based on Eq. (5.76) is used. Figure 5.8 indicates errors of

$$\frac{2.983 - 2.6}{2.983} \times 100 = 13\% \text{ in } \beta \quad \text{and} \quad \frac{2.084 - 2}{2.084} \times 100 = 4\% \text{ in } K$$

if a core length of 2.54 cm instead of the representative core length of 30 cm is used.

This exercise demonstrates that the core length average permeability and inertial flow coefficient are dependent on the core length, and therefore, correct values can only be obtained by extrapolation to the representative core length. The core length dependency is more pronounced for tight porous formations because of more pressure variation, and it is negligible for high-permeability formations. The results shown in Figures 5.6 and 5.7 indicate that the pseudopressure function is a better linear representation of experimental data than the pressure-squared function.

5.8 FLOW DEMARCATION CRITERIA

Ergun (1952) interpreted Forchheimer's equation in the following form:

$$-\frac{dp}{dx} = A \frac{(1-\phi)^2}{\phi^3} \frac{\mu u}{d^2} + B \frac{(1-\phi)}{\phi^3} \frac{\rho u^2}{d},$$
(5.93)

where d denotes the characteristic length of the porous medium, such as the diameter of spherical particles forming the porous media, and A and B denote the dimensionless fitting constants. Ergun (1952) estimated by means of experimental data that $A = 150$ and $B = 1.75$. However, the later works indicate some variations from these values. Nevertheless, Kececioglu and Jiang (1994) facilitate Ergun's equation to develop the flow demarcation criteria for the determination of transition between the various types of flows as described schematically in Figure 5.9. Kececioglu and Jiang (1994) accomplished this objective by expressing Ergun's equation in dimensionless form. Their expression is modified here as the following by substituting the mean hydraulic flow tube diameter

$$D = 4\sqrt{\frac{F_s \tau K}{\phi}},$$

where F_s is a geometric shape factor ($F_s = 2$ for cylindrical flow tubes):

$$\tilde{P} \equiv \frac{K}{\mu u} \frac{dp}{dx} = \tilde{A} + \tilde{B} \text{Re}_K,$$
(5.94)

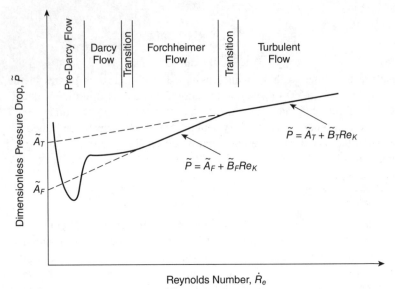

Figure 5.9 Description of the flow regimes for fluid flow through porous media (after Kececioglu and Jiang, 1994; © 1994, reprinted by permission of the American Society of Mechanical Engineers).

where

$$\tilde{A} = \frac{A}{16F_s\tau}\left(\frac{1-\phi}{\phi}\right)^2, \tilde{B} = \frac{B}{4\sqrt{F_s\tau}}\frac{1-\phi}{\phi}. \tag{5.95}$$

The pore flow Reynolds number is given by

$$\mathrm{Re}_K = \frac{\rho u}{\mu\phi}\sqrt{\frac{K}{\phi}}. \tag{5.96}$$

Kececioglu and Jiang (1994) determined the coefficient of Eq. (5.94) empirically for the different flow regimes described in Figure 5.9. They proposed the following flow demarcation criteria:

Pre-Darcy flow regime: $\mathrm{Re}_K < 0.06$

Darcy flow regime: $0.06 < \mathrm{Re}_K < 0.12$, $\tilde{A} = 1.0$, $\tilde{B} = 0$

Post-Darcy laminar Forchheimer flow regime: $0.34 < \mathrm{Re}_K < 2.3$, $\tilde{A} = 1.0$, $\tilde{B} = 0.7$

Turbulent flow regime: $\mathrm{Re}_K > 3.4$, $\tilde{A} = 1.9$, $\tilde{B} = 0.22$

Bear et al. (1968) emphasized the presence of a threshold pressure gradient in porous media and a deviation from Darcy's law in the pre-Darcy flow regime. This phenomenon is indicated also in the classification of the flow regimes presented in Figure 5.10 modified after Basak (1977). By analyzing various experimental data, Longmuir (2004) determined that pre-Darcy flow may occur in compacted

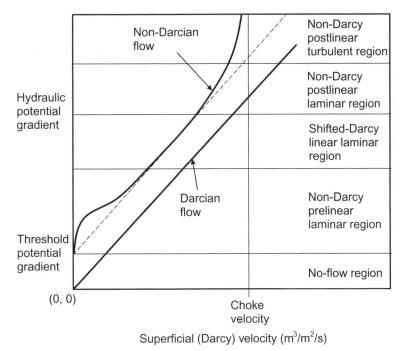

Figure 5.10   Relationship of the potential gradient versus flow for fluid flow through porous media (modified after Basak, 1977; prepared by the author).

porous media having a large amount of specific surface areas, low permeability, shale formations, or upon injection of water having different salinities than brine present in porous reservoir formations. Under such condition and relatively low flow rates, interactions between the fluids and porous media may be sufficient enough to interfere with the flow appreciably and may cause a significant deviation from the Darcy flow behavior. Based on their experimental results, Kececioglu and Jiang (1994) conclude that the pressure drop in the pre-Darcy flow regime is inversely proportional to the Reynolds number.

5.9   ENTROPY GENERATION IN POROUS MEDIA

As stated by Civan and Tiab (1989), "The second law of thermodynamics determines the practical limits of operating systems. However, within these limits, conditions leading to optimum performance can be found." As explained by Parvatiyar (1998), various mass, momentum, and energy transfer processes occurring between the fluids and their boundaries present within the porous media are irreversible and therefore consume the energy of the fluid by entropy generation. Civan and Tiab (1989) add that "The efficiency of real thermodynamic systems decreases as the entropy generation increases. Therefore, to maximize the utilization of energy, systems should be designed and operated such that the entropy generation is minimized."

Porous media can be represented by a bundle of tortuous hydraulic tubes. Frequently, as the fluids flow through porous media, there is a possibility of a variety of rock–fluid interactions at the pore surface. As the fluid flows through hydraulic tubes, nonequilibrium conditions arise because of the viscous dissipation of the fluid energy and the transfer of species from the pore fluids to the pore surface or vice versa due to rock–fluid interactions. As a result, nonequilibrium conditions are created and the entropy production increases, meaning that the thermodynamic efficiency of the overall system decreases. A mathematical analysis of the flow and rock–fluid interactions is carried here to determine the optimal conditions leading to minimum entropy production during the process similar to Civan and Tiab (1989) and Parvatiyar (1998).

5.9.1   Flow through a Hydraulic Tube

A representative differential section of a hydraulic tube is considered as the control volume for the analysis. Then, the formulation of the entropy production during horizontal, isothermal, and steady flow of a fluid through a hydraulic tube can be presented according to Parvatiyar (1998) in a form consistent with the rest of the chapter.

The length of the differential element of the hydraulic tube is dl. The hydraulic tube diameter is given by (Janna, 1993)

$$D_h = 4A/\wp, \tag{5.97}$$

where A is the cross-sectional area of the tube and $\wp$ is its perimeter. For circular tubes of diameter D,

$$A = \pi D^2/4 \tag{5.98}$$

and

$$\wp = \pi D. \tag{5.99}$$

Hence, substituting Eqs. (5.98) and (5.99) into Eq. (5.97) yields

$$D_h = D. \tag{5.100}$$

The mass and volumetric flow rates of the flowing fluid phase are given, respectively, by

$$\dot{m} = \rho v A \tag{5.101}$$

and

$$\dot{V} = \dot{m}/\rho = vA, \tag{5.102}$$

where ρ and v are the density and velocity of the fluid. Next, consider a species transferred from the wall of the hydraulic tube (or pore surface) to the fluid phase flowing through the hydraulic tube as a result of rock–fluid interactions. Let C_w denote the molar concentration of the species at the tube wall and C_b denote the bulk volume average molar concentration of the species in the fluid. Both are given in kilomole per cubic meter units. Let $\dot{m}_w \left(kg/m^2\text{-}s \right)$ and $\dot{n}_w$ (kmol/m²-s) denote the mass and molar fluxes of the species transferred from the tube wall to the fluid phase. Because the driving force for the species transfer is the concentration difference between the wall and bulk fluids, the flux of the species transfer from the wall to the bulk fluid can be expressed as being proportional to the concentration difference as

$$\dot{n}_w = \dot{m}_w / M = k\left(C_w - C_b \right), \tag{5.103}$$

where M is the molecular weight of the species in kilomole per kilogram and k is an empirically determined mass transfer coefficient in meter per second.

For a relatively small concentration change in the fluid phase, the change in the species chemical potential can be approximated as (Bejan, 1982; Parvatiyar, 1998)

$$\mu_w - \mu_b = RT\ell n\left(\frac{f_w}{f_b} \right) \approx RT\left(\frac{f_w}{f_b} - 1 \right) \approx RT\left(\frac{C_w}{C_b} - 1 \right) \tag{5.104}$$

because $\ell n\left(1 + x \right) = x - x^2/2 + x^3/3 - + \cdots \approx x$ for small x values. f is fugacity; R is the universal gas constant; and T is the absolute temperature.

Combining Eqs. (5.103) and (5.104), the following expression can be obtained:

$$\mu_w - \mu_b \cong RT\left(C_p - 1 \right) = \dot{m}_w / \left(Mk_1 C_b \right), \tag{5.105}$$

where C_p is the wall-to-bulk concentration ratio or the concentration polarization given by

$$C_p = C_w / C_b. \tag{5.106}$$

For the differential control volume element of length $d\ell$ and cross-sectional area A, the following mass (continuity equation), enthalpy (first law of thermodynamics), and entropy (second law of thermodynamics) balance equations can be written, respectively, as

$$Ad\dot{m} = \dot{m}_w \wp d\ell = Md\dot{n}_w, \tag{5.107}$$

$$d\dot{H} = h_w \dot{m}_w \wp d\ell, \tag{5.108}$$

and

$$d\dot{S}_p = d\dot{S} - s_w m_w \wp d\ell. \tag{5.109}$$

In Eqs. (5.105)–(5.107), $\dot{S}_p$ (J/kg-s) is the rate of entropy production; $\dot{H}$ (J/s) and $\dot{S}$ (J/s-K) denote the total enthalpy and entropy rates of the fluid phase, respectively; and h_w (J/kg), and s_w (J/kg-K) represent the specific enthalpy and entropy of the species transferred from the wall to the fluid, respectively. The Gibbs equation can be written to express the rate of change of the enthalpy as

$$d\dot{H} = Td\dot{S} + \dot{V}dp + \mu_b d\dot{n}_w. \tag{5.110}$$

For a pure species leaving the tube wall, the chemical potential is given by

$$\mu_w = \left(h_w - Ts_w \right) M. \tag{5.111}$$

Thus, combining Eqs. (5.102) and (5.105)–(5.111) yields the following expression for the rate of entropy production per length of the hydraulic tube:

$$\frac{d\dot{S}_p}{d\ell} = -\frac{\dot{m}}{\rho T}\frac{dp}{d\ell} + \pi DkR^2 TC_b\left(C_p - 1\right)^2. \tag{5.112}$$

The frictional pressure loss is related to the wall shear stress, τ_w, by (Janna, 1993)

$$\tau_w = \frac{D_h}{4}\left(-\frac{dp}{d\ell}\right)_f. \tag{5.113}$$

The Fanning friction factor is defined by

$$f_F = \frac{\tau_w}{\frac{1}{2}\rho v^2}. \tag{5.114}$$

Thus, Eqs. (5.100), (5.101), (5.113), and (5.114) can be combined to obtain

$$\left(-\frac{dp}{d\ell}\right)_f = \frac{2f_F\rho v^2}{D} = \frac{32\dot{m}^2 f_F}{\pi^2\rho D^5}. \tag{5.115}$$

The Reynolds number is given by

$$\text{Re} = \rho v D/\mu = 4\dot{m}/(\pi D\mu). \tag{5.116}$$

Hence, using Eqs. (5.115) and (5.116), Eq. (5.112) can be expressed as

$$\frac{d\dot{S}_p}{d\ell} = \frac{\pi\mu^3 f_F \text{Re}^3}{2T\rho^2 D^2} + \pi DkR^2 TC_b\left(C_p - 1\right)^2, \tag{5.117}$$

which implies that the entropy production is zero (the minimum) when the flow rate is zero, and there is no concentration polarization, if all other variables remain constant. Using $\rho = 1010 \text{ kg/m}^3$, $\mu = 1.3\times10^{-3} \text{ Pa·s}$, $C_b = 1.0\times10^{-6} \text{ kmol/m}^3$, $R = 8314 \text{ Pa-m}^3/\text{kmol-K}$, $T = 298 \text{ K}$, and $M = 10^6 \text{ kg/kmol}$ in Eq. (5.117), Parvatiyar (1998) predicted the entropy production rate as functions of the Reynolds number, hydraulic tube diameter, and concentration polarization as depicted in Figures 5.11–5.13.

5.9.2   Flow through Porous Media

The pressure loss during horizontal flow through porous media is given by Forchheimer's equation:

$$-\frac{dp}{d\ell} = \frac{\mu}{K}u + \beta\rho u^2. \tag{5.118}$$

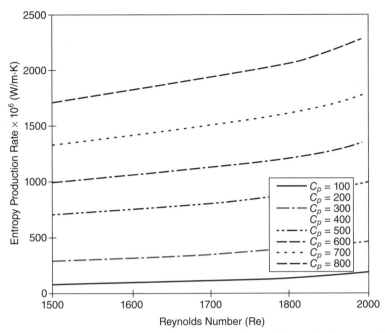

Figure 5.11   Entropy production as a function of Reynolds number for $D_m = 0.016\,\text{m}$ (after Parvatiyar, 1998, *Journal of Membrane Science*; reprinted by permission of Elsevier).

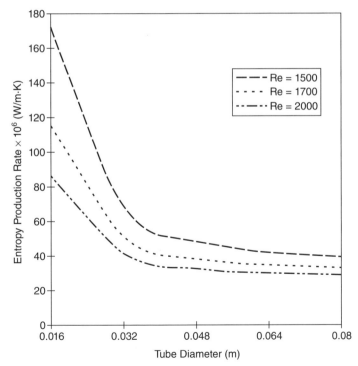

Figure 5.12   Entropy production as a function of tube diameter for $C_p = 100$ (after Parvatiyar, 1998, *Journal of Membrane Science*; reprinted by permission of Elsevier).

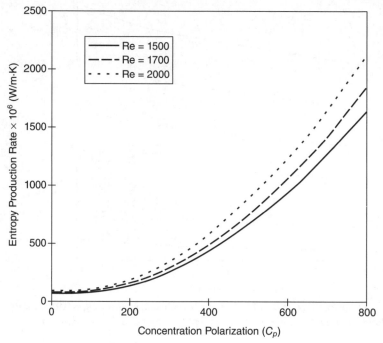

Figure 5.13 Entropy production as a function of concentration polarization for $D_m = 0.032\,\text{ID}$ (after Parvatiyar, 1998, *Journal of Membrane Science*; reprinted by permission of Elsevier).

Neglecting the rock–fluid interactions, in Eq. (5.112), the entropy production is given by

$$\frac{d\dot{S}_p}{d\ell} = -\frac{\dot{m}}{\rho T}\frac{dp}{d\ell}. \tag{5.119}$$

The superficial velocity, $u\,(\text{m}^3/\text{m}^2\text{-s})$, and the mass flow rate, $\dot{m}\,(\text{kg/s})$, are related by

$$\dot{m} = \rho u A. \tag{5.120}$$

where A denotes the cross-sectional area of porous material.

Therefore, combining Eqs. (5.118)–(5.120) yields

$$\frac{d\dot{S}_p}{d\ell} = \frac{\mu}{\rho^2 TKA}\dot{m}^2 + \frac{\beta}{\rho^2 TA^2}\dot{m}^3. \tag{5.121}$$

Alternatively, substituting the porous media Reynolds number

$$\text{Re} = \frac{\beta K\dot{m}}{A\mu}, \tag{5.122}$$

Eq. (5.121) can be expressed as

$$\frac{d\dot{S}_p}{d\ell} = \frac{A\mu^3}{\rho^2 TK^3\beta^2}(1+\text{Re})\,\text{Re}^2, \tag{5.123}$$

which implies that the entropy production increases with increasing flow rate and zero (the minimum) when the flow rate is zero, if all other variables remain constant.

5.10   VISCOUS DISSIPATION IN POROUS MEDIA

The term $(\tau : \nabla \mathbf{u})$ involved in the conservation of energy equation denotes the viscous dissipation of energy by frictional forces. Al-Hadhrami et al. (2003) derived an expression of the viscous dissipation term, which honors the asymptotic limiting conditions of low permeability ($K \rightarrow 0$ and thus $\tilde{\mu} \rightarrow 0$) and high permeability ($K \rightarrow \infty$ and thus $\tilde{\mu} \rightarrow \mu_f$). Here, μ_f denotes the clear fluid viscosity and $\tilde{\mu}$ is the effective fluid viscosity. Consider the velocity vector is given by $\mathbf{v} = U\mathbf{i} + V\mathbf{j} + W\mathbf{k}$, where U, V, and, W are the components of the velocity vector, and $\mathbf{i}$, $\mathbf{j}$, and $\mathbf{k}$ denote the unit vectors in the X-, Y-, and Z-Cartesian coordinates, respectively.

Al-Hadhrami et al. (2003) derived the following expressions for the work done (W.D.) by frictional forces for three special cases

- Motion of clear fluid described by the Navier–Stokes equation:

$$W.D. = \rho \frac{D}{Dt}\left(\frac{U^2 + V^2 + W^2}{2}\right) - p\left(\frac{\partial U}{\partial x} + \frac{\partial V}{\partial y} + \frac{\partial W}{\partial z}\right)$$
$$+ \mu_f\left[2\left(\frac{\partial U}{\partial x}\right)^2 + 2\left(\frac{\partial V}{\partial y}\right)^2 + 2\left(\frac{\partial W}{\partial z}\right)^2 + \left(\frac{\partial U}{\partial y} + \frac{\partial V}{\partial x}\right)^2 + \left(\frac{\partial U}{\partial z} + \frac{\partial W}{\partial x}\right)^2$$
$$+ \left(\frac{\partial V}{\partial z} + \frac{\partial W}{\partial y}\right)^2\right],$$

$$(5.124)$$

where the substantial derivative is defined by

$$\frac{Dp}{Dt} \equiv \frac{\partial p}{\partial t} + \mathbf{u} \cdot \nabla p. \tag{5.125}$$

The first term on the right of Eq. (5.124) denotes the amount of work transformed into kinetic energy; the second term denotes the work involved in fluid compression; and the third term denotes the amount of work lost by dissipation into heat, respectively (Beckett, 1980; Beckett and Friend, 1984).

- Motion of fluid in porous media described by the Darcy equation:

$$W.D. = -p\left(\frac{\partial U}{\partial x} + \frac{\partial V}{\partial y} + \frac{\partial W}{\partial z}\right) + \frac{\mu_f}{K}\left(U^2 + V^2 + W^2\right) \tag{5.126}$$

The first term on the right denotes the work involved in fluid compression and the second term denotes the amount of work lost by dissipation into heat, respectively (Ingham et al., 1990).

- Motion of fluid in porous media described by the Brinkman equation (Al-Hadhrami et al., 2003):

$$W.D. = -p\left(\frac{\partial U}{\partial x} + \frac{\partial V}{\partial y} + \frac{\partial W}{\partial z}\right) + \frac{\mu_f}{K}\left(U^2 + V^2 + W^2\right)$$

$$+ \tilde{\mu}\left[2\left(\frac{\partial U}{\partial x}\right)^2 + 2\left(\frac{\partial V}{\partial y}\right)^2 + 2\left(\frac{\partial W}{\partial z}\right)^2 + \left(\frac{\partial U}{\partial y} + \frac{\partial V}{\partial x}\right)^2 + \left(\frac{\partial U}{\partial z} + \frac{\partial W}{\partial x}\right)^2\right.$$

$$\left. + \left(\frac{\partial V}{\partial z} + \frac{\partial W}{\partial y}\right)^2\right] \tag{5.127}$$

Hence, Al-Hadhrami et al. (2003) determined that the viscous dissipation (V.D.) for $K \to 0\,(\tilde{\mu} \to 0)$ is expressed by the following, as in the case of the motion of fluid in porous media described by the Darcy equation:

$$V.D. \equiv (\mathbf{\tau}\!:\!\nabla\mathbf{u}) = \frac{\mu_f}{K}\left(U^2 + V^2 + W^2\right). \tag{5.128}$$

For $K \to \infty\,(\tilde{\mu} \to \mu)$, it is expressed by as in the case of the motion of clear fluid described by the Navier–Stokes equation:

$$V.D. = \tilde{\mu}\left[2\left(\frac{\partial U}{\partial x}\right)^2 + 2\left(\frac{\partial V}{\partial y}\right)^2 + 2\left(\frac{\partial W}{\partial z}\right)^2 + \left(\frac{\partial U}{\partial y} + \frac{\partial V}{\partial x}\right)^2 + \left(\frac{\partial U}{\partial z} + \frac{\partial W}{\partial x}\right)^2\right.$$

$$\left. + \left(\frac{\partial V}{\partial z} + \frac{\partial W}{\partial y}\right)^2\right]. \tag{5.129}$$

Note Eq. (5.128) can be written as

$$V.D. \equiv (\mathbf{\tau}\!:\!\nabla\mathbf{u}) = \mu_f \mathbf{u} \cdot \mathbf{K}^{-1} \cdot \mathbf{u} = -\mathbf{u} \cdot \nabla\mathbf{p}. \tag{5.130}$$

5.11 GENERALIZED DARCY'S LAW BY CONTROL VOLUME ANALYSIS

The present formulation given according to Civan (2008d) combines the application of two different concepts as depicted in Figures 5.14 and 5.15. The interactions of the effective pore fluid (mobile fluid portion excluding the immobile fluids such as connate water and residual oil) with the surrounding are handled across the areas open for flow over the exterior surfaces of the control volume as shown in Figure 5.14 (modified after Das, 1999). In a heterogeneous medium, different open flow surfaces may be established in different Cartesian directions. Figure 5.15 shows the internal flow structure represented by the capillary orifice model of the preferential flow paths (bundle of tortuous flow tubes having frequent pore-throat constrictions) formed in a given Cartesian direction (modified after Blick, 1966 and Blick and Civan, 1988 for the tortuosity effect).

In an anisotropic medium, different preferential paths may be established in other Cartesian directions. The possibility of cross-flow between bundles of capillary tubes is neglected. The internal resistive forces acting upon fluid are considered by means of the capillary orifice model according to Blick and Civan (1988). The capil-

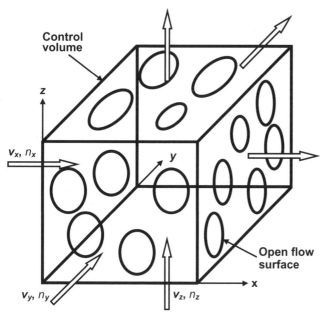

Figure 5.14   External open flow surfaces represented by the control volume model of interactions with the surrounding media. Different open flow surfaces are established in the different Cartesian directions (modified after Civan, 2008d; © 2008 SPE, reprinted by permission of the Society of Petroleum Engineers).

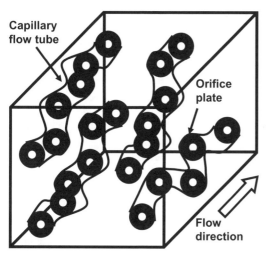

Figure 5.15   Internal flow structure represented by the tortuous capillary orifice model of preferential flow paths formed in a given Cartesian direction. Different preferential paths are established in the different Cartesian directions (modified after Civan, 2008d; © 2008 SPE, reprinted by permission of the Society of Petroleum Engineers).

lary orifice model attempts at the incorporation of the effect of frequent pore-throat acceleration and pore-body deceleration of fluid during flow through porous media into the Kozeny–Carman bundle of capillary tube model (Kozeny, 1927; Carman, 1938, 1956) by placing orifice plates along the flow paths as schematically depicted in Figure 5.15. The capillary tube wall acts as the pore surface and the orifice plate acts as the pore throat. Hence, this model provides convenience in characterizing the internal forces resisting the fluid flow through porous media.

The surface forces acting through the various surfaces of the fluid present inside the control volume of the skin friction and pressure types are considered. The orifice plates provide a mechanism for an estimation of the drag force exerted by the pore-throat constrictions. Blick and Civan (1988) derived the capillary orifice model only for a single straight capillary tube. Civan (2008d) extended and generalized this approach by considering the tortuosity effect for a bundle of capillary tubes present in a representative elementary bulk volume according to Civan (2001, 2002d, 2007b). The porous media is assumed to be heterogeneous and anisotropic, and fluid and porous media properties are variable. Therefore, the parameters of the preferential capillary flow paths formed in different Cartesian coordinate directions have different values. There is a critical threshold pressure gradient that must be overcome to initiate flow in porous media (Prada and Civan, 1999; Wang et al., 2006).

Note that the formulation elaborated here assumes a fully elastic porous matrix (Biot's coefficient $\alpha = 0$ according to Section 5.3). It needs modification for other cases as described in Section 5.3.

5.11.1 General Formulation

In the following formulation u, μ, p, and ρ represent the volume flux, viscosity, pressure, and density of fluid, respectively. K and ϕ are the permeability and porosity of porous media; g is the gravitational acceleration; and A is the area normal to flow. The x-, y- and z-Cartesian coordinates are selected in the principal directions of the permeability tensor. Hence, the diagonal elements K_x, K_y, and K_z of the permeability tensor are nonzero and the off-diagonal elements are taken zeros. The nonbold, lowercase bold, and uppercase bold symbols denote the scalar, vector, and tensor properties, respectively.

Consider the control volume defined by the volume of the mobile pore fluid contained within interconnected (effective) pore space present inside the elemental bulk volume of porous media, described in Figures 5.14 and 5.15.

The equation of motion for this fluid will be derived by applying the following equation of linear momentum balance (Crowe et al., 2001):

$$\sum_{CV} \mathbf{F} = \sum_{CV} \mathbf{F}_{\text{External}} + \sum_{CV} \mathbf{F}_{\text{Internal}} = \sum_{CS} (\mathbf{n} \cdot \mathbf{A}) \rho \mathbf{v} + \frac{\partial}{\partial t} (V \rho \mathbf{v})_{CV}, \qquad (5.131)$$

where CV and CS refer to the control volume and control surface of the mobile pore fluid; $\mathbf{F}$ is the net force vector acting upon fluid; $\mathbf{A}$ is the surface area vector acting outward and normal to the area open for flow; t is time; ρ is the density; $\mathbf{v}$ is the velocity vector; and V is the volume of the fluid. Applying Eq. (5.131) in the x-direction for the control volume shown in Figure 5.14 yields

$$
\begin{cases}
\left(p_E \Delta y \Delta z \phi\right)_x - \left(p_E \Delta y \Delta z \phi\right)_{x+\Delta x} \\
+ \left(T_{xx} \Delta y \Delta z \phi\right)_x - \left(T_{xx} \Delta y \Delta z \phi\right)_{x+\Delta x} \\
+ \left(T_{yx} \Delta x \Delta z \phi\right)_y - \left(T_{yx} \Delta x \Delta z \phi\right)_{y+\Delta y} \\
+ \left(T_{zx} \Delta x \Delta y \phi\right)_z - \left(T_{zx} \Delta x \Delta y \phi\right)_{z+\Delta z}
\end{cases}_{\text{External}}
$$

$$
+ \left\{ F_{Bx}\left(\Delta x \Delta y \Delta z \phi\right) - F_{Sx}\left(\Delta x \Delta y \Delta z \phi\right) - F_{Ox}\left(\Delta x \Delta y \Delta z \phi\right) \right\}_{\text{Internal}} \tag{5.132}
$$

$$
= -\left[\left(n_x \Delta y \Delta z \phi\right)\rho v_x\right]_x + \left[\left(n_x \Delta y \Delta z \phi\right)\rho v_x\right]_{x+\Delta x}
$$

$$
- \left[\left(n_y \Delta x \Delta z \phi\right)\rho v_x\right]_y + \left[\left(n_y \Delta x \Delta z \phi\right)\rho v_x\right]_{y+\Delta y}
$$

$$
- \left[\left(n_z \Delta x \Delta y \phi\right)\rho v_x\right]_z + \left[\left(n_z \Delta x \Delta y \phi\right)\rho v_x\right]_{z+\Delta z}
$$

$$
+ \frac{\partial}{\partial t}\left[\left(\Delta x \Delta y \Delta z \phi\right)\rho v_x\right],
$$

where Δx, Δy, and Δz denote the dimensions of the bulk volume element, and the subscripts indicate the components in the various Cartesian directions. The other terms appearing in this equation are subsequently described and elaborated. Dividing Eq. (5.132) by $(\Delta x \Delta y \Delta z)$ and then taking a limit as the control volume shrinks to the representative elementary volume (REV) yields

$$
-\frac{\partial}{\partial x}\left(p_E \phi\right) + \phi F_{Bx} - \phi F_{Sx} - \phi F_{Ox}
$$

$$
-\frac{\partial}{\partial x}\left(T_{xx}\phi\right) - \frac{\partial}{\partial y}\left(T_{yx}\phi\right) - \frac{\partial}{\partial z}\left(T_{zx}\phi\right) \tag{5.133}
$$

$$
= \frac{\partial}{\partial x}\left(\phi n_x \rho v_x\right) + \frac{\partial}{\partial y}\left(\phi n_y \rho v_x\right) + \frac{\partial}{\partial z}\left(\phi n_z \rho v_x\right) + \frac{\partial}{\partial t}\left(\phi \rho v_x\right).
$$

However, by definition of derivatives, the limit should be taken to zero volume. But this violates the fact that REV is a finite quantity and cannot be collapsed to a point of zero volume. Nevertheless, the application of Eq. (5.133) may be justified if the *extrapolated limit* concept introduced by Hubbert (1956) for defining macroscopic (porous media volume averaged) quantities is adapted also for derivatives. The details of the above-mentioned derivation are shown here to clarify the volume averaging of the pressure and stress terms, which have been the subject of many debates. As can be seen by Eq. (5.133), these terms should include the porosity associated with them under the conditions explained in Section 5.3. The various terms appearing in Eq. (5.133) can be resolved as described in the following.

The effective or net pressure p_E that can put fluid into motion is given by

$$
p_E = p - p_{th}, \tag{5.134}
$$

where p is the fluid pressure and p_{th} is the threshold pressure defined as the minimum pressure above which fluid can begin flow. Fluid can flow through porous media only when the fluid force can overcome the threshold pressure gradient. Prada and Civan (1999) correlated the threshold pressure gradient by

$$
-\left(\partial p / \partial x\right)_{th} = 16 M_x^{-0.8}, \; 57 < M_x < 2.8 \times 10^4, \tag{5.135}
$$

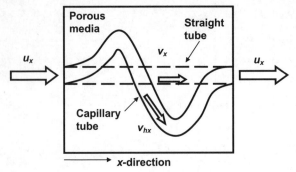

Figure 5.16 Schematic realization of the fluid capillary velocity (interstitial fluid velocity) v_{hx}, the bulk volume average velocity (superficial fluid velocity) u_x, and the fluid volume average velocity (pore fluid velocity) v_x, defined over the representative elementary bulk volume of porous media (modified after Civan, 2008d; © 2008 SPE, reprinted by permission of the Society of Petroleum Engineers).

where the threshold pressure gradient $(\partial p/\partial x)_{th}$ is in the units of pounds per square inch per centimeter, and the x-direction fluid mobility $M_x \equiv K_x/\mu$ is in millidarcy per centipoise. A similar correlation is also applicable in other coordinate directions. The threshold pressure gradient decreases by increasing fluid mobility. Thus, the threshold pressure gradient may be neglected for gases but important for liquids.

The body force acting on the fluid is considered to be of the gravitational type, given by

$$F_{Bx} = \rho g_x, \tag{5.136}$$

where g_x is the gravitational acceleration in the x-direction.

Consider now the schematic of a preferential flow path given in Figure 5.16. For convenience, only one of the capillary flow tubes is shown. The straight tube represents the flow averaged over the mobile pore fluid volume (pore average velocity). The skin (pore surface) friction force per unit fluid volume is given by (Blick and Civan, 1988)

$$F_{Sx} = \frac{n_x \pi D_{hx} L_{hx}}{V_p} T_{wx}, \tag{5.137}$$

where

$$V_p = n_x \frac{\pi D_{hx}^2}{4} L_{hx}$$

denotes the mobile pore fluid volume (excluding the immobile fluids such as connate water and residual oil); n_x, D_{hx}, and L_{hx} are the number, mean diameter, and length of the preferential capillary flow tubes formed by flow in the x-direction, and T_{wx} is the wall shear stress, given by

$$T_{wx} = c_f \frac{1}{2} \rho v_{hx}^2, \tag{5.138}$$

where v_{hx} denotes the fluid interstitial velocity as it moves along the tortuous capillary tube according to Figure 5.16, and c_f denotes the skin friction coefficient that can be expressed in terms of the friction factor f as

$$c_f = f / 4. \tag{5.139}$$

Laminar flow conditions are assumed to prevail near the pore surface, and therefore (Blick and Civan, 1988)

$$f = \frac{64}{\text{Re}_x}, \text{Re}_x \equiv \frac{\rho v_{hx} D_{hx}}{\mu} < 2000, \tag{5.140}$$

where Re_x is the Reynolds number for flow in the x-direction. Thus, applying Eqs. (5.139) and (5.140) to Eq. (5.138) yields

$$T_{wx} = \frac{8\mu v_{hx}}{D_{hx}}. \tag{5.141}$$

Express the bulk volume average (superficial) velocity u_x in terms of the fluid volume average velocity (pore average velocity) v_x according to Figure 5.16 by

$$u_x = \phi v_x. \tag{5.142}$$

Express the capillary tube (interstitial) fluid velocity v_{hx} in terms of the fluid volume average velocity v_x according to Figure 5.16 by

$$v_{hx} = \tau_{hx} v_x. \tag{5.143}$$

The tortuosity τ_{hx} of the capillary tubes representing the preferential flow paths in the x-direction is given by

$$\tau_{hx} = L_{hx} / L_{bx}, \tag{5.144}$$

where L_{bx} is the length of bulk porous media in the x-direction. Thus, the relationship between the fluid interstitial velocity, v_{hx}, superficial velocity, u_x, tortuosity, τ_{hx}, and porosity, ϕ, is given by (Dupuit, 1863)

$$v_{hx} = \frac{\tau_{hx} u_x}{\phi}. \tag{5.145}$$

Utilizing the above-mentioned equations, Eq. (5.137) becomes

$$F_{Sx} = \frac{\mu}{K_x} \phi v_x = \frac{\mu}{K_x} u_x, \tag{5.146}$$

where K_x is defined as a lump parameter referring to the permeability of porous medium in the x-principal direction, expressed by

$$K_x = \frac{\phi V_p}{8n_x \pi L_{hx} \tau_{hx}} = \frac{\phi D_{hx}^2}{32\tau_{hx}}. \tag{5.147}$$

The orifice (pore throat) drag force per unit mobile pore fluid volume is given by Blick (1966) and Blick and Civan (1988):

$$F_{Ox} = n_x \frac{L_{hx}}{D_{hx}} \frac{F_{Dx}}{V_p},$$ (5.148)

where

$$\frac{L_{hx}}{D_{hx}}$$

represents the number of orifices placed at an equal distance of the mean pore diameter D_{hx} from each other along the capillary tubes. The drag force for one orifice plate is given by

$$F_{Dx} = c_{Dx} \frac{1}{2} \rho |v_{hx}| v_{hx} \frac{\pi D_{hx}^2}{4},$$ (5.149)

where c_{Dx} denotes the drag coefficient in the x-direction. The velocity terms were separated as shown in the above-mentioned equation, one taken with its absolute value, to consider the effect of flow direction on the drag force direction. Then, utilizing the above equations in Eq. (5.148) yields

$$F_{Ox} = \phi^2 \rho \beta_x |v_x| v_x = \rho \beta_x |u_x| u_x,$$ (5.150)

where β_x is defined as a lump parameter referring to the inertial flow coefficient of porous medium in the x-principal direction, expressed by

$$\beta_x = \frac{c_{Dx} n_x \pi D_{hx} L_{hx} \tau_{hx}^2}{8 \phi^2 V_p} = \frac{c_{Dx} \tau_{hx}^2}{2 \phi^2 D_{hx}}.$$ (5.151)

Combining Eqs. (5.147) and (5.151) according to Geertsma (1974) yields

$$\beta_x K_x^{1/2} \phi^{3/2} = \frac{c_{Dx} D_{hx} \tau_{hx}^{3/2}}{16} \sqrt{\frac{n_x \pi L_{hx}}{2 V_p}} = \frac{c_{Dx} \tau_{hx}^{3/2}}{8 \sqrt{2}}.$$ (5.152)

Substituting $D_{hx} = 4 \sqrt{2 \tau_{hx}} \sqrt{K_x / \phi}$ for the mean diameter of the capillary flow tubes (Civan, 2005a) or Eqs. (5.147) and (5.151) into the following dimensionless group also yields a similar expression as

$$\frac{\beta_x K_x \phi}{D_{hx}} = \frac{\beta_x K_x^{1/2} \phi^{3/2}}{4 \sqrt{2 \tau_{hx}}} = \frac{c_{Dx} \tau_{hx}}{64} = \text{dimensionless}.$$ (5.153)

Therefore, the Liu et al. (1995) correlation yields a dimensional constant as

$$\frac{\beta_x K_x \phi}{\tau_{hx}} = \frac{c_{Dx} D_{hx}}{64} = 8.91 \times 10^8,$$

containing the relevant unit conversion factors, where the units of β_x, K_x, and ϕ are per foot, millidarcy, and fraction. Eq. (5.152) indicates that the value of the right side is variable for Re < 4000 because the drag coefficient c_{Dx} varies with the Reynolds number (Crowe et al., 2001). The value of the right side becomes constant for $c_{Dx} = 2.3$ under the flow conditions of Re > 4000. The empirical correlation given by Geertsma (1974) for high rate flow suggests that the right side has a value of about 0.180–0.205 (Blick and Civan, 1988). This evidence supports the applicability of the capillary tube model.

In view of the preceding developments, Eq. (5.133) can be now written as

$$-\frac{\partial}{\partial x}(p\phi)+\frac{\partial}{\partial x}(p_{th}\phi)+\rho\phi g_x-\phi^2\frac{\mu}{K_x}v_x-\phi^3\rho\beta_x|v_x|v_x$$

$$-\frac{\partial}{\partial x}(T_{xx}\phi)-\frac{\partial}{\partial y}(T_{yx}\phi)-\frac{\partial}{\partial z}(T_{zx}\phi)$$

$$=\frac{\partial}{\partial x}(\phi n_x\rho v_x)+\frac{\partial}{\partial y}(\phi n_y\rho v_x)+\frac{\partial}{\partial z}(\phi n_z\rho v_x)+\frac{\partial}{\partial t}(\phi\rho v_x).$$

(5.154)

Eq. (5.154) can be generalized in the vector tensor notation as

$$-\nabla(p\phi)+\nabla(p_{th}\phi)+\rho\phi\mathbf{g}-\phi^2\mu\mathbf{K}^{-1}\cdot\mathbf{v}-\phi^3\rho\boldsymbol{\beta}\cdot|\mathbf{v}|\,\mathbf{v}$$

$$-\nabla\cdot(\mathbf{T}\phi)=\nabla\cdot(\phi\mathbf{n}\rho\mathbf{v})+\frac{\partial}{\partial t}(\phi\rho\mathbf{v}),$$

(5.155)

where ∇ is the gradient operator and $\mathbf{T}$ is the shear stress tensor. Note that

$$\phi\mathbf{n}\rho\mathbf{v}=\phi\mathbf{v}\rho\mathbf{v}-\mathbf{D}\cdot\nabla(\phi\rho\mathbf{v})$$

(5.156)

and

$$-\nabla(p\phi)+\nabla(p_{th}\phi)+\rho\phi\mathbf{g}-\phi^2\mu\mathbf{K}^{-1}\cdot\mathbf{v}-\phi^3\rho\boldsymbol{\beta}\cdot|\mathbf{v}|\,\mathbf{v}$$

$$-\nabla\cdot(\mathbf{T}\phi)=\nabla\cdot(\phi\mathbf{v}\rho\mathbf{v})-\nabla\cdot[\mathbf{D}\cdot\nabla(\phi\rho\mathbf{v})]+\frac{\partial}{\partial t}(\phi\rho\mathbf{v}).$$

(5.157)

Define a flow potential as

$$\Psi=\frac{1}{\phi}\left\{\int_{P_o}^{P}\frac{d[(p-p_{th})\phi]}{\rho}+g(\phi z-\phi_o z_o)\right\}.$$

(5.158)

Thus, Eq. (5.115) can be written as

$$-\rho\nabla(\Psi\phi)-\phi^2\mu\mathbf{K}^{-1}\cdot\mathbf{v}-\phi^3\rho\boldsymbol{\beta}\cdot|\mathbf{v}|\,\mathbf{v}$$

$$-\nabla\cdot(\mathbf{T}\phi)=\nabla\cdot(\phi\mathbf{v}\rho\mathbf{v})-\nabla\cdot[\mathbf{D}\cdot\nabla(\phi\rho\mathbf{v})]+\frac{\partial}{\partial t}(\phi\rho\mathbf{v}).$$

(5.159)

Substituting Eq. (5.142) into Eq. (5.115) yields

$$-\nabla(p\phi)+\nabla(p_{th}\phi)+\rho\phi\mathbf{g}-\phi\mu\mathbf{K}^{-1}\cdot\mathbf{u}-\phi\rho\boldsymbol{\beta}\cdot|\mathbf{u}|\,\mathbf{u}$$

$$-\nabla\cdot(\mathbf{T}\phi)=\nabla\cdot\left(\frac{\rho\mathbf{u}\mathbf{u}}{\phi}\right)-\nabla\cdot[\mathbf{D}\cdot\nabla(\rho\mathbf{u})]+\frac{\partial}{\partial t}(\rho\mathbf{u}).$$

(5.160)

or

$$-\rho\nabla(\Psi\phi)-\phi\mu\mathbf{K}^{-1}\cdot\mathbf{u}-\phi\rho\boldsymbol{\beta}\cdot|\mathbf{u}|\,\mathbf{u}$$

$$-\nabla\cdot(\mathbf{T}\phi)=\nabla\cdot\left(\frac{\rho\mathbf{u}\mathbf{u}}{\phi}\right)-\nabla\cdot[\mathbf{D}\cdot\nabla(\rho\mathbf{u})]+\frac{\partial}{\partial t}(\rho\mathbf{u}).$$

(5.161)

The terms on the left side of Eq. (5.161) denote the momentum owing to flow potential gradient, shear force, pore-throat orifice drag, and viscous diffusion, respectively. The terms on the right side of Eq. (5.161) denote the inertial effects (negligible for creeping-flow), momentum dispersion, and rate of momentum change, respectively.

Applying the control volume approach (Civan, 2008d) similar to the derivation of the equation of motion, the porous media mass balance equation can be derived as

$$\nabla \cdot (\rho \phi \mathbf{v}) + \frac{\partial}{\partial t}(\rho \phi) = \dot{m}\phi, \tag{5.162}$$

or by substituting Eq. (5.142),

$$\nabla \cdot (\rho \mathbf{u}) + \frac{\partial}{\partial t}(\rho \phi) = \dot{m}\phi, \tag{5.163}$$

where $\dot{m}$ denotes the rate of mass added per unit volume of the pore fluid. Ordinarily, it is equal to zero when there is no internal source present. By means of Eq. (5.162), Eq. (5.115) can be simplified as

$$-\nabla(p\phi) + \nabla(p_{th}\phi) + \rho\phi\mathbf{g} - \phi^2\mu\mathbf{K}^{-1} \cdot \mathbf{v} - \phi^3\rho\boldsymbol{\beta} \cdot |\mathbf{v}|\,\mathbf{v}$$
$$-\nabla \cdot (\mathbf{T}\phi) = \dot{m}\phi\mathbf{v} + \rho\phi\left(\mathbf{v} \cdot \nabla\mathbf{v} + \frac{\partial \mathbf{v}}{\partial t}\right) - \nabla \cdot [\mathbf{D} \cdot \nabla(\phi\rho\mathbf{v})]. \tag{5.164}$$

Alternatively, substituting Eq. (5.142) into Eq. (5.164) yields

$$-\nabla(p\phi) + \nabla(p_{th}\phi) + \rho\phi\mathbf{g} - \phi\mu\mathbf{K}^{-1} \cdot \mathbf{u} - \phi\rho\boldsymbol{\beta} \cdot |\mathbf{u}|\,\mathbf{u}$$
$$-\nabla \cdot (\mathbf{T}\phi) = \dot{m}\mathbf{u} + \rho\phi\left[\frac{\mathbf{u}}{\phi} \cdot \nabla\left(\frac{\mathbf{u}}{\phi}\right) + \frac{\partial}{\partial t}\left(\frac{\mathbf{u}}{\phi}\right)\right] - \nabla \cdot [\mathbf{D} \cdot \nabla(\rho\mathbf{u})]. \tag{5.165}$$

In the limit, as fluid ceases to flow (i.e., $\mathbf{u} = \mathbf{0}$), then Eq. (5.165) simplifies as

$$-\nabla(p\phi) + \nabla(p_{th}\phi) + \rho\phi\mathbf{g} = 0. \tag{5.166}$$

This means that fluid can flow (i.e., $\mathbf{u} > \mathbf{0}$), only when the following condition is satisfied under prevailing threshold pressure gradient:

$$-\nabla(p\phi) + \nabla(p_{th}\phi) + \rho\phi\mathbf{g} > 0. \tag{5.167}$$

5.11.2 Simplified Equations of Motion for Porous Media Flow

The equations derived up to this point are general and are applicable for variable fluid and porous media properties. Some special equations are obtained in the following.

When the porosity is constant, then Eq. (5.165) simplifies to

$$-\nabla p + \nabla p_{th} + \rho\mathbf{g} - \mu\mathbf{K}^{-1} \cdot \mathbf{u} - \rho\boldsymbol{\beta} \cdot |\mathbf{u}|\,\mathbf{u}$$
$$-\nabla \cdot \mathbf{T} = \frac{\dot{m}}{\phi}\mathbf{u} + \rho\left[\frac{\mathbf{u}}{\phi^2} \cdot \nabla\mathbf{u} + \frac{1}{\phi}\frac{\partial \mathbf{u}}{\partial t}\right] - \frac{1}{\phi}\nabla \cdot [\mathbf{D} \cdot \nabla(\rho\mathbf{u})]. \tag{5.168}$$

This means that fluid can flow only when

$$-\nabla p + \nabla p_{th} + \rho\mathbf{g} \geq 0. \tag{5.169}$$

Further, for Newtonian fluids according to Slattery (1972),

$$-\nabla \cdot \mathbf{T} = (\lambda + \mu)\nabla(\nabla \cdot \mathbf{v}) + \mu\nabla \cdot (\nabla\mathbf{v}), \tag{5.170}$$

where ρ, μ and λ are the mass density, shear coefficient of viscosity, and bulk coefficient of viscosity, respectively.

The Blick and Civan (1988) equation is given in vector–tensor notation as

$$-\nabla p + \rho\mathbf{g} - \mu\mathbf{K}^{-1} \cdot \mathbf{u} - \rho\boldsymbol{\beta} \cdot |\mathbf{u}|\mathbf{u} = \rho\left[\frac{\mathbf{u}}{\phi^2} \cdot \nabla\mathbf{u} + \frac{1}{\phi}\frac{\partial\mathbf{u}}{\partial t}\right]. \tag{5.171}$$

A comparison of Eqs. (5.168) and (5.171) reveals that the threshold pressure gradient, viscous, and source terms have been omitted by Blick and Civan (1988). This is justifiable, however, as they were concerned essentially with applications to gases. Eq. (5.171) may be simplified to obtain the Darcy and Darcy–Forchheimer equations, respectively, as the following:

$$-\nabla p + \rho\mathbf{g} - \mu\mathbf{K}^{-1} \cdot \mathbf{u} = 0 \tag{5.172}$$

and

$$-\nabla p + \rho\mathbf{g} - \mu\mathbf{K}^{-1} \cdot \mathbf{u} - \rho\boldsymbol{\beta} \cdot |\mathbf{u}|\mathbf{u} = 0. \tag{5.173}$$

This, in turn, reveals that the applicability of Eqs. (5.171)–(5.173) is subject to the assumption of constant porosity.

For incompressible fluids, constant porosity, and constant or no internal source, Eqs. (5.162), (5.168), and (5.170) can be combined as

$$-\nabla p + \nabla p_{th} + \rho\mathbf{g} - \mu\mathbf{K}^{-1} \cdot \mathbf{u} - \rho\boldsymbol{\beta} \cdot |\mathbf{u}|\mathbf{u} + \frac{\mu}{\phi}\nabla \cdot (\nabla\mathbf{u})$$
$$= \frac{\dot{m}}{\phi}\mathbf{u} + \rho\left[\frac{\mathbf{u}}{\phi^2} \cdot \nabla\mathbf{u} + \frac{1}{\phi}\frac{\partial\mathbf{u}}{\partial t}\right] - \frac{1}{\phi}\nabla \cdot [\mathbf{D} \cdot \nabla(\rho\mathbf{u})]. \tag{5.174}$$

The Brinkman and the Darcy–Brinkman–Forchheimer equations can be derived from Eq. (5.174), respectively, as

$$-\nabla p + \rho\mathbf{g} - \mu\mathbf{K}^{-1} \cdot \mathbf{u} + \frac{\mu}{\phi}\nabla \cdot (\nabla\mathbf{u}) = 0. \tag{5.175}$$

and

$$-\nabla p + \rho\mathbf{g} - \mu\mathbf{K}^{-1} \cdot \mathbf{u} - \rho\boldsymbol{\beta} \cdot |\mathbf{u}|\mathbf{u} + \frac{\mu}{\phi}\nabla \cdot (\nabla\mathbf{u}) = 0 \tag{5.176}$$

The applicability of Eqs. (5.175) and (5.176) is subject to the same conditions of Eq. (5.174).

5.12   EQUATION OF MOTION FOR NON-NEWTONIAN FLUIDS

Present knowledge and capability of predicting the flow of non-Newtonian fluids and the associated frictional drag in porous media are insufficient, especially in terms of the effect of pore (or porous matrix attributes), flow regimes, viscoelasticity of non-Newtonian fluids, fluid entrapment, and fluid–pore wall interactions (Chhabra et al., 2001). The following sections present various formulations according to Chhabra et al. (2001).

5.12.1   Frictional Drag for Non-Newtonian Fluids

In general, a fluid is referred to as behaving as a Newtonian fluid in porous media when the average wall shear stress $\langle \tau_w \rangle$ at the pore wall is proportional to the nominal shear rate $\langle \dot{\gamma}_w \rangle_n$ at the pore wall, that is, linearly related by

$$\langle \tau_w \rangle = \mu \langle \dot{\gamma}_w \rangle_n, \tag{5.177}$$

where the proportionality coefficient μ is ordinarily defined as the fluid viscosity, and the convention $\langle \ \rangle$ denotes an average over the circumference of the hydraulic flow tube (Chhabra et al., 2001). The following can be written for laminar viscous flow through packed grains of bulk length L (Kemblowski et al., 1989):

$$\langle \tau_w \rangle = \frac{D_h}{4} \left(\frac{\Delta p}{L} \right) \tag{5.178}$$

and

$$\langle \dot{\gamma}_w \rangle_n = K_o \left(\frac{4v}{D_h} \right) \tag{5.179}$$

where K_o is a shape factor ($K_o = 2$ in the case of circular capillary tubes); D_h denotes the mean hydraulic tube diameter (hydraulic radius is $R_h = D_h/4$) and v denotes the average pore fluid interstitial velocity, given by (duPuit, 1863)

$$v = \frac{u \tau_h}{\phi}, \tag{5.180}$$

where u denotes the superficial velocity or volumetric flux; ϕ is porosity; and τ_h is the tortuosity of preferential flow paths, given by

$$\tau_h = \frac{L_h}{L} \tag{5.181}$$

where L_h and L denote the lengths of flow paths and porous media, respectively.

When the effect of the walls or boundaries confining the porous media is neglected, the hydraulic diameter D_h for a pack of grains having a porosity of ϕ is given by

$$D_h = \frac{4\phi}{(1-\phi)\Sigma_g} \tag{5.182}$$

where Σ_g denotes the specific grain surface expressed per unit grain volume. When the grain shape is approximated as a sphere, then it is given by

$$\Sigma_g = \frac{6}{D_g} \tag{5.183}$$

where D_g is the grain diameter. Thus, substituting Eqs. (5.180)–(5.182) into Eqs. (5.178) and (5.179) yields

$$\langle \tau_w \rangle = \frac{\phi}{(1-\phi)\Sigma_g}\left(\frac{\Delta p}{L}\right), \tag{5.184}$$

and

$$\langle \dot{\gamma}_w \rangle_n = \Sigma_g K_o \tau_h \left(\frac{1-\phi}{\phi^2}\right) u. \tag{5.185}$$

On the other hand, a fluid is referred to as a non-Newtonian fluid when its behavior cannot be described by Eq. (5.177). Frequently, the behavior of inelastic non-Newtonian fluids is described by an empirical power-law relationship between the wall shear stress and the wall nominal shear rate according to the Rabinowitsch and Mooney equation (Metzner, 1956; Chhabra et al., 2001):

$$\langle \tau_w \rangle = m'\left(\langle \dot{\gamma}_w \rangle_n\right)^{n'}, \tag{5.186}$$

where m' [Pa·s^n] denotes the apparent fluid consistency index or power-law consistency coefficient and n' is the apparent flow behavior index. For circular and noncircular tubes having $0.83 \leq K_o \leq 3$ (Miller, 1972),

$$m' = m\left(\frac{3n+1}{4n}\right)^n, \qquad n' = n. \tag{5.187}$$

For Newtonian fluids, $n = 1$ and $m = \mu$.

Among the various attempts, Chhabra et al. (2001) state that the capillary tube model incorporating the pore wall and grain shape effects (Comiti and Renaud, 1989, Sabiri and Comiti, 1995) and the hybrid method (Machac and Dolejs, 1981) perform well in predicting the frictional pressure loss during flow of time-independent non-Newtonian fluids through unconsolidated packs of grains.

5.12.2 Modified Darcy's Law for Non-Newtonian Fluids

Substituting Eqs. (5.184) and (5.185) into Eq. (5.186) yields

$$u^n = \frac{K}{\mu_e}\frac{\Delta p}{L}, \tag{5.188}$$

where

$$\frac{K}{\mu_e} = \frac{\phi^{2n+1}}{m'\left(K_o \tau_h\right)^n\left[(1-\phi)\Sigma_g\right]^{n+1}}, \tag{5.189}$$

where μ_e is an effective viscosity and K is permeability of porous media. Eq. (5.188) is the modified Darcy's law of power-law non-Newtonian fluids (Bird et al., 1960; Pascal, 1986, 1990; Pascal and Pascal, 1989). In the limit as $L \to 0$, Eq. (5.188) can be written as

$$-\frac{\partial p}{\partial x} = \frac{\mu_e}{K} u^n. \tag{5.190}$$

For Newtonian fluids, $\mu_e = \mu, m' = \mu$, and $n' = n = 1$. Therefore, Eq. (5.189) simplifies to the well-known Kozeny–Carman equation, given by (Kozeny, 1927; Carman, 1938, 1956)

$$K = \frac{\phi^3}{K_o \tau_h \left[(1-\phi) \Sigma_g \right]^2} = \frac{\phi D_h^2}{16 K_o \tau_h}. \tag{5.191}$$

Consequently, substituting Eq. (5.191) into Eq. (5.189) reveals that

$$\mu_e = m' \left[\frac{K_o \tau_h (1-\phi) \Sigma_g}{\phi^2} \right]^{n-1} = m' \left(\frac{4 K_o \tau_h}{\phi D_h} \right)^{n-1}. \tag{5.192}$$

Sabiri and Comiti (1995) first determined the tortuosity τ and the specific surface area of the porous media grains Σ_g by measuring the pressure loss during flow of a Newtonian fluid. Caution that the surface area measured in this way may be less than the actual surface area of individual grains because of the reduction of surface area in a pack of grains by overlapping such as by embedment and fusing (Chhabra et al., 2001). Then, for laminar flow of pore fluid, they express the wall shear rate $\langle \dot{\gamma}_w \rangle_n$ by

$$\langle \dot{\gamma}_w \rangle_n = \left(\frac{3n+1}{2n} \right) \Sigma_g \tau_h \left(\frac{1-\phi}{\phi^2} \right) u. \tag{5.193}$$

The Fanning pore friction factor f_F is given by

$$f_{pF} = \frac{\dfrac{D_h}{4} \left(\dfrac{\Delta p}{L} \right)}{\dfrac{1}{2} \rho v^2} = \frac{2 \phi^3}{\rho u^2 \tau_h^2 (1-\phi) \Sigma_g} \left(\frac{\Delta p}{L} \right). \tag{5.194}$$

Note that the Moody pore friction factor is $f_M = 4 f_F$.
The pore Reynolds number is given by

$$\text{Re}_p = \frac{\rho \phi^{2n-2} (\tau_h u)^{2-n}}{2^{n-3} m \left(\dfrac{3n+1}{4n} \right)^n \left[(1-\phi) \Sigma_g \right]^n}. \tag{5.195}$$

Sabiri and Comiti (1997a) developed a dimensionless correlation as

$$f_{pF} = \frac{16\alpha}{\text{Re}_p} + 0.194\beta, \tag{5.196}$$

where the coefficients α and β indicate the wall effect of confined porous media, given by

$$\alpha = 1 + \frac{4}{D(1-\phi)\Sigma_g} \tag{5.197}$$

and

$$\beta = \left(1 - \frac{D_g}{D}\right)^2 + 0.427\left[1 - \left(1 - \frac{D_g}{D}\right)^2\right]. \tag{5.198}$$

Here, D denotes the diameter of confined porous media and D_g denotes the equivalent grain diameter. Thus, for an infinite-size porous media ($D \gg D_g$, $\alpha = \beta = 1$), Eq. (5.196) simplifies as

$$f_{pF} = \frac{16}{Re_p} + 0.194. \tag{5.199}$$

Successful correlation of data shown by Comiti et al. (2000) verifies the validity of Eq. (5.199) for $Re_p \leq 600$ in the range of $0.31 \leq \phi \leq 0.46$ porosity, $0.27 \leq n' \leq 0.91$ flow behavior index, and including data for $D_g/D < 0.15$ wall effects for various shape grains.

5.12.3 Modified Forchheimer Equation for Non-Newtonian Fluids

Substituting Eqs. (5.194) and (5.195) into Eq. (5.199) yields a modified Forchheimer equation for non-Newtonian fluids in the limit as $L \to 0$:

$$-\frac{\partial p}{\partial x} = \frac{\mu_e}{K}u^n + \beta\rho u^2, \tag{5.200}$$

where $\dfrac{\mu_e}{K}$ is given by Eq. (5.189) and the inertial flow coefficient is given by

$$\beta = \frac{0.194\tau_h^2(1-\phi)\Sigma_g}{2\phi^3} = \frac{0.388\tau_h^2}{\phi^2 D_h}. \tag{5.201}$$

The product of K and β according to Eqs. (5.191) and (5.201) yields a dimensionless group as

$$\frac{\beta K\phi}{\tau_h D_h} = \frac{0.388}{16K_o}. \tag{5.202}$$

Civan (2008d) has shown that

$$\frac{\beta K\phi}{\tau_h D_h} = \frac{c_D}{64}, \tag{5.203}$$

where c_D is the drag coefficient. Comparing Eqs. (5.202) and (5.203) yields

$$c_D = \frac{1.552}{K_o}. \tag{5.204}$$

For example, $K_o = 2$ and therefore $c_D = 0.776$ for cylindrical flow tubes.

5.13 EXERCISES

1. A density formulation of Eqs. (5.80)–(5.82) may be convenient when the real gas deviation factor, z, is given as a function of density, ρ, such as by Dranchuk et al., (1974).
Carry out a density formulation of the Forchheimer equation as the following. First
differentiate the real gas equation of state given by Eq. (5.4) to obtain (Civan and Evans,
1996)

$$\frac{dp}{dx} = \left(\frac{RT}{M}\right)\left(z + \rho\frac{dz}{d\rho}\right)\frac{d\rho}{dx}. \tag{5.205}$$

Then, substitute Eqs. (5.60) and (5.205) into Eq. (5.80) to derive the following density
formulation:

$$\frac{d\rho}{dx} = -\frac{wM\mu}{ART\rho}\left(z + \rho\frac{dz}{d\rho}\right)^{-1}\left(\frac{1}{K} + \beta\frac{w}{A\mu}\right) \tag{5.206}$$

subject to

$$\rho = \rho(p_1), x = 0 \tag{5.207}$$

and

$$\rho = \rho(p_2), x = L. \tag{5.208}$$

2. Show that, for the calculation of the average viscosity, $\bar{\mu}$, the average real gas deviation
factor, $\bar{z}$, and the pseudopressure differential, $\Delta m = m(p_1) - m(p_2)$, the differential
forms of Eqs. (5.63), (5.64), and (5.75) are given as

$$dI_1/dp = \mu/(p_1 - p_2), \tag{5.209}$$
$$dI_2/dp = z/(p_1 - p_2), \tag{5.210}$$

and

$$dI_3/dp = \rho/\mu. \tag{5.211}$$

In Eqs. (209)–(5.211), I_1, I_2, and I_3 represent the integrals according to Eqs. (5.63),
(5.64), and (5.75), respectively. Thus, when Eqs. (5.209)–(5.210) are integrated from
p_1 to p_2 beginning with the conditions given by

$$I_1 = 0, I_2 = 0, \quad \text{and} \quad I_3 = 0, \text{ at } p = p_1, \tag{5.212}$$

the average values are obtained, respectively, as

$$\bar{\mu} = I_1, \bar{z} = I_2 \quad \text{and} \quad \Delta m = I_3, \text{ at } p = p_2. \tag{5.213}$$

Integrate Eqs. (5.209)–(5.211) numerically by the Runge–Kutta method from the inlet to the outlet pressures, that is, from p_1 to p_2 for pressure values of your choice.

3. Transform the Liu et al. (1995) equation given in field units so that it can be used in the International System of Units (SI).

4. The generalized Darcy's equation of Das (1999), applicable for a fully elastic porous matrix ($\alpha = 1$ according to Section 5.3), can be written as

$$-\nabla(p\phi) + \rho\phi\mathbf{g} - \phi^2\mu\mathbf{K}^{-1}\cdot\mathbf{v} = \dot{m}\phi\mathbf{v} + \rho\phi\frac{\partial\mathbf{v}}{\partial t}. \tag{5.214}$$

Determine the terms missing in the equation derived by Das (1999) by comparing Eqs. (5.164) and (5.214), and justify under which conditions they can be neglected.

5. Consider heterogeneous porous media with $\phi = \phi(x, y, z)$ and $\phi \neq \phi(t)$, and a slightly compressible fluid described by the isothermal compressibility coefficient

$$c = \frac{1}{\rho}\frac{\partial\rho}{\partial p}.$$

Substitute $\dot{m}\phi \equiv q$ in Eq. (5.162) and

$$\rho\phi\frac{\partial\mathbf{v}}{\partial t} \cong 0$$

in Eq. (5.214) according to Das (1999) and eliminate the velocity $\mathbf{v}$ between these equations. Then, show that expanding the resulting equation yields (Civan, 2008d)

$$\phi c\frac{\partial p}{\partial t} = \nabla\cdot\left(\frac{\phi^2\nabla p + p\phi\nabla\phi}{q + \phi^2\mu K^{-1}}\right) + c\nabla p\cdot\left(\frac{\phi^2\nabla p + p\phi\nabla\phi}{q + \phi^2\mu K^{-1}}\right) + \frac{q}{\rho}. \tag{5.215}$$

Hint: Substituting $\dot{m}\phi \equiv q$ in Eq. (5.162) according to Das yields

$$\nabla\cdot(\rho\phi\mathbf{v}) + \frac{\partial}{\partial t}(\rho\phi) = q. \tag{5.216}$$

Substituting

$$\rho\phi\frac{\partial\mathbf{v}}{\partial t} \cong 0$$

in Eq. (5.214) according to Das yields

$$-\nabla(p\phi) + \rho\phi\mathbf{g} - \phi^2\mu\mathbf{K}^{-1}\cdot\mathbf{v} = q\mathbf{v}. \tag{5.217}$$

For horizontal flow in an isotropic porous media having the same permeability K in all coordinate directions, Eq. (5.217) is solved for pore fluid velocity as

$$\mathbf{v} = \frac{-\phi\nabla p - p\nabla\phi}{q + \phi^2\mu K^{-1}}. \tag{5.218}$$

Thus, substituting Eq. (5.218) into the equation of continuity given by Eq. (5.216) yields

$$\frac{\partial}{\partial t}(\rho\phi) = \nabla\cdot\left[\rho\left(\frac{\phi^2\nabla p + p\phi\nabla\phi}{q + \phi^2\mu K^{-1}}\right)\right] + q \tag{5.219}$$

Next, consider heterogeneous porous media with $\phi = \phi(x, y, z)$ and $\phi \neq \phi(t)$, and a slightly compressible fluid described by the following isothermal compressibility coefficient:

$$c = \frac{1}{\rho} \frac{\partial \rho}{\partial p}. \tag{5.220}$$

Thus, expanding Eq. (5.219) and then applying Eq. (5.220) yield

$$\phi c \frac{\partial p}{\partial t} = \nabla \cdot \left(\frac{\phi^2 \nabla p + p \phi \nabla \phi}{q + \phi^2 \mu K^{-1}} \right) + c \nabla p \cdot \left(\frac{\phi^2 \nabla p + p \phi \nabla \phi}{q + \phi^2 \mu K^{-1}} \right) + \frac{q}{\rho}. \tag{5.221}$$

6. For conventional practice omitting the source term q erroneously in the equation of motion while retaining it in the equation of continuity, assuming constant properties and dropping the second term on the right of Eq. (5.215), and then applying a first-order finite difference approximation to the time derivative of error Δp in the pressure calculation, show that an equation is obtained as (Civan, 2008d)

$$\Delta p = \frac{\Delta t}{\phi c} \left(\frac{\phi^2}{q + \phi^2 \mu K^{-1}} - \frac{K}{\mu} \right) \nabla^2 p. \tag{5.222}$$

The term $\nabla^2 p$ is evaluated by a second-order accurate five-point spatial finite difference approximation. Using the data of Das (1999) presented as the following, show that Eq. (5.222) provides an estimate of the error given by $\Delta p(Pa) = -6.4 \Delta t(s)$. Data: fluid density, $\rho = 1000 \text{ kg/m}^3$; fluid viscosity, $\mu = 0.001 \text{ Pa·s}$; fluid compressibility, $c = 1.422 \times 10^{-9} \text{ Pa}^{-1}$; mass flow rate per unit bulk volume of porous media, $q = 9.74 \times 10^{-4} \text{ kg/m}^3\text{-s}$; average porosity, $\phi = 20\%$; average permeability, $K = 10^{-13} \text{ m}^2$; pay thickness, $h = 6.096 \text{ m}$; grid size in the x-direction, $\Delta x = 30.48 \text{ m}$; grid size in the y-direction, $\Delta y = 30.48 \text{ m}$. The pressure and porosity values at various grid points: $(i, j + 1)$: 1399 kPa, 22%; $(i - 1, j)$: 1410 kPa, 23%; (i, j): 1406 kPa, 20%; $(i + 1, j)$: 1417 kPa, 18%; and $(i, j - 1)$: 1417 kPa, 19%.
Hint: Assuming constant properties and dropping the second term on the right of Eq. (5.215) yields

$$\phi c \frac{\partial p}{\partial t} = \left(\frac{\phi^2}{q + \phi^2 \mu K^{-1}} \right) \nabla^2 p + \frac{q}{\rho}. \tag{5.223}$$

For conventional practice, omitting the source term q erroneously in the equation of motion while retaining it in the equation of continuity yields

$$\phi c \frac{\partial p}{\partial t} = \frac{K}{\mu} \nabla^2 p + \frac{q}{\rho}. \tag{5.224}$$

Hence, subtracting Eq. (5.224) from Eq. (5.223) and then applying a first-order finite difference approximation to the time derivative of error Δp in pressure calculation,

$$\Delta p = \frac{\Delta t}{\phi c} \left(\frac{\phi^2}{q + \phi^2 \mu K^{-1}} - \frac{K}{\mu} \right) \nabla^2 p, \tag{5.225}$$

where $\Delta p \equiv p$ (Eq. 5.223) $- p$ (Eq. 5.224) represents the error made if the conventional Eq. (5.224) is used instead of the present Eq. (5.223). Following Das (1999), the term $\nabla^2 p$ indicated in Eq. (5.222) is evaluated by applying a second-order accurate five-point spatial finite difference approximation as

$$\nabla^2 p = \frac{p_{i-1,j} - 2p_{i,j} + p_{i+1,j}}{\Delta x^2} + \frac{p_{i,j-1} - 2p_{i,j} + p_{i,j+1}}{\Delta y^2}. \tag{5.226}$$

7. Assuming constant permeability and viscosity, no source effect, neglecting the term

$$\left[\nabla \left(\frac{p}{\phi} \right) \right] \cdot \nabla \phi$$

and dropping the second term on the right of Eq. (5.215), and then applying a first-order finite difference approximation to the time derivative of error Δp in the pressure calculation, show that an equation is obtained as

$$\Delta p = \frac{\Delta t K}{\phi^2 c \mu} \left(p \nabla^2 \phi \right). \tag{5.227}$$

The term $\nabla^2 \phi$ is evaluated by a second-order accurate five-point spatial finite difference approximation. Using the data of Das (1999) presented earlier in Exercise 6, show that Eq. (5.227) provides an estimate of the error given by $\Delta p(Pa) = 53.2 \Delta t(s)$.

Hint: Assuming constant permeability and viscosity, no source effect, neglecting the term

$$\left[\nabla \left(\frac{p}{\phi} \right) \right] \cdot \nabla \phi,$$

and dropping the second term on the right of Eq. (5.215) can be updated as

$$\phi c \frac{\partial p}{\partial t} = \frac{K}{\mu} \left[\nabla^2 p + \frac{p}{\phi} \nabla^2 \phi \right]. \tag{5.228}$$

Then, the conventional form of this equation is obtained by omitting the term $\nabla^2 \phi$ involving the heterogeneity effect as the following:

$$\phi c \frac{\partial p}{\partial t} = \frac{K}{\mu} \nabla^2 p. \tag{5.229}$$

Hence, subtracting Eq. (5.229) from Eq. (5.228) and then applying a first-order finite difference approximation to the time derivative of error Δp in pressure calculation,

$$\Delta p = \frac{\Delta t K}{\phi^2 c \mu} \left(p \nabla^2 \phi \right), \tag{5.230}$$

where $\Delta p \equiv p$ (Eq. 5.228) $- p$ (Eq. 5.229) represents the error made if the conventional Eq. (5.229) is used instead of the present Eq. (5.228) according to the terms included in the derivation.

8. Assuming constant properties, no source effect, and neglecting the heterogeneity ($\nabla\phi = 0$) in Eq. (5.215), show that the conventional equation used in well testing can be derived as

$$\phi c \frac{\partial p}{\partial t} = \frac{K}{\mu} \nabla^2 p. \tag{5.231}$$

9. Note that Eqs. (5.142), (5.143), and (5.145) can be written in vector–tensor form, respectively, as

$$\mathbf{u} = \phi \mathbf{v}, \tag{5.232}$$

and

$$\mathbf{v}_h = \mathbf{T}_h \cdot \mathbf{v}, \tag{5.233}$$

$$\mathbf{v}_h = \frac{1}{\phi}(\mathbf{T}_h \cdot \mathbf{u}), \tag{5.234}$$

where $\mathbf{T}_h$ denotes the tortuosity tensor. Express the generalized Darcy's equations derived in this section in terms of $\mathbf{v}_h$.

10. Note that the formulation elaborated in Section 5.11 for derivation of the generalized Darcy's law considers a fully elastic porous matrix ($\alpha = 0$ according to Section 5.3).

(a) Repeat this derivation for other cases ($\alpha \neq 0$) according to the definition of pressure as described in Section 5.3.

(b) The present formulation accommodates for various factors and mechanisms but ignores the effect of cross-flow. Investigate the possibility of considering the effect of cross-flow in the formulation.

(c) Extend the present formulation to flow of multiphase fluid systems through porous media.

11. Consider the Forchheimer equation given in a consistent unit system of your choice. Convert the units of the variables in this equation so that this equation can be used directly with the following units:

$$[p] = \text{atm}, [x] = \text{cm}, [\mu] = \text{cP}, [K] = \text{md}, [\rho] = \text{g/cm}^3,$$
$$[u] = \text{cm/s, and } [\beta] = \text{cm}^{-1}$$

12. Consider a porous core plug of 1-in. diameter, 2-in. length, 20% porosity, and 1.7 tortuosity. Determine the actual velocity of oil flowing through the tortuous flow paths if oil is injected into this core plug at a rate of 5 mL/s.

13. What will be the interstitial velocity in centimeter per second of the fluid flowing at a rate of 10 mL/min through a 1.0-in. diameter core having a 1.5 tortuosity and 20% porosity?

14. Estimate the threshold pressure gradient for an effective fluid permeability of 50 md and viscosity of 0.1 cP.

15. Consider the correlation of the threshold pressure gradient given by Prada and Civan (1999), where p is in pounds per square inch, x is in centimeter, and λ_m is in millidarcy per centipoise. Convert the units of the variables in this equation so that it can be used with the desired units prescribed as the following: p is in pascal, x is in meter, and λ_m is in square meter per pascal per second.

CHAPTER 6

GAS TRANSPORT IN TIGHT POROUS MEDIA

6.1   INTRODUCTION

Description of various gaseous flow regimes through tight porous media has drawn considerable attention because the conventional Darcy's law cannot realistically describe the variety of the relevant flow regimes other than the viscous flow regime.* For example, Javadpour et al. (2007) and Javadpour (2009) have determined that gas flow in shales (extremely low-permeability porous rock) deviates from behavior described by the conventional Fick's and Darcy's laws. Therefore, many attempts have been made in describing the transfer of gas through tight porous media under various regimes.

Description of gas flow in tight porous media, such as sequestration of CO_2 into tight geological porous media, and gas production from coal seams, and sandstone and Devonian shale formations, requires a different approach than using the conventional Darcy's equation (Ertekin et al., 1986; King and Ertekin, 1991). The same can be observed in nanoporous materials and asymmetric porous membranes made of ceramic, zeolite, and organic and inorganic materials, which are frequently used for gas separation (Rangarajan et al., 1984; Finol and Coronas, 1999). This is because the fluid present in the interconnected pores of porous media of various sizes may have different flow regimes depending on the prevailing pore and fluid conditions and the mean free path of molecules relative to the pore size (Rangarajan et al., 1984). Consequently, given the pore size distribution of porous media, the interconnected pores of porous media can be classified into several groups in which different

---

* Parts of this chapter have been reproduced with modifications from the following:

Civan, F. 2002a. A triple-mechanism fractal model with hydraulic dispersion for gas permeation in tight reservoirs. Paper SPE 74368, SPE Intl. Petroleum Conference and Exhibition, Villahermosa, Mexico, February 10–12, 2002, © 2002 SPE, with permission from the Society of Petroleum Engineers;

Civan, F. 2010a. Effective correlation of apparent gas permeability in tight porous media. Transport in Porous Media, 82(2), pp. 375–384, with permission from Springer; and

Civan, F. 2010d. A review of approaches for describing gas transfer through extremely tight porous media. In: Porous Media and Its Applications in Science, Engineering, and Industry, K. Vafai, ed., Proceedings of the Third ECI International Conference on Porous Media and its Applications in Science, Engineering and Industry (June 20–25, 2010), Montecatini Terme, Italy, pp. 53–58, with permission from the American Institute of Physics.

Porous Media Transport Phenomena, First Edition. Faruk Civan.
© 2011 John Wiley & Sons, Inc. Published 2011 by John Wiley & Sons, Inc.

TABLE 6.1 Classification of Flow Conditions in Pipes and Their Modeling Approaches according to the Knudsen Number Limits (Modified after Roy et al., 2003; Schaaf and Chambre, 1961; Reproduced with Permission from the American Institute of Physics after Civan, 2010d)

Flow Regimes	Models
Continuum flow (Kn ≤ 0.001)	Boltzmann equation, Euler equations, and no-slip Navier-Stokes equations
Slip flow (0.001 < Kn < 0.1)	Boltzmann equation, slip Navier-Stokes equations
Transition flow (0.1 < Kn < 10)	Boltzmann equation, Burnett equation
Free molecular flow (Kn ≥ 10)	Boltzmann equation

flow regimes takes place based on the prevailing Knudsen number (Rangarajan et al., 1984). This allows the determination of a representative apparent permeability of porous media, whose value depends on the prevailing fluid and flow conditions, and the pore size distribution. Further, Bravo (2007) proposed that the velocity profile should be expressed in a piecewise continuous manner encompassing the various flow regimes occurring over the cross-sectional area of flow.

Table 6.1 classifies the flow conditions and proper modeling approaches according to the Knudsen number determined by Schaaf and Chambre (1961) by experiments conducted in pipes (Roy et al., 2003; Civan, 2010d). The fundamental modeling approaches available for describing gas transfer through low-permeability porous media are the following: lattice–Boltzmann (LB), molecular dynamics (MD), direct simulation Monte Carlo (DSMC), Burnett equation (BE), and Navier–Stokes equation (NSE) (Schaaf and Chambre, 1961; Loyalka and Hamoodi, 1990; Bird, 1994; Beskok and Karniadakis, 1999).

Beskok and Karniadakis (1999) developed a unified Hagen–Poiseuille-type equation covering the fundamental flow regimes in tight porous media, including continuum fluid flow, slip flow, transition flow, and free molecular flow conditions. Civan (2010a) demonstrated that an accurate correlation of of the data of the dimensionless rarefaction coefficient can be accomplished using a simple inverse power-law function and a Vogel–Tammann–Fulcher (VTF)-type equation.

This chapter demonstrates that using the apparent permeability in a Darcy-type equation (Civan, 2010a) and Civan's (2002a,e) macroscopic mass conservation equation incorporating transport by hydraulic dispersion allow for proper physical description of gas flow in tight porous formations. These equations are solved analytically for one-dimensional, horizontal, and steady flow in laboratory core tests and radial flow in the near-wall bore region.

6.2 GAS FLOW THROUGH A CAPILLARY HYDRAULIC TUBE

Consider the flow of gas through a capillary tube according to Civan (2010a,d), denoted by the subscript h, representing a hydraulic flow path in porous media. The length and hydraulic diameter of the capillary tube are denoted by L_h and D_h. The molecular weight, density, temperature, pressure, and the capillary tube cross-sectional average velocity of the flowing gas are denoted by M, ρ, T, p, and v_h (meter per second).

The volumetric flow rate (cubic meter per second) is given by

$$q_h = \left(\pi D_h^2 / 4 \right) v_h, \tag{6.1}$$

where $A_h = \pi D_h^2 / 4$ (square meter) denotes the cross-sectional area of flow tubes. The mass flow rate $\dot{m}_h$ (kilogram per second) is given by

$$\dot{m}_h = \rho q_h. \tag{6.2}$$

The mole flow rate $\dot{n}_h$ (kilomole per second) is given by

$$\dot{n}_h = \dot{m}_h / M. \tag{6.3}$$

The density of gas is given by the following real gas equation of state:

$$\rho = \frac{Mp}{ZRT}, \tag{6.4}$$

where Z and R denote the real gas deviation factor and the universal gas constant. Eqs. (6.1)–(6.4) can be combined to obtain:

$$\dot{n}_h = \frac{\dot{m}_h}{M} = \frac{\rho q_h}{M} = \frac{\rho \pi D_h^2 v_h}{4M} = \frac{\pi D_h^2 v_h p}{4ZRT}. \tag{6.5}$$

The flow potential is defined by

$$\psi = p + \int_0^z \rho g dz. \tag{6.6}$$

The wall shear stress is given by

$$\tau_w = \frac{\left(\pi D_h^2 / 4 \right) \Delta \psi}{\pi D_h L_h} = \frac{D_h}{4} \frac{\Delta \psi}{L_h}. \tag{6.7}$$

6.3 RELATIONSHIP BETWEEN TRANSPORTS EXPRESSED ON DIFFERENT BASES

The porosity is denoted by ϕ and the ratio of the length of the hydraulic tube L_h and the bulk length of porous media $L = L_b$ is referred to as the tortuosity, given by

$$\tau_h = L_h / L. \tag{6.8}$$

The relationships between the fluid volume average velocity (pore average velocity) v_ϕ, the bulk volume average velocity (superficial or Darcy velocity) u, and the interstitial fluid velocity v_h as fluid moves along the tortuous preferential flow paths are given by (Civan, 2008d)

$$u = \phi v_\phi \tag{6.9}$$

and

$$v_h = \tau_h v_\phi.$$ (6.10)

The duPuit (1863) relationship is obtained from Eqs. (6.9) and (6.10) as

$$v_h = \frac{\tau_h u}{\phi}.$$ (6.11)

The following relationship can be written for the molar flux (kilomole per square meter per second) across the surface area of bulk volume:

$$J = \frac{\dot{n}}{A} = \frac{\rho q}{MA} = \frac{\rho u}{M}.$$ (6.12)

Similarly, the molar fluxes over the cross-sectional areas of pores and tortuous hydraulic tubes can be written, respectively, as

$$J_\phi = \frac{\rho v_\phi}{M}$$ (6.13)

and

$$J_h = \frac{\dot{n}_h}{A_h} = \frac{\rho q_h}{MA_h} = \frac{\rho v_h}{M}.$$ (6.14)

Consequently, it can be shown that

$$J = \phi J_\phi,$$ (6.15)

$$J_h = \tau_h J_\phi,$$ (6.16)

and

$$J_h = \frac{\tau_h J}{\phi}.$$ (6.17)

Now, consider porous media having pores described by a pore size distribution function $f(H)$ expressed in the units of cubic meter per kilogram-meter, where H denotes the pore width. From Eq. (6.17),

$$J = \frac{1}{\tau_h} (\phi J_h)_{\text{cumulative}} = \frac{1}{\tau_h} \int_{\sigma_{SS}}^{\infty} J_h(p, H) \rho_p f(H) dH.$$ (6.18)

where σ_{ss} denotes the collision diameter of the solid atom.

Next, assume that the flux over bulk porous media surface area and the hydraulic tube cross-sectional surface area can be described by the following gradient models:

$$J = -B(p, H) \frac{\partial p}{\partial z}$$ (6.19)

and

$$J_h = -B_h(p, H) \frac{\partial p}{\partial z},$$ (6.20)

where B represents the total permeability and B_h the permeability of the tortuosity flow paths. Hence, invoking Eqs. (6.19) and (6.20) into Eq. (6.18) yields

$$B_h(p,H) = \frac{\rho_p}{\tau_h} \int\limits_{\sigma_{SS}}^{\infty} B_h(p,H) f(H) dH. \tag{6.21}$$

6.4 THE MEAN FREE PATH OF MOLECULES: FHS VERSUS VHS

The word "fluid" refers to a continuum description of an assembly of molecules, which are in reality discrete entities. As observed by Knudsen (1909) and Radhakrishnan et al. (2000a,b, 2002a,b), the molecules of fluids behave differently in tightly bounded space present within the pores of porous media than in an unconfined space. The proximity of the pore walls limits the distance available for free motion of the fluid molecules and increases the probability of the interactions of molecules with the pore surface. The molecules of sufficiently low-density fluids, such as gas, in an unbounded space can move freely and randomly along straight paths until colliding with other molecules or the capillary tube wall (Knudsen, 1909). The mean free path of molecules is the average distance followed by the molecule along a straight path until collision occurs with another molecule or the pore wall. Under certain pressure and temperature conditions, the mean free path of molecules is a characteristic value, denoted by λ. Based on the kinetic theory of molecular motion, the mean free path of molecules is given by (Metz, 1976)

$$\lambda = \frac{k_B T}{\sqrt{2}\pi d^2 \bar{p}}, \tag{6.22}$$

where $k_B = 1.381 \times 10^{-23}$ J/K is the Boltzmann constant, T is the absolute temperature, $\bar{p}$ is the absolute average pressure, and d is the collision diameter of the gas molecules. The mean free path decreases as temperature decreases and/or pressure increases.

Bird (1983) emphasizes that the frequently used fixed cross-section hard-sphere model (FHS) applies for an ideal gas, but a variable cross-section hard-sphere model (VHS) should be considered for real gases.

Hence, Bird (1983) develops a VHS model, which involves two basic assumptions. The cross-sectional area of collision σ for a gas molecule is inversely proportional to its translational energy:

$$\sigma \propto \frac{1}{T^\zeta}, \tag{6.23}$$

where ζ is an exponent. The real gas viscosity is described by a power-law function of temperature T:

$$\mu \propto T^\omega, 0.6 \leq \omega \leq 0.9, \tag{6.24}$$

where ω is an empirically determined exponent. Note that $\omega = \frac{1}{2} + \zeta$.

Consequently, Bird (1983) provides the following modified expressions for the mean free path and the Chapman–Enskog viscosity, respectively:

$$\lambda = \frac{1}{(2-\zeta)^\zeta \, \Gamma(2-\zeta)\sqrt{2}n\sigma} \tag{6.25}$$

and

$$\mu = \frac{15M}{8} \frac{\sqrt{\pi R_g T / M}}{(2-\zeta)^\zeta \, \Gamma(4-\zeta)\sigma}, \tag{6.26}$$

where n (kilomole per cubic meter) is the number density of molecules in the gas, that is, the moles of gas available per unit volume of gas. The gamma function, where k is a dummy variable, is given by

$$\Gamma(k) = (k-1)! \tag{6.27}$$

When the average cross section σ is eliminated between Eqs. (6.25) and (6.26), the following expression is obtained (Bird, 1983):

$$\lambda = \frac{2\mu(7-2\omega)(5-2\omega)}{15} \frac{1}{\rho\sqrt{2\pi R_g T / M}}. \tag{6.28}$$

Note that

$$\rho = nM. \tag{6.29}$$

Thus, the following expression used by Roy et al. (2003) is derived by substituting $\omega = 0.5$ into Eq. (6.28) for the FHS:

$$\lambda = \frac{16\mu}{5} \frac{1}{\rho\sqrt{2\pi R_g T / M}}. \tag{6.30}$$

Further, the expression given by Loeb (1934) can be derived by considering an ideal gas and by substituting $16/5 \cong \pi$ in Eq. (6.30):

$$\lambda = \frac{\mu}{\rho}\sqrt{\frac{\pi}{2R_g T / M}} = \frac{\mu}{P}\sqrt{\frac{\pi R_g T}{2M}}. \tag{6.31}$$

6.5 THE KNUDSEN NUMBER

The Knudsen number Kn is used as a measure of rarefaction or the density lowness of a gas. It is expressed by

$$\text{Kn} = \frac{\lambda}{l_h}, \tag{6.32}$$

where l_h is the characteristic length and λ denotes the mean free path of molecules.

Frequently, the characteristic length for flow through capillary tubes is taken as the hydraulic diameter $D_h = 2R_h$ in definition of the Knudsen number, although many literatures, including Beskok and Karniadakis (1999), used the hydraulic radius R_h as indicated later in Eq. (6.91) for the characteristic length. Therefore, care should be paid to the particular form of the Knudsen number used in various studies.

The hydraulic radius R_h of flow paths is usually approximated by (Carman, 1938; Civan, 2010a)

$$R_h = 2\sqrt{2\tau_h}\sqrt{\frac{K_\infty}{\phi}} \quad \text{or} \quad K_\infty = \frac{\phi R_h^2}{8\tau_h}, \tag{6.33}$$

where τ_h is the tortuosity, K_∞ is the intrinsic permeability, and ϕ is the porosity of porous media.

ρ (kilogram per cubic meter) denotes the gas density given by the following real gas equation of state:

$$\rho = \frac{\bar{M}P}{ZR_gT}, \tag{6.34}$$

where P is the absolute gas pressure (pascal), T is the absolute temperature (kelvin), $\bar{M}$ is the average molecular mass (kilogram per kilomole), $R_g = 8314\,\text{J/kmol/K}$ is the universal gas constant, and Z is the real gas deviation factor.

Roy et al. (2003) express the Knudsen number by the ratio of the Mach to Reynolds numbers as

$$\text{Kn} = \sqrt{\frac{\pi\gamma}{2}}\frac{\text{Ma}}{\text{Re}} = \frac{\mu}{l_h}\sqrt{\frac{\pi}{2\rho P}} = \sqrt{\frac{\pi\gamma}{2}}\frac{\mu}{\rho l_h\sqrt{\frac{\gamma R_g T}{M}}} = \frac{\mu}{Pl_h}\sqrt{\frac{\pi R_g T}{2M}}. \tag{6.35}$$

The Mach (Ma) and Reynolds (Re) numbers are given as the following:

$$\text{Ma} = \frac{u}{c} = u\sqrt{\frac{\rho}{\gamma P}} = \frac{u}{\sqrt{\frac{\gamma R_g T}{M}}}, \tag{6.36}$$

where the speed of sound is given by

$$c = \sqrt{\frac{\gamma P}{\rho}} \tag{6.37}$$

and

$$\text{Re} = \frac{\rho u l_h}{\mu}, \tag{6.38}$$

where u (m³/m²·s) is the volumetric flux, μ (pascal-second) is the viscosity of gas, and $\gamma = C_p/C_v$ is the ratio of the specific heat capacity at constant pressure and the specific heat capacity at constant volume.

If D denotes the average diameter of the interconnected pore space, through which the fluid is flowing, a dimensionless number comparing the mean free paths of molecules with the pore diameter, called the Knudsen (1909) number, can be defined as

$$Kn = \lambda / D. \tag{6.39}$$

Shapiro and Wesselingh (2008) defined the Knudsen dimensionless number as

$$Kn = \lambda S, \tag{6.40}$$

where the pore size D is expressed by the reciprocal specific internal surface of porous media $1/S$.

Eqs. (6.22) and (6.39) indicate that

$$\lambda \to 0, Kn \to 0, \text{ as } p \to \infty \text{ and } / \text{ or } T \to 0. \tag{6.41}$$

Hence, the state of the molecules approaches a liquid and ultimately a solid. Conversely, a state of gas is attained for

$$\lambda \to \infty, Kn \to \infty, \text{ as } p \to 0 \text{ and/or } T \to \infty. \tag{6.42}$$

Gas flow through porous media may occur by several mechanisms, depending on the prevailing pore fluid conditions and the mean pore diameter (Liepmann, 1961; Stahl, 1971; Rangarajan et al., 1984; Kaviany, 1991). Schaaf and Chambre (1961) classified the flow regimes as continuum fluid flow ($Kn \le 0.001$), slip flow ($0.001 < Kn < 0.1$), transition flow ($0.1 < Kn < 10$), and free molecular flow ($Kn \ge 10$) (Table 6.1, Roy et al., 2003, Civan, 2010d). However, Beskok and Karniadakis (1999) emphasize that the Knudsen number limits given in this classification are based on pipe flow experiments and may vary by the geometry of other cases.

6.6   FLOW REGIMES AND GAS TRANSPORT AT ISOTHERMAL CONDITIONS

Transport of gas molecules through tight porous media at a prescribed temperature may occur by various mechanisms depending upon the pore size H and the bulk pressure p. The fundamental transport processes of gaseous molecules in porous media are identified as the following (Schaaf and Chambre, 1961; Do et al., 2001; Bae and Do, 2005):

1. Knudsen diffusion, referred to as pore diffusion (dominant in micropores)
2. Transition flow
3. Slip flow
4. Gaseous viscous flow, referred to as Darcy flow (dominant in macropores)
5. Adsorbed-phase diffusion, referred to as surface diffusion (dominant in micropores), which involves only the uppermost adsorbed molecules and depends on the concentration gradient of the uppermost molecules acting as the driving force.

6. Liquid viscous or condensate flow, referred to as capillary condensation flow (occurs when reduced pressure [pore pressure-to-vapor pressure ratio p_p/p_o] approaches the unity)

Thus, the total permeability of porous media B is expressed by

$$B = B_K + B_T + B_S + B_V + B_\mu + B_C, \tag{6.43}$$

where B_K, B_T, B_S, B_V, B_μ, and B_C denote the Knudsen diffusion (pore diffusion), transition, slip, viscous flow (Darcy flow), adsorbed-phase diffusion (surface diffusion), and condensate flow (capillary condensation flow), respectively.

Let p_c denote the critical pressure required for filling of pores with adsorbate. Then, the pore pressure p_p is calculated by

$$p_p = p \exp\left(-\frac{\alpha_c E_o}{R_g T}\right), \tag{6.44}$$

where p is the bulk pressure, α_c is a correction factor necessary for the overestimation of pore pressure in very small pores, E_o is the energy required for compressing molecules within the pore space, and R_g is the universal gas constant.

Deposition of gas molecules inside the pore space may lead to the pore-layering and pore-filling processes as depicted in Figure 6.1 by Bae and Do (2005).

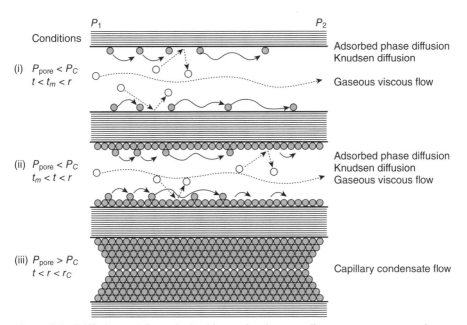

Figure 6.1   Diffusion and flow of adsorbing molecules according to pore pressure and statistical thickness in a pore of activated carbon (after Bae and Do, 2005; reproduced with permission from John Wiley & Sons).

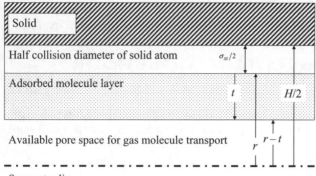

Symmetry line

Figure 6.2   Parts of a half pore space (prepared by the author).

As indicated in Figure 6.2, $H/2$ is the half pore width, $\sigma_{ss}/2$ denotes the collision radius of solid atom, t is the thickness of the adsorbed molecule layer, r is the accessible pore half-width, and $(r - t)$ denotes the pore width that is available for gas transport. The thickness of a monolayer is denoted by t_m, given by

$$t_m = \left(\frac{v_M}{N_A} \right)^{1/3},$$ (6.45)

where v_M is the liquid molar volume of the adsorbed phase and N_A is the Avogadro number.

Three types of transport modes occur as described in Figure 6.1 by Bae and Do (2005):

1. If conditions are $p_p < p_c$ and $t < t_m < r$, then adsorbed-phase diffusion (surface diffusion), Knudsen diffusion (pore diffusion), and gaseous viscous flow (Darcy flow) occur simultaneously.

2. If conditions are $p_p < p_c$ and $t_m < t < r$, then, as in the first case, adsorbed-phase diffusion (surface diffusion), Knudsen diffusion (pore diffusion), and gaseous viscous flow (Darcy flow) occur simultaneously.

3. If conditions are $p_p > p_c$ and $t < r < r_c$, then only the capillary condensate flow occurs, where r_c denotes the half collision diameter of the adsorbing molecules.

The mathematical descriptions of gas flow through a preferential flow path or tube under various flow regimes are presented in the following.

6.6.1 Knudsen Regime

At sufficiently low pressures, the mean free path of molecules may exceed the mean pore diameter (gas molecules–wall surface collisions). Therefore, $Kn \gg 1.0$ and the wall shear stress is negligible, $\tau_w \approx 0.0$. Thus, the Knudsen (1909) flow regime prevails. Knudsen empirically described the nonlinear flow behavior with respect to pressure in this region. Knudsen's (1909) equation can be expressed as

$$v_h = \left[A + \frac{D_K}{p} \frac{1+Bp}{1+Cp} \right] \frac{\Delta\psi}{L}, \tag{6.46}$$

where A, B, and C are empirical fitting parameters, and D_K is the Knudsen diffusivity coefficient. Hence, substituting Eq. (6.46) into Eq. (6.5) yields

$$\dot{n}_{Kh} = \frac{\pi D_h^4}{4} \left[A + \frac{D_K}{p} \frac{1+Bp}{1+Cp} \right] \frac{p}{ZRT} \frac{\Delta\psi}{L_h}$$

$$= \frac{\pi D_h^4}{4} \left[A + \frac{D_K}{p} \frac{1+Bp}{1+Cp} \right] \frac{p}{ZRT} \frac{1}{\tau_h} \frac{\Delta\psi}{L} \tag{6.47}$$

and

$$J_{Kh} = \frac{\dot{n}_{Kh}}{A_h} = \left[A + \frac{D_K}{p} \frac{1+Bp}{1+Cp} \right] \frac{p}{ZRT} \frac{1}{\tau_h} \frac{\Delta\psi}{L}, \tag{6.48}$$

whereas Present (1958) derived a different equation, given as

$$v_{Kh} \equiv \bar{c} = \frac{D\bar{c}}{3p} \frac{\Delta\psi}{L_h}. \tag{6.49}$$

Hence, substituting Eq. (6.49) into Eq. (6.5) yields

$$\dot{n}_{Kh} = \frac{\pi D^3 \bar{c}}{12p} \frac{p}{ZRT} \frac{\Delta\psi}{L_h} \tag{6.50}$$

and

$$J_{Kh} = \frac{\dot{n}_{Kh}}{A_h} = \frac{D_h \bar{c}}{3p} \frac{p}{ZRT} \frac{1}{\tau_h} \frac{\Delta\psi}{L}. \tag{6.51}$$

The Knudsen diffusivity (pore diffusion) of subcritical gases is given by

$$D_K = \frac{4K_o(r)}{3} \sqrt{\frac{8R_gT}{\pi M}} \left(\frac{2-f_p}{f_p} \right) = \frac{16R_gTK_o(r)}{3\sqrt{2\pi MR_gT}} \left(\frac{2-f_p}{f_p} \right), \tag{6.52}$$

where f_p is the collision reflection factor expressing the fraction of molecules colliding with the pore surface, which are bouncing back from the surface. $K_o(r)$ is the Knudsen diffusion parameter given by (Nicholson and Petropoulos, 1985)

$$K_o(r) = r/2 \text{ for cylindrical pores of } r \text{ radius} \quad \text{and}$$
$$K_o(r) = 3r/8 \text{ for slit pores of } r \text{ half-width.} \tag{6.53}$$

Bae and Do (2005) express the Knudsen diffusion permeability of all the pores by

$$B_K(p) = \frac{\rho_p}{\tau_K} \frac{16}{3\sqrt{2\pi MR_gT}} \int\limits_{r_c(p)}^{\infty} K_o(r-t) \left(\frac{2-f_p}{f_p} \right) R_o(r,t) f(r) dr \text{ for } r > t, \tag{6.54}$$

where the pore volume reduction by adsorbed molecule layering is given by

$$R_o(r,t) = \left(\frac{r-t}{r}\right)^2 \text{ for cylindrical pores and}$$

$$R_o(r,t) = \frac{r-t}{r} \text{ for slit pores.}$$

(6.55)

It is more convenient to express the integral as a sum over various pore size ranges when discrete pore size distribution data are used, as the following:

$$B_K(p) = \frac{\rho_p}{\tau_K} \frac{16}{3\sqrt{2\pi MR_g T}} \sum_{i=r_c}^{\infty} K_o(r_i - t_i)\left(\frac{2-f_{pi}}{f_{pi}}\right) R_o(r_i,t_i) f(r_i)\Delta r_i \text{ for } r > t. \quad (6.56)$$

6.6.2   Slip/Transition Regime

A region of slip or transition flow regime occurs between the Knudsen and viscous flow regions. The intermediate pressure conditions prevailing in this region allow for slip of molecules over the pore surface, represented by the following slip boundary condition approximating the wall shear stress as being proportional to the cross-sectional average fluid velocity as (Kaviany, 1991)

$$\tau_w = \alpha_f v_{STh}, \quad (6.57)$$

in which α_f is an empirical friction coefficient. Thus, equating Eqs. (6.7) and (6.57), an expression for the slip velocity can be derived as (Kaviany, 1991)

$$v_{STh} = \frac{D_h}{4\alpha} \frac{\Delta\psi}{L_h}. \quad (6.58)$$

Hence, substituting Eq. (6.58) into Eq. (6.5) yields an expression for the molar gas flow rate as

$$\dot{n}_{STh} = \frac{\pi D_h^3}{16\alpha_f} \frac{p}{ZRT} \frac{\Delta\psi}{L_h}. \quad (6.59)$$

However, it is more rigorous to consider the wall shear stress as being proportional to the kinetic energy of the flowing fluid according to (Civan, 2002a)

$$\tau_w = \alpha_f \tfrac{1}{2} \rho v_{STh}^2, \quad (6.60)$$

in which α_f is an empirical drag coefficient. Thus, equating Eqs. (6.7) and (6.60) yields

$$v_{STh} = \sqrt{\frac{D_h}{2\alpha_f \rho} \frac{\Delta\psi}{L_h}}. \quad (6.61)$$

Hence, substituting Eq. (6.61) into Eq. (6.5) yields

$$\dot{n}_{STh} = \frac{\pi D_h^2}{4} \sqrt{\frac{D_h}{2\alpha_f M} \frac{p}{ZRT} \frac{\Delta\psi}{L_h}}. \quad (6.62)$$

Alternatively, Eq. (6.61) can be written as

$$v_{STh} = \frac{D_h}{2\alpha_f \rho v_{STh}} \frac{\Delta \psi}{L_h}.$$ (6.63)

Hence, substituting Eq. (6.63) into Eq. (6.5) yields

$$\dot{n}_{STh} = \frac{\pi D_h^3}{8} \frac{1}{\alpha_f M v_{STh}} \frac{\Delta \psi}{L_h}.$$ (6.64)

As a simplification, assume $\alpha_f = 1$ and the mean gas speed is given by

$$v_{STh} = \bar{c} = \sqrt{\frac{8RT}{\pi M}}.$$ (6.65)

Then, substituting Eq. (6.65) into Eq. (6.64) yields (Metz, 1976)

$$\dot{n}_{STh} = \frac{\pi D_h^3}{8} \frac{1}{M\bar{c}} \frac{\Delta \psi}{L_h}.$$ (6.66)

Alternatively, Eqs. (6.65) and (6.66) can be combined as

$$\dot{n}_{STh} = \frac{\pi^2 D_h^3 \bar{c}}{64p} \frac{p}{ZRT} \frac{\Delta \psi}{L_h} = \frac{\pi^2 D_h^3 \bar{c}}{64p} \frac{p}{ZRT} \frac{1}{\tau_h} \frac{\Delta \psi}{L}$$ (6.67)

and

$$J_{STh} = \frac{\pi D_h \bar{c}}{16p} \frac{p}{ZRT} \frac{1}{\tau_h} \frac{\Delta \psi}{L}.$$ (6.68)

6.6.3   Viscous Regime

Gas flows as a continuous phase in the viscous regime (Shapiro and Wesselingh, 2008). Temperature gradient effect on flow is negligible. At elevated pressures, the fluid density is high and therefore the interactions between the fluid molecules and pore surface increase. As a result, the fluid motion is subject to intermolecular (molecule–molecule collisions) and molecule and pore surface friction effects. As high-density fluids flow through capillary tubes or hydraulic flow paths in porous media, the wall shear stress due to friction between the fluid and solid wall may be significantly high. Hence the fluid molecules may stick to the tube wall and the fluid may not have slip over the tube wall. Thus, the fluid velocity at the pore wall is zero, $v_w = 0$. The Hagen–Poiseuille equation, given as follows, applies under these conditions:

$$v_h = \frac{D_h^2}{32\mu} \frac{\Delta \psi}{L_h}.$$ (6.69)

Hence, substituting Eq. (6.69) into Eq. (6.5) yields

$$\dot{n}_{Vh} = \frac{\pi D_h^4}{128\mu} \frac{p}{ZRT} \frac{\Delta \psi}{L_h} = \frac{\pi D_h^4}{128\mu} \frac{p}{ZRT} \frac{1}{\tau_h} \frac{\Delta \psi}{L}$$ (6.70)

and

$$J_h = \frac{\dot{n}_{vh}}{A_h} = \frac{\rho}{M} v_h = \frac{D_h^2}{32\mu} \frac{p}{ZRT} \frac{1}{\tau_h} \frac{\Delta\psi}{L}. \tag{6.71}$$

Bae and Do (2005) express the gaseous viscous flow permeability of all the pores by

$$B_V(p) = \frac{\rho_p}{\tau_V} \frac{p}{\mu_g R_g T} \int_{r_c(p)}^{\infty} B_o(r-t) R_o(r,t) f(r) dr \text{ for } r > t, \tag{6.72}$$

where the pore volume reduction by adsorbed molecule layering is given by

$$R_o(r,t) = \left(\frac{r-t}{r}\right)^2 \text{ for cylindrical pores } \text{ and}$$

$$R_o(r,t) = \frac{r-t}{r} \text{ for slit pores.} \tag{6.73}$$

The viscous flow parameter $B_o(r)$ is given by (Nicholson and Petropoulos, 1985)

$$B_o(r,t) = r^2/8 \text{ for cylindrical pores } \text{ and}$$

$$B_o(r,t) = r^2/3 \text{ for slit pores.} \tag{6.74}$$

It is more convenient to express the integral as a sum over various pore size ranges when discrete pore size distribution data are used, as the following:

$$B_V(p) = \frac{\rho_p}{\tau_V} \frac{p}{\mu_g R_g T} \sum_{i=r_c(p)}^{\infty} B_o(r_i - t_i) R_o(r_i,t_i) f(r_i) \Delta r_i \text{ for } r > t. \tag{6.75}$$

6.6.4   Adsorbed-Phase Diffusion

Transport by adsorbed-phase diffusion (or surface diffusion) is expressed by a Fickian-type equation (Do et al., 2001):

$$J_\mu = \frac{\rho_p}{\tau_\mu} D_\mu^* \frac{C_\mu}{p} \frac{\Delta\psi}{L_h} = \frac{\rho_p}{\tau_\mu} D_\mu^* \frac{C_\mu}{p} \frac{1}{\tau_h} \frac{\Delta\psi}{L}, \tag{6.76}$$

where C_μ denotes the adsorbed concentration (mole per gram), given by

$$C_\mu(p) = \int_{\sigma_{SS}}^{\infty} \Psi^E(p,r) f(r) dr, \tag{6.77}$$

where $\Psi^E(p,r)$ is the effective local isotherm function given by

$$\Psi^E(p,r) = t/r \text{ for } t \le t_m \text{ and}$$

$$\Psi^E(p,r) = t_m/r \text{ for } t > t_m. \tag{6.78}$$

Bae and Do (2005) express the adsorbed-phase diffusion (surface diffusion) permeability of all the pores by

$$B_\mu(p) = \frac{\rho_p}{\tau_\mu v_M} \int_{r_c(p)}^{\infty} D_\mu^*(p,r) \frac{\Psi^E(p,r)}{p} f(r)dr, \tag{6.79}$$

where τ_μ is the tortuosity for surface diffusion, v_M is the liquid molar volume (cubic meter per kilomole), and $D_\mu^*(p,r)$ is the corrected surface diffusivity. It is more convenient to express the integral as a sum over various pore size ranges when discrete pore size distribution data are used, as the following:

$$B_\mu(p) = \frac{\rho_p}{\tau_\mu v_M} \sum_{i=r_c(p)}^{\infty} D_\mu^*(p,r_i) \frac{\Psi^E(p,r_i)}{p} f(r_i)\Delta r_i. \tag{6.80}$$

6.6.5 Liquid Viscous or Capillary Condensate Flow

When $p_p > p_c$, then capillary condensation occurs, leading to the pore-filling process. Assume that the condensate behaves as a Newtonian fluid. Thus, the viscosity is constant across the pore cross-sectional area ($\mu = $ ct.) and no slip occurs over the pore surface ($v_z[r] = 0$). Then, the capillary condensate flow permeability is given by

$$B_C(p) = \frac{\rho_p}{\tau_C} \left(\frac{R_g T}{3\mu_L v_M^2} \right) \int_1^{r_c(p)} \frac{r^2}{p} f(r)dr, \tag{6.81}$$

where τ_c is the tortuosity for capillary condensate flow and μ_L is the viscosity of the liquid (condensate). Bae and Do (2005) express the integral as a sum over various pore size ranges when discrete pore size distribution data are used, as the following:

$$B_C(p) = \frac{\rho_p}{\tau_C} \left(\frac{R_g T}{3\mu_L v_M^2} \right) \sum_1^{r_c(p)} \frac{r_i^2}{p} f(r_i)\Delta r_i. \tag{6.82}$$

6.7 GAS TRANSPORT AT NONISOTHERMAL CONDITIONS

Shapiro and Wesselingh (2008) analyzed the flow of gas in tight porous media under nonisothermal conditions. Based on the gas kinetic theory, Shapiro and Wesselingh (2008) derived the equation of flow of gas in tight porous media in the kinetic regime. They included the Knudsen dilute gas kinetic regime and the intermediate or dense gas kinetic regime; the second is also referred to as the slip/transition flow regime. They assumed that all molecules colliding with the surface are adsorbed and that porous media are isotropic. Under these conditions, they expressed the flow occurring because of the pressure and temperature gradients ∇p and ∇T as

$$\mathbf{u} = \frac{\phi v}{S} \beta \left(\beta_T \frac{\nabla T}{T} - \frac{\nabla p}{p} \right), \tag{6.83}$$

where S is the specific pore surface (square meter per cubic meter) and v is the characteristic average molecular velocity given by

$$v = \sqrt{\frac{kT}{M}}, \tag{6.84}$$

where k is the Boltzmann constant, M is the molecular mass (usually called weight), and T is the absolute temperature. The β constant is given by

$$\beta = \frac{24}{13}\sqrt{\frac{2}{\pi}} \cong 1.5 \text{ for the Knudsen regime (dilute gas regime)} \quad \text{and}$$

$$\beta = \frac{6\sqrt{2\pi}}{16+\pi} \cong 0.8 \text{ for the intermediate regime (dense gas regime).} \tag{6.85}$$

The parameter β_T is given by

$$\beta_T = 1/2 \text{ for the Knudsen regime (dilute gas regime)} \quad \text{and}$$

$$\beta_T = \frac{8\lambda S}{15\sqrt{2\pi}vRN\phi} \tag{6.86}$$

$$\cong 0.07\frac{\gamma Kn}{\phi} \text{ for the intermediate regime (dense gas regime),}$$

where N is the gas molecule number density or concentration, the Knudsen number Kn is given by Eq. (6.40), and κ is the heat conductivity, given by

$$\kappa = \Psi Nc_V v\lambda, \tag{6.87}$$

where c_V is the specific heat capacity of gas at constant volume, $\gamma = c_V/R$, and Ψ is a dimensionless collision parameter ($\Psi \cong 1/3$). Thus,

$$\beta_T = \frac{8\Psi vKn}{15\sqrt{2\pi}\phi} \cong 0.07\frac{\gamma Kn}{\phi} \text{ for the intermediate regime (dense gas regime).} \tag{6.88}$$

Under constant pressure conditions, the gas flows from the low- to high-temperature direction in the Knudsen regime, according to Eq. (6.83).

6.8   UNIFIED HAGEN–POISEUILLE-TYPE EQUATION FOR APPARENT GAS PERMEABILITY

The Hagen–Poiseuille-type equation of Beskok and Karniadakis (1999) considers a single-pipe flow. When the bundle of tortuous tube realization of the preferential flow paths in porous media (Carman, 1956) is considered, the number and tortuosity of the preferential flow paths formed in porous media should be considered as important parameters. In the following sections, the derivation of the apparent gas permeability equation and the correlation of the rarefaction coefficient and the Klinkenberg gas slippage factor are presented according to Civan's (2010a).

6.8.1 The Rarefaction Coefficient Correlation

All the fundamental gaseous flow regimes can be described using the following unified Hagen–Poiseuille-type equation for volumetric gas flow q_h through a single pipe (Beskok and Karniadakis, 1999):

$$q_h = f(\text{Kn})\frac{\pi R_h^4}{8\mu}\frac{\Delta p}{L_h}, \tag{6.89}$$

where the flow condition function $f(\text{Kn})$ is given by

$$f(\text{Kn}) = (1+\alpha\text{Kn})\left(1+\frac{4\text{Kn}}{1-b\text{Kn}}\right), \tag{6.90}$$

where Kn is the Knudsen number given by

$$\text{Kn} = \frac{\lambda}{R_h}, \tag{6.91}$$

where R_h and L_h denote the hydraulic radius and length of flow tube, and λ denotes the mean free path of molecules, given by (Loeb, 1934)

$$\lambda = \frac{\mu}{p}\sqrt{\frac{\pi R_g T}{2M}}, \tag{6.92}$$

where p is the absolute gas pressure in pascal, T is the absolute temperature in kelvin, M is the molecular mass in kilogram per kilomole, $R_g = 8314\,\text{J/kmol/K}$ is the universal gas constant, and μ is the viscosity of gas in pascal-second.

The parameter α appearing in Eq. (6.90) is a dimensionless rarefaction coefficient, which varies in the range of $0 < \alpha < \alpha_o$ over $0 \le \text{Kn} < \infty$. Beskok and Karniadakis (1999) provide an empirical correlation as

$$\alpha = \alpha_o\frac{2}{\pi}\tan^{-1}\left(\alpha_1\text{Kn}^{\alpha_2}\right), \tag{6.93}$$

where $\alpha_1 = 4.0$, $\alpha_2 = 0.4$, and α_o is an asymptotic upper limit value of α as $\text{Kn} \to \infty$ (representing free molecular flow condition), calculated by

$$\alpha_o \equiv \alpha_{\text{Kn}\to\infty} = \frac{64}{3\pi\left(1-\dfrac{4}{b}\right)}. \tag{6.94}$$

Here, b denotes a slip coefficient. For example, $b = -1$ is substituted for a fully developed slip flow through tubes for all gases. They indicate that $\alpha = 0$ and $b = -1$ in the slip flow condition, and therefore Eq. (6.94) becomes

$$\alpha_o = \frac{64}{15\pi}. \tag{6.95}$$

However, Civan (2010a) provides a simple inverse power-law expression as a much more accurate and practical alternative to Eq. (6.93) in the range of data analyzed by Beskok and Karniadakis (1999), given as

$$\frac{\alpha_o}{\alpha} - 1 = \frac{A}{\mathrm{Kn}^B}, A > 0, B > 0, \tag{6.96}$$

where A and B are empirical fitting constants. Note that Eq. (6.96) honors the limiting conditions of $0 < \alpha < \alpha_o$ over $0 \leq \mathrm{Kn} < \infty$.

Using Eq. (6.96), Civan (2010a) accurately correlated the data of both Loyalka and Hamoodi (1990) using the theoretically predicted upper limit value of $\alpha_o = 1.358$ and Tison and Tilford (1993) using an adjusted upper limit value of $\alpha_o = 1.205$. Consequently, the data of Loyalka and Hamoodi (1990) are correlated as

$$\log(1.358/\alpha - 1) = -0.4348 \log \mathrm{Kn} - 0.7496, R^2 = 0.9871. \tag{6.97}$$

Thus, $A = 0.1780$ and $B = 0.4348$. Similarly, the data of Tison and Tilford (1993) are correlated as

$$\log(1.205/\alpha - 1) = -0.365 \log \mathrm{Kn} - 0.7011, R^2 = 0.9486. \tag{6.98}$$

Thus, $A = 0.199$ and $B = 0.365$.

6.8.2 The Apparent Gas Permeability Equation

The Beskok and Karniadakis (1999) unified Hagen–Poiseuille-type equation (Eq. 6.89) derived for flow q_h through a single pipe can now be applied for the volumetric gas flow through a bundle of tortuous flow paths, as

$$q = nq_h = nf(\mathrm{Kn})\frac{\pi R_h^4}{8\mu}\frac{\Delta p}{L_h}, \tag{6.99}$$

where L_h denotes the length of the tortuous flow paths and n denotes the number of preferential hydraulic flow paths formed in porous media. The latter can be approximated by rounding the value calculated by the following equation to the nearest integer (Civan, 2007a):

$$n = \frac{\phi A_b}{\pi R_h^2}, \tag{6.100}$$

where ϕ is porosity and A_b is the bulk surface area of porous media normal to flow direction. The symbol q denotes the total volumetric flow through porous media. It can be described macroscopically by a Darcy-type gradient law of flow, where the flow is assumed proportional to the pressure gradient, given by

$$q = \frac{KA_b}{\mu}\frac{\Delta p}{L}, \tag{6.101}$$

where K denotes the apparent gas permeability of tight porous media and L is the length of bulk porous media.

Note that Eq. (6.101) is used frequently, although it is not rigorously correct. The corrections required on Eq. (6.101), such as the effect of the threshold pressure gradient (Prada and Civan, 1999), are provided elsewhere by Civan (2008d), but are ignored here to avoid unnecessary complications for purposes of the present discussion and derivation. Nevertheless, Civan (2008d) argued that such corrections are usually negligible for gaseous flow, although the validity of such claim for tight porous media needs detailed investigation.

The tortuosity factor τ_h of hydraulic preferential flow paths in porous media is defined by

$$\tau_h = \frac{L_h}{L}. \tag{6.102}$$

Hence, the following expression can be derived for the apparent gas permeability by combining Eqs. (6.99)–(6.102):

$$K = K_\infty f(\mathrm{Kn}), \tag{6.103}$$

where K_∞ denotes the intrinsic liquid permeability of porous media, given by

$$K_\infty = \frac{\phi R_h^2}{8\tau_h}. \tag{6.104}$$

Eq. (6.104) can be rearranged to express the hydraulic tube radius as

$$R_h = 2\sqrt{2\tau_h}\sqrt{\frac{K_\infty}{\phi}}. \tag{6.105}$$

Alternatively, it can be shown for a pack of porous media grains that (Civan, 2007a)

$$R_h = \frac{2}{\Sigma_g}\left(\frac{\phi}{1-\phi}\right), \tag{6.106}$$

where Σ_g denotes the specific grain surface in porous media. Hence, equating Eqs. (6.105) and (6.106) yields the well-known Kozeny–Carman equation of permeability as (Carman, 1956; Civan, 2007a)

$$\sqrt{\frac{K_\infty}{\phi}} = \frac{1}{\Sigma_g\sqrt{2\tau_h}}\left(\frac{\phi}{1-\phi}\right). \tag{6.107}$$

The function $f(\mathrm{Kn})$ does not appear in Eq. (6.107) because the intrinsic permeability K_∞ of porous media is only a property of porous media and does not depend on the fluid type and flow condition.

6.8.3 The Klinkenberg Gas Slippage Factor Correlation

Under slip flow conditions, $\alpha = 0$ and $b = -1$, and therefore Eq. (6.103) combined with Eq. (6.90) can be written as

$$K = K_\infty\left(1+\frac{4\mathrm{Kn}}{1+\mathrm{Kn}}\right). \tag{6.108}$$

Florence et al. (2007) approximate this equation for $\mathrm{Kn}\ll 1$, upon substitution of the Kn number expression Eq. (6.91), as

$$K \cong K_\infty(1+4\mathrm{Kn}) = K_\infty\left(1+\frac{4\lambda}{R_h}\right), \tag{6.109}$$

whereas the equation of Klinkenberg (1941) is given by

$$K = K_\infty \left(1 + \frac{b_K}{p} \right),$$ (6.110)

where b_K is the slippage factor. Comparing Eqs. (6.109) and (6.110) yields an expression as

$$b_K = 4p\text{Kn} = \frac{4p\lambda}{R_h}.$$ (6.111)

Substituting Eqs. (6.92) and (6.105) into Eq. (6.111) results in

$$b_K = \beta \left(\frac{K_\infty}{\phi} \right)^{-1/2},$$ (6.112)

where the coefficient β is defined by

$$\beta = p\lambda \sqrt{\frac{2}{\tau_h}}, \text{ or } \frac{\beta}{\mu} = \sqrt{\frac{\pi R_g T}{\tau_h M}}.$$ (6.113)

Using Eq. (6.113), Civan (2010a) provides an accurate correlation for the data of Florence et al. (2007) involving the flow of various gases (hydrogen, helium, nitrogen, air, and carbon dioxide) in a given porous medium (sandstone) under isothermal conditions (assumed as 298 K) as the following:

$$\beta = \mu \sqrt{\frac{\pi R_g T}{\tau_h M}} = \frac{2.79 \times 10^3 \mu}{\sqrt{M}},$$ (6.114)

where β is in pascal-meter, μ is in pascal-second, and M is in kilogram per kilomole.

Note that Eqs. (6.111)–(6.113) can be used to derive an expression for the Knudsen number as

$$\text{Kn} = \frac{b_K}{4p} = \frac{\mu}{4} \sqrt{\frac{\pi R_g T}{\tau_h M}} \frac{1}{p} K_\infty^{-1/2} \phi^{1/2}.$$ (6.115)

For example, applying the correlation given by Eq. (6.114) to Eq. (6.115) yields

$$\text{Kn} = \frac{698\mu}{\sqrt{M}} \frac{1}{p} K_\infty^{-1/2} \phi^{1/2},$$ (6.116)

and therefore

$$b_k = \frac{2.79 \times 10^3 \mu}{\sqrt{M}} K_\infty^{-1/2} \phi^{1/2},$$ (6.117)

where b_k is in pascal, μ is in pascal-second, M is in kilogram per kilomole, p is in pascal, K_∞ is in square meter, and ϕ is a fraction. When applied for the N_2 gas, Eq. (6.117) yields

$$b_k = 0.0094 \left(\frac{K_\infty}{\phi} \right)^{-1/2}. \tag{6.118}$$

The tortuosity τ_h of the preferential hydraulic flow paths in the porous medium can be estimated using Eq. (6.114) as the following, expressed in the consisted SI units:

$$\tau_h = \frac{\pi R_g T}{7.78 \times 10^6}, \tag{6.119}$$

where R_g is 8314 J/kmol-K and T is in kelvin. Eq. (6.119) may be used in determining the tortuosity of porous media.

6.9 SINGLE-COMPONENT GAS FLOW

The equation of total mass conservation for gas flowing through porous media is expressed by (Civan, 2002d, 2010b)

$$\frac{\partial(\rho\phi)}{\partial t} + \frac{\partial(\rho u)}{\partial x} = \frac{\partial}{\partial x}\left[D_m \frac{\partial(\rho\phi)}{\partial x} \right] - \dot{m}, \tag{6.120}$$

where t and x are the time and distance, respectively; $\dot{m}$ is the rate of gas mass removed from the flowing gas per unit bulk volume of porous media by various processes including reaction, absorption, or adsorption; and D_m denotes the bulk mass dispersion coefficient.

Eq. (6.120) implies that the total gas mass flux j_T is given by

$$j_T = j_A + j_D, \tag{6.121}$$

where j_A denotes the advective mass flux given by the following Hagen–Poiseuille-type equation (Civan, 2010b):

$$j_A = \rho u = -F \frac{\rho\phi}{\tau_h} \frac{R_h^2}{8\mu} \frac{\partial P}{\partial x}. \tag{6.122}$$

Here, P is the absolute total pressure and F is a wall slip correction factor, determined by (Brown et al., 1946)

$$F = 1 + 4\left(\frac{2}{\beta} - 1 \right) \frac{\mu}{R_h P} \sqrt{\frac{\pi R_g T}{2M}}, \tag{6.123}$$

where β is an empirical parameter representing the fraction of gas molecules returning after striking at the pore wall.

The j_D term represents the gas-dispersive mass flux given by the following expression, applying Eq. (6.34) and the Knudsen diffusivity D_K for the bulk diffusivity:

$$j_{Dm} = -D_m \frac{\partial(\rho\phi)}{\partial x} = -\frac{D_K}{\tau_h} \frac{\partial}{\partial x}\left(\frac{\bar{M}P\phi}{R_g T} \right). \tag{6.124}$$

The Knudsen diffusivity D_K is given by (Roy et al., 2003)

$$D_K = \frac{2R_h}{3}\sqrt{\frac{8R_g T}{\pi \bar{M}}}.$$ (6.125)

Eq. (6.124) yields the equation of Roy et al. (2003) as the following for $\tau_h = 1.0$ and constant properties:

$$j_{Dm} = -D_K \frac{\bar{M}\phi}{R_g T}\frac{\partial P}{\partial x}.$$ (6.126)

Invoking Eq. (6.125) into Eq. (6.124) and assuming constant properties yield

$$j_{Dm} = -\frac{2R_h \phi}{3\tau_h}\sqrt{\frac{8\bar{M}}{\pi R_g T}}\frac{\partial P}{\partial x}.$$ (6.127)

A substitution of Eqs. (6.122) and (6.127) into Eq. (6.121) yields (Javadpour, 2009)

$$j_T = -\left[F\frac{\rho}{\mu}\frac{\phi R_h^2}{8\tau_h} + \frac{2R_h \phi}{3\tau_h}\sqrt{\frac{8\bar{M}}{\pi R_g T}}\right]\frac{\partial P}{\partial x}.$$ (6.128)

On the other hand, a Darcy-type expression can be written as

$$j_T = -\frac{\rho}{\mu}K\frac{\partial P}{\partial x}.$$ (6.129)

Then, the following expression can be derived for the apparent permeability K scaled by the intrinsic permeability given by Eq. (6.105), K_∞, by equating Eqs. (6.128) and (6.129) and applying Eq. (6.149), $\mathrm{Kn} = \lambda / R_h$:

$$\frac{K}{K_\infty} = F + \frac{8}{R_h}\frac{2}{3}\sqrt{\frac{8R_g T}{\pi \bar{M}}}\frac{\mu}{P} = F + \frac{8}{R_h}\frac{2}{3}\frac{\mu}{\rho}\sqrt{\frac{8\bar{M}}{\pi R_g T}} = F + \frac{20}{3}\mathrm{Kn}.$$ (6.130)

6.10 MULTICOMPONENT GAS FLOW

Now, let M denote the molecular weight, p the absolute partial pressure, c the mass concentration, C the mol concentration, n the number concentration, w the weight fraction, X the mole fraction of the gas species, P the total absolute pressure of the gas mixture, and $N_A = 6.022 \times 10^{23}$ molecules/mol the Avogadro's number.

The concentration of gas species present in the gas mixture can be expressed in various forms as (Civan, 2010d)

$$w = \frac{c}{\rho}, c = MC = \frac{Mp}{R_g T} = \frac{MPX}{R_g T}, n = N_A C, w = \frac{XM}{\bar{M}},$$

$$\bar{M} = \sum_i (XM)_i, C = \frac{XP}{R_g T} = \frac{p}{R_g T}, X = \frac{p}{P}.$$ (6.131)

The equation of gas species mass conservation for gas flowing through porous media is expressed by (Civan, 2002d, 2010d)

$$\frac{\partial(\rho w \phi)}{\partial t} + \frac{\partial(\rho w u)}{\partial x} = \frac{\partial}{\partial x}\left[D\frac{\partial(\rho w \phi)}{\partial x}\right] - \dot{m}, \tag{6.132}$$

where $\dot{m}$ denotes the rate of gas species mass removed from the flowing gas phase per unit bulk volume of porous media by a process including reaction, absorption, or adsorption.

Eq. (6.132) implies that the species gas mass flux j_T is given by

$$j_T = j_A + j_D, \tag{6.133}$$

where j_A is the advective mass flux of gas species, given by

$$j_A = \rho u w = u c = u M C = u M n / N_A. \tag{6.134}$$

The volumetric flux of the flowing gas mixture u is given by a Darcy-type equation as

$$u = -\frac{1}{\mu}K\frac{\partial P}{\partial x}. \tag{6.135}$$

Hence, the following expression can be derived (Bravo, 2007):

$$j_A = cu = MC\left(-\frac{1}{\mu}K\frac{\partial P}{\partial x}\right) = \frac{MPX}{R_gT}\left(-\frac{1}{\mu}K\frac{\partial P}{\partial x}\right). \tag{6.136}$$

The dispersive mass flux j_D of gas species is given by the modified Fick's law (Civan, 2002e, 2010d):

$$j_{Db} = -D_b\frac{\partial(\rho w \phi)}{\partial x} = -\phi\rho\frac{D_K}{\tau_h}\left[\frac{\partial w}{\partial x} + \frac{w}{\phi\rho}\frac{\partial(\rho\phi)}{\partial x}\right], \tag{6.137}$$

where D_K is the Knudsen diffusivity, given by (Roy et al., 2003)

$$D_K = \frac{2R_h}{3}\sqrt{\frac{8R_gT}{\pi M}}. \tag{6.138}$$

Invoking Eqs. (6.34) and (6.131) into Eq. (6.137) gives

$$j_{Db} = -\phi\frac{\bar{M}P}{R_gT}\frac{D_K}{\tau_h}\left(\frac{\partial w}{\partial x} + \frac{w}{P}\frac{\partial P}{\partial x}\right). \tag{6.139}$$

The following equation given by Bravo (2007) can be derived by substituting $\tau_h = 1.0$ and neglecting the second term in Eq. (6.139):

$$j_D = -\frac{\phi D_K MP}{R_gT}\frac{\partial X}{\partial x}. \tag{6.140}$$

Consequently, when Eqs. (6.136) and (6.140) are substituted into Eq. (6.133), the following equation given by Bravo (2007) can be derived:

$$j_T = -\frac{MPXK}{R_g T \mu}\frac{\partial P}{\partial x} - \frac{\phi D_K MP}{R_g T}\frac{\partial X}{\partial x}. \qquad (6.141)$$

6.11   EFFECT OF DIFFERENT FLOW REGIMES IN A CAPILLARY FLOW PATH AND THE EXTENDED KLINKENBERG EQUATION

The formulation given by Bravo (2007) is presented in the following with some modifications. Bravo (2007) extended the formulation of Klinkenberg (1941) by representing the velocity profile over the cross-sectional area of a capillary tube in a piecewise continuous manner. This allowed the determination of the pressure dependence of the Klinkenberg parameter b_K by considering the molecule–molecule and molecule–capillary tube wall interactions occurring during flow. Let R_h denote the radius of the capillary flow tube, R_o the radius of the transition/Knudsen region, and R_b the radius of the continuous flow region.

Bravo (2007) relates R_h and R_b by

$$R_b = R_h - w_1 \lambda, \qquad (6.142)$$

where w_1 is an empirical parameter. Thus, the dimensionless radius is given by

$$f = 1 - \frac{w_1 \lambda}{R_h} = 1 - w_1 \mathrm{Kn}. \qquad (6.143)$$

Similarly, Bravo (2007) relates R_h and R_o by

$$R_o = R_h - w_o \lambda, \qquad (6.144)$$

where w_o is an empirical parameter. Thus, the dimensionless radius is given by

$$f_o = 1 - \frac{w_o \lambda}{R_h} = 1 - w_o \mathrm{Kn}. \qquad (6.145)$$

The velocity profiles for various flow regimes are described in the following manner.

The gas transport under continuum flow regime prevailing over the inner region $(0 < r < R_b)$ is described by the Navier-Stokes equation (NSE) considering the pressure and viscous dissipation forces resulting from molecule–molecule interactions:

$$\frac{dp}{dx} = \mu \frac{1}{r}\frac{d}{dr}\left(r\frac{dv}{dr}\right). \qquad (6.146)$$

The solution of this equation provides the Poiseuille velocity profile as

$$v(r) = \frac{R_b^2 - r^2}{4\mu}\left(-\frac{dp}{dx}\right), 0 < r < R_b. \qquad (6.147)$$

The intermediate transition/Knudsen region ($R_b \le r \le R_o$) involving gas transport under viscous dissipation forces occurs because of the molecule–capillary tube wall interactions. The inner region velocity profile is connected to the near-wall region slip velocity by a linear velocity profile, given by

$$v(r) = \frac{v_o - v_b}{R_o - R_b}(r - R_b) + v_b, \ R_b < r < R_o, \tag{6.148}$$

where $v_b = nv_o$, n is an empirical constant, and v_o is the slip velocity estimated in the following.

The free molecular gas transport in the near-wall region ($R_o < r \leq R_w$) represented by a constant slip velocity, given by

$$v(r) = v_o = \frac{D^K}{p}\left(-\frac{dp}{dx}\right), 0 \leq r \leq R_o, \tag{6.149}$$

where D^K is the Knudsen diffusivity, given by

$$D^K = \frac{4K_o}{3}\sqrt{\frac{8k_B T}{\pi M}}, \tag{6.150}$$

where K_o is a Klinkenberg diffusion parameter.

Bravo (2007) developed the following correlation, where D^K is in square meter per second and K_∞ is in square meter:

$$D^K = 1.86 K_\infty^{0.70}. \tag{6.151}$$

The mean interstitial pore fluid velocity in the capillary tube is calculated by

$$\bar{v} = \frac{2}{R_h^2}\int_0^{R_h} v(r) r \, dr. \tag{6.152}$$

Hence, using the velocity profiles presented earlier, the mean interstitial pore fluid velocity in the capillary tube is calculated by

$$\bar{v} = \frac{R_h^2 f^4}{8\mu}\left(-\frac{dp}{dx}\right) + v_o\left[\frac{n-1}{3}(f^2 + f_o f + f_o^2) + 1\right], \text{ when } f, f_o > 0$$

$$\bar{v} = v_o\left[\frac{n-1}{3}\left(\frac{f_o^3}{f_o - f}\right) + 1\right], \text{ when } f \leq 0, f_o > 0 \tag{6.153}$$

$$\bar{v} = v_o, \text{ when } f, f_o \leq 0.$$

Note the interstitial pore fluid velocity $\bar{v}$ (meter per second) is given by (Dupuit, 1863)

$$\bar{v} = \frac{\tau_h u}{\phi}. \tag{6.154}$$

Darcy's law is given by

$$u = \frac{K}{\mu}\left(-\frac{dp}{dx}\right). \tag{6.155}$$

The Klinkenberg equation is given by

$$K = K_\infty\left(1 + \frac{b}{P}\right),$$ (6.156)

where K_∞ is given by Eq. (6.105).

Thus, combining Eqs. (6.105) and (6.153)–(6.156) yields

$$b = \frac{\mu D^K}{K_\infty}\left[\frac{n-1}{3}(f^2 + f_o f + f_o^2) + 1\right] - (1 - f^4)P, \text{ when } f, f_o > 0$$

$$b = \frac{\mu D^K}{K_\infty}\left[\frac{n-1}{3}\left(\frac{f_o^3}{f_o - f}\right) + 1\right] - P, \text{ when } f \leq 0, f_o > 0$$ (6.157)

$$b = \frac{\mu D^K}{K_\infty} - P, \text{ when } f, f_o \leq 0.$$

6.12   EFFECT OF PORE SIZE DISTRIBUTION ON GAS FLOW THROUGH POROUS MEDIA

Consider the flow of gas through a porous medium in which the preferred flow paths can be realized as a bundle of tortuous capillary tubes. The number of tubes of various diameters, D, can be expressed by means of a suitable frequency distribution function, $F(D)$, such that

$$\int_{D_{\min}}^{D_{\max}} F(D)dD = N_t,$$ (6.158)

where $D_{\min} \leq D \leq D_{\max}$ represents the capillary tube diameter range and N_t denotes the total number of hydraulic tubes formed by the fluid flowing through the preferential flow paths in porous media. For this purpose, the following normal distribution function can be used (Rangarajan et al., 1984):

$$F(D) = \frac{N_t}{\sigma\sqrt{2\pi}}\exp\left[-\frac{1}{2}\left(\frac{D - \bar{D}}{\sigma}\right)^2\right],$$ (6.159)

where σ and $\bar{D}$ denote the standard deviation of the tube diameter distribution and the mean tube diameter. The following fractal distribution function can also be adopted (Perrier et al., 1996):

$$F(D) = N_t\beta(e - d)D^{e-d-1},$$ (6.160)

where $e = 3$ denotes the Euclidian dimension, $2 < d < 3$ represents the fractal dimension range, and β is a coefficient. Eq. (6.160) can be written as

$$F(D) = cD^{\tilde{d}},$$ (6.161)

where

$$c = N_t\beta(e - d), \tilde{d} = e - d - 1.$$ (6.162)

Following Rangarajan et al. (1984), the hydraulic tubes of porous media can be classified into three groups as $D_{\min} \leq D \leq D_{K-S}$, $D_{K-S} \leq D \leq D_{S-V}$, and $D_{S-V} \leq D \leq D_{\max}$, corresponding to the Knudsen, slip, and viscous flow regimes, respectively, based on the prevailing Knudsen number. Applying the parallel flow analogy of flow through porous media, then the total mole gas flow rate through porous media is given by

$$\dot{n} = \dot{n}_K + \dot{n}_S + \dot{n}_V, \tag{6.163}$$

where $\dot{n}_K, \dot{n}_S, \dot{n}_V$ are the mole gas flow rates through the tubes involving the Knudsen, slip, and viscous flow regimes, given, respectively, by

$$\dot{n}_K = \sum_{i=1}^{N_{Kt}} F(D_i)\dot{n}_K(D_i), \tag{6.164}$$

$$\dot{n}_S = \sum_{i=1}^{N_{St}} F(D_i)\dot{n}_S(D_i), \tag{6.165}$$

and

$$\dot{n}_V = \sum_{i=1}^{N_{Vt}} F(D_i)\dot{n}_V(D_i). \tag{6.166}$$

where N_{Kt}, N_{St}, N_{Vt} denote the total number of capillary tubes undergoing the Knudsen, slip, and viscous flow regimes. Hence, the total number of all flow tubes in porous media is given by

$$N_t = N_{Kt} + N_{St} + N_{Vt}. \tag{6.167}$$

When a continuous distribution function such as Eqs. (6.159) or (6.161) is used to express the number frequency of hydraulic tubes, the following expressions can be used to express the molar flow rates:

$$\dot{n}_K = \int_{D_{\min}}^{D_{K-S}} F(D)\dot{n}_K(D)dD, \tag{6.168}$$

$$\dot{n}_S = \int_{D_{K-S}}^{D_{S-V}} F(D)\dot{n}_S(D)dD, \tag{6.169}$$

and

$$\dot{n}_V = \int_{D_{S-V}}^{D_{\max}} F(D)\dot{n}_V(D)dD. \tag{6.170}$$

On the other hand, the total mole gas flow rate through porous media can be expressed by

$$\dot{n} = \frac{\dot{m}}{M} = \frac{\rho q}{M} = \frac{\rho A u}{M} = \frac{A u p}{ZRT}, \tag{6.171}$$

where A denotes the area of porous media normal to flow and u is the Darcy velocity or the superficial flow through porous media, given by

$$u = \frac{\dot{n}M}{\rho A} = \frac{\dot{n}}{A}\frac{ZRT}{p}. \tag{6.172}$$

The Darcy-type gradient law is given by

$$u = \frac{K_a}{\mu} \frac{\Delta \psi}{L_b},$$ (6.173)

where L_b and K_a are length and apparent permeability of porous media.

Thus, substituting Eq. (6.163) with Eqs. (6.168)–(6.170) into Eq. (6.172) and then comparing with Eq. (6.173) yields the following expression for the apparent permeability of porous media:

$$K_a(p) = \frac{\mu}{\tau_h A} \frac{N_t}{\sigma \sqrt{2\pi}} \frac{p}{ZRT}$$

$$\left\{ \begin{array}{l} \displaystyle\int_{D_{\min}}^{D_{K-S}} \exp\left[-\frac{1}{2}\left(\frac{D - \bar{D}}{\sigma} \right)^2 \right] \frac{\pi D^3 \bar{c}}{12p} dD + \\[3mm] \displaystyle\int_{D_{K-S}}^{D_{S-V}} \exp\left[-\frac{1}{2}\left(\frac{D - \bar{D}}{\sigma} \right)^2 \right] \frac{\pi^2 D^3 \bar{c}}{64p} dD + \\[3mm] \displaystyle\int_{D_{S-V}}^{D_{\max}} \exp\left[-\frac{1}{2}\left(\frac{D - \bar{D}}{\sigma} \right)^2 \right] \frac{\pi D^4}{128\mu} dD \end{array} \right\}.$$ (6.174)

The integral appearing in Eq. (6.174) can be evaluated numerically by means of an appropriate method, such as the Gaussian quadrature. If the fractal equation, Eq. (6.161), is used, the apparent permeability expression is obtained as

$$K_a(p) = \frac{\mu}{\tau_h A} \frac{p}{ZRT} c \left\{ \begin{array}{l} \displaystyle\int_{D_{\min}}^{D_{K-S}} D^{\tilde{d}} \frac{\pi D^3 \bar{c}}{12p} dD + \\[3mm] \displaystyle\int_{D_{K-S}}^{D_{S-V}} D^{\tilde{d}} \frac{\pi^2 D^3 \bar{c}}{64p} dD + \\[3mm] \displaystyle\int_{D_{S-V}}^{D_{\max}} D^{\tilde{d}} \frac{\pi D^4}{128\mu} dD \end{array} \right\},$$ (6.175)

which can be integrated analytically to obtain

$$K_a(p) = \frac{\mu}{\tau_h A} \frac{p}{ZRT} c \left\{ \begin{array}{l} \displaystyle\frac{\pi \bar{c}}{12p} \frac{D_{K-S}^{\tilde{d}+4} - D_{\min}^{\tilde{d}+4}}{\tilde{d}+4} + \\[3mm] \displaystyle\frac{\pi^2 \bar{c}}{64p} \frac{D_{S-V}^{\tilde{d}+4} - D_{K-S}^{\tilde{d}+4}}{\tilde{d}+4} + \\[3mm] \displaystyle\frac{\pi}{128\mu} \frac{D_{\max}^{\tilde{d}+5} - D_{S-V}^{\tilde{d}+5}}{\tilde{d}+5} \end{array} \right\}.$$ (6.176)

This formulation used the mean tube tortuosity. However, the formulation can be extended using a tortuosity distribution function. Analogous to Knudsen's experimental studies with capillary tubes, Klinkenberg (1941) empirically determined that the apparent permeability tensor of porous media could be expressed by (Wa'il Abu-El-Sha'r and Abriola, 1997)

$$\mathbf{K}_a(p) = \mathbf{K}_\infty + \mathbf{I}\frac{D_K \mu}{p}, \tag{6.177}$$

where $\mathbf{K}_\infty$ is the intrinsic or liquid permeability tensor of porous media and $\mathbf{I}$ denotes the unit tensor. The data of the Knudsen diffusivity (square meter per second) versus the intrinsic permeability (square meter) for dry clayey soil, provided by Olivella et al. (2000), can be correlated as

$$D_K = aK_\infty^b, a = 1.0 \times 10^9, b = 1.0. \tag{6.178}$$

Next, for illustration purposes, consider the macroscopic equation of continuity for flow through porous media, given by (Civan, 2002a,e)

$$\frac{\partial}{\partial t}(\phi\rho) + \nabla \cdot (\rho\mathbf{u}) = \nabla \cdot [\mathbf{D}_b \cdot \nabla(\phi\rho)], \tag{6.179}$$

where $\mathbf{D}_b$ denotes the porous media hydraulic dispersion tensor.

Assume a steady-state one-dimensional horizontal flow of gas through a porous core plug. Substituting Eqs. (6.177) and (6.4) into Eq. (6.179) yields

$$\frac{d}{dx}\left[(\delta + \gamma p)\frac{dp}{dx}\right] = 0, \tag{6.180}$$

where

$$\delta = D_K + D_b\phi \tag{6.181}$$

and

$$\gamma = K_\infty / \mu. \tag{6.182}$$

The boundary conditions are prescribed as

$$p = p_0, x = 0 \tag{6.183}$$

and

$$p = p_L, x = L. \tag{6.184}$$

The analytical solution of Eqs. (6.180)–(6.184) is obtained as

$$\delta p + (\gamma/2)p^2 = C_1 x + C_2, \tag{6.185}$$

where

$$C_1 = (\delta p_L + (\gamma/2)p_L^2 - C_2)/L \tag{6.186}$$

and

$$C_2 = \delta p_0 + (\gamma/2)p_0^2. \tag{6.187}$$

In the radial coordinate, Eq. (6.180) becomes

$$\frac{1}{r}\frac{d}{dr}\left[(\delta + \gamma p)r\frac{dp}{dr}\right] = 0. \tag{6.188}$$

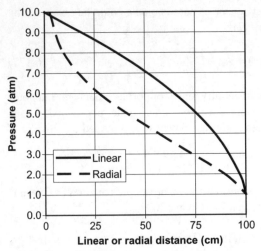

Figure 6.3   Comparison of solutions for linear and radial flows (after Civan, 2002a; © 2002 SPE, reproduced with permission of the Society of Petroleum Engineers).

The boundary conditions are prescribed as

$$p = p_w, r = r_w \tag{6.189}$$

and

$$p = p_e, r = r_e. \tag{6.190}$$

The analytical solution of Eqs. (6.180)–(6.184) is obtained as

$$\delta p + (\gamma/2)p^2 = C_3 \ln r + C_4, \tag{6.191}$$

where

$$C_3 = \frac{\delta(p_e - p_w) + (\delta/2)(p_e^2 - p_w^2)}{\ln(r_e/r_w)} \tag{6.192}$$

and

$$C_4 = \delta p_w + (\delta/2)p_w^2 - C_3 \ln r_w. \tag{6.193}$$

Garrison et al. (1992, 1993) used Eq. (6.161) and derived $\tilde{d}$ as the apparent surface fractal dimension from the pore diameter distribution. Figure 6.3 presents a comparison of the linear and radial flow problems using the data of Garrison et al. (1993) for the Norphlet sandstone parameters.

6.13   EXERCISES

1. Javadpour (2009) accomplished the best fit of Eq. (6.128) to the experimental data of Pong et al. (1994) by adjusting the value of the β parameter to 0.8. The conditions of the gas flow tests were the following: Argon gas was flown through a membrane made

of an alumina filter. The pores consist of straight cylindrical-shaped nanotubes. The thickness of the membrane is $L = 60\,\mu m$. The radius of the nanotubes is $R_h = 100\,nm$, and the tortuosity is $\tau_h = 1.0$. The porosity is $\phi \cong 0.2 - 0.3$. Make reasonable assumptions for any missing data. Prepare a plot of values predicted by Eq. (6.128) against the experimental data to determine the accuracy of this equation.

2. Prepare a plot of K/K_∞ versus pressure P using Eq. (6.130) using the parameter values given in the previous problem.

3. Prepare a plot of α versus Kn over the range of $0 \le Kn < \infty$ using Eqs. (6.93), (6.97), and (6.98), and compare the curves obtained by these equations.

4. Prepare a plot of β/μ versus $\sqrt{M}$ using Eq. (6.113) for $\tau_h = \sqrt{2}$ at various temperatures.

5. Determine the tortuosity τ_h using Eq. (6.119) for the sandstone sample used in the tests assuming a temperature of 298 K according to Florence et al. (2007). Note, however, that Florence et al. (2007) mentioned that the actual temperature was unknown for the data involved in their studies and they simply assumed the value of 298 K for their calculations.

6. Bravo (2007) correlated the data of Klinkenberg (1941) and Li et al. (2004) satisfactorily using Eqs. (6.156) and (6.157) using the best estimate parameter values given as the following: case 1: $K_\infty = 0.024 \times 10^{-15}\,m^2$, $n = 14$, $w_1 = 10.8$; case 2: $K_\infty = 2.36 \times 10^{-15}\,m^2$, $n = 15$, $w_1 = 3.56$; and case 3: $K_\infty = 23 \times 10^{-15}\,m^2$, $n = 80$, $w_1 = 40.7$. Develop plots of K versus P using these parameter values.

7. The data of Garrison et al. (1993) for Norphlet sandstone parameters are presented as $\phi = 5.5\%$, $K = 0.20\,mD$, $D_{min} = 1.01\mu$, $D_{max} = 70.85\mu$, $D_{mean} = 3.83\mu$, $c = 3.8 \times 10^4$, and $d = 2.087$. The data of Rangarajan et al. (1984) are $D_{K-S} = 0.1\lambda\ \mu$ and $D_{S-V} = 100\lambda\ \mu$. The other required data assumed for calculatrion are (Civan, 2002a) $p_w = 10\,atm$, $p_e = 1\,atm$, $r_w = 2.54\,cm$, and r_e or $L = 100\,cm$. Calculate and plot the curves given in Figure 6.3 for the linear and radial flow problems using these data.

8. Prepare a plot of the flow condition function $f\,(Kn)$ versus the Knudsen number Kn in the range of $0 \le Kn < 1000$ using $\alpha_o = 1.358$, $b = -1$, $A = 0.1780$, and $B = 0.4348$ for the data of Loyalka and Hamoodi (1990). What would be the intrinsic (actual) permeability of the core plug in the units of square meter and millidarcy if the apparent permeability of a core plug was measured as $2.5 \times 10^{-20}\,m^2$ for the Knudsen number value of $Kn = 10$?

9. Estimate the mean free path of the methane molecule in meter at 25°C and 1-atm pressure conditions using the equation given by Loeb (1934).

10. If the Knudsen number is equal to 0.1 for gas flowing through tight porous media, what will be the apparent to intrinsic (true) permeability ratio (correction factor)? Apply the correlation obtained for the rarefaction coefficient α for the data of Tison and Tilford (1993).

11. Consider a continuous flow of a real gas through a horizontal porous core sample under isothermal (25°C) and steady-state conditions. The length of this core plug is 100 cm and the diameter is 2.54 cm. The porosity is 20%. The tortuosity is 1.5. The gas pressures at the inlet and outlet sections of the core are 5 and 1 atm, respectively. A correction on the intrinsic (actual) permeability of the core $K_\infty = 2.7\,mD$ is necessary according to the following Klinkenberg equation given by

$$K = K_\infty \left(1 + \frac{1.3}{p}\right),$$

where the pressure p is in the atmosphere unit. For simplification purposes, assume the following constant average values for the real gas deviation factor $Z = 0.8$ and the real gas viscosity $\mu = 0.012\,cP$.

(a) Write down the equations required for the description of gas flow in the core sample.

(b) Obtain an analytical solution of the resulting pressure equation (a simple equation) for pressure.

(c) Obtain the expressions for calculation of the Darcy and actual fluid velocities using the above-obtained pressure solution.

(d) Calculate the velocity of the gas just before and after it leaves the porous core sample at the outlet section.

12. Is the apparent permeability greater or smaller than the actual permeability of a tight porous rock?

13. Concerning the formulation of Civan et al. (2010, 2011) for shale-gas permeability and diffusivity determination:

(a) Develop a finite difference numerical solution scheme.

(b) Compare your numerical solutions with those of Civan et al. (2010, 2011) for the applications presented in their studies.

CHAPTER 7

FLUID TRANSPORT THROUGH POROUS MEDIA

7.1 INTRODUCTION

Single- and multiphase mass and momentum balance equations are coupled under isothermal conditions to derive a hydraulic diffusion equation, and its applications are demonstrated by several examples.*

First, a generalized leaky-tank reservoir model is developed, including the non-Darcy and generalized fluid effects (Civan, 2002e). The resulting model equations are applied for a typical case. It is demonstrated that special plotting schemes can provide a practical technique for the determination of the thickness, permeability, and the water-drive strength of petroleum reservoirs from deliverability test data, including the non-Darcy effect due to converging flow pattern in the near-wellbore region, showing the effect of convective acceleration.

Next, convenient formulations of the immiscible displacement in porous media are presented and applied for waterflooding. The macroscopic equation of continuity for immiscible displacement is derived. Richardson's (1961) approach and the fractional flow formulation are extended and generalized for anisotropic and heterogeneous porous media. The integral transformations according to Douglas et al. (1958) and the coordinate transformations lead to differential equations, which do not involve the variable fluid and porous media properties explicitly in the differential operators. Fractional flow and unit end-point mobility ratio formulations are also derived for specific applications to reduce the computational requirements and to accomplish rapid simulation of waterflooding of petroleum reservoirs. The resulting equations can be discretized and solved more conveniently and accurately

* Parts of this chapter have been reproduced with modifications from the following:

Civan, F. 1996b. Convenient formulation for immiscible displacement in porous media. SPE 36701, Proceedings of the 71st SPE Annual Tech Conference and Exhibition (October 6–9, 1996), Denver, Colorado, pp. 223–236, © 1996 SPE, with permission from the Society of Petroleum Engineers;

Civan, F., 2000e. Leaky-tank reservoir model including the non-Darcy effect. Journal of Petroleum Science and Engineering, 28(3), pp. 87–93, with permission from Elsevier; and

Penuela, G. and Civan, F. 2001. Two-phase flow in porous media: Property identification and model validation, Letter to the Editor. AIChE Journal, 47(3), pp. 758–759, © 2001 AIChE, with permission from the American Institute of Chemical Engineers.

Porous Media Transport Phenomena, First Edition. Faruk Civan.
© 2011 John Wiley & Sons, Inc. Published 2011 by John Wiley & Sons, Inc.

than the conventional formulation, which requires cumbersome discretization formulae for mixed derivatives involving the fluid and porous media properties. The convenient formulations offer potential advantages over the usual formulation used in the simulation of waterflooding, such as improved accuracy and reduced computational effort.

Finally, models facilitating the streamline and stream tube or flow channel concepts, which are frequently used for description of flow in porous media, are described. It is shown that these models offer certain advantages such as convenience, practicality, reduced computational effort, and insight into flow patterns. Suitable flow problems involving porous media are formulated as potential flow problems and are solved analytically by means of the methods of superposition, images, and front tracking.

7.2 COUPLING SINGLE-PHASE MASS AND MOMENTUM BALANCE EQUATIONS

Consider the equation of mass conservation given by (Civan, 2002e)

$$\frac{\partial(\phi\rho)}{\partial t} + \nabla \cdot (\rho\mathbf{u}) = \nabla \cdot [\mathbf{D} \cdot \nabla(\phi\rho)] + r, \tag{7.1}$$

where r denotes the source term and $\mathbf{D}$ is the hydraulic dispersion coefficient tensor. Eq. (7.1) can be rearranged as

$$\frac{\partial(\phi\rho)}{\partial t} + (\mathbf{v} - \nabla \cdot \mathbf{D}) \cdot \nabla(\phi\rho) + \phi\rho\nabla \cdot \mathbf{v} = \mathbf{D} \cdot \nabla^2(\phi\rho) + r, \mathbf{v} = \frac{\mathbf{u}}{\phi}. \tag{7.2}$$

If $\rho = $ ct., then Eq. (7.2) becomes

$$\frac{\partial\phi}{\partial t} + (\mathbf{v} - \nabla \cdot \mathbf{D}) \cdot \nabla\phi + \phi\nabla \cdot \mathbf{v} = \mathbf{D} \cdot \nabla^2\phi + \frac{r}{\rho}, \mathbf{v} = \frac{\mathbf{u}}{\phi}. \tag{7.3}$$

If $\phi = $ ct., then Eq. (7.2) becomes

$$\frac{\partial\rho}{\partial t} + (\mathbf{v} - \nabla \cdot \mathbf{D}) \cdot \nabla\rho + \rho\nabla \cdot \mathbf{v} = \mathbf{D} \cdot \nabla^2\rho + \frac{r}{\phi}, \mathbf{v} = \frac{\mathbf{u}}{\phi}. \tag{7.4}$$

If $\phi = $ ct. and $\rho = $ ct., then Eq. (7.2) becomes

$$\nabla \cdot \mathbf{v} = \frac{r}{\phi\rho}, \mathbf{v} = \frac{\mathbf{u}}{\phi}. \tag{7.5}$$

Now, neglect the dispersion term in Eq. (7.1) so that it can be written as

$$\frac{\partial(\phi\rho)}{\partial t} + \nabla \cdot (\rho\mathbf{u}) = r. \tag{7.6}$$

Consider Darcy's law for horizontal flow, given by

$$u = -\frac{K}{\mu}\nabla p. \tag{7.7}$$

The isothermal fluid and pore compressibility factors are defined by

$$c_\rho = \frac{1}{\rho}\frac{\partial \rho}{\partial p}, c_\phi = -\frac{1}{\phi}\frac{\partial \phi}{\partial p}, \rho = \rho_o \exp[c_\rho (p - p_o)], c_T = c_\rho + c_\phi. \qquad (7.8)$$

Combining Eqs. (7.6)–(7.8) yields

$$\frac{\partial p}{\partial t} = \alpha_h \nabla^2 p + c_\rho \alpha_h (\nabla p)^2 + \frac{1}{\phi c_T}[\nabla \cdot (\phi \rho c_T)]\nabla p$$

$$+ \frac{r}{\phi c_T \rho_o \exp[c_\rho (p - p_o)]}, \alpha_h = \frac{K}{\phi \rho c_T}, \qquad (7.9)$$

where α_h denotes the hydraulic diffusivity.

If all the parameters are constants, there is no source, and $(\nabla p)^2 \cong 0$, then Eq. (7.9) simplifies as

$$\frac{\partial p}{\partial t} = \alpha_h \nabla^2 p. \qquad (7.10)$$

This equation is referred to as the hydraulic diffusion equation.

7.3 CYLINDRICAL LEAKY-TANK RESERVOIR MODEL INCLUDING THE NON-DARCY EFFECT

Frequently, the cylindrical tank model, as shown in Figure 7.1, is resorted to describe the flow within the reservoir drainage area of wells, for convenience in the formulation of the governing equations of flow and to generate analytic solutions. The resulting simplified radial flow model is assumed to approximately represent the reservoir fluid conditions within the drainage area and leads to simplified analytic

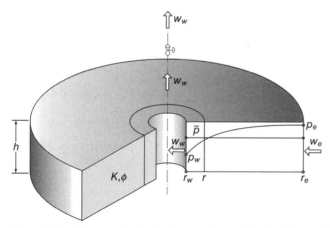

Figure 7.1   Schematic leaky-tank reservoir model (after Civan, 2000e; reprinted by permission from Elsevier).

solutions, which reduce the complexity of the reservoir engineering analysis and are justifiable in view of the uncertainties involving in most reservoir conditions.

Kumar (1977a,b) formulated the radial flow equations necessary for the determination of the water-drive strength from transient well test data using Darcy's law. Civan and Tiab (1991) extended Kumar's (1977a,b) formulation for the non-Darcy effect. Civan (2000e) generalized and extended their methodologies and formulations as described in the following. This leads to improved and practical inflow deliverability equations, considering the non-Darcy effects associated with the converging/diverging flow around wellbores.

Here, the leaky-tank model by Civan (2000e) is described. Its defining equations and the conditions of solutions are derived for generalized applications, irrespective of the type of the reservoir fluid, which may be classified as incompressible, slightly compressible, or compressible. The assumptions and their implications are expressed quantitatively. The potential effects of the non-Darcy flow and partial water-drive conditions are also considered by the application of Forchheimer's equation of motion. The leaky-tank model provides a valuable tool for the determination of reservoir formation permeability and thickness, and the strength of partial water drive from deliverability test data.

Consider the cylindrical leaky-tank model shown schematically in Figure 7.1. This model is intended to approximate the drainage area of a well completed in a reservoir, undergoing production by a partial water-drive mechanism. Although the following derivation is carried out in the radial coordinate, the results can be readily transformed to linear and elliptic flow conditions. The porosity, ϕ, permeability, K, and thickness, h, of the reservoir formation are assumed constant. A partially acting aquifer surrounding the reservoir is considered. A piston (or unit mobility ration) displacement of the reservoir fluid by the incoming water from the surrounding aquifer is assumed. The reservoir formation and fluid are assumed isothermal. The reservoir is producing at a constant terminal rate. The shrinkage of the radius of the outer reservoir boundary due to water influx is neglected because it is much greater compared with the wellbore radius.

The formulations are carried out according to Civan (2000e) in terms of the quantities expressed in mass units to avoid complications involving the varying fluid properties and to maintain generality irrespective of the fluid types.

The bulk cylindrical area normal to flow at a radius r from the well centerline is given by

$$A = 2\pi rh. \tag{7.11}$$

The volumetric and mass flow rates can be expressed, respectively, as

$$q = Au \tag{7.12}$$

and

$$w = \rho q, \tag{7.13}$$

where ρ and u are the density and volumetric flux of the fluid.

Frequently, it is convenient to express the flow rates in terms of the volumes expressed at appropriate base conditions, such as the standard, usually taken as

14.7 psia (1 atm) and 60°F (15.5°C) in the petroleum industry, or appropriate reference temperature and pressure conditions, denoted here by the subscript b. Hence, Eqs. (7.12) and (7.13) can be written in alternative forms, respectively, as

$$q_b = Au_b \tag{7.14}$$

and

$$w = \rho_b q_b. \tag{7.15}$$

Applying Eqs. (7.12)–(7.15), the following derivations can be readily expressed in conventional terms, that is, in pressure and volumetric flow.

The mass balance of the fluid for the cylindrical leaky-tank model shown in Figure 7.1 is given by

$$\frac{d}{dt}\left[\pi\left(r_e^2 - r_w^2\right)h\phi\bar{\rho}\right] = w_e - w_w, \tag{7.16}$$

in which t is time, r_w and r_e denote the wellbore and drainage area radii, and $\bar{\rho}$ is the average density of the fluid in the reservoir. w_e denotes the mass flow rate of the reservoir fluid displaced due to water influx at the outer reservoir boundary, and w_w denotes the constant terminal mass production rate of the reservoir fluid at the wellbore.

Kumar (1977a) defined the ratio of the reservoir influx and efflux rates as the water-drive strength factor by

$$f = w_e/w_w = q_{eb}/q_{wb}, \tag{7.17}$$

where q_{eb} and q_{wb} are the volumetric water influx and well production flow rates expressed at the base conditions.

Although the radius of the leaky boundary shrinks during water influx, that is, $r_e = r_e(t)$, the variation of the outer radius is negligible at the actual field conditions. Also, note that $r_w \ll r_e$, and h and ϕ are constants. Consequently, substituting Eq. (7.17) into Eq. (7.16) leads to the following simplified mass balance equation:

$$\pi r_e^2 h\phi\frac{d\bar{\rho}}{dt} = -(1-f)w_w, \tag{7.18}$$

subject to the initial condition given by

$$p = p_i, \quad \bar{\rho} = \rho(p_i), \quad t = 0. \tag{7.19}$$

Therefore, the analytic solution of Eqs. (7.18) and (7.19) yields

$$t = \frac{\pi r_e^2 h\phi}{(1-f)w_w}[\bar{\rho}(p_i) - \bar{\rho}(\bar{p})], \tag{7.20}$$

where $\bar{p}$ is the average reservoir fluid pressure in the leaky-tank model given by (Dake, 1978)

$$\bar{p} = \int_{r_w}^{r_e} p2\pi rh\phi \, dr \Big/ \left[\pi\left(r_e^2 - r_w^2\right)h\phi\right]. \tag{7.21}$$

The equation of continuity in the radial coordinate is given by

$$\frac{\partial(\phi\rho)}{\partial t} + \frac{1}{r}\frac{\partial(r\rho u)}{\partial r} = 0. \tag{7.22}$$

ϕ is constant and substituting Eqs. (7.11)–(7.13) into Eq. (7.22) results in

$$2\pi h\phi r\frac{\partial\rho}{\partial t} + \frac{\partial w}{\partial r} = 0. \tag{7.23}$$

This equation can be solved subject to the following conditions.

For all practical purposes, the reservoir fluid conditions in a very large reservoir can be assumed stabilized under the influence of the outer boundary conditions of the reservoir, if the reservoir has been producing for a sufficiently long time (Dake, 1978). Therefore, the stabilized state condition with respect to time can be approximated by Eq. (7.18) as

$$\frac{\partial\rho}{\partial t} \cong \frac{d\bar{\rho}}{dt} = -\frac{(1-f)w_w}{\pi r_e^2 h\phi} = \text{constant}, \tag{7.24}$$

where (Kumar, 1977b)

$$f = 0 \text{ for semi-steady-state (closed reservoir,} \atop \text{zero water influx) condition,} \tag{7.25}$$

$$0 < f < 1 \text{ for intermediate-state (leaky external} \atop \text{boundary, partial water influx) condition,} \tag{7.26}$$

$$f = 1 \text{ for steady-state (constant external boundary} \atop \text{pressure, full water influx) condition, and} \tag{7.27}$$

$$f > 1 \text{ for excess fluid injection condition.} \tag{7.28}$$

The conditions with respect to the radial distance or the boundary conditions at the wellbore and external radii are given, respectively, by

$$p = p_w, w = -w_w, r = r_w \tag{7.29}$$

and

$$p = p_e, w = -w_e = -fw_w, r = r_e. \tag{7.30}$$

Eliminating $\partial\rho/\partial t$ between Eqs. (7.23) and (7.24) yields

$$\frac{dw}{dr} = (1-f)w_w\frac{2r}{r_e^2}, \tag{7.31}$$

which, upon integration, subject to the leaky-boundary condition given by Eq. (7.30), yields

$$w = -w_w\left[1-(1-f)(r/r_e)^2\right]. \tag{7.32}$$

The minus sign in Eqs. (7.19), (7.20), and (7.22) indicates that the flow direction is opposite of the radial coordinate direction.

The equation of motion in the radial coordinate is given by (Forchheimer, 1901, 1914)

$$-\frac{\partial p}{\partial r} = \frac{\mu}{K}u + \beta\rho|u|u. \tag{7.33}$$

Following Civan and Evans (1998), consider a pseudopressure function defined by

$$m(p) = \int_{p_b}^{p} \frac{\rho}{\mu} dp. \tag{7.34}$$

Hence, substituting Eqs. (7.11)–(7.15) and (7.34) into Eq. (7.33) yields an alternative form of the Forchheimer equation as

$$-\frac{dm(p)}{dr} = \frac{w}{2\pi rh}\left[\frac{1}{K} + \left(\frac{\beta}{\mu}\right)\frac{|w|}{2\pi rh}\right]. \tag{7.35}$$

For all practical purposes, the effect of pressure on the fluid viscosity is small. Thus, invoking Eq. (7.32) and the average fluid viscosity defined by (Civan and Evans, 1998)

$$\mu \approx \bar{\mu} = \int_{p_1}^{p_2} \mu\,dp / (p_2 - p_1), \tag{7.36}$$

where $p_1 < p < p_2$ is some representative range of pressure variation of the reservoir fluid, such as $p_1 = p_w$ and $p_2 = p_e$; Eq. (7.35) can be solved, subject to the wellbore boundary condition given by Eq. (7.29), to obtain an analytic solution as

$$
\begin{aligned}
m(p) - m(p_w) &= \frac{w_w}{2\pi hK}\left[\ell n\left(\frac{r}{r_w}\right) - (1-f)\frac{r^2 - r_w^2}{2r_e^2}\right] \\
&\quad + \frac{\beta}{\bar{\mu}}\left(\frac{w_w}{2\pi h}\right)^2\left[-\frac{1}{r} + \frac{1}{r_w} - 2(1-f)\frac{r - r_w}{r_e^2} + (1-f)^2\frac{r^3 - r_w^3}{3r_e^4}\right].
\end{aligned}
\tag{7.37}
$$

Applying Eq. (7.30) and rearranging Eq. (7.37) leads to the following linear form expressing the reciprocal pseudo-productivity index as a linear function of the production rate:

$$\frac{m(p_e) - m(p_w)}{w_w} = a + bw_w, \tag{7.38}$$

where, considering $r_w \ll r_e$,

$$a \cong \left[\ell n(r_e/r_w) - (1/2)(1-f)\right]/(2\pi hK) \tag{7.39}$$

and

$$b \cong \frac{\beta}{r_w\bar{\mu}(2\pi h)^2}. \tag{7.40}$$

Thus, f and (Kh) can be determined from a straight-line plot of

$$\frac{[m(p_e) - m(p_w)]}{w_w} \text{ versus } w_w.$$

The preceding formulations are general, irrespective of the fluid types. The present formulations can be readily applied for incompressible, slightly compressible, and compressible (gas) fluids by substituting the following expressions for the fluid density, respectively:

$$\rho = \rho_o = \text{constant}, \tag{7.41}$$

$$\rho = \rho_b \exp[c(p - p_b)], \tag{7.42}$$

and

$$\rho = \frac{Mp}{ZRT}, \tag{7.43}$$

where c is the compressibility coefficient, M is the molecular weight of the gas, Z is the real gas deviation factor, T is the absolute temperature, and R is the universal gas constant.

The data considered for the example are $h = 10\,\text{m}$, $K = 9.0 \times 10^{-15}\,\text{m}^2$, $\beta = 5.0 \times 10^{10}\,\text{m}^{-1}$, $\phi = 0.20$, $r_w = 0.0508\,\text{m}$, $r_e = 9100\,\text{m}$, $T = 410°\text{K}$, $p_i = 4.83 \times 10^7\,\text{Pa}$ (or kg/m·s^2), and $\bar{\mu} = 1.25 \times 10^{-5}\,\text{kg/m·s}$. The reservoir fluid is a 0.65-gravity natural gas. Using these data, the typical dimensionless average fluid density versus dimensionless time, dimensionless mass flow rate versus dimensionless radial distance, and the reciprocal pseudo-productivity index versus mass production rate trends for $f = 0$, 0.5, 1.0, and 1.5 are plotted in Figures 7.2–7.4, respectively.

When the actual field measurements of flowing bottom-hole pressure, p_w, versus production rate, q_{wb}, of a well and the fluid properties data are available, a straight-line plot of the reciprocal pseudo-productivity index versus mass production rate similar to Figure 7.4 can be constructed by means of the least squares linear regression method. The values of the a and b parameters of Eq. (7.38) are determined

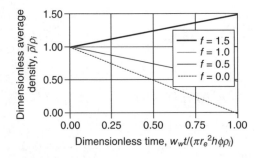

Figure 7.2   Dimensionless average density versus dimensionless time (after Civan, 2000e; reprinted by permission from Elsevier).

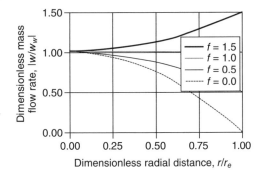

Figure 7.3 Dimensionless mass flow rate versus dimensionless radial distance (after Civan, 2000e; reprinted by permission from Elsevier).

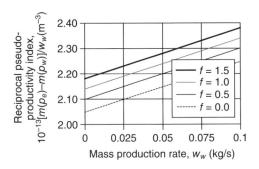

Figure 7.4 Reciprocal pseudo-productivity index versus mass production rate (after Civan, 2000e; reprinted by permission from Elsevier).

by the intercept and slope of this line, respectively. These values are substituted into Eqs. (7.39) and (7.40), which are then expressed for h and Kh as

$$Kh = \left[\ln\left(r_e/r_w\right) - (1/2)(1-f)\right]/(2\pi a) \tag{7.44}$$

and

$$h = \frac{1}{2\pi}\sqrt{\frac{\beta}{r_w \bar{\mu} b}}. \tag{7.45}$$

Thus, eliminating h between Eqs. (7.44) and (7.45) yields

$$f = 1 + 2\left[Ka\sqrt{\frac{\beta}{r_w \bar{\mu} b}} - \ln\left(\frac{r_e}{r_w}\right)\right]. \tag{7.46}$$

The β value can be estimated using an appropriate empirical correlation, such as by Liu et al. (1995) (See Chapter 5).

Note that Eqs. (7.44)–(7.46) require an implicit solution approach for the determination of the h, Kh, and f values, because the Liu et al. (1995) or other available β correlations are given as functions of permeability K, as well as other parameters, such as porosity ϕ and tortuosity τ, whose values should be acquired by appropriate means.

7.4 COUPLING TWO-PHASE MASS AND MOMENTUM BALANCE EQUATIONS FOR IMMISCIBLE DISPLACEMENT

Formulations of the governing equations and boundary conditions for immiscible displacement in porous media are presented according to Civan (1996b). Neglecting the capillary pressure effect, Buckley and Leverett (1942) obtained a simplified formulation. This equation can be solved analytically using Welge's (1952) method in an isotropic homogeneous media for one-dimensional linear, radial, and spherical flow. The formulation results in a convection-dispersion type of equation when the capillary pressure effect is included.

The formulations are applied to water/oil systems, and the generalized boundary conditions are derived. Convenient formulations are presented by transforming the governing equations by means of an integral transformation of the function and transformation of the coordinates. The fractional flow and end-point mobility ratio formulations are also presented for specific applications.

7.4.1 Macroscopic Equation of Continuity

The mass balance of phase j can be written as (Civan, 2002e)

$$\frac{\partial}{\partial t}(\varepsilon_j \rho_j) + \nabla \cdot (\rho_j \mathbf{u}_j) - \nabla \cdot (\varepsilon_j \mathbf{D}_j \cdot \nabla \rho_j) + r_j = 0, \qquad (7.47)$$

where $\mathbf{D}$ denotes the hydraulic diffusivity tensor and ε_j is the volume fraction of the flowing phase j in porous media given by

$$\varepsilon_j = \phi(S_j - S_{ji}). \qquad (7.48)$$

ϕ is the porosity and S_j is the saturation of phase j and S_{ji} is the immobile phase j saturation. r_j denotes the loss of fluid mass per unit bulk volume of porous media per unit time.

Adding Eq. (7.47) over all the fluid phases yields the total mass balance equation as

$$\frac{\partial(\phi\rho)}{\partial t} + \nabla \cdot (\rho \mathbf{u}) - \nabla \cdot (\phi \mathbf{D} \cdot \nabla \rho) + r = 0, \qquad (7.49)$$

in which the following average quantities have been used:

$$\rho = \sum_j S_j \rho_j, \qquad (7.50)$$

$$\rho \mathbf{u} = \sum_j \rho_j \mathbf{u}_j, \qquad (7.51)$$

$$\mathbf{D} \cdot \nabla \rho = \sum_j S_j \mathbf{D}_j \cdot \nabla \rho_j, \qquad (7.52)$$

and

$$r = \sum_j r_j. \qquad (7.53)$$

For incompressible fluids, invoking Eq. (7.49) into Eq. (7.47) yields a volumetric balance equation as

$$\frac{\partial(\phi S_j)}{\partial t} + \nabla \cdot \mathbf{u}_j + q_j = 0, \tag{7.54}$$

where

$$q_j = r_j / \rho_j. \tag{7.55}$$

Thus, adding Eq. (7.54) over all the fluid phases results in the following total volumetric balance equation:

$$\frac{\partial \phi}{\partial t} + \nabla \cdot \mathbf{u} + q = 0, \tag{7.56}$$

in which

$$\mathbf{u} = \sum_j \mathbf{u}_j \tag{7.57}$$

and

$$q = \sum_j q_j. \tag{7.58}$$

7.4.2 Application to Oil/Water Systems

The macroscopic equation of continuity for the water phase can be obtained from Eqs. (7.47) and (7.48) as

$$\frac{\partial}{\partial t}(\phi S_w \rho_w) + \nabla \cdot (\rho_w \mathbf{u}_w) - \nabla \cdot (\phi S_w \mathbf{D}_w \cdot \nabla \rho_w) + r_w = 0. \tag{7.59}$$

In Eq. (7.59), t is time, ∇ is the divergence operator, ϕ is the porosity, S_w is the water saturation in the porous media, ρ_w is the density of water, $\mathbf{u}_w$ is the volumetric flux of the water phase, $\mathbf{D}_w$ is the coefficient of hydraulic dispersion due to spatial variations in porous media, and r_w is the mass rate of the water lost.

The volumetric fluxes of the water and oil phases through porous media are given, respectively, by

$$\mathbf{u}_w = -\frac{k_{r_w}}{\mu_w} \mathbf{K} \cdot (\nabla p_w + \rho_w g \nabla z) \tag{7.60}$$

and

$$\mathbf{u}_o = -\frac{k_{r_o}}{\mu_o} \mathbf{K} \cdot (\nabla p_o + \rho_o g \nabla z) = -\frac{k_{r_o}}{\mu_o} \mathbf{K} \cdot (\nabla p_w + \nabla p_c + \rho_o g \nabla z), \tag{7.61}$$

because the capillary pressure is defined as

$$p_c = p_o - p_w. \tag{7.62}$$

In Eqs. (7.60) and (7.61), k_{rw} and k_{ro}, μ_w and μ_o, and ρ_w and ρ_o denote the relative permeability, viscosity, density, and pressure of the water and oil phases, respectively. $\mathbf{K}$, g, and z are the permeability tensor, the gravitational acceleration, and the distance in the gravity direction.

The total volumetric flux is given by

$$\mathbf{u} = \mathbf{u}_w + \mathbf{u}_o. \tag{7.63}$$

Substitution of Eqs. (7.60) and (7.61) into Eq. (7.63) yields

$$\mathbf{u} = -\left(\frac{k_{r_w}}{\mu_w} + \frac{k_{r_o}}{\mu_o}\right)\mathbf{K} \cdot \nabla p_w - \frac{k_{r_o}}{\mu_o}\frac{dp_c}{dS_w}\mathbf{K} \cdot \nabla S_w - \left(\frac{k_{r_w}}{\mu_w}\rho_w + \frac{k_{r_o}}{\mu_o}\rho_o\right)g\mathbf{K} \cdot \nabla z. \tag{7.64}$$

Eliminating ∇p_w between Eqs. (7.60) and (7.64) and rearranging leads to

$$\mathbf{u}_w = F_w\left(\mathbf{u} + \frac{k_{r_o}}{\mu_o}\frac{dp_c}{dS_w}\mathbf{K} \cdot \nabla S_w - \frac{k_{r_o}}{\mu_o}\Delta\rho g\mathbf{K} \cdot \nabla z\right), \tag{7.65}$$

in which

$$\Delta\rho = \rho_w - \rho_o \tag{7.66}$$

and

$$F_w = \left(1 + \frac{k_{r_o}\mu_w}{\mu_o k_{r_w}}\right)^{-1}, \tag{7.67}$$

where M is the mobility ratio given by

$$M = \frac{k_{r_o}\mu_w}{k_{r_w}\mu_o}. \tag{7.68}$$

For formulation of the boundary conditions, the total mass flux of water is defined as the sum of the transport by bulk flow (convection) and hydraulic dispersion according to Eq. (7.59) as

$$\mathbf{m}_w = \rho_w\mathbf{u}_w - \phi S_w\mathbf{D} \cdot \nabla\rho_w. \tag{7.69}$$

Then, the water injection end boundary condition can be written as

$$\rho_w\mathbf{u}_w - \phi S_w\mathbf{D} \cdot \nabla\rho_w = (\rho_w\mathbf{u}_w)_{\text{in}}. \tag{7.70}$$

Note that the right of Eq. (7.70) is prescribed by the conditions of the injected water. Similarly, the outlet (or production) end boundary condition is obtained as

$$\phi S_w\mathbf{D} \cdot \nabla\rho_w = 0. \tag{7.71}$$

For incompressible fluids, Eq. (7.70) simplifies to

$$\mathbf{u}_w = (\mathbf{u}_w)_{\text{in}}, \tag{7.72}$$

and Eq. (7.71) drops out.

7.4.2.1 Pressure and Saturation Formulation
Consider incompressible fluids and formation. The immiscible displacement of oil by water is described by the following differential equations, respectively, by invoking Eq. (7.64) into Eq. (7.56) and Eq. (7.60) into Eq. (7.59):

$$\nabla \cdot \left[\begin{array}{l} \left(\dfrac{k_{rw}}{\mu_w} + \dfrac{k_{ro}}{\mu_o} \right) \mathbf{K} \cdot \nabla p_w + \dfrac{k_{ro}}{\mu_o} \dfrac{dp_c}{dS_w} \mathbf{K} \cdot \nabla S_w \\[2mm] + \left(\dfrac{k_{rw}}{\mu_w} \rho_w + \dfrac{k_{ro}}{\mu_o} \rho_o \right) g \mathbf{K} \cdot \nabla z \end{array} \right] = q. \tag{7.73}$$

and

$$\phi \frac{\partial S_w}{\partial t} - \nabla \cdot \left[\frac{k_{rw}}{\mu_w} \mathbf{K} \cdot (\nabla p_w + \rho_w g \nabla z) \right] + q_w = 0. \tag{7.74}$$

7.4.2.2 Saturation Formulation

Assuming incompressible fluids and substituting Eq. (7.65) into Eq. (7.59) according to Richardson (1961) and Peters et al. (1993) yields

$$\frac{\partial (\phi S_w)}{\partial t} + \nabla \cdot \left[F_w \left(\mathbf{u} + \frac{k_{ro}}{\mu_o} \frac{dp_c}{dS_w} \mathbf{K} \cdot \nabla S_w - \frac{k_{ro}}{\mu_o} \Delta \rho g \mathbf{K} \cdot \nabla z \right) \right] + q_w = 0, \tag{7.75}$$

where q_w is defined according to Eq. (7.55). Thus, substituting Eq. (7.56) into Eq. (7.75) (or eliminating $(\nabla \cdot \mathbf{u})$) yields

$$\phi \frac{\partial S_w}{\partial t} + (S_w - F_w) \frac{\partial \phi}{\partial t} + \frac{dF_w}{dS_w} \mathbf{u} \cdot \nabla S_w + \nabla \cdot \left[F_w \frac{k_{ro}}{\mu_o} \frac{dp_c}{dS_w} \mathbf{K} \cdot \nabla S_w \right]$$
$$- \nabla \cdot \left[F_w \frac{k_{ro}}{\mu_o} \Delta \rho g \mathbf{K} \cdot \nabla z \right] - F_w q + q_w = 0. \tag{7.76}$$

When $q = 0$, $q_w = 0$, and an isotropic and homogeneous porous media is considered, Eq. (7.76) simplifies to Eq. (16.101) given by Richardson (1961). Eq. (7.76) is interesting because it reveals a transient state, convection, dispersion, with source nature of the water phase transport. Eq. (7.76) can also be written in the following compact form:

$$\phi \frac{\partial S_w}{\partial t} + (S_w - F_w) \frac{\partial \phi}{\partial t} + \frac{dF_w}{dS_w} \mathbf{u} \cdot \nabla S_w - \nabla \cdot [D_w (S_w) \mathbf{K} \cdot \nabla S_w]$$
$$- \nabla \cdot [T_w (S_w) \mathbf{K} \cdot \nabla z] - F_w q + q_w = 0, \tag{7.77}$$

which, upon expansion, yields the following alternative form:

$$\phi \frac{\partial S_w}{\partial t} + (S_w - F_w) \frac{\partial \phi}{\partial t} + \left[\frac{dF_w}{dS_w} \mathbf{u} - \frac{dD_w}{dS_w} \cdot (\mathbf{K} \cdot \nabla S_w) - \frac{dT_w}{dS_w} (\mathbf{K} \cdot \nabla z) \right] \cdot \nabla S_w$$
$$- D_w \nabla \cdot (\mathbf{K} \cdot \nabla S_w) - T_w \nabla \cdot (\mathbf{K} \cdot \nabla z) - F_w q + q_w = 0, \tag{7.78}$$

in which D_w and T_w can be referred to as the capillary and gravitational dispersion coefficients defined, respectively, as

$$D_w (S_w) = F_w \frac{k_{ro}}{\mu_o} \left(-\frac{dp_c}{dS_w} \right) \tag{7.79}$$

and

$$T_w (S_w) = F_w \frac{k_{ro}}{\mu_o} \Delta \rho g. \tag{7.80}$$

7.4.2.3 Boundary Conditions   The conditions at permeable boundaries can be formulated from the continuity of the water phase and total phase fluxes across permeable boundaries, respectively, expressed by

$$(\mathbf{u}_w \cdot \mathbf{n})^- = (\mathbf{u}_w \cdot \mathbf{n})^+ \tag{7.81}$$

and

$$(\mathbf{u} \cdot \mathbf{n})^- = (\mathbf{u} \cdot \mathbf{n})^+. \tag{7.82}$$

Note that − and + denote the inside and outside of the permeable boundary. The outside conditions are assumed to be prescribed. Therefore, by substituting Eqs. (7.63) and (7.65) into Eqs. (7.81) and (7.82), the following general boundary condition is obtained:

$$\left[\left(\frac{dp_c}{dS_w} \mathbf{K} \cdot \nabla S_w - \Delta \rho g \mathbf{K} \cdot \nabla z \right) \cdot \mathbf{n} \right]^- = \left(\frac{\mu_w}{k_{r_w}} \right)^- (\mathbf{u}_w \cdot \mathbf{n})^+ - \left(\frac{\mu_o}{k_{r_o}} \right)^- (\mathbf{u}_o \cdot \mathbf{n})^+. \tag{7.83}$$

7.4.3 One-Dimensional Linear Displacement

For illustration purposes, consider the classical problem of one-dimensional flow of oil/water immiscible displacement in a homogeneous core (Civan, 1996b). Assume that the core is initially saturated with oil up to the connate water saturation and the flow begins with injecting water at a constant rate to displace oil.

For this case, Eq. (7.76) simplifies as

$$\phi \frac{\partial S_w}{\partial t} + \frac{dF_w}{dS_w} u \frac{\partial S_w}{\partial x} + \frac{\partial}{\partial x} \left(F_w \frac{k_{r_o}}{\mu_o} \frac{dp_c}{dS_w} K \frac{\partial S_w}{\partial x} \right) = 0, 0 \leq x \leq L, t > 0. \tag{7.84}$$

The initial condition is given by

$$S_w = S_{wc}, 0 \leq x \leq L, t = 0. \tag{7.85}$$

At the inlet boundary, because only water is injected, $(u_o)^+ = 0$ and $(u_w)^+ \equiv (u_w)_{in}$. Thus, Eq. (7.83) simplifies to

$$\frac{dp_c}{dS_w} \frac{\partial S_w}{\partial x} = \frac{\mu_w (u_w)_{in}}{Kk_{r_w}}, x = 0, t > 0. \tag{7.86}$$

At the outlet, three separate conditions need to be formulated.

1. Until the water break through time, t^*, $(u_w)^+ = 0$ and the overall volume balance dictates for incompressible fluids that $(u_o)^+ = (u_o)_{out} = (u_w)_{in}$. Thus,

$$\frac{dp_c}{dS_w} \frac{\partial S_w}{\partial x} = -\frac{\mu_o (u_w)_{in}}{Kk_{r_o}}, x = L, 0 < t \leq t^*. \tag{7.87}$$

Note that Eqs. (7.84), (7.86), and (7.87) can be shown to be the same as the equations of Douglas et al. (1958).

2. Between the breakthrough and infinite water injection, both $(u_w)^+ = (u_w)_{out} \neq 0$ and $(u_o)^+ = (u_o)_{out} \neq 0$. Thus, Eq. (7.83) is written as

$$\frac{dp_c}{dS_w}\frac{\partial S_w}{\partial x} = \frac{\mu_w (u_w)_{out}}{Kk_{r_w}} - \frac{\mu_o (u_o)_{out}}{Kk_{r_o}}, \; x = L, t^* < t < \infty. \tag{7.88}$$

In addition, the overall balance for the core dictates that

$$(u_w)_{in} = (u_w)_{out} + (u_o)_{out}. \tag{7.89}$$

3. During the infinite water injection $(u_o)^+ = (u_o)_{out} = 0$ and $(u_w)^+ = (u_w)_{out} = (u_w)_{in}$. Thus, Eq. (7.83) simplifies to

$$\frac{dp_c}{dS_w}\frac{\partial S_w}{\partial x} = \frac{\mu_w (u_w)_{in}}{Kk_{r_w}}, \; x = L, t \to \infty. \tag{7.90}$$

At later times, Eqs. (7.88) and (7.90) can be replaced by the following simplified boundary condition:

$$S_w = 1 - S_{or}. \tag{7.91}$$

7.4.4 Numerical Solution of Incompressible Two-Phase Fluid Displacement Including the Capillary Pressure Effect

For the description of a two-phase flow in porous core plugs, Penuela and Civan (2001) considered that the core, water, and oil properties remain constant; the core is initially saturated with water; the flow begins by injecting oil at a constant rate to displace the water; the pressure at the production outlet face of the core is constant; and the flow is one dimensional and horizontal. Under these conditions, the immiscible displacement in a core plug can be described by the following equations (Richardson, 1961; Dullien, 1992; Civan, 1996b):

$$\phi\frac{\partial S_w}{\partial t} + \frac{dF_w}{dS_w}u\frac{\partial S_w}{\partial x} + \frac{\partial}{\partial x}\left[F_w \frac{k_{ro}}{\mu_o}\frac{dP_c}{dS_w}K\frac{\partial S_w}{\partial x} \right] = 0, 0 \le x \le L, t > 0, \tag{7.92}$$

subject to the initial condition given by

$$S_w = S_w^*(x), 0 \le x \le L, t = 0, \tag{7.93}$$

and the injection and production face boundary conditions given, respectively, by

$$\frac{\partial S_w}{\partial x} = -\frac{\mu_o (u_o)_{inj}}{Kk_{ro}}\left[\frac{dP_c}{dS_w} \right]^{-1}, x = 0, t > 0 \tag{7.94}$$

and

$$S_w = S_w\big|_{P_c=0}, x = L, t > 0. \tag{7.95}$$

The zero capillary pressure and zero gravity fractional water function is given by Eq. (7.67).

In Eqs. (7.92)–(7.95), x and t denote the distance and time; ϕ and K are the porosity and permeability of the core plug; u is the total volumetric flux of the pore fluids, which is equal to the injection fluid volumetric flux under the considered conditions; L is the length of the core plug; μ_o and k_{ro} denote the viscosity and relative permeability of the oil; μ_w and k_{rw} are the viscosity and relative permeability of

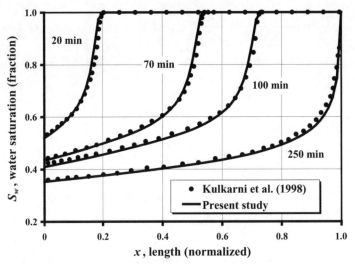

Figure 7.5 Comparison of the numerical solutions obtained by Kulkarni et al. (1998) and Penuela and Civan (2001) (after Penuela and Civan, 2001; © 2001 AIChE, reprinted by permission from the American Institute of Chemical Engineers).

the water; P_c denotes the capillary pressure; S_w is the water phase saturation in the core; $S_w^*(x)$ represents the initial water saturation distribution in the core; and $S_w|_{P_c=0}$ defines the water saturation corresponding to zero capillary pressure. For the problem considered by Kulkarni et al. (1998), $S_w^*(x) = 1$ and $S_w|_{P_c=0} = 1 - S_{or}$, where the residual oil saturation is taken as zero, $S_{or} = 0$. $(u_o)_{inj}$ is the volumetric flux of the oil injected into the core plug.

Figure 7.5 shows a comparison of the fully implicit numerical solutions of Eqs. (7.92)–(7.95) obtained by Penuela and Civan (2001), using a first-order accurate temporal and second-order accurate spatial finite difference approximations, with the solutions given by Kulkarni et al. (1998). As can be seen, the Penuela and Civan (2001) solutions and those given by Kulkarni et al. (1998) are very close.

7.4.5 Fractional Flow Formulation

Although it is not necessary, it has been customary to express the water phase volumetric flux, $\mathbf{u}_w$, in terms of the fractional flow of water, f_w. For one-dimensional problems, however, this formulation provides some convenience. Civan (1994b) defined a scalar fractional water, f_w, according to

$$\mathbf{u}_w = f_w \mathbf{u}. \tag{7.96}$$

Thus, equating Eqs. (7.65) and (7.96) leads to the following expression:

$$f_w = F_w \left(1 + \frac{k_{r_o}}{u^2 \mu_o} \frac{dp_c}{dS_w} \mathbf{u} \cdot \mathbf{K} \cdot \nabla S_w - \frac{k_{r_o}}{u^2 \mu_o} \Delta\rho g \mathbf{u} \cdot \mathbf{K} \cdot \nabla z \right), \tag{7.97}$$

in which

$$u = |\mathbf{u}| = \left(u_x^2 + u_y^2 + u_z^2 \right)^{1/2},$$

(7.98)

where u_x, u_y, and u_z denote the volumetric flux components in the x-, y-, and z-direction. Consequently, assuming incompressible fluids and invoking Eqs. (7.56), (7.96), and (7.98) into Eq. (7.54) for the water phase yields

$$\phi \frac{\partial S_w}{\partial t} + \mathbf{u} \cdot \nabla f_w - q f_w + q_w + S_w \frac{\partial \phi}{\partial t} = 0.$$

(7.99)

7.4.6 The Buckley–Leverett Analytic Solution Neglecting the Capillary Pressure Effect

For this purpose, the Buckley and Leverett (1942) equation expressing the velocity of phase 1 at a given saturation can be expressed as the following (Marle, 1981). Consider

$$\phi \frac{\partial S_1}{\partial t} + u \frac{\partial f_1}{\partial x} = 0,$$

(7.100)

where

$$S_1 = S_1 (x, t).$$

(7.101)

Thus, for a prescribed value of saturation S_1,

$$dS_1 = 0 = \frac{\partial S_1}{\partial t} dt + \frac{\partial f_1}{\partial x} dx,$$

(7.102)

we obtain

$$\frac{\partial S_1}{\partial t} = -\frac{\partial f_1}{\partial x} \left(\frac{dx}{dt} \right)_{S_1}.$$

(7.103)

Consequently, substituting Eq. (7.103) into Eq. (7.100) yields

$$(dx/dt)_{S_1} = (u/\phi)(df_1/dS_1)_{S_1}.$$

(7.104)

However, the superficial flow u and cumulative volume Q are related by

$$u = \frac{1}{A} \frac{dQ}{dt},$$

(7.105)

where A denotes the cross-sectional area of porous material normal to flow. Substituting Eq. (7.105), Eq. (7.104) becomes

$$dx_{S_1} = (df_1/dS_1)_{S_1} dQ/(A\phi).$$

(7.106)

Because $f_1 = f_1(S_1)$ only, then $(df_1/dS_1)_{S_1}$ is a fixed value. Thus, integrating and applying the initial condition that

$$x_{S_1} = 0, \quad Q = 0, \quad t = 0.$$

(7.107)

Eq. (7.106) leads to an expression for the location of point with a given saturation value as

$$x_{S_1} = \frac{Q}{A\phi} \left(\frac{df_1}{dS_1} \right)_{S_1}. \tag{7.108}$$

This equation is used to generate a semianalytic solution, which provides the locations of prescribed fluid saturations for various cumulative volumes of injected fluid.

7.4.7 Convenient Formulation

As stated by Civan and Sliepcevich (1985a), the transformations of transport equations, in which "the variable properties do not appear explicitly," may lead to numerical solutions of "better accuracy with less computational effort." Therefore, Eq. (7.77) will be converted to a form involving direct applications of the various mathematical operations to the function.

For this purpose, first, define an integral transformation according to Douglas et al. (1958) as

$$r(S_w) = \frac{1}{Z} \int_{S_{wc}}^{S_w} D_w (S_w) \, dS_w, \tag{7.109}$$

where S_{wc} and S_{or} denote the connate water and residual oil saturations, respectively, and

$$Z = \int_{S_{wc}}^{(1-S_{or})} D_w (S_w) \, dS_w. \tag{7.110}$$

Thus, applying Eq. (7.109), Eq. (7.77) can be transformed into

$$\frac{\phi}{D_w} \frac{\partial r}{\partial t} + \frac{1}{Z} \left[\frac{dF_w}{dr} \mathbf{u} + \frac{dT_w}{dr} (\mathbf{K} \cdot \nabla z) \right] \cdot \nabla r + \nabla \cdot (\mathbf{K} \cdot \nabla r)$$

$$+ \frac{T_w}{Z} \nabla \cdot (\mathbf{K} \cdot \nabla z) + \frac{1}{Z} \left(\frac{r_w}{\rho_w} - F_w \frac{r}{\rho} \right) + (S_w - F_w) \frac{\partial \phi}{\partial t} = 0. \tag{7.111}$$

Next, consider coordinate transformation from (x, y, z) to (ξ, η, ζ) such that the mapping transformation is given by

$$\begin{bmatrix} \dfrac{\partial r}{\partial \xi} \\[2mm] \dfrac{\partial r}{\partial \eta} \\[2mm] \dfrac{\partial r}{\partial \zeta} \end{bmatrix} = \begin{bmatrix} \dfrac{\partial x}{\partial \xi} & \dfrac{\partial y}{\partial \xi} & \dfrac{\partial z}{\partial \xi} \\[2mm] \dfrac{\partial x}{\partial \eta} & \dfrac{\partial y}{\partial \eta} & \dfrac{\partial z}{\partial \eta} \\[2mm] \dfrac{\partial x}{\partial \zeta} & \dfrac{\partial y}{\partial \zeta} & \dfrac{\partial z}{\partial \zeta} \end{bmatrix} \begin{bmatrix} \dfrac{\partial r}{\partial x} \\[2mm] \dfrac{\partial r}{\partial y} \\[2mm] \dfrac{\partial r}{\partial z} \end{bmatrix} \tag{7.112}$$

or simply by

$$\tilde{\nabla} r = \mathbf{J} \cdot \nabla r, \tag{7.113}$$

where $\mathbf{J}$ denotes the Jacobian matrix as shown in Eq. (7.112) and ∇ and $\tilde{\nabla}$ are the gradient operators in the original and the transformed coordinates, respectively. To make Eq. (7.111) free of properties in the various operators, define

$$\tilde{\nabla}r = \mathbf{K} \cdot \nabla r. \tag{7.114}$$

Thus, comparing Eqs. (7.113) and (7.114) leads to the following Jacobian matrix, which can be used to develop the mapping functions for the desired transformation:

$$\mathbf{J} = \mathbf{K}. \tag{7.115}$$

This method of transformation can be readily extended to other coordinate systems. Consequently, applying Eq. (7.114), Eq. (7.111) is transformed to the following convenient form:

$$\frac{\phi}{D_w}\frac{\partial r}{\partial t} + \frac{1}{Z}\left[\frac{dF_w}{dr}\mathbf{u} + \frac{dT_w}{dr}\tilde{\nabla}z\right] \cdot \mathbf{K}^{-1} \cdot \tilde{\nabla}r + \mathbf{K}^{-1} \cdot \tilde{\nabla} \cdot \tilde{\nabla}r$$
$$+ \frac{T_w}{Z}\mathbf{K}^{-1} \cdot \tilde{\nabla} \cdot \tilde{\nabla}z + \frac{1}{Z}(q_w - F_w q) + (S_w - F_w)\frac{\partial\phi}{\partial t} = 0. \tag{7.116}$$

For example, for horizontal one-dimensional flow in an isotropic homogeneous core, with no source terms, Eq. (7.116) simplifies to

$$\frac{\phi}{D_w}\frac{\partial r}{\partial t} + \frac{1}{Z}\frac{dF_w}{dr}uK^{-1}\frac{\partial r}{\partial \xi} + K^{-1}\frac{\partial^2 r}{\partial \xi^2} = 0. \tag{7.117}$$

Eq. (7.117) is more convenient for numerical discretization than Eq. (7.84) because it involves simple direct derivatives of the function.

7.4.8 Unit End-Point Mobility Ratio Formulation

Inferred by Collins (1961), Craig (1980), and Dake (1978), the following simplifying assumptions are considered for the unit mobility ratio formulation (Civan, 1993, 1996a,b). The fluid properties (density and viscosity) are assumed constant. The capillary pressure effect is neglected:

$$p_c = 0, \; p_o = p_w = p. \tag{7.118}$$

The relative permeabilities are approximated by linear functions of the normalized saturations (Yokoyama and Lake, 1981):

$$k_{rw} = k'_{rw}\hat{S}_w \tag{7.119}$$

and

$$k_{ro} = k'_{ro}\left(1 - \hat{S}_w\right), \tag{7.120}$$

where the normalized mobile water saturation is given by

$$\hat{S}_w = \frac{S_w - S_{wc}}{1 - S_{wc} - S_{or}}. \tag{7.121}$$

The fluid densities are replaced by the arithmetic average value as

$$\rho_w = \rho_o = \rho_{avg},$$ (7.122)

where

$$\rho_{avg} = (\rho_w + \rho_o)/2.$$ (7.123)

The end-point relative mobilities are replaced by the arithmetic average value as

$$\frac{k'_{rw}}{\mu_w} = \frac{k'_{ro}}{\mu_o} = \left(\frac{k_r}{\mu}\right)_{avg},$$ (7.124)

where

$$\left(\frac{k_r}{\mu}\right)_{avg} = \frac{1}{2}\left(\frac{k'_{rw}}{\mu_w} + \frac{k'_{ro}}{\mu_o}\right).$$ (7.125)

A flow potential for incompressible fluids is defined as

$$\Psi = p + \rho g z.$$ (7.126)

Based on these considerations, Eqs. (7.64) and (7.97) simplify, respectively as

$$\mathbf{u} = -(k_r/\mu)\mathbf{K}\cdot\nabla\Psi$$ (7.127)

and

$$f_w = \hat{S}_w.$$ (7.128)

Therefore, substituting Eq. (7.127) into Eq. (7.56) leads to the following flow potential equation:

$$-(k_r/\mu)\nabla\cdot(\mathbf{K}\cdot\nabla\Psi) + q + \partial\phi/\partial t = 0.$$ (7.129)

Substituting Eqs. (7.121) and (7.118) into Eq. (7.99) results in the following water saturation equation:

$$(1 - S_{or} - S_{wc})\phi\frac{\partial\hat{S}_w}{\partial t} + \mathbf{u}\cdot\nabla\hat{S}_w - \hat{S}_w q + q_w + \left[S_{wc} - (S_{wc} + S_{or})\hat{S}_w\right]\partial\phi/\partial t = 0.$$ (7.130)

Note that Eq. (7.129) can be more conveniently solved after transforming into the following form by applying the transformation given by Eq. (7.114)

$$-(k_r/\mu)\mathbf{K}^{-1}\cdot\tilde{\nabla}\cdot\tilde{\nabla}\Psi + q + \partial\phi/\partial t = 0.$$ (7.131)

Note that Eq. (7.131) is solved once if all the parameters are independent of time.

7.4.8.1 *Example 1* This example deals with the time–space numerical solution of the following problem (Civan, 1996a,b):

$$\frac{\partial S}{\partial t} + \frac{u_x}{\phi}\frac{\partial S}{\partial x} + \frac{u_y}{\phi}\frac{\partial S}{\partial y} = 0,$$ (7.132)

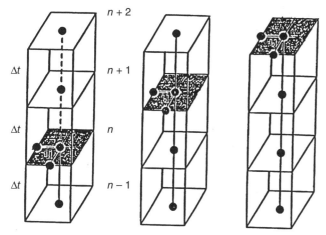

Figure 7.6   Schematic of the computational molecule having three time steps and three space points (after Civan, 1996b; © 1996 SPE, with permission from the Society of Petroleum Engineers).

subject to

$$S = 0, 0 \leq x, y \leq 1, t = 0, \tag{7.133}$$

$$S = 1, x = 0, 0 \leq y \leq 1, t > 0, \tag{7.134}$$

and

$$S = 1, 0 \leq x \leq 1, y = 0, t > 0. \tag{7.135}$$

A finite difference discretization of Eq. (7.132) at n, $n + 1$, and $n + 2$ time levels according to Figure 7.6 yields, respectively:

$$\left(-2S_{ij}^{n-1} - 3S_{ij}^{n} + 6S_{ij}^{n+1} - S_{ij}^{n+2}\right)/(6\Delta t) + \left(u_x/\phi\right)_{ij}^{n}\left(S_{ij}^{n} - S_{i-1,j}^{n}\right)/\Delta x$$
$$+ \left(u_y/\phi\right)_{ij}^{n}\left(S_{ij}^{n} - S_{i,j-1}^{n}\right)/\Delta y = 0, \tag{7.136}$$

$$\left(S_{ij}^{n-1} - 6S_{ij}^{n} + 3S_{ij}^{n+1} + 2S_{ij}^{n+2}\right)/(6\Delta t) + \left(u_x/\phi\right)_{ij}^{n+1}\left(S_{ij}^{n+1} - S_{i-1,j}^{n+1}\right)/\Delta x$$
$$+ \left(u_y/\phi\right)_{ij}^{n+1}\left(S_{ij}^{n+1} - S_{i,j-1}^{n+1}\right)/\Delta y = 0, \tag{7.137}$$

and

$$\left(-2S_{ij}^{n-1} + 9S_{ij}^{n} - 18S_{ij}^{n+1} + 11S_{ij}^{n+2}\right)/(6\Delta t) + \left(u_x/\phi\right)_{ij}^{n+2}\left(S_{ij}^{n+2} - S_{i-1,j}^{n+2}\right)/\Delta x$$
$$+ \left(u_y/\phi\right)_{ij}^{n+2}\left(S_{ij}^{n+2} - S_{i,j-1}^{n+2}\right)/\Delta y = 0. \tag{7.138}$$

Eqs. (7.136)–(7.138) are solved analytically for S_{ij}^{n}, S_{ij}^{n+1}, and S_{ij}^{n+2}. Figure 7.7a–c show typical solutions obtained using a 10×10 equally spaced spatial grid ($\Delta x = \Delta y$) over a square domain and $\Delta t/\Delta x = 0.1$. The solutions using 0.5, 1.0, 5.0, and 10.0 for $\Delta t/\Delta x$ have also been carried out. The accuracy of the numerical solutions degenerates by increasing values of Δt. This is due to the low-order discretization

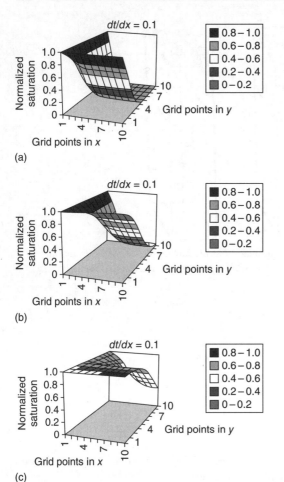

Figure 7.7   (a–c) Saturation profiles after 5, 15, and 30 time steps for $\Delta x / \Delta t = 0.1$ (after Civan, 1996b; © 1996 SPE, with permission from the Society of Petroleum Engineers).

of the spatial derivatives used in this specific example and can be readily alleviated by using higher-order discretization according to Civan (1994a, 2009).

7.4.8.2   Example 2 Consider the following problem describing a waterflooding in a homogeneous, anisotropic, and rectangular domain without any wells inside, as shown. Eq. (7.129) simplifies as (Civan, 1996a)

$$k_x \frac{\partial^2 p}{\partial x^2} + k_y \frac{\partial^2 p}{\partial y^2} = 0. \tag{7.139}$$

Assuming $S_{wc} = S_{or} = 0$, Eq. (7.130) simplifies as

$$\phi \frac{\partial S_w}{\partial t} + v_x \frac{\partial S_w}{\partial x} + v_y \frac{\partial S_w}{\partial y} = 0, \tag{7.140}$$

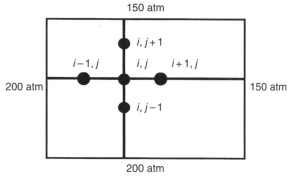

Figure 7.8   Schematic of a rectangular computational region and a general five-point computational molecule (prepared by the author).

where

$$v_x = -\frac{k_x}{\mu}\frac{\partial p}{\partial x} \tag{7.141}$$

and

$$v_y = -\frac{k_y}{\mu}\frac{\partial p}{\partial y}. \tag{7.142}$$

The initial condition is assumed as

$$S_w = 0, \text{ everywhere, } t = 0. \tag{7.143}$$

The boundary pressures are prescribed as shown in Figure 7.8.
The saturation boundary condition is prescribed as

$$S_w = 1 \text{ along the } x \text{ and } y\text{-axes.} \tag{7.144}$$

This problem is solved for the pressure and saturation profiles as described in the following.

The finite difference solution can be obtained by discretizing Eqs. (7.139)–(7.144) respectively as the following:

$$k_x \frac{p_{i+1,j} - 2p_{ij} + p_{i-1,j}}{(\Delta x)^2} + k_y \frac{p_{i,j+1} - 2p_{ij} + p_{i,j-1}}{(\Delta y)^2} = 0, \tag{7.145}$$

$$v_{x_{ij}} = -\frac{k_x}{\mu}\frac{p_{i+1,j} - p_{i-1,j}}{2\Delta x}, \tag{7.146}$$

$$v_{y_{ij}} = -\frac{k_y}{\mu}\frac{p_{i,j+1} - p_{i,j-1}}{2\Delta y}, \tag{7.147}$$

$$v = \sqrt{v_x^2 + v_y^2}, \tag{7.148}$$

and

$$\phi\frac{S_i^n - S_i^{n-1}}{\Delta t} + v_{x_{ij}}\frac{S_{i,j}^n - S_{i-1,j}^n}{\Delta x} + v_{y_{ij}}\frac{S_{i,j}^n - S_{i,j-1}^n}{\Delta y} = 0. \tag{7.149}$$

First, the pressure equation, Eq. (7.145), is solved. Then, the velocity components are calculated using Eqs. (7.146) and (7.147). Finally, Eq. (7.149) is solved explicitly for the saturation.

The sweep efficiency can be calculated by

$$E_D = \int\int_{x\ y} \frac{S_{oi} - S_o}{S_{oi}} \, dydx = \int\int_{x\ y} \frac{S_w - S_{wc}}{1 - S_{wc}} \, dydx \qquad (7.150)$$

where the double integration can be evaluated by an appropriate numerical procedure, such as the trapezoidal rule.

7.5 POTENTIAL FLOW PROBLEMS IN POROUS MEDIA

Irrotational flow in porous media can be formulated conveniently in terms of the flow potential, Ψ, defined by

$$\Psi = \int_{p_o}^{p} \frac{dp}{\rho} + g(z - z_o), \qquad (7.151)$$

and using the fluid flux expressed by Darcy's law, given by

$$\mathbf{u} = -\frac{\rho}{\mu} \mathbf{K} \cdot \nabla\Psi. \qquad (7.152)$$

The potential flow problems can be solved analytically under certain simplifying conditions. Analytic solutions can provide an accurate representation of intricate flow problems, such as preferential flow through fractures, cross-flow across discontinues, and converging/diverging flow around wellbores, in heterogeneous reservoirs involving anisotropy and/or stratification and wells and fractures. Hence, for such complicated flow situations, analytic solutions may be more advantageous than numerical solutions, the accuracy of which often suffers from insufficient levels of resolution of numerical grid systems (Shirman and Wojtanowicz, 1996). Frequently, the inherent grid point orientation effect and numerical dispersion and instability problems associated with numerical solution methods for differential equations render inaccurate numerical solutions. Under these conditions, well-conditioned analytic solutions with accuracies limited by simplifying assumptions may be favored over numerical solutions with inherent difficulties.

7.5.1 Principle of Superposition

The superposition principle expresses the solution of linear equations as a weighted linear sum of their special solutions. For example, the following Laplace equation is a linear differential equation:

$$\nabla^2\Psi = 0. \qquad (7.153)$$

If N special solutions denoted by $\Psi_i : i = 1, 2, ..., N$ can be found for this equation, then each will satisfy Eq. (7.153). Thus,

$$\nabla^2 \Psi_i = 0 : i = 1, 2, ..., N. \tag{7.154}$$

Consequently, multiplying Eq. (7.154) by some constant weighting coefficients $C_i : i = 1, 2, ..., N$ and then adding yields

$$\nabla^2 \sum_{i=1}^{N} C_i \Psi_i = 0. \tag{7.155}$$

Therefore, by comparing Eqs. (7.153) and (7.155), the solution of Eq. (7.153) can be expressed as a weighted linear sum of the special solutions, $\Psi_i : i = 1, 2, ..., N$, as follows:

$$\Psi = \sum_{i=1}^{N} C_i \Psi_i. \tag{7.156}$$

The weighting coefficients $C_i : i = 1, 2, ..., N$ are determined in a manner to satisfy the prescribed conditions of specific problems, such as the consistency and compatibility conditions and the source/sink conditions.

As an example, consider the three-dimensional potential flow involving multiple-point sinks in an infinite size homogeneous porous media. When only a single well i producing at a rate of q_i is present, the flow field around the well is spherical and the solution of the potential flow problem is given by the integration of Darcy's law in radial distance for constant flow rate q_i:

$$\Psi_i (r) = \Psi_\infty - \frac{q_i \mu}{4 \pi K r}, \tag{7.157}$$

where the radial distance r is given by

$$r = \sqrt{(x - x_i)^2 + (y - y_i)^2 + (z - z_i)^2}, \tag{7.158}$$

where x, y, and z denote the general Cartesian coordinates, and x_i, y_i, and z_i denote the Cartesian coordinates of well i. For convenience, the potential can be expressed relative to the potential at the infinite distance as

$$\Psi_{ri} (r) \equiv \Psi_i (r) - \Psi_\infty = -\frac{q_i \mu}{4 \pi K r}. \tag{7.159}$$

When there are N wells present, producing at the rates of $q_i : i = 1, 2, ..., N$, then the analytic solution can be readily obtained by applying Eq. (7.156) as

$$\Psi_r = \sum_{i=1}^{N} C_i \Psi_{ri} = -\sum_{i=1}^{N} C_i \frac{q_i \mu}{4 \pi K r}. \tag{7.160}$$

Note that Eq. (7.160) constitutes a solution regardless of the values of the weighting coefficients $C_i : i = 1, 2, ..., N$. For example, they all might as well be assigned a value of unity; that is, $C_i = 1 : i = 1, 2, ..., N$.

7.5.2 Principle of Imaging

The method of images is a powerful technique used to generate analytic solutions for potential flow problems in intricate domains involving discontinuities by utilizing analytic solutions of problems considering infinite size domains. Here, the method of images is described according to Shirman and Wojtanowicz (1996), who explained this method by analogy to an optical problem. The following presentation involves some modifications on the treatise presented by Shirman and Wojtanowicz (1996) for consistency with the materials covered in this chapter.

Shirman and Wojtanowicz (1996) explained the principle of imaging by means of two lamps and two photometers symmetrically placed in a box with respect to its center. Thus, the photometers indicate the same values when the strengths of the lamps are exactly the same. However, if the lamp on the right is turned off, the light available in the box will be less and will be unevenly distributed. Therefore, both photometers will read lower values than the previous case where both lamps are turned on. Further, the photometer closer to the lamp on the left will read a higher value than that on the right and farther away from the lamp. However, when a perfect mirror, capable of reflecting with 100% efficiency, is placed in the middle of the box, the photometer on the left side of the mirror now reads the full value because of the superposition of the lights coming from the lamp and its mirror image. However, the photometer on the right of the mirror reads zero because the mirror does not permit any light to pass to the other side on the right behind the mirror and therefore acts as a barrier to light transmission, or becomes an analog to a no-flow-type boundary.

As a next exercise, Shirman and Wojtanowicz (1996) assume that the 100% reflecting mirror is now replaced by a 50% reflecting and therefore 50% transparent mirror. Consequently, in contrast to the previous example, the photometer on the right of the mirror now indicates some value instead of zero and the photometer on the left of the mirror indicates a reading less than the previous 100% reflecting mirror case. Hence, the 50% reflecting mirror acts as an analog to a cross-flow boundary.

The lamps or light sources considered in the preceding example by Shirman and Wojtanowicz (1996) correspond to the wells or fluid sinks in flow through porous media. The mirror corresponds to a discontinuity, no-flow, or cross-flow boundaries, depending on its transmissibility/reflectivity conditions. The photometer readings correspond to the flow potential.

7.5.3 Basic Method of Images

By analogy to the above optical problem, Shirman and Wojtanowicz (1996) derived the following rules for applications on flow through porous media, as quoted below:

> "**1.** The image well affects only the zone it was reflected from, and its strength is equal to the product of the strength of a real well and a coefficient of reflection C, and
>
> **2.** the strength of the real well across the discontinuity is reduced by the refraction coefficient D."

As an example of the application of these rules, Shirman and Wojtanowicz (1996) consider a single-point sink located in the upper layer near a planar discontinuity separating two layers of different permeability having K_1 permeability in the upper layer and K_2 permeability in the lower layer. A real well producing at a constant rate, q, acting as a point sink is placed in the upper layer at a distance, z_R, from the planar cross-flow boundary. Thus, the image well is placed in the lower layer at an equal distance, $z_I = z_R$, from the planar cross-flow boundary.

If the real well existed alone in an infinite-size porous medium of uniform permeability K, the potential distribution around the well would have been given by applying Eq. (7.157):

$$\Psi = \Psi_e - \frac{q\mu}{4\pi K}\left(\frac{1}{r} - \frac{1}{r_e}\right), \tag{7.161}$$

where

$$r = \sqrt{(x - x_R)^2 + (y - y_R)^2 + (z - z_R)^2}. \tag{7.162}$$

Ψ_e denotes the flow potential at a radius of r_e. When there is a porous media discontinuity, however, the problem is analogous to the above-described partial reflecting and therefore partial transparent mirror case. Thus, applying the superposition principle and rule 2, the potential distribution in the upper layer of permeability K_1 containing the real well is given by the sum of the full contribution of the real well and the partial contribution of its image with respect to the planar discontinuity according to

$$\Psi_1 = \Psi_R + C\Psi_I, \tag{7.163}$$

whereas the potential distribution in the lower layer of permeability K_2 is given by the partial contribution of the real well as the following, applying rule 2:

$$\Psi_2 = D\Psi_R. \tag{7.164}$$

In Eqs. (7.163) and (7.164), C and D denote the reflection and refraction coefficients of the planar discontinuity. The values of the coefficients are determined to satisfy the compatibility and consistency conditions at the planar discontinuity given, respectively, by

$$\Psi_1 = \Psi_2, z = 0 \tag{7.165}$$

and

$$u_1 = u_2 \text{ or } -\frac{K_1}{\mu}\frac{\partial\Psi_1}{\partial z} = -\frac{K_2}{\mu}\frac{\partial\Psi_2}{\partial z}, z = 0. \tag{7.166}$$

Ψ_R and Ψ_I denote the flow potentials of the real and image wells, when they are present alone in a uniform permeability media K in general, given, respectively, by Eqs. (7.167) and (7.168)

$$\Psi_R = \Psi_e - \frac{q\mu}{4\pi K}\left(\frac{1}{\sqrt{(x - x_R)^2 + (y - y_R)^2 + (z - z_R)^2}} - \frac{1}{r_e}\right) \tag{7.167}$$

and

$$\Psi_I = \Psi_e - \frac{q\mu}{4\pi K}\left(\frac{1}{\sqrt{(x-x_R)^2+(y-y_R)^2+(z-z_R)^2}}-\frac{1}{r_e}\right). \tag{7.168}$$

As a result, substituting Eqs. (7.167) and (7.168) into Eqs. (7.163)–(7.166) yields, respectively,

$$\Psi_1 = \Psi_e - \frac{q\mu}{4\pi K_1}\left(\frac{1}{\sqrt{(x-x_R)^2+(y-y_R)^2+(z-z_R)^2}}-\frac{1}{r_e}\right)$$
$$-\frac{Cq\mu}{4\pi K_1}\left(\frac{1}{\sqrt{(x-x_R)^2+(y-y_R)^2+(z-z_R)^2}}-\frac{1}{r_e}\right), \tag{7.169}$$

$$\Psi_2 = \Psi_e - \frac{Dq\mu}{4\pi K_2}\left(\frac{1}{\sqrt{(x-x_R)^2+(y-y_R)^2+(z-z_R)^2}}-\frac{1}{r_e}\right), \tag{7.170}$$

$$\frac{1+C}{K_1}=\frac{D}{K_2}, \tag{7.171}$$

and

$$1-C=D. \tag{7.172}$$

Consequently, Eqs. (7.171) and (7.172) can be solved to obtain the coefficients of reflection and refraction as functions of the permeability of the two layers as

$$C=\frac{K_1-K_2}{K_1+K_2}, 0 \leq C \leq 1 \tag{7.173}$$

and

$$D=\frac{2K_2}{K_1+K_2}, 0 \leq D \leq 1. \tag{7.174}$$

Thus, the potential distributions in the K_1 and K_2 permeability layers can be determined by invoking Eqs. (7.173) and (7.174) into Eqs. (7.169) and (7.170).

Shirman and Wojtanowicz (1996) point out several interesting features of this solution. First, Eq. (7.172) confirms that the fractional energies passing through and reflected off the planar discontinuity surface add up to the unity, $C + D = 1$. Second, when the planar discontinuity is a full no-flow boundary or 100% reflector, then $C = 1$ and $D = 0$. Third, if a uniform media having no planar discontinuity is considered, then $K_1 = K_2 = K$ and $C = 0$ and $D = 1$. Fourth, $K_2 \rightarrow \infty$, then the planar discontinuity represents a constant potential boundary and Eqs. (7.166)–(7.168) yield $C = -1$, $D = 2$, and $\Psi_2 = \Psi_e$.

Shirman and Wojtanowicz (1996) present solution for another interesting case dealing with a single well dissected by the planar cross-flow boundary located between the porous layers of K_1 and K_2 permeability. In this case, both the real and image wells are located at the same point $(x_I = x_R, y_I = y_R, z_I = z_R = 0)$ on the cross-flow boundary, Hence, Eqs. (7.169) and (7.170) simplify, respectively, as

$$\Psi_1 = \Psi_e - \frac{q\mu}{4\pi K_1}\left(\frac{1}{\sqrt{x^2 + y^2 + z^2}} - \frac{1}{r_e}\right)$$
$$- \frac{Cq\mu}{4\pi K_1}\left(\frac{1}{\sqrt{x^2 + y^2 + z^2}} - \frac{1}{r_e}\right)$$

(7.175)

and

$$\Psi_2 = \Psi_e - \frac{Dq\mu}{4\pi K_2}\left(\frac{1}{\sqrt{x^2 + y^2 + z^2}} - \frac{1}{r_e}\right),$$

(7.176)

where C and D are given by Eqs. (7.173) and (7.174).

7.5.4 Expanded Method of Images

As described by Shirman and Wojtanowicz (1996), when a single-point sink is located between two parallel cross-flow boundaries, an infinite number of images with respect to both boundaries need to be considered. Shirman and Wojtanowicz (1996) explain that the contribution of each new image well due to the presence of a reflecting boundary is determined by a product of its rate of production and coefficient of reflection C. Therefore, the contribution of the higher-generation images rapidly diminishes, especially after the third image. When one or both boundaries are the no-flow types, then the coefficient of perfect reflection of $C = 1$ is applied for all images with respect to the no-flow boundary.

Shirman and Wojtanowicz (1996) described that when a single-point sink is placed in a layer of a multilayer porous media, the strengths of the signals approaching the parallel cross-flow boundaries, having the reflection coefficients of $C_i : i = 1$, $2, \ldots, N$, will split into the reflecting and refracting fractions consecutively. For instance, the strength of the image source resulting from the reflection of a signal approaching the first cross-flow boundary is qC_1 and the strength of the refracting signal crossing the first cross-flow boundary will be $q(1 - C_1)$. Similarly, the signal of strength $q(1 - C_1)$ approaching the second cross-flow boundary will split into the reflected strength $[q(1 - C_1)]C_2$ for the second image and the refracted strength $[q(1 - C_1)](1 - C_2)$ crossing the second cross-flow boundary. This procedure is repeated for an infinite number of images, each having a rapidly diminishing contribution.

7.6 STREAMLINE/STREAM TUBE FORMULATION AND FRONT TRACKING

The development and implementation of models based on streamline/stream tube formulation and front tracking have been attempted for variety of applications. The fundamentals of the streamline/stream tube formulation are presented according to Martin and Wegner (1979), and Akai (1994). Consider Figure 7.9 showing a porous medium confined inside a region with sealed boundaries. A fluid is flowing through this medium because of the injection and production of the fluid at diagonally

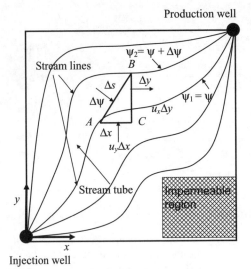

Figure 7.9   Schematic of two streamlines and a stream tube (prepared by the author).

opposite corners. The paths of the fluid particles are depicted by continuous lines extending from the injection to the production corners. These are called the streamlines. The regions present in between the consecutive streamlines can be viewed as stream tubes or flow channels. Streamlines form no-flow boundaries for the stream tubes because the streamlines are tangent or parallel curves to flow directions. Therefore, stream tubes can be viewed as a nonleaky hose whose width is determined by the flow pattern. The volumetric rate of a fluid flowing through a stream tube can be expressed by a scalar function called the stream function. The value of this function remains constant along each streamline, and the difference between the stream function values of the two streamlines forming a stream tube is equal to the volumetric rate of fluid flowing through that stream tube.

7.6.1   Basic Formulation

Consider the pair of streamlines as shown in Figure 7.9 in an x-y domain. The fluid is assumed incompressible and the flow is two dimensional for this case. The stream function in a two-dimensional domain is a prescribed scalar function, represented by

$$\Psi = \Psi(x, y). \tag{7.177}$$

The exact differential of the stream function is given by

$$d\Psi = \frac{\partial \Psi}{\partial x} dx + \frac{\partial \Psi}{\partial y} dy. \tag{7.178}$$

On the other hand, a volumetric balance over the unit-thick triangular element shown in Figure 7.9 leads to (Martin and Wegner, 1979)

$$dq \equiv \Delta \Psi = u_y \Delta x - u_x \Delta y. \tag{7.179}$$

The limit of this equation as Δx and Δy approach zero is

$$d\Psi = u_y dx - u_x dy. \tag{7.180}$$

Thus, comparing Eqs. (7.178) and (7.180) yields

$$u_x = -\frac{\partial \Psi}{\partial y} \tag{7.181}$$

and

$$u_y = \frac{\partial \Psi}{\partial x}. \tag{7.182}$$

Substituting Eqs. (7.181) and (7.182) into Eq. (7.180) yields

$$d\Psi = 0. \tag{7.183}$$

This indicates that the stream function value remains constant along a streamline. Further, for a fluid flowing through a stream tube confined between two streamlines, for which the stream function values are denoted by Ψ_1 and Ψ_2; the volume flow rate is given by (Akai, 1994)

$$q = |\Psi_2 - \Psi_1|. \tag{7.184}$$

In addition, Eqs. (7.181) and (7.182), and therefore the stream function, satisfy the incompressible fluid equation of continuity in an incompressible porous media, given by

$$\frac{\partial u_x}{\partial x} + \frac{\partial u_y}{\partial y} = 0. \tag{7.185}$$

If s denotes the length along a streamline, the parametric representation of a streamline in two dimensions is given by

$$x = x(s) \tag{7.186}$$

and

$$y = y(s), \tag{7.187}$$

whereas the velocity components in the x-, y-, and s-directions are defined, respectively, by

$$u_x = \frac{dx}{dt}, \tag{7.188}$$

$$u_y = \frac{dy}{dt}, \tag{7.189}$$

and

$$u = \frac{ds}{dt}. \tag{7.190}$$

Then, the equation of a streamline can be obtained by eliminating the time variable between Eqs. (7.188) and (7.189) as

$$\frac{dy}{dx}\bigg|_{\Psi=ct.} = \frac{u_y}{u_x}.$$ (7.191)

The same can be obtained by applying Eq. (7.183) to Eq. (7.180) (Akai, 1994). It is also true from Eqs. (7.188)–(7.190) that

$$\frac{dx}{ds}\bigg|_{\Psi=ct.} = \frac{u_x}{u}$$ (7.192)

and

$$\frac{dy}{ds}\bigg|_{\Psi=ct.} = \frac{u_y}{u}.$$ (7.193)

For horizontal two-dimensional flow, Darcy's law expresses the components of the fluid fluxes in the x- and y-directions by

$$u_x = -\lambda_x \frac{\partial p}{\partial x}$$ (7.194)

and

$$u_y = -\lambda_y \frac{\partial p}{\partial y},$$ (7.195)

where the directional fluid mobility are given by

$$\lambda_x = K_x/\mu$$ (7.196)

and

$$\lambda_y = K_y/\mu,$$ (7.197)

where μ is the viscosity of the fluid and K_x and K_y are the directional permeability of porous media.

Substituting Eqs. (7.194) and (7.195) into Eq. (7.185) yields

$$\frac{\partial}{\partial x}\left(\lambda_x \frac{\partial p}{\partial x}\right) + \frac{\partial}{\partial y}\left(\lambda_y \frac{\partial p}{\partial y}\right) = 0.$$ (7.198)

Similarly, substituting Eqs. (7.181) and (7.182) into Eq. (7.185) yields

$$\frac{\partial}{\partial x}\left(\frac{\partial \Psi}{\partial y}\right) = \frac{\partial}{\partial y}\left(\frac{\partial \Psi}{\partial x}\right).$$ (7.199)

This conforms to Schwartz's rule of exactness and indicates that the stream function is exact.

Previously, the fluid was assumed incompressible. Further assume inviscid flow, that is, flow involving no viscosity or frictional effects. Then, the flow obeys the following irrotationality condition (Bertin, 1987; Akai, 1994):

$$\frac{\partial u_y}{\partial x} - \frac{\partial u_x}{\partial y} = 0.$$ (7.200)

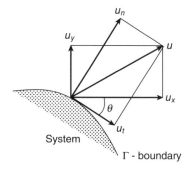

Figure 7.10   Decomposition of the volume flux at a boundary (prepared by the author).

Thus, substituting Eqs. (7.194) and (7.195) into Eq. (7.200) yields

$$\frac{\partial}{\partial x}\left(\lambda_y \frac{\partial p}{\partial y}\right) = \frac{\partial}{\partial y}\left(\lambda_x \frac{\partial p}{\partial x}\right). \tag{7.201}$$

Similarly, substituting Eqs. (7.181) and (7.182) into Eq. (7.200) yields the Laplace equation:

$$\frac{\partial^2 \Psi}{\partial x^2} + \frac{\partial^2 \Psi}{\partial y^2} = 0. \tag{7.202}$$

Consider a boundary of the system as shown in Figure 7.10. The volumetric flux vector can be decomposed into the components in the x- and y-directions, denoted as u_x and u_y, respectively. It is also possible to decompose the volume flux vector into the tangential and normal components relative to the boundary position. The latter is convenient for prescription of the boundary conditions for Eq. (7.202) and can be related to the former by (Marathe et al., 1995)

$$u_n = u_x \sin\theta - u_y \cos\theta \tag{7.203}$$

and

$$u_t = u_x \cos\theta + u_x \sin\theta, \tag{7.204}$$

where θ is the angle between the tangent and the horizontal direction in x.

Typical boundary conditions for Eq. (7.202) can be specified as follows (Shikhov and Yakushin, 1987; Marathe et al., 1995):

1. *Impermeable, no-flow, or sealed boundaries.* The normal flows are zero along sealed boundaries Γ:

$$u_n|_\Gamma = 0, \frac{\partial \Psi}{\partial t}\bigg|_\Gamma = 0, \Psi|_\Gamma = \text{ct.} \tag{7.205}$$

For example, for a rectangular domain,

$$\frac{\partial \Psi}{\partial y} = -u_x = 0 \text{ along the } y\text{-boundary } \Gamma_y \tag{7.206}$$

and

$$\frac{\partial \Psi}{\partial x} = u_y = 0 \text{ along the } x\text{-boundary } \Gamma_x. \quad (7.207)$$

Therefore, Eqs. (7.206) and (7.207) imply, respectively, that

$$\Psi = \text{ct. along the } y\text{-direction} \quad (7.208)$$

and

$$\Psi = \text{ct. along the } x\text{-direction}. \quad (7.209)$$

For simplicity, the constant can be chosen as zero for reference purposes.

2. *Equipotential boundaries.* The pressures are constant and the tangential flows are zero along the equipotential boundaries. Thus,

$$p = \text{ct.}, u_t|_\Gamma = 0, \left.\frac{\partial \Psi}{\partial n}\right|_\Gamma = 0. \quad (7.210)$$

For example, for a rectangular domain,

$$u_x = -\lambda_x \frac{\partial p}{\partial x} = -\frac{\partial \Psi}{\partial y} = 0 \text{ along the } x\text{-boundary } \Gamma_x \quad (7.211)$$

and

$$u_y = -\lambda_y \frac{\partial p}{\partial y} = \frac{\partial \Psi}{\partial x} = 0 \text{ along the } y\text{-boundary } \Gamma_y. \quad (7.212)$$

3. *Sources and sinks.* Along the boundaries across which fluid flows into and out of the system, such as wellbores and leaky boundaries, the difference between the stream function values of two streamlines is equal to the rate of fluid flowing into or out of the system. Hence, the strength of the source/sink is equal to the discontinuity in the stream function value. Thus,

$$\Psi_1 - \Psi_2 = q, \text{ where } q > 0 \text{ for production,}$$
$$q = 0 \text{ for no flow, and } q < 0 \text{ for injection.} \quad (7.213)$$

For example, consider Figure 7.11 showing several streamlines emanating from an injection well. The normal volume flux for axisymmetric flow is given by

$$u_n = \frac{\partial \Psi}{\partial \Gamma} = \frac{q}{2\pi r h}, \quad (7.214)$$

where Γ denotes distance along the wellbore boundary. An integration of Eq. (7.214) between any pair of i and j streamlines yields

$$\Psi_j - \Psi_i = q_{ij} = \frac{q}{2\pi r h}(\Gamma_j - \Gamma_i) \quad (7.215)$$

and

$$q = \sum q_{ij}. \quad (7.216)$$

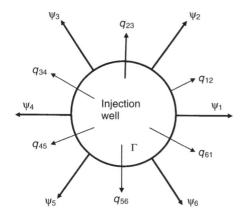

Figure 7.11 Streamlines emanating from an injection well (prepared by the author).

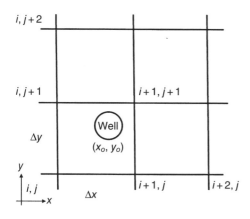

Figure 7.12 Schematic of a wellbore contained within a grid block (prepared by the author).

7.6.2 Finite Analytic Representation of Wells in Porous Media

The finite analytic method generates finite difference-type numerical schemes for governing differential equations by using local analytic solutions (Civan, 1995, 2009).

Consider the schematic of a wellbore contained within a grid block as shown in Figure 7.12. Because a wellbore diameter is generally much smaller than a typical reservoir grid block size considered for numerical solution in reservoir simulation, accurate representation of wells requires special attention in the numerical modeling of such wells. Among others, Shikhov and Yakushin (1987), and Peaceman (1990) offered effective procedures that incorporate near-wellbore radial flow analytic solutions into numerical solution schemes. Here, the method of Shikhov and Yakushin (1987) is presented.

Shikhov and Yakushin (1987) developed a finite analytic numerical scheme for isotropic flow in the vicinity of a well. For this purpose, by substituting $\lambda_x = \lambda_y = \lambda = $ ct. into Eq. (7.198), the following Laplace equation is obtained:

$$\frac{\partial^2 p}{\partial x^2} + \frac{\partial^2 p}{\partial y^2} = \frac{1}{r}\frac{d}{dr}\left(r\frac{dp}{dr}\right) = 0. \tag{7.217}$$

The analytic solution of this equation is given by

$$p = p_o + \frac{q\mu}{K}\ln r = p_o + \frac{q\mu}{K}\ln\sqrt{(x-x_o)^2 + (y-y_o)^2}. \tag{7.218}$$

Therefore,

$$u_x = -\frac{K}{\mu}\frac{\partial p}{\partial x} = \frac{q}{2\pi h}\frac{x-x_o}{(x-x_o)^2 + (y-y_o)^2} \tag{7.219}$$

and

$$u_y = -\frac{K}{\mu}\frac{\partial p}{\partial y} = \frac{q}{2\pi h}\frac{y-y_o}{(x-x_o)^2 + (y-y_o)^2}. \tag{7.220}$$

On the other hand,

$$u_x = -\frac{\partial\Psi}{\partial y} \tag{7.221}$$

and

$$u_y = \frac{\partial\Psi}{\partial x}. \tag{7.222}$$

Now, consider the four grid points associated with the grid block containing the well as shown in Figure 7.12. The central finite difference approximations of Eqs. (7.221) and (7.222) are given, respectively, by

$$(u_x)_{i,j+1} = -\frac{\Psi_{i,j+2} - \Psi_{i,j}}{2\Delta y} \tag{7.223}$$

and

$$(u_y)_{i+1,j} = \frac{\Psi_{i+2,j} - \Psi_{i,j}}{2\Delta x}. \tag{7.224}$$

These equations can be solved for $\Psi_{i,j}$, respectively, as

$$\Psi_{i,j} = \Psi_{i,j+2} - 2\Delta y(u_x)_{i,j+1} \tag{7.225}$$

and

$$\Psi_{i,j} = \Psi_{i+2,j} + 2\Delta x(u_y)_{i+1,j}. \tag{7.226}$$

Then, the above two expressions can be averaged arithmetically as

$$\Psi_{i,j} = \frac{1}{2}\left[\Psi_{i,j+2} + \Psi_{i+2,j} - 2\Delta y(u_x)_{i,j+1} + 2\Delta x(u_y)_{i+1,j}\right]. \tag{7.227}$$

In this equation, u_x and u_y are calculated using Eqs. (7.219) and (7.220). A formula can be generated for $\Psi_{i+1,j}$ also from Eq. (7.227) by progressing the index. Note that Eq. (7.227) is slightly different from the expression given by Shikhov and Yakushin (1987).

The central finite difference approximations of Eqs. (7.221) and (7.222) at the other two points are given, respectively, by

$$(u_x)_{i,j} = -\frac{\Psi_{i,j+1} - \Psi_{i,j-1}}{2\Delta y} \tag{7.228}$$

and

$$(u_y)_{i+1,j+1} = \frac{\Psi_{i+2,j+1} - \Psi_{i,j+1}}{2\Delta x}. \tag{7.229}$$

These equations can be solved for $\Psi_{i,j+1}$ and then averaged arithmetically to obtain

$$\Psi_{i,j+1} = \frac{1}{2}\Big[\Psi_{i,j-1} + \Psi_{i+2,j+1} + 2\Delta y(u_x)_{i,j} - 2\Delta x(u_y)_{i+1,j+1}\Big]. \tag{7.230}$$

In this equation, u_x and u_y are calculated using Eqs. (7.219) and (7.220). A formula can be generated for $\Psi_{i+1,j+1}$ also from Eq. (7.230) by progressing the index.

7.6.3 Streamline Formulation of Immiscible Displacement in Unconfined Reservoirs

An unconfined reservoir is an infinitely large theoretical system in which the flow potential in the vicinity of the production/injection wells is only influenced by the flow rates of these wells, while the potential at infinite distance remains constant because there are no acting boundary effects.

Consider $i = 1, 2, \ldots, N$ wells in an unconfined reservoir that is an areal, uniform-thick, and isotropic porous medium. A unit mobility ratio displacement of the resident fluid by the injected fluids under steady-state flow conditions is considered for simplicity.

The potential distribution around the i well, when it is present alone in an unconfined medium, is given by the following analytic solution:

$$\Psi_i = \Psi_w - \frac{q_i\mu}{2\pi Kh}\ln\left(\frac{r}{r_w}\right), \tag{7.231}$$

where Ψ_w denotes the flow potential at the wellbore. The radial distance from the well center is given by

$$r = \sqrt{(x - x_i)^2 + (y - y_i)^2}, \tag{7.232}$$

where x_i and y_i are the Cartesian coordinates of the center point of the i well. Applying the superposition principle of potential flow, the potential distribution incorporating the influence of all wells present in an unconfined medium can be expressed by

$$\Psi(x, y) = \sum_{i=1}^{N}\Psi_i(x, y) = \Psi_w - \frac{\mu}{4\pi Kh}\sum_{i=1}^{N} q_i \ln\left[(x - x_i)^2 + (y - y_i)^2\right]. \tag{7.233}$$

Applying Darcy's law, the directional volume fluxes (superficial velocities) are given by

$$u_x = -\frac{K}{\mu}\frac{\partial \Psi}{\partial x} = \frac{1}{2\pi h}\sum_{i=1}^{N}\frac{q_i(x-x_i)}{(x-x_i)^2+(y-y_i)^2} \tag{7.234}$$

and

$$u_y = -\frac{K}{\mu}\frac{\partial \Psi}{\partial y} = \frac{1}{2\pi h}\sum_{i=1}^{N}\frac{q_i(y-y_i)}{(x-x_i)^2+(y-y_i)^2}. \tag{7.235}$$

The pore fluid directional velocities are given by

$$v_x = \frac{u_x}{\phi\left(1-\sum_{r} S_r\right)} \tag{7.236}$$

and

$$v_y = \frac{u_y}{\phi\left(1-\sum_{r} S_r\right)}, \tag{7.237}$$

where S_r denotes the immobile fluid saturations, such as the connate water and residual oil saturations involved in the waterflooding of oil reservoirs.

Figure 7.13 of Marathe et al. (1995) shows the typical streamline patterns involving an injection well surrounded by four equally spaced production wells present in an unconfined reservoir.

7.6.4   Streamline Formulation of Immiscible Displacement Neglecting Capillary Pressure Effects in Confined Reservoirs

Marathe et al. (1995) developed a calculation scheme for the determination of flood patterns resulting from multiple injection and production wells present in irregularly

Figure 7.13   Typical streamline patterns involving an injection well surrounded by four equally spaced production wells present in an unconfined reservoir (after Marathe et al., 1995; © 1995 SPE, with permission from the Society of Petroleum Engineers).

bounded reservoirs. They approximated the immiscible fluid displacement process as a unit mobility ratio displacement, leading to a homogeneous reservoir treatment. Their approach for modeling immiscible displacement to determine the potential and streamline distributions and for displacing fluid front positions is presented here with some modifications for consistency with the rest of the presentation of this chapter.

Consider Figure 7.14 by Marathe et al. (1995), showing an irregularly bounded, uniform-thick areal reservoir containing a number of injection and production wells. This reservoir may be referred to as the "actual" system. For simplification, Marathe et al. (1995) first approximated the reservoir boundaries by several straight-line segments as depicted in Figure 7.15. This reservoir may be referred to as the "model" system. Second, place N image wells outside, a distance away from and along the reservoir boundary as shown in Figure 7.16 to represent the effect of the boundaries

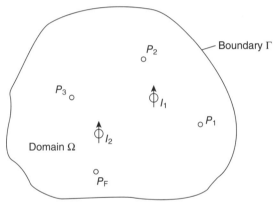

Figure 7.14 An irregularly bounded, uniform-thick areal reservoir containing a number of injection and production wells (after Marathe et al., 1995; © 1995 SPE, with permission from the Society of Petroleum Engineers).

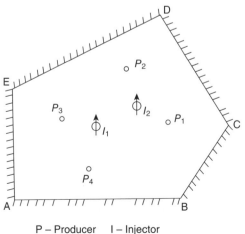

Figure 7.15 Reservoir under pressure-maintenance no-flow boundaries (after Marathe et al., 1995; © 1995 SPE, with permission from the Society of Petroleum Engineers).

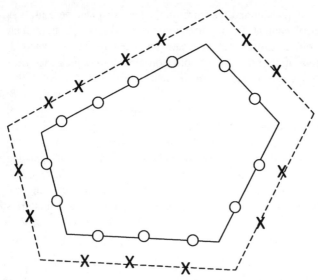

Figure 7.16   Schematic of N image wells and M boundary points (prepared by the author).

in a confined reservoir. Third, sample a sufficiently large number of boundary points M ($M > N$) along the boundary segments and identify them with associated boundary conditions, such as constant potential or no-flow.

Applying the superposition principle for potential flow, the directional volume flux components are expressed as a sum of the contributions by the N_R-real and N_I-image wells as

$$u_x = \frac{1}{2\pi h} \sum_{i=1}^{N_T} \frac{q_i (x - x_i)}{(x - x_i)^2 + (y - y_i)^2} \qquad (7.238)$$

and

$$u_y = \frac{1}{2\pi h} \sum_{i=1}^{N_T} \frac{q_i (y - y_i)}{(x - x_i)^2 + (y - y_i)^2}, \qquad (7.239)$$

where the total number of real and image wells is given by

$$N_T = N_R + N_I. \qquad (7.240)$$

Because the streamlines are parallel to no-flow boundaries, the normal fluid flux vanishes. Thus,

$$u_n = u_x \sin \theta - u_y \cos \theta = 0. \qquad (7.241)$$

Substituting Eqs. (7.238) and (7.239) into Eq. (7.241) and then applying for the sample points located along the no-flow boundaries, and rearranging and composing the resulting system of M linear equations lead to a matrix equation as

$$Aq_I = b, \qquad (7.242)$$

where the elements of the q_I column vector are the unknown flow rates q_{Ii} of $i = 1$, $2, ..., N_I$ image wells placed outside the reservoir boundaries; the elements of the b column vector are given as follows in terms of the known flow rates q_{Ri} of $i = 1$, $2, ..., N_R$ real wells present inside the reservoir boundaries:

$$b_j = \sum_{k=1}^{N_R} \frac{q_{Ri}\left[(x_j - x_k)\sin\theta - (y_j - y_k)\cos\theta\right]}{(x_j - x_k)^2 + (y_j - y_k)^2}, j = 1, 2, ..., M. \tag{7.243}$$

The elements of the coefficient matrix A are given by

$$a_{ji} = \frac{(x_j - x_i)\sin\theta - (y_j - y_i)\cos\theta}{(x_j - x_i)^2 + (y_j - y_i)^2}, j = 1, 2, ..., M \text{ and } i = 1, 2, ..., N. \tag{7.244}$$

Because the streamlines are normal to the equipotential boundaries, the tangential fluid flux vanishes. Thus,

$$u_t = u_x \cos\theta + u_y \sin\theta = 0. \tag{7.245}$$

An analysis similar to no-flow boundaries can be carried out for equipotential boundaries to determine the flow rates for the image wells placed along and outside the equipotential boundaries. The resulting linear equations are also incorporated into Eq. (7.242), which can then be solved by an appropriate numerical method, such as the singular value decomposition method (Forsythe et al, 1977), for the unknown flow rates q_{Ii} of $i = 1, 2, ..., N_I$ image wells.

Now that the flow rates of the real and image wells and their positions are known, then Eqs. (7.238) and (7.239) can be used to determine the front positions and streamlines according to the procedure described in this chapter.

Figures 7.17 and 7.18 by Marathe et al. (1995) show the typical streamline patterns, and areal sweep efficiency and water cut versus pore volume involving an

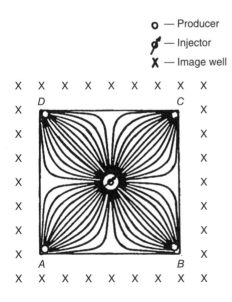

o — Producer

♂ — Injector

X — Image well

Figure 7.17 Typical streamline patterns for the five-spot pattern confined case with image wells (after Marathe et al., 1995; © 1995 SPE, with permission from the Society of Petroleum Engineers).

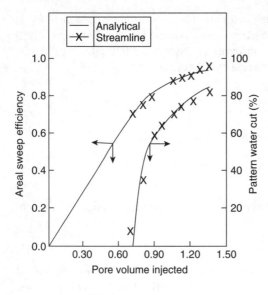

Figure 7.18 Areal sweep efficiency and water cut versus pore volume involving an injection well surrounded by four equally spaced production wells in a confined reservoir. The streamline solution matches well with the analytic solution (after Marathe et al., 1995; © 1995 SPE, with permission from the Society of Petroleum Engineers).

injection well surrounded by four equally spaced production wells in a confined reservoir. Figure 7.18 indicates that the streamline solution matches well with the analytic solution of the same problem, confirming the validity of the streamline solution method. Figure 7.19a–c by Marathe et al. (1995) show the streamline patterns for two injection and four production wells in an unconfined reservoir, a confined reservoir with no-flow boundaries, and a confined reservoir with mixed equipotential and no-flow boundaries, respectively. Figure 7.20 by Marathe et al. (1995) depicts the areal sweep efficiency and water cut versus pore volume calculated for the bounded reservoir case.

7.7 EXERCISES

1. Construct a plot of the dimensionless pressure versus dimensionless time for a real gas to replace the plot given in Figure 7.2.
2. What form would Eq. (7.9) take if the derivation was carried out by including the dispersion term?
3. Obtain a Buckley–Leverett solution of saturation S_1 versus distance x_{S1} at various times using Eq. (7.108) for typical assumed values.
4. Repeat the Penuela and Civan (2001) numerical solution approach using the relative permeability and capillary pressure data given by Kulkarni et al. (1998) after replacing the outlet boundary condition given by Eq. (7.95) by the general outlet boundary condition formulated by Civan (1996b) given by Eq. (7.83) so that a numerical solution can be obtained until infinite fluid throughput at which condition all the mobile fluid will have been displaced by the injection fluid.

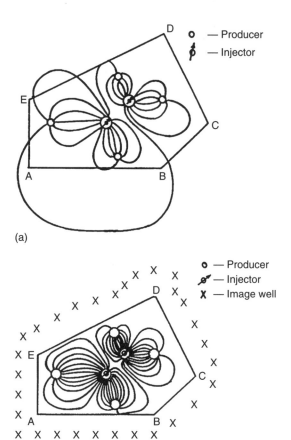

(a)

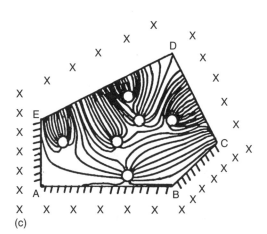

(b)

(c)

Figure 7.19 Streamline patterns for two injection and four production wells in (a) an unconfined reservoir, (b) a confined reservoir with no-flow boundaries, and (c) a confined reservoir with mixed equipotential and no-flow boundaries (after Marathe et al., 1995; © 1995 SPE, with permission from the Society of Petroleum Engineers).

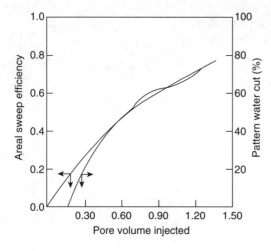

Figure 7.20   Areal sweep efficiency and water cut versus pore volume for the bounded reservoir (after Marathe et al., 1995; © 1995 SPE, with permission from the Society of Petroleum Engineers).

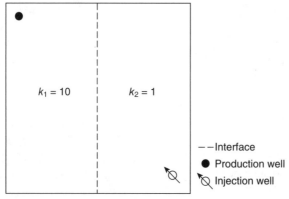

Figure 7.21   Reservoir's geometry for the validation of the numerical method (after Guevara-Jordan and Rodriguez-Hernandez, 2001; © 2001 SPE, with permission from the Society of Petroleum Engineers).

5. Obtain a numerical solution of the problem given in Example 1 using the numerical method described there.

6. Obtain a numerical solution of the problem given in Example 2 using the numerical method described there.

7. Guevara-Jordan and Rodriguez-Hernandez (2001) derived the analytic solutions for pressure distribution over a model areal reservoir of a unit square shape and containing equal size, contrasting permeability porous formations, one having an injection well and the other having a production well with unit strengths, as illustrated in Figure 7.21.

   (a) Derive and confirm their analytic solution given next for the special case of an infinite size reservoir:

$$p(\mathbf{x}) = \frac{1}{4\pi K_1} \left[\ln|\mathbf{x} - \mathbf{x}_{R1}|^2 + \left(\frac{K_1 - K_2}{K_1 + K_2} \right) \ln|\mathbf{x} - \mathbf{x}_{I1}|^2 \right]$$

$$- \frac{1}{2\pi} \left(\frac{1}{K_1 + K_2} \right) \ln|\mathbf{x} - \mathbf{x}_{I2}|^2 \text{ in } K_1 - \text{region,}$$

(7.246)

and

$$p(\mathbf{x}) = \frac{-1}{4\pi K_2} \left[\ln|\mathbf{x} - \mathbf{x}_{R2}|^2 - \left(\frac{K_1 - K_2}{K_1 + K_2} \right) \ln|\mathbf{x} - \mathbf{x}_{I2}|^2 \right]$$

$$+ \frac{1}{2\pi} \left(\frac{1}{K_1 + K_2} \right) \ln|\mathbf{x} - \mathbf{x}_{I1}|^2 \text{ in } K_2 - \text{region,}$$

(7.247)

where the subscripts R and I denote the real and image wells, and 1 and 2 denote the production and injection wells, shown in Figure 7.21. $K_1 = 1$ and $K_2 = 10\,\text{md}$, although no units were provided by the authors.

(b) Using the analytic solution given in (a), determine the streamline distribution for the infinite-size reservoir and compare this with those given in Figure 7.22.

(c) Applying the method of images and superposition with the analytic solutions given in (a) according to the mirror imaging scheme depicted in Figure 7.23, derive the corresponding analytic solutions for the unit square reservoir with no-flow boundaries, shown in Figure 7.24.

(d) Using the analytic solution derived in part c, determine the streamline distribution for the unit square reservoir with no-flow boundaries and compare this with those given in Figure 7.25.

8. Consider Figure 7.26 showing a horizontal well completed inside a slab-shaped homogeneous and isotropic reservoir containing a plane discontinuity. Determine the transient-state analytic solution and plot the isopotential lines for the flow potential distribution in this reservoir if this well begins production at a constant rate at a given time.

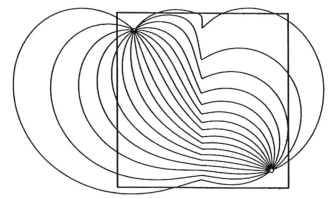

Figure 7.22   Streamline distribution without imaginary wells around the reservoir (after Guevara-Jordan and Rodriguez-Hernandez, 2001; © 2001 SPE, with permission from the Society of Petroleum Engineers).

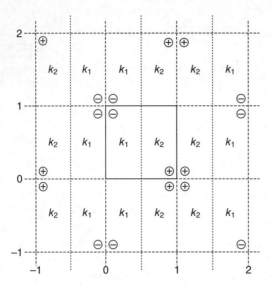

Figure 7.23   Mirror imaging of the reservoir in Figure 7.24 used in the analytic solution by the method of images (after Guevara-Jordan and Rodriguez-Hernandez, 2001; © 2001 SPE, with permission from the Society of Petroleum Engineers).

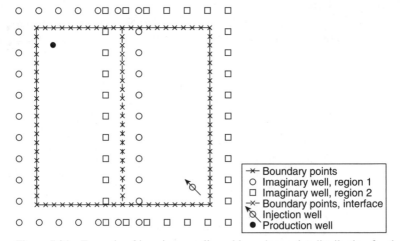

Figure 7.24   Example of imaginary wells and boundary point distribution for the reservoir considered in the validation (after Guevara-Jordan and Rodriguez-Hernandez, 2001; © 2001 SPE, with permission of the Society of Petroleum Engineers).

9. Carry out an analysis similar to Shikhov and Yakushin (1987) for a well in an anisotropic porous media.

10. Carry out an analysis similar to Marathe et al. (1995) for an anisotropic reservoir.

11. The analytic solution of the leaky-tank model was plotted as straight lines for the dimensionless average fluid density versus the dimensionless time by Civan (2000e). Construct a plot of the average reservoir fluid pressure versus the dimensionless time corresponding to the dimensionless density and dimensionless time range indicated by

Figure 7.25 Streamline distribution computed with the new method for the validation problem (after Guevara-Jordan and Rodriguez-Hernandez, 2001; © 2001 SPE, with permission from the Society of Petroleum Engineers).

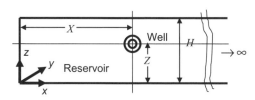

Figure 7.26 Cross-sectional view of a semi-infinite slab reservoir containing a horizontal well (prepared by the author).

Figure 7.2 for typical slightly compressible and compressible fluids using the data presented by Civan (2000e).

12. Solve the following problem using the data given as porosity $\phi = 0.25$, $K_x = 250\,\text{mD} = 250 \times 10^{-15}\,\text{m}^2$, $K_y = 150\,\text{mD} = 150 \times 10^{-15}\,\text{m}^2$, formation thickness $= 30\,\text{ft} = 9.14\,\text{m}$, $r_w = 0.25 = 0.0762\,\text{m}$, oil production rate $q_w = 200\,\text{stb/day} = 4.4 \times 10^{-4}\,\text{cm}^3/\text{s}$, $B_o = 1.2\,\text{bbl/stb}$, oil gravity $= 35\,\text{API}$, viscosity $= 3.0\,\text{cp}$, pressure at the external boundary $= 3000\,\text{psia} = 204\,\text{atm}$, reservoir half-length $(x_e) = 4000\,\text{ft} = 1219\,\text{m}$, reservoir half-width $(y_e) = 2500\,\text{ft} = 762\,\text{m}$, number of grid blocks in the x-direction $(N) = 40$, and number of grid blocks in the y-direction $(M) = 25$. Prepare the two-dimensional (x, y) contour plots of the reservoir fluid pressure, Darcy velocities in the x- and y-Cartesian directions, the resultant Darcy velocity, and its direction angle with respect to the x-axis over the quadrant of a rectangular-shaped reservoir of unit thickness based on the finite difference numerical solution method (Fig. 7.27). Make reasonable assumptions for any missing data. Explore ways of improving the solution method and accuracy. You may use any computing tool of your preference.

13. Consider the quadrant of a rectangular-shaped reservoir having a uniform thickness in which a well in the center is hydraulically fractured as shown in Figure 7.28. Half-fracture length $(x_f) = 500\,\text{ft}$, fracture width $(W_f) = 1.0\,\text{in.}$, half-reservoir length $(x_e) = 2500\,\text{ft}$, half-reservoir width $(y_e) = 1500\,\text{ft}$, reservoir permeability $(K) = 0.1\,\text{darcy}$, fracture permeability $(K_f) = 1.0\,\text{darcy}$, wellbore pressure $(p_w) = 1500\,\text{psia}$, fluid viscosity $(\mu) = 0.7\,\text{cP}$, number of grid points in the x-direction $N = 30$, and number of grid

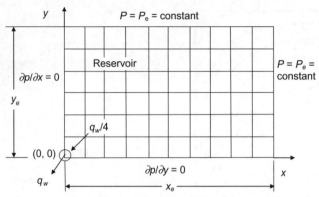

Figure 7.27   Production well in the quadrant of a rectangular-shaped reservoir having a uniform thickness (prepared by the author).

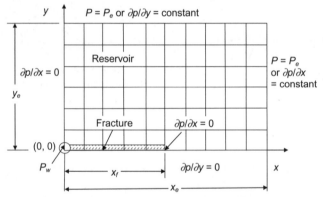

Figure 7.28   Hydraulically fractured well in the quadrant of a rectangular-shaped reservoir having a uniform thickness (prepared by the author).

points in the y-direction $M = 20$. Assume a no-flow condition at the fracture tip. Repeat the solution using exactly the same data but assuming (a) constant pressure Dirichlet boundary conditions along the external reservoir boundaries prescribed by $p = p_e = 3000$ psia, (b) constant flow Neumann boundary conditions along the external reservoir boundaries specified by $\partial p/\partial x = 3.0$ psia/ft, and $\partial p/\partial y = 3.0$ psia/ft. (c) Compare the pressure profiles along the fracture obtained in parts a and b. (d) Compare the pressure profiles (isobar contours) over the reservoir domain obtained in parts a and b.

14. Write down the equation of continuity with a constant source term for one-dimensional transient-state radial flow of a single-component gas through a homogeneous and isotropic porous media. Substitute Darcy's law to obtain a pressure equation. Solve this equation for flow of an ideal gas in the radial domain of $r_w < r < r_e$ under the steady-state conditions, where r_w and r_e denote the wellbore and external reservoir boundaries along which the constant pressures are specified as p_w and p_e, respectively.

15. The water-to-oil viscosity ratio is

$$\frac{\mu_w}{\mu_o} = 1.0.$$

The water-to-oil relative permeability ratio correlation is given by

$$\frac{k_{ro}}{k_{rw}} = a\exp(-bS_w),$$

where $a = 1.0$ and $b = 2.0$. At what water saturation value does the fractional water versus water saturation curve has an inflection point?

16. If the water-drive strength factor f of the leaky-tank boundary is greater than 1.0, what does it indicate about the condition of the leaky external boundary condition?

CHAPTER **8**

PARAMETERS OF FLUID TRANSFER IN POROUS MEDIA

8.1 INTRODUCTION

The methods required for defining and determining the essential parameters affecting fluid transport through porous media is presented.* Wettability, wettability index, capillary pressure, relative permeability, and their measurement and temperature dependence, and wall drag and interfacial drag are discussed.

The wettability state-dependent properties of porous material may vary with temperature and chemical and physicochemical processes, such as decomposition, precipitation and dissolution, and adsorption and desorption over the pore surface (Madden and Strycker, 1989; Buckley, 2002). When fluids having temperatures different from pore fluids are introduced into porous media, the wettability of porous material may assume different wettability states because of the effect of temperature on the material and fluid properties, and the interactions of pore fluids with the porous material (Civan, 2004). Civan (2004) applied the Arrhenius equation for correlation of the temperature effect of the wettability-related properties of porous media and verified with experimental data, including computed tomography (CT)

---

* Parts of this chapter have been reproduced with modifications from the following:

Civan, F. 2004. Temperature dependence of wettability-related rock properties correlated by the Arrhenius equation. Petrophysics, 45(4), pp. 350–362, © 2004 SPWLA, with permission from the Society of Petrophysicists and Well Log Analysts;

Civan, F. and Evans, R.D. 1991. Non-Darcy flow coefficients and relative permeabilities for gas/brine systems. Paper SPE 21516, Proceedings of the Gas Technology Symposium (January 23–25, 1991), Houston, TX, pp. 341–352, © 1991 SPE; with permission from the Society of Petroleum Engineers;

Civan, F. and Evans, R.D. 1993. Relative permeability and capillary pressure data from non-Darcy flow of gas/brine systems in laboratory cores. Paper SPE 26151, Proceedings of the Gas Technology Symposium (June 28–30, 1993), Calgary, Canada, pp. 139–153, © 1993 SPE, with permission from the Society of Petroleum Engineers;

Tóth, J., Bódi, T., Szücs, P., and Civan, F. 2002. Convenient formulae for determination of relative permeability from unsteady-state fluid displacements in core plugs. Journal of Petroleum Science and Engineering, 36(1–2), pp. 33–44, with permission from Elsevier; and

Ucan, S., Civan, F., and Evans, R.D. 1997. Uniqueness and simultaneous predictability of relative permeability and capillary pressure by discrete and continuos means. Journal of Canadian Petroleum Technology, 36(4), pp. 52–61, © 1997 SPE, with permission from the Society of Petroleum Engineers.

---

number (relating to the pore fluid saturation), capillary pressure, unfrozen water content, and wettability index, according to the theory presented in the following sections. The correlation by the Arrhenius equation provides useful information about the activation energy requirements associated with the imbibition and drainage processes involving the flow of immiscible fluids in porous media.

As stated by Ucan et al. (1997),

> Petrophysical properties of multiphase flow systems in porous rock are complex functions of the morphology and topology of the porous medium, interactions between rock and fluids, phase distribution and flow patterns and regimes. It is impractical to deconvolute the effect of the individual factors and forces from the macroscopic petrophysical properties. Therefore, the effect of these properties on the flow behavior is lumped in the form of empirically determined relative permeability and capillary pressure functions which are used as the primary flow parameters for the macroscopic description of multiphase flow in porous media. However, development of reliable methods for simultaneous determination of relative permeability and capillary pressure data from laboratory core flood tests is a challenging task and is of continuing interest to the oil and gas industry.

Most of the previous methods developed for extracting relative permeability and capillary pressure data are indirect and based on reservoir core history matching, which requires iterative numerical solutions of inverse problems until a satisfactory match to experimental core flow test data is obtained. The direct methods are based on integral formulations of the equations describing the fluid pressures and saturations for laboratory cores, and they do not require numerical solution methods for differential equations involved in the indirect methods. Therefore, the numerical stability and accuracy and the history matching problems associated with the indirect methods are eliminated.

First, the direct methods for the determination of relative permeability and capillary pressure data from unsteady-state non-Darcy fluid displacement in laboratory cores are presented. These methods take advantage of the internal fluid displacement data that can be obtained by noninvasive techniques in addition to the usual external fluid rate and pressure data. The mathematical interpretation methods developed by Civan and Donaldson (1989) and Civan and Evans (1991, 1993) enable the determination of relative permeability and capillary pressure simultaneously without the inherent limitations of the previous methods. These are also computationally more convenient, based on a semianalytic solution of an algebraic formulation of the flow through porous media. Because of algebraic formulation, these methods also provide a numerically more accurate interpretation of laboratory core data.

In general, the type of experiments to be conducted and the data to be measured depend on the theory available to interpret the data. Previously reported methods for interpretation of immiscible fluid displacement data to generate relative permeability and capillary pressure curves relied upon external fluid data such as the effluent fluid rates and differential fluid pressure between the fluid influx and efflux. Examples of such methods are given by Johnson et al. (1959), Jones and Roszelle (1978), Marle (1981), and Civan and Donaldson (1989). Ramakrishnan and Cappiello (1991) have designed a special experimental method to guarantee that the capillary pressure is zero for the boundary condition at the core outlet so that a simplified theory can be used to predict only the nonwetting-phase relative permeability. These methods allow for a direct calculation of relative permeability and involve some limitations due to the inherent assumptions. There are also some indirect methods given by Kerig and Watson (1987) and Ouenes et al. (1992), which rely on the history matching methods. Indirect methods are prone to errors resulting from the numerical solution of the governing differential equations and the nonuniqueness of the estimated parameter values.

Several methods take advantage of the advancements in noninvasive techniques such as nuclear magnetic resonance imaging (NMRI) (Chen et al., 1992), ultrasonic imaging (Chardaire-Riviere et al., 1992), and X-ray tomography (Vinegar, 1986) to infer interior fluid data such as the fluid saturation distribution over the core length. However, most of these studies still rely upon the indirect method of history matching for interpretation, although such internal data are particularly suitable for direct interpretation.

Enwere and Archer (1992a,b) have shown that the saturation profiles dynamically measured by NMRI in conjunction with capillary pressure and relative permeability curves can be used to determine the capillary pressure gradients in a core. But the real objective is to determine capillary pressure and relative permeability from the saturation profiles.

Second, the indirect methods are discussed. Ucan et al. (1997) addressed the issue of uniqueness in the determination of relative permeability and capillary pressure functions by means of the history matching of unsteady-state displacement data obtained from laboratory core flow tests. Ucan et al. (1997) point out that history matching (the inverse problem) requires a reliable porous media averaged, macroscopic flow description model (the forward problem) to predict the values of the observable parameters such as cumulative production, pressure drop histories, and saturation history profiles so that the best estimates of the relative permeability and capillary pressure functions can be determined. However, some model parameters may be overdetermined while leaving the others underdetermined, unless the interpretation method is designed to assimilate a proper set of experimental data. Lack of intrinsic data, experimental uncertainties, and an accurate physical representation of the complex flow affect the reliability of the predictions.

When a problem is ill-posed, its solution may not necessarily be unique and, therefore, perturbation of any model parameters may lead to arbitrary variations of the solutions. Ucan et al. (1997) demonstrated that observed values are quite sensitive to the flow functions. In simulation of flow in subsurface porous formations, the flow functions are usually chosen as the first model parameters to be tuned for history matching. It is of continuing interest to develop satisfactory general theoretical flow functions for this purpose. For the prediction of two-phase relative permeability alone, many different empirical models have been proposed (Honarpoor et al., 1986, and Siddiqui et al., 1993). Although these empirical models are applicable only under specific conditions, they have been selected arbitrarily in many applications without any particular basis. Ucan et al. (1997) have shown that the application of the empirical flow functions globally over the whole saturation range does not always yield a satisfactory history match, and better results can be achieved when these empirical models are applied locally in a piecewise continuous manner. Consequently, the parameters of these empirical models assume different values over the various segments.

Numerous investigations have been reported for the simultaneous estimation of capillary pressure and relative permeability functions from laboratory core data. Efforts have been made to gather internal core data such as saturation and pressure history profiles, which can be measured along the core by various techniques including gamma-ray attenuation, CT scanning, or NMRI, in addition to the overall pressure drop across the core and the core production data. Various functional representations of relative permeability and capillary pressure curves with a variety of interpretation methods and optimization techniques have been investigated.

Kerig and Watson (1987) used a regression-based method to estimate the relative permeability curves using a spline function representation. Richmond and Watson (1990) extended this regression-based method for the simultaneous estimation of a functional representation of capillary pressure and relative permeability curves. Chardaire et al. (1989, 1990, 1992) used piecewise linear functions to determine the relative permeability and capillary pressure simultaneously by using a multiscale representation of parameters. Chardaire et al. (1989, 1990, 1992) have proposed an automatic adjustment method to determine the capillary pressure and relative permeability by using optimal control theory. Ouenes et al. (1992) used a discrete representation of the relative permeability and capillary pressure using simulated annealing for the Darcy flow of incompressible fluids. These studies have not incorporated internal core data information into the history match.

It has been shown that it is possible to match the saturation profiles and recovery curves from a laboratory test by means of different forms of relative permeability and capillary pressure curves. Firoozabadi and Aziz (1991) have pointed out that entirely different relative permeability models could match the same recovery performance. Thus, using only the recovery curves and overall pressure differential data, referred to as the external data, to represent flow functions by empirical global functions does not guarantee a unique set of relative permeability and capillary pressure curves. Ucan et al. (1997) have demonstrated that internal core data (saturation and/or pressure profiles) along with the conventional external data (overall pressure and effluent fluid production data) can help achieve unique solutions provided that sufficient experimental data are available for the fluid/rock system of interest.

The wettability, relative permeability, and capillary pressure issues are reviewed in the following sections.

8.2 WETTABILITY AND WETTABILITY INDEX

Wettability is a generic term expressing the relative affinity of solid to various fluid phases due to intermolecular interactions, and therefore it is a measure of the spreading tendency of fluids over solid surfaces (Hirasaki, 1991). The wettability of porous materials can be generally related to the activation energy required for immiscible fluid displacement. This energy depends on the relative affinities of the solid to the fluids involved in immiscible displacement (Sharma, 1985).

In principle, the wettability of a solid surface is a well-defined property. It can be determined in terms of several basic parameters, such as the surface roughness, contact angle of fluid interface, and surface tension of fluids involved in competition with each other for spreading over a solid surface. However, the wettability of porous material is a macroscopic property averaged over the representative bulk volume element of porous material.

Wettability is an important property of porous materials. It affects the macroscopic properties of porous material such as distribution and saturation, relative permeability, and capillary pressure of the pore fluids in a complicated manner. Therefore, the wettability of porous material can be quantified and expressed in terms of the relationship of relative permeability or capillary pressure to pore fluid saturation. However, because relative permeability and capillary pressure vary with saturation, a practical measure of wettability is considered as an integral effect over the mobile fluid saturation, bounded by the end-point or immobile fluid saturations, and in terms of the work done during complete immiscible displacement of one fluid phase by another.

There are various approaches available for the assessment of the wettability of petroleum-bearing reservoir formations. These approaches include the Amott–Harvey wettability index (WI_{AH}) (Amott, 1959), the United States Bureau of Mines wettability index (WI_{USBM}) (Donaldson et al., 1969, 1980a,b), nuclear magnetic resonance relaxation techniques (WI_{NMR}) (Guan et al., 2002), and inferring wettability via the imbibition rate measurements (Zhou et al., 2000; Matejka et al., 2002). However, the most practical approach is to express the wettability of porous materials using the USBM wettability index as defined in the following.

The wettability of porous materials can be generally related to the energy required for immiscible fluid displacement. This energy depends on the relative affinities of the solid to the fluids involved in immiscible displacement (Sharma, 1985). The wettability index provides a measure of the comparison of the works associated with the drainage and imbibition processes. Therefore, the wettability index can be defined by the following equation:

$$WI_{USBM} = \log_{10}(W_{drainage} / W_{imbibition}), \tag{8.1}$$

where $W_{drainage}$ and $W_{imbibition}$ denote the works associated with the forced drainage and imbibition processes, indicated by the forced drainage and imbibition capillary pressure curves, respectively. $WI > 0$ for wetting fluids and $WI < 0$ for nonwetting fluids. The wettability index is an indicator of the wetting characteristics of porous materials, which affect the end-point saturations, range of the immobile and mobile fluid saturations, capillary pressure and relative permeability of fluids, fluid displacement efficiency, and fluid transport effectiveness (Civan, 2000a).

8.3 CAPILLARY PRESSURE

The capillary pressure of immiscible fluids in porous media is defined as the difference between the pressures of the nonwetting and wetting fluid phases at the interface. Here, the discussion is limited to the oil/water systems in a uniformly water-wet porous media.

Frequently, the capillary pressure is correlated with the normalized fluid saturation using the Brooks and Corey (1966) empirical power-law function, given by

$$p_c = \frac{p_e}{\overline{S}_w^{1/\lambda}}, \tag{8.2}$$

where p_e and λ denote the entry capillary pressure and the pore size distribution index, respectively. The normalized fluid saturation for a nonwetting/wetting fluid system, such as an oil/water system in a uniformly water-wet material, is given by

$$\overline{S}_w = \frac{S_w - S_{wr}}{(1 - S_{nr}) - S_{wr}}, \tag{8.3}$$

where S_w denotes the wetting-phase fluid saturation of material. S_{wr} and S_{nr} denote the irreducible saturations of the wetting and nonwetting fluids.

Leverett (1941) proposed the following dimensionless number, referred to as the Leverett J-function, for convenient correlation of the capillary pressure of immiscible fluids in porous media against fluid saturation over a representative bulk volume element. It is expressed as

$$J(S_w) = \frac{p_c \sqrt{K/\phi}}{\sigma \cos\theta}, \tag{8.4}$$

where σ represents the interfacial tension between the fluid phases. θ denotes the contact angle of the fluid interface with the solid. The contact angle in porous media is a macroscopic property expressing the average value of the contact angles associated with multiphase pore fluids over a representative bulk volume element (Hirasaki, 1991; Robin et al., 1995). The wetting coefficient is defined by (Grant and Salehzadeh, 1996)

$$\kappa = \cos\theta. \tag{8.5}$$

Based on the J-function analogy, the following transformation between the capillary pressures at two different conditions, referred to as conditions A and B, can be written (Li and Horne, 2000; Civan, 2000a) as

$$p_c^A(S_w) = \frac{\sigma_A \cos\theta_A \sqrt{K_B/\phi_B}}{\sigma_B \cos\theta_B \sqrt{K_A/\phi_A}} p_c^B(S_w). \tag{8.6}$$

Lacking sufficient data on the temperature dependency of the various parameters of Eq. (8.6), Li and Horne (2000) simplified Eq. (8.6) as follows by attributing the effect of temperature on capillary functions essentially via the dependency of the surface tension on temperature:

$$p_c^T(S_w) \cong \frac{\sigma_T}{\sigma_{T_o}} p_c^{T_o}(S_w), \tag{8.7}$$

where T and T_o are the temperatures at two different conditions. They showed that the resulting theory matched the experimental data satisfactorily for the steam/water capillary pressures in Berea sandstone and geothermal materials.

Grant and Salehzadeh (1996) estimated temperature effects on capillary pressure functions based on the temperature dependency of the interfacial tension and the wetting coefficient according to

$$p_c \equiv p_c(\sigma, \kappa, S_w) = \frac{2\sigma\kappa}{r(S_w)}, \tag{8.8}$$

where $r(S_w)$ is the effective mean radius of the capillary hydraulic tubes in a partially saturated porous medium depending on the water saturation. Therefore, it follows that

$$\left(\frac{\partial p_c}{\partial T}\right)_{S_w} = \frac{p_c}{\sigma}\left(\frac{\partial \sigma}{\partial T}\right)_{S_w} + \frac{p_c}{\kappa}\left(\frac{\partial \kappa}{\partial T}\right)_{S_w}, \tag{8.9}$$

where the variation of the wetting coefficient with temperature was given by

$$\frac{\partial \kappa}{\partial T} = \frac{1}{\sigma}\left(\frac{\kappa\sigma + \Delta_G^L h^S}{T} - \kappa\frac{\partial \sigma}{\partial T}\right). \tag{8.10}$$

In Eq. (8.10), T is the absolute temperature and $\Delta_G^L h^S$ denotes the enthalpy of immersion per unit area. Grant and Salehzadeh (1996) show that the experimental data regarding interfacial tension correlate linearly with temperature. However, the surface tensions measured at different temperatures can be correlated more accurately using the following asymptotic power-law expression (Rowlinson and Widom, 1982):

$$\sigma = \sigma_o(1 - T/T_c)^m, \tag{8.11}$$

where the exponent is $m = 1.26$ and σ_o is a coefficient. T_c is the critical absolute temperature. Nevertheless, the value of $m = 1.0$ used by Grant and Salehzadeh, (1996) is close to $m = 1.26$. Therefore, their correlation is satisfactory in view of the measurement errors involved in experimental data. Furthermore, Grant and Salehzadeh (1996) showed that temperature effects on capillary pressure functions could be expressed as follows when both the wetting coefficient and the enthalpy of immersion are assumed also to be linear functions of temperature:

$$\frac{p_c(S_w)}{\left[\dfrac{\partial p_c(S_w)}{\partial T}\right]_{S_w}} = \frac{-C_1 + C_2 T + C_3 T \ln T}{C_2 + C_3 + C_3 \ln T}, \tag{8.12}$$

where C_1, C_2, and C_3 are empirical constants. Consequently, the integration of Eq. (8.12) from a reference temperature, T_r, to an observational temperature, T_f, yields

$$p_c(S_w, T_f) = p_c(S_w, T_r)\left[\frac{-C_1 + C_2 T_f + C_3 T_f \ln T_f}{-C_1 + C_2 T_r + C_3 T_r \ln T_r}\right]. \tag{8.13}$$

Provided that T_0 is a suitable scaling temperature value, a series expansion can be considered as

$$-\ln T = -\ln\left(1 - \frac{T_0 - T}{T_0}\right) = 1 - \frac{T}{T_0} + \frac{1}{2}\left(1 - \frac{T}{T_0}\right)^2 + \frac{1}{3}\left(1 - \frac{T}{T_0}\right)^3 + \dots. \tag{8.14}$$

Thus, extending the formulation by Grant and Salehzadeh (1996), Eq. (8.13) can be approximated by

$$p_c\left(S_w, T_f\right) = p_c\left(S_w, T_r\right)\left[\frac{\beta_0 + \beta_1 T_f + \beta_2 T_f^2 + \dots}{\beta_0 + \beta_1 T_r + \beta_2 T_r^2 + \dots}\right]. \tag{8.15}$$

The studies of Grant and Salehzadeh (1996) involving the drainage phenomena concerning Plano silt loam and Elkwood sandy loam indicated that the coefficients of the power series are given by

$$-398K < \beta_0 < -346K, \beta_1 \cong 1.0, \text{ and } \beta_i = 0.0 : i = 2, 3, \dots \text{ at } T_r = 298.15K. \tag{8.16}$$

Therefore, the following expression given by Grant and Salehzadeh (1996) can be used with reasonable accuracy:

$$p_c\left(S_w, T_f\right) \cong p_c\left(S_w, T_r\right)\left[\frac{\beta_0 + \beta_1 T_f}{\beta_0 + \beta_1 T_r}\right]. \tag{8.17}$$

8.4   WORK OF FLUID DISPLACEMENT

The energy required for immiscible fluid displacement in porous media depends on the relative affinities of the solid to the fluids involved in the immiscible displacement (Sharma, 1985). The work of immiscible displacement in porous materials can be expressed by

$$W = \int_{S_{wr}}^{1-S_{nr}} p_c dV_w, \tag{8.18}$$

where S_{wr} and S_{nr} denote the irreducible saturations of the wetting and nonwetting fluids (referred to as the end-point saturations), and p_c is the capillary pressure between the nonwetting and wetting phases. V_w is the wetting-phase volume, given by

$$V_w = V_b \phi S_w, \tag{8.19}$$

where V_b, ϕ, and S_w denote the bulk volume, porosity, and wetting-phase fluid saturation of the material, respectively. Hence, substituting Eq. (8.19) into Eq. (8.18), the work of immiscible fluid displacement per unit bulk volume of a homogeneous material is expressed by (Yan et al., 1997)

$$W/V_b = \phi \int_{S_{wr}}^{1-S_{nr}} p_c dS_w. \tag{8.20}$$

The integral term appearing in Eq. (8.20) is equal to the area indicated by the capillary pressure curve over the mobile fluid saturation range bounded by the end-point saturations.

As indicated by Eq. (8.4), the capillary pressure of the pore fluids is a macroscopic property, which is related to the surface tension of fluids, contact angle of fluid interface, and fluid saturation, as well as the porosity and permeability of material, in a complicated manner. Therefore, the effect of temperature variation is due to the combined effect of temperature effects on the material and fluid properties, and on the interaction forces between them. Consequently, Eq. (8.4) is a lumped representation of the temperature dependence of the various parameters involved in the immiscible displacement work.

The wettability state is defined in terms of the integrated effect of the capillary pressure variation with pore fluid saturation for the complete immiscible displacement of one fluid phase by another phase. This amounts to the work involved in changing the pore fluid condition from the immobile saturation of one fluid phase to the immobile saturation of the other. The required effort or energy can be expressed in terms of the area below the capillary pressure curve, given by

$$\int_{S_{wr}}^{1-S_{nr}} p_c dS_w \cong \left[\int_{S_{wr}}^{1-S_{nr}} p_c\left(S_w, T_r\right) dS_w \right] \left[\frac{\beta_0 + \beta_1 T_f}{\beta_0 + \beta_1 T_r} \right]. \tag{8.21}$$

Using a suitable scaling temperature value, T_*, Eq. (8.21) can be approximated, by means of a truncated series approximation, as

$$\ln \int_{S_{wr}}^{1-S_{nr}} p_c dS_w \cong \ln \left[\frac{\int_{S_{wr}}^{1-S_{nr}} p_c\left(S_w, T_r\right) dS_w}{\beta_0 + \beta_1 T_r} \right] + \ln \left\{ T_* \left[1 - \left(1 - \frac{\beta_0 + \beta_1 T_f}{T_*} \right) \right] \right\}$$

$$\cong \ln \left[\frac{T_* \int_{S_{wr}}^{1-S_{nr}} p_c\left(S_w, T_r\right) dS_w}{\beta_0 + \beta_1 T_r} \right] - \left[1 - \frac{\beta_0 + \beta_1 T_f}{T_*} \right] = a + b T_f. \tag{8.22}$$

where a and b are empirical parameters.

8.5 TEMPERATURE EFFECT ON WETTABILITY-RELATED PROPERTIES OF POROUS MEDIA

The wettability index and other related properties of porous formations can be correlated with temperature by means of the Arrhenius (1889) equation (Civan, 2004). The Arrhenius equation is an empirical equation originally introduced to express the dependency of the reaction rate coefficient (usually incorrectly called a constant) on temperature, given by

$$\ln f = \ln f_\infty - \frac{E_a}{R_g T}, \tag{8.23}$$

where the reaction rate coefficient and its high-temperature limit ($T \to \infty$) or the pre-exponential coefficient are denoted by f and f_∞, respectively. T is the absolute temperature, E_a is the activation energy, and R_g is the universal gas constant. The activation energy is the least amount of energy necessary for reactants to be able to undergo a chemical reaction. In practical words, this energy barrier is a measure of the effort required to overcome the inertia of a system in order to initiate a process. Although the Arrhenius equation was originally proposed for chemical reactions, Arrhenius-type expressions and its extension, known as the Vogel–Tammann–Fulcher (VTF) equation, have also been used successfully for temperature correlation of other parameters (Civan, 2008b), such as viscosity (Civan and Weers, 2001), emulsion stability (Civan et al., 2004), and diffusion coefficient (Callister, 2000; Civan and Rasmussen, 2003). Zhang et al. (2003) estimated the temperature effects on water flow in variably saturated soils using the activation energy concept and the Arrhenius equation.

Civan (2000d, 2004, 2008b) and Zhang et al. (2003) estimated the temperature effects on porous media processes using an Arrhenius-type equation. For example, Civan (2004) obtained good correlations of the various experimental data using Eq. (8.23) with coefficients of linear regressions very close to one by substituting $f = W$ for the work of immiscible fluid displacement; $f = W_{drainage}/W_{imbibition}$ for the wettability index WI;

$$f = \frac{1 - \phi_a}{\phi_u - \phi_a}$$

for the unfrozen water content, where ϕ_a and ϕ_u denote the mass fractions of the adsorbed and unfrozen waters of the total water/ice system; $f = V$ for the volume of unfrozen water per dry mass of porous media; and

$$f = \frac{1}{L} \int_0^S S dx$$

for the average fluid saturation, where L denotes the core length.

For example, the fraction of unfrozen water in wet soils at temperatures below the freezing temperature of water (e.g., 0°C at 1 atm) depends on the properties of soil, water, and ice (Civan, 2000d) and the interactions between these phases. Such properties determine the affinity of soil to water and ice, and hence the distribution and saturation of water and ice. Therefore, it is assumed that the dependence of the unfrozen water saturation in the pore volume should relate to the temperature dependence of wettability of the pore surface and vice versa. The wettability can be expressed by Eqs. (8.1) and (8.18) in terms of the capillary pressure, whereas the capillary pressure is related to fluid saturation, surface tension, permeability, and porosity according to Eq. (8.4).

Freezing and thawing of water in moist soils occur gradually below the freezing point of water (Civan and Sliepcevich, 1984, 1985b, 1987). Civan (2000d) analyzed unpublished data reported by Nakano and Brown (1971) for the unfrozen water content of the Point Barrow silt. Civan (2004) correlated these data according to Eq. (8.23) by plotting

$$\ln\left(\frac{1 - \phi_a}{\phi_u - \phi_a}\right)$$

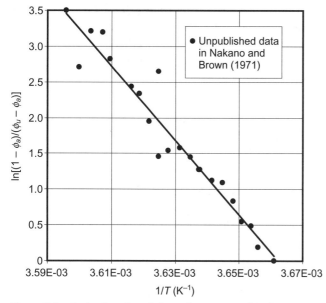

Figure 8.1   Arrhenius plot of the unfrozen water fraction versus temperature below the freezing point of water (unpublished data in Nakano and Brown, 1971) for the system of water/ice in soil (after Civan, 2004; © 2004 SPWLA, reprinted with permission from the Society of Petrophysicists and Well Log Analysts).

versus T (kelvin) with coefficients of linear regressions very close to one as shown in Figure 8.1. ϕ_a and ϕ_u denote the mass fractions of the adsorbed and unfrozen waters of the total water/ice system present in the Point Barrow silt, respectively.

In a study concerning the calculation of steam and water saturations in Berea sandstone, Li and Horne (2001) demonstrated the temperature dependence of the CT numbers obtained using an X-ray CT scanner. Li and Horne (2001) expressed the effect of temperature on the steam saturation of a steam/water pore fluid system according to the equation

$$S_{\text{steam}} = \frac{CT_{\text{wet}}(T) - CT_{\text{partial}}(T)}{CT_{\text{wet}}(T) - CT_{\text{dry}}(T)},$$
(8.24)

where $CT_{\text{dry}}(T)$ and $CT_{\text{wet}}(T)$ denote the CT numbers of the completely steam- and water-saturated material, and $CT_{\text{partial}}(T)$ denotes the CT number for the partially steam-saturated material. Because the CT number of fluid-saturated material depends on the composite density of the material and fluid system, the effect of temperature on the CT number is due to the temperature dependence of density (Li and Horne, 2001). Furthermore, because wettability determines the pore fluid distribution in a material sample, it is reasonable to assume that the temperature dependence of fluid saturation should also indicate a similar dependence trend on temperature as wettability. The validity of this issue is demonstrated in the following (Civan, 2004). The CT numbers measured by Li and Horne (2001) over the core length at 20, 60,

75, and 92°C temperatures are integrated over the core length and are then correlated with temperature. The area below the measured curve of the CT numbers versus the distance along the core plug was calculated at the four test temperatures, and the logarithmic values of the areas were plotted against temperature according to Eq. (8.23) with coefficients of linear regressions very close to one.

8.6 DIRECT METHODS FOR THE DETERMINATION OF POROUS MEDIA FLOW FUNCTIONS AND PARAMETERS

In this section, two direct methods are presented for the immiscible displacement of non-Darcy flow of variable density and viscosity fluids to determine the relative permeability and capillary pressure according to Civan and Evans (1993). The first method only uses the internal core data on saturation and saturation-weighted fluid pressure. The second method uses only the saturation-weighted fluid pressure and effluent fluid rates. The effects of fingering, porous formation compressibility, and gas solubility are neglected. The properties of realistic cores are usually heterogeneous and anisotropic. Hence, the flow characteristics of fluids during flow through the cores are not uniform through the cross-sectional area of cores due to fingering and other pertinent phenomena. In addition, the fluids do not always move as continuous streams and the flow regime for a given phase may vary from Darcy to non-Darcy over a particular cross-sectional area. The measured flow rates at the core outlet end and the pressure and saturation profiles along the core bear the effects of the irregularities that occur inside the core during fluid displacement. Hence, the Darcy and non-Darcy relative permeability, the capillary pressure, and the interfacial drag calculated by the methods presented here are considered to be the apparent properties of the actual core represented as being homogeneous and isotropic.

8.6.1 Direct Interpretation Methods for the Unsteady-State Core Tests

The equations required for the interpretation of unsteady-state immiscible displacement data are derived in the following for determination of the two-phase relative permeability data from core flow tests.

8.6.1.1 Basic Relationships One-dimensional, isothermal, and unsteady rapid flow of two immiscible and compressible fluids in an inclined, homogeneous, and isotropic porous media is considered, and the relationships essential for determination of relative permeability, non-Darcy flow coefficients, capillary pressure, and the interfacial drag force are presented according to Civan and Evans (1993). The core length is L, the cross-sectional area A, the porosity ϕ, and the permeability k. The calculation procedure is described. In the following, phases 1 and 2 refer to the wetting and nonwetting fluids, respectively. The relative permeabilities of the respective fluid phases are k_{r_1} and k_{r_2}.

The mass balance equations for phases 1 and 2 are given, respectively, by

$$\frac{\partial(\rho_1 u_1)}{\partial x} + \phi \frac{\partial(\rho_1 S_1)}{\partial t} = 0 \tag{8.25}$$

and

$$\frac{\partial(\rho_2 u_2)}{\partial x} + \phi \frac{\partial(\rho_2 S_2)}{\partial t} = 0. \tag{8.26}$$

The saturations add up to one; that is,

$$S_1 + S_2 = 1.0. \tag{8.27}$$

The momentum balance equations are given by (Tutu et al., 1983; Schulenberg and Muller, 1987)

$$-\frac{\partial p_1}{\partial x} + \rho_1 g \sin\theta = \frac{\mu_1}{kk_{r_1}} u_1 + \frac{\rho_1}{\eta\eta_{r_1}} u_1^2 + \frac{F_{12}}{S_1} \tag{8.28}$$

and

$$-\frac{\partial p_2}{\partial x} + \rho_2 g \sin\theta = \frac{\mu_2}{kk_{r_2}} u_2 + \frac{\rho_2}{\eta\eta_{r_2}} u_2^2 + \frac{F_{21}}{S_2}. \tag{8.29}$$

In Eqs. (8.178) and (8.179), K is the absolute permeability and η is the reciprocal of the inertial flow coefficient, β. F_{12} and F_{21} are the interfacial drag forces experienced by phases 1 and 2, which are the opposite and equal of each other; that is,

$$F_{21} = -F_{12}. \tag{8.30}$$

The interfacial drag force, F_{12}, is correlated as the following. Schulenberg and Muller (1987) derived an empirical correlation for the interfacial drag force, F_{12}, considering that the relevant quantities are the buoyant force, $(\rho_1 - \rho_2)g$; the viscous force in the liquid phase with respect to the relative velocity of the phases, $(\mu_1/k)(u_2/S_2 - u_1/S_1)$; the inertial force in the liquid phase with respect to the phases, $(\rho_1/\eta)(u_2/S_2 - u_1/S_1)^2$; and the capillary force, σ/k. Their equation can be written, using Eqs. (8.44) and (8.45), as

$$F_{12} = u^2 \tilde{F}_{12}, \tag{8.31}$$

where

$$\tilde{F}_{12} = \frac{kg}{\eta\sigma}\left(1 - \frac{\rho_2}{\rho_1}\right)\left(\frac{f_2}{S_2} - \frac{f_1}{S_1}\right)^2 W(S_1). \tag{8.32}$$

$W(S_1)$ is an empirically determined dimensionless function of the liquid phase saturation S_1. Schulenberg and Muller (1987) determined this function for air with different liquids and porous media of packed grains to be

$$W(S_1) = W_o S_1^m (1 - S_1), \tag{8.33}$$

with $W_o = 350$ and $m = 7$. k_r is the relative permeability and η_r is the relative reciprocal inertial flow coefficient.

The capillary pressure is given by

$$p_c(S_1) = p_2 - p_1.\tag{8.34}$$

Applying Eqs. (8.31) and (8.34) and the chain rule, Eq. (8.29) can be expressed as

$$-\frac{\partial p_1}{\partial x} + \rho_2 g \sin\theta = \frac{\mu_2}{kk_{r_2}} u_2 + \frac{\rho_2}{\eta\eta_{r_2}} u_2^2 - \frac{F_{12}}{S_2} + \frac{\partial p_c}{\partial S_1}\frac{\partial S_1}{\partial x}.\tag{8.35}$$

The relationships between volumetric fluxes and cumulative volumes of phases 1 and 2 and the total fluid are given, respectively, by

$$u_1 = \frac{1}{A}\frac{dQ_1}{dt},\tag{8.36}$$

$$u_2 = \frac{1}{A}\frac{dQ_2}{dt},\tag{8.37}$$

and

$$u = \frac{1}{A}\frac{dQ}{dt}.\tag{8.38}$$

The total volumetric flux and total cumulative volume are given by

$$u = u_1 + u_2\tag{8.39}$$

and

$$Q = Q_1 + Q_2.\tag{8.40}$$

The relationship between the volumetric flow rate and the volumetric fluxes are given by

$$q_1 = Au_1\tag{8.41}$$

and

$$q_2 = Au_2.\tag{8.42}$$

The total volumetric flow rate is given by

$$q = q_1 + q_2 = Au.\tag{8.43}$$

The fractional flow of the phases are given by

$$f_1 = u_1/u = dQ_1/dQ\tag{8.44}$$

and

$$f_2 = u_2/u = dQ_2/dQ.\tag{8.45}$$

Thus, the fractional flows add up to one:

$$f_1 + f_2 = 1.\tag{8.46}$$

Equations (8.28) and (8.35) can be manipulated for convenience. First, subtracting Eq. (8.28) from Eq. (8.35), using Eq. (8.27), and rearranging yields

$$-\frac{\partial p_c}{\partial S_1}\frac{\partial S_1}{\partial x}+(\rho_2-\rho_1)g\sin\theta=\frac{\mu_2}{kk_2}u_2-\frac{\mu_1}{kk_n}u_1+\frac{\rho_2}{\eta\eta_{r_2}}u_2^2-\frac{\rho_1}{\eta\eta_n}u_1^2-\frac{F_{12}}{S_1S_2}. \tag{8.47}$$

By means of Eqs. (8.39) and (8.44)–(8.46), Eq. (8.47) can be transformed into a fractional flow equation as

$$-\frac{\partial p_c}{\partial S_1}\frac{\partial S_1}{\partial x}+(\rho_2-\rho_1)g\sin\theta=\frac{\mu_2}{kk_2}u-\left(\frac{\mu_1}{k_n}+\frac{\mu_2}{k_{r_2}}\right)f_1$$
$$+\frac{u_2}{\eta}\left[-\frac{\rho_1 f_1^2}{\eta_n}+\frac{\rho_2(1-f_1)^2}{\eta_{r_2}}\right]-\frac{F_{12}}{S_1S_2}. \tag{8.48}$$

Second, by rearranging Eqs. (8.28) and (8.35), one obtains

$$\left(-\frac{\partial p_1}{\partial x}+\rho_1 g\sin\theta\right)\frac{kk_n}{\mu_1}=u_1+\frac{kk_n\rho_1}{\eta\eta_n\mu_1}u_1^2+\frac{kk_n}{\mu_1}\frac{F_{12}}{S_1} \tag{8.49}$$

and

$$\left(-\frac{\partial p_1}{\partial x}+\rho_2 g\sin\theta\right)\frac{kk_{r_2}}{\mu_2}=u_2+\frac{kk_{r_2}\rho_2}{\eta\eta_{r_2}\mu_2}u_2^2+\frac{kk_{r_2}}{\mu_2}\left(-\frac{F_{12}}{S_2}+\frac{\partial p_c}{\partial S_1}\frac{\partial S_1}{\partial x}\right). \tag{8.50}$$

Adding Eqs. (8.49) and (8.50) and substituting Eqs. (8.39) and (8.44)–(8.46) yields the following pressure equation:

$$-\frac{\partial p_1}{\partial x}\left(\frac{k_n}{\mu_1}+\frac{k_{r_2}}{\mu_2}\right)+g\sin\theta\left(\frac{k_n}{v_1}+\frac{k_{r_2}}{v_2}\right)=\frac{u}{k}+\frac{u^2}{\eta}\left[\frac{k_n f_1^2}{\eta_n v_1}+\frac{k_{r_2}(1-f_1)^2}{\eta_{r_2}v_2}\right], \tag{8.51}$$

in which $v_1=\mu_1/\rho_1$ and $v_2=\mu_2/\rho_2$. Note that Eqs. (8.48) and (8.51) collapse to Marle's (1981) equations when $u\to 0$, $p_c=0$, and $F_{12}=0$.

Inferred by Chung and Catton (1988) and by Schulenberg and Muller (1987), first we assume

$$\eta_r=k_r. \tag{8.52}$$

By substituting Eqs. (8.39), (8.44)–(8.46), and (8.52) into Eqs. (8.28) and (8.35), one obtains

$$k_n=\frac{u}{k}\mu_1 f_1\left(1+\frac{\rho_1 u f_1 k}{\mu_1\eta}\right)\div\left[-\frac{\partial p_1}{\partial x}+\rho_1 g\sin\theta-\frac{F_{12}}{S_1}\right] \tag{8.53}$$

and

$$k_{r_2}=\frac{u}{k}\mu_2(1-f_1)\left(1+\frac{\rho_2 u(1-f_1)k}{\mu_2\eta}\right)\div\left[-\frac{\partial p_1}{\partial x}-\frac{\partial p_c}{\partial S_1}\frac{\partial S_1}{\partial x}+\rho_2 g\sin\theta-\frac{F_{12}}{S_2}\right]. \tag{8.54}$$

Similarly, Eqs. (8.48) and (8.51) yield the following alternative expressions, respectively:

$$k_n=\frac{u}{k}\left[-\mu_1 f_1\left(1+\frac{\rho_1 u f_1 k}{\mu_1\eta}\right)+\left(\frac{k_n}{k_{r_2}}\right)\mu_2(1-f_1)\left(1+\frac{\rho_2 u(1-f_1)k}{\mu_2\eta}\right)\right]$$
$$\div\left[-\frac{\partial p_c}{\partial S_1}\frac{\partial S_1}{\partial x}+(\rho_2-\rho_1)g\sin\theta+\frac{F_{12}}{S_1S_2}\right] \tag{8.55}$$

and

$$
\frac{1}{k_{r_2}} = \left\{ -\frac{\partial p_1}{\partial x} \left[\left(\frac{k_{r_1}}{k_{r_2}} \right) \frac{1}{\mu_1} + \frac{1}{\mu_2} \right] + g \sin \theta \left[\left(\frac{k_{r_1}}{k_{r_2}} \right) \frac{1}{v_1} + \frac{1}{v_2} \right] \right.
$$
$$
\left. - F_{12} \left[\left(\frac{k_{r_1}}{k_{r_2}} \right) \frac{1}{\mu_1 S_1} - \frac{1}{\mu_2 S_2} \right] - \frac{1}{\mu_2} \frac{\partial p_c}{\partial S_1} \frac{\partial S_1}{\partial x} \right\} \div \left\{ \frac{u}{k} + \frac{\mu^2}{\eta} \left[\frac{f_1^2}{v_1} + \frac{(1-f_1)^2}{v_2} \right] \right\}.
$$

(8.56)

Eqs. (8.55) and (8.56) can be solved simultaneously for k_{r_1} and k_{r_2}, provided that the other data are available.

8.6.1.2 Solution Neglecting the Capillary End Effect for Constant Fluid Properties

Constant Injection Rate By substituting $p_c = 0$ and $p_1 = p_2 = p$ and $F_{12} = u^2 F_{12}$ (Eq. 8.33), Eq. (8.55) simplifies as

$$
-\frac{\partial p}{\partial x} = \left\{ \frac{u}{k} - g \sin \theta \left(\frac{k_{r_1}}{v_1} + \frac{k_{r_2}}{v_2} \right) + u^2 \left[\frac{1}{\eta} \left(\frac{f_1^2}{v_1} + \frac{f_2^2}{v_2} \right) \right. \right.
$$
$$
\left. \left. + \tilde{F}_{12} \left(\frac{k_{r_1}}{\mu_1 S_1} - \frac{k_{r_2}}{\mu_2 S_2} \right) \right] \right\} \div \left(\frac{k_{r_1}}{\mu_1} - \frac{k_{r_2}}{\mu_2} \right).
$$

(8.57)

On the other hand, the overall pressure differential for a core length of L is given by

$$
\Delta p = \int_0^L (-\partial p / \partial x) \, dx,
$$

(8.58)

in which $(-\partial p/\partial x)$ is given by Eq. (8.57). For constant physical properties, ρ_1, ρ_2, μ_1, μ_2, and $p_c = 0$, the saturation is the only variable in Eq. (8.57). Therefore, we will change from variable x to S_1 in Eq. (8.57). For this purpose, consider the following expression (Marle, 1981):

$$
(dx/dt)_{S_1} = (u/\phi)(df_1/dS_1)_{S_1}.
$$

(8.59)

Substituting Eq. (8.38), Eq. (8.59) becomes

$$
dx_{S_1} = (df_1/dS_1)_{S_1} \, dQ/(A\phi).
$$

(8.60)

Since $f_1 = f_1(S_1)$ only, then $(df_1/dS_1)_{S1}$ is a fixed value. Thus, integrating and applying the initial condition that

$$
x_{S_1} = 0, Q = 0, t = 0,
$$

(8.61)

Eq. (8.60) leads to an expression for the location of the point with a given saturation value as

$$
x_{S_1} = \frac{Q}{A\phi} \left(\frac{df_1}{dS_1} \right)_{S_1},
$$

(8.62)

from which

$$
dx_{S_1} = \frac{Q}{A\phi} d\left(\frac{df_1}{dS_1} \right)_{S_1}.
$$

(8.63)

If u is a constant, then Eq. (8.38) simplifies to

$$Q = uAt. \tag{8.64}$$

Substituting Eq. (8.64) into Eqs. (8.62) and (8.63) yields

$$x_{s_1} = \frac{ut}{\phi} \left(\frac{df_1}{dS_1} \right)_{s_1} \tag{8.65}$$

and

$$dx_{s_1} = \frac{ut}{\phi} d\left(\frac{df_1}{dS_1} \right)_{s_1}. \tag{8.66}$$

Substituting Eqs. (8.65) and (8.66) into Eq. (8.58) and dividing by t leads to

$$\frac{\Delta p}{t} = \int_0^{\frac{df_1}{dS_1} = \frac{\phi L}{ut}} \left(-\frac{\partial p}{\partial x} \right) \frac{u}{\phi} d\left(\frac{df_1}{dS_1} \right). \tag{8.67}$$

Taking a derivative with respect to t and applying the Leibniz rule and substituting Eq. (8.57) into Eq. (8.67) yield the following expression, which applies at $x = L$:

$$\frac{1}{L}\left(\Delta p - t\frac{d\Delta p}{dt} \right) = \left\{ \frac{u}{k} - g\sin\theta \left(\frac{k_{r_1}}{v_1} + \frac{k_{r_2}}{v_2} \right) + u^2 \left[\frac{1}{\eta}\left(\frac{f_1^2}{v^1} + \frac{f_2^2}{v^2} \right) \right. \right.$$
$$\left. \left. + \tilde{F}_{12}\left(\frac{k_{r_1}}{\mu_1 S_1} + \frac{k_{r_2}}{\mu_2 S_2} \right) \right] \right\} \div \left(\frac{k_{r_1}}{\mu_1} + \frac{k_{r_2}}{\mu_2} \right). \tag{8.68}$$

Eq. (8.68) has two unknown relative permeabilities, k_{r1} and k_{r2}.

Eq. (8.54) provides the second equation needed to obtain solutions for k_{r1} and k_{r2}.

Substituting $\eta \to \infty$, $p_c = 0$, and $\tilde{F}_{12} = 0$ into Eqs. (8.54) and (8.68), and then solving analytically, yields

$$\frac{1}{k_{r_1}} = \frac{K}{L\mu_1 f_1 u}\left[\Delta p + \rho_1 g\sin\theta L - t\frac{d\Delta p}{dt} \right] \tag{8.69}$$

and

$$\frac{1}{k_{r_2}} = \frac{\mu_1 f_1 + \frac{K}{u}(\rho_2 - \rho_1)g\sin\theta k_{r_1}}{\mu_2(1 - f_1)k_{r_1}}. \tag{8.70}$$

For constant rate q, injection $t/dt = Q/dQ$ because $Q = qt$. Note that Eqs. (8.69) and (8.70) simplify to Marle's (1981) equations for $\theta = 90°$.

Variable Injection Rate   The fractional flows are given by

$$f_1 = \frac{dQ_1}{dQ} \tag{8.71}$$

and

$$f_2 = \frac{dQ_2}{dQ}. \tag{8.72}$$

For horizontal flow, $\theta = 0°$ and (Marle, 1981):

$$\frac{k_{r_1}}{k_{r_2}} = \frac{\mu_1 f_1}{\mu_2 (1 - f_1)} = \frac{\mu_1}{\mu_2} \frac{\dfrac{dQ_1}{dQ}}{1 - \dfrac{dQ_1}{dQ}}. \tag{8.73}$$

When the total volumetric flux u is a variable, a solution similar to the preceding one can be obtained for a horizontal core, $\theta = 0$. By extending the procedure by Marle (1981) and substituting Eqs. (8.38), (8.57), (8.62), and (8.63) into Eq. (8.58) and rearranging, one obtains

$$\frac{\Delta p}{Q \dfrac{dQ}{dt}} = \frac{1}{A^2 k \phi} \int_{0}^{\frac{df_1}{dS_1} = \frac{A \phi L}{Q}} \varphi d \left(\frac{df_1}{dS_1} \right) + \frac{1}{A^3 \phi} \frac{dQ}{dt} \int_{0}^{\frac{df_1}{dS_1} = \frac{A \phi L}{Q}} \Psi d \left(\frac{df_1}{dS_1} \right)_{S_1}, \tag{8.74}$$

in which

$$\varphi = \left(\frac{k_{r_1}}{\mu_1} + \frac{k_{r_2}}{\mu_2} \right)^{-1} \tag{8.75}$$

and

$$\Psi = \left[\frac{1}{\eta} \left(\frac{f_1^2}{v_1} + \frac{f_2^2}{v_2} \right) + \tilde{F}_{12} \left(\frac{k_{r_1}}{\mu_1 S_1} + \frac{k_{r_2}}{\mu_2 S_2} \right) \right] \varphi. \tag{8.76}$$

By taking a derivative with respect to Q and applying the Leibniz rule and rearranging, Eq. (8.74) yields

$$\zeta = \int_{0}^{\frac{df_1}{dS_1} = \frac{A \phi L}{Q}} \Psi d \left(\frac{df_1}{dS_1} \right), \tag{8.77}$$

in which

$$\zeta = A^3 \phi \frac{d}{dQ} \left[\frac{\Delta p}{Q \dfrac{dQ}{dt}} \right] + \frac{A^2 \phi L}{k} \frac{1}{Q^2} \left[\varphi + \frac{k}{A} \frac{dQ}{dt} \Psi \right] \div \frac{d}{dQ} \left(\frac{dQ}{dt} \right). \tag{8.78}$$

By taking another derivative with respect to Q, Eq. (8.77) yields

$$\frac{d\zeta}{dQ} = \Psi \left(-\frac{A \phi L}{Q^2} \right). \tag{8.79}$$

In Eq. (8.78),

$$\frac{d}{dQ}\left(\frac{dQ}{dt}\right) = \frac{dt}{dQ}\frac{d}{dt}\left(\frac{dQ}{dt}\right) = \left(\frac{dQ}{dt}\right)^{-1}\frac{d^2Q}{dt^2}. \tag{8.80}$$

Eqs. (8.54) and (8.53) apply at the core outlet end (i.e., $x = L$) and can be solved numerically for k_{r1} and k_{r2}.

When $\eta \to \infty$ and $\tilde{F}_{12} = 0$, Eqs. (8.77) and (8.78) can be combined and simplified to lead to Marle's (1981) equation:

$$AK\frac{d}{dQ}\left[\frac{\Delta p}{Q\,\dfrac{dQ}{dt}}\right] = -\frac{L}{Q^2}\frac{1}{\dfrac{k_{r1}}{\mu_1} + \dfrac{k_{r2}}{\mu_2}}. \tag{8.81}$$

Thus, solving Eqs. (8.73) and (8.81) yields Marle's equations:

$$\frac{1}{k_{r1}} = \frac{AK}{\mu_1 L\dfrac{dQ_1}{dt}}\left[\Delta p\left(1 + Q\frac{\dfrac{d^2Q}{dt^2}}{\left(\dfrac{dQ}{dt}\right)^2}\right) - Q\frac{d\Delta p}{dQ}\right] \quad \text{and} \quad \frac{1}{k_{r2}} = \frac{\mu_1}{\mu_2}\frac{\dfrac{dQ_1}{dQ}}{1 - \dfrac{dQ_1}{dQ}}\frac{1}{k_{r1}}. \tag{8.82}$$

8.6.1.3 Inferring Function and Function Derivative Values from Average Function Values

The following is an extension of the procedure given by Jones and Roszelle (1978) to determine $\partial f/\partial x$ from f.

According to the mean value theorem

$$\bar{f} = \frac{1}{x}\int_0^x f\,dx, \tag{8.83}$$

from which we obtain by differentiating

$$f = \bar{f} + x\frac{\partial \bar{f}}{\partial x} \tag{8.84}$$

and

$$\frac{\partial f}{\partial x} = \frac{\partial}{\partial x}\left(\bar{f} + x\frac{\partial \bar{f}}{\partial x}\right), \tag{8.85}$$

Eq. (8.84) can be reformulated as follows:

$$f = \bar{f} + x\frac{\partial \bar{f}}{\partial x}\frac{\partial(Q/x)}{\partial(Q/x)}$$
$$f = \bar{f} + x\frac{\partial \bar{f}}{\partial(Q/x)}\frac{\partial(Q/x)}{\partial x}. \tag{8.86}$$

Similarly, Eq. (8.85) leads to

$$\frac{\partial f}{\partial x} = \frac{\partial (Q/x)\partial}{\partial (Q/x)\partial x}\left(\overline{f} + x\frac{\partial \overline{f}}{\partial x}\frac{\partial (Q/x)}{\partial (Q/x)}\right)$$
$$= \frac{\partial (Q/x)}{\partial x}\frac{\partial}{\partial (Q/x)}\left(\overline{f} + x\frac{\partial \overline{f}}{\partial (Q/x)}\frac{\partial (Q/x)}{\partial x}\right). \tag{8.87}$$

Now consider the following:

$$\partial\left(\frac{Q}{x}\right) = \frac{1}{x}\partial Q - \frac{Q}{x^2}\partial x. \tag{8.88}$$

Thus,

$$\frac{\partial (Q/x)}{\partial x} = \frac{1}{x}\frac{\partial Q}{\partial x} - \frac{Q}{x^2}. \tag{8.89}$$

Since $Q = Q(t)$ only, Eq. (8.89) yields

$$\frac{\partial (Q/x)}{\partial x} = -\frac{Q}{x^2}. \tag{8.90}$$

By substituting Eq. (8.90), Eqs. (8.86) and (8.87), respectively, become

$$f = \overline{f} - \frac{Q}{x}\frac{\partial \overline{f}}{\partial (Q/x)} \tag{8.91}$$

and

$$\frac{\partial f}{\partial x} = -\frac{Q}{x^2}\frac{\partial}{\partial (Q/x)}\left[\overline{f} - \frac{Q}{x}\frac{\partial \overline{f}}{\partial (Q/x)}\right]. \tag{8.92}$$

Eq. (8.92) can be expanded to obtain

$$\frac{\partial f}{\partial x} = \frac{Q^2}{x^3}\frac{\partial^2 \overline{f}}{\partial (Q/x)^2}. \tag{8.93}$$

Applying Eq. (8.88) at $x = L$ yields

$$\partial (Q/x)_{x=L} = \frac{1}{L}\partial Q. \tag{8.94}$$

Thus, by substituting Eq. (8.94), Eqs. (8.91) and (8.93) at $x = L$, respectively, become

$$f|_{x=L} = \overline{f}_L - Q\frac{d\overline{f}}{dQ} \tag{8.95}$$

and

$$\left.\frac{\partial f}{\partial x}\right|_{x=L} = \frac{Q^2}{L^2}\frac{d^2 \overline{f}}{dQ^2}. \tag{8.96}$$

8.6.1.4 Relationships for Processing Experimental Data   The calculation of relative permeability and capillary pressure values from fluid displacement data requires additional relationships derived in this section. For this purpose, we consider two cases of measurements. In the first case, we assume that the dynamic saturation profiles along the core can be measured by well-established methods such as NMRI, X-ray tomography, or ultrasonic imaging methods. In the second case, we assume that only the effluent fluid volumes can be measured. In both cases, we assume that the dynamic total fluid pressures along the core are measurable by appropriate methods such as by means of pressure transducers mounted along the core.

Evaluation of S_1 and $\partial S_1/\partial x$ If S_1 versus x are measured, then $\partial S_1/\partial x$ can be directly evaluated from these data. If only the efflux fluid rates are measured, then S_1 and $\partial S_1/\partial x$ can be evaluated indirectly by the method described as follows.

Define average density and saturation over the core length according to the mean value theorem, respectively, by

$$\bar{\rho}_1 = \frac{1}{x}\int_0^x \rho_1 dx \tag{8.97}$$

and

$$\bar{\rho}_1\bar{S} = \frac{1}{x}\int_0^x \rho_1 S_1 dx. \tag{8.98}$$

For an incompressible fluid, Eq. (8.98) becomes

$$\bar{S}_1 = \frac{1}{x}\int_0^x S_1 dx. \tag{8.99}$$

Differentiating Eq. (8.98) twice, with respect to x, gives

$$\rho_1 S_1 = \bar{\rho}_1\bar{S}_1 + x\frac{\partial\left(\bar{\rho}_1\bar{S}_1\right)}{\partial x} \tag{8.100}$$

and

$$\frac{\partial\left(\rho_1 S_1\right)}{\partial x} = 2\frac{\partial\left(\bar{\rho}_1\bar{S}_1\right)}{\partial x} + x\frac{\partial^2\left(\bar{\rho}_1\bar{S}_1\right)}{\partial x^2}. \tag{8.101}$$

Note that ρ_1 is calculated at the pressure p_1 determined by Eq. (8.122) given later. Following the procedure given earlier, we can express Eqs. (8.100) and (8.101) at $x = L$ in terms of the cumulative injection Q^* of the displacing phase at $x = 0$, where * indicates the conditions at $x = 0$, as

$$\rho_1 S_1 = \bar{\rho}_1\bar{S}_1 - Q^*\frac{d\left(\bar{\rho}_1\bar{S}_1\right)}{dQ^*} \tag{8.102}$$

and

$$\frac{\partial\left(\rho_1 S_1\right)}{\partial x} = \left(\frac{Q^*}{L}\right)^2\frac{d^2\left(\bar{\rho}_1\bar{S}_1\right)}{dQ^{*2}}. \tag{8.103}$$

For an incompressible fluid, Eqs. (8.102) and (8.103) becomes

$$S_1 = \bar{S}_1 - Q^* \frac{d\bar{S}_1}{dQ^*} \tag{8.104}$$

and

$$\frac{\partial S_1}{\partial x} = \left(\frac{Q^*}{L} \right)^2 \frac{d^2 \bar{S}_1}{dQ^{*2}}. \tag{8.105}$$

To express the second-order derivative on the right of Eq. (8.103), we use Eq. (8.25). Integrating Eq. (8.25) over the core length yields

$$\rho_1 u_1 - \rho_1^* u_1^* + \phi \frac{d}{dt} \int_0^L \rho_1 S_1 dx = 0. \tag{8.106}$$

By substituting Eqs. (8.97) and (8.98) into Eq. (8.106), one obtains

$$\rho_1 u_1 - \rho_1^* u_1^* + \phi L \frac{d(\bar{\rho}_1 \bar{S}_1)}{dt} = 0. \tag{8.107}$$

By substituting Eq. (8.36) for u_1 and u_1^* and dividing by dQ^*, Eq. (8.107) leads to

$$\frac{d(\bar{\rho}_1 \bar{S}_1)}{dQ^*} = \frac{1}{A\phi L} \left[\rho_1^* \frac{dQ_1^*}{dQ^*} - \rho_1 \frac{dQ_1}{dQ^*} \right]. \tag{8.108}$$

Integrating Eq. (8.108) for a cumulative injection from zero to Q^* yields

$$\bar{\rho}_1 \bar{S}_1 = \bar{\rho}_1 \bar{S}_1 \big|_{\substack{Q^*=0 \\ (t=0)}} + \frac{1}{A\phi L} \int_0^{Q^*} \left(\rho_1^* \frac{dQ_1^*}{dQ^*} - \rho_1 \frac{dQ_1}{dQ^*} \right) dQ^*. \tag{8.109}$$

Differentiating, Eq. (8.108) yields

$$\frac{d^2(\bar{\rho}_1 \bar{S}_1)}{dQ^{*2}} = \frac{1}{A\phi L} \frac{d}{dQ^*} \left[\rho_1^* \frac{dQ_1^*}{dQ^*} - \rho_1 \frac{dQ_1}{dQ^*} \right]. \tag{8.110}$$

For an incompressible fluid, $Q = Q^*$ and Eqs. (8.108)–(8.110) become

$$\frac{d\bar{S}_1}{dQ^*} = \frac{1}{A\phi L} \left[\frac{dQ_1^*}{dQ^*} - \frac{dQ_1}{dQ^*} \right] = \frac{f_1^* - f_1}{A\phi L}, \tag{8.111}$$

$$\bar{S}_1 = \bar{S}_1 \big|_{\substack{Q^*=0 \\ (t=0)}} + \frac{1}{A\phi L} (Q_1^* - Q_1), \tag{8.112}$$

and

$$\frac{d^2 \bar{S}_1}{dQ^{*2}} = \frac{1}{A\phi L} \left[\frac{df_1^*}{dQ^*} - \frac{df_1}{dQ^*} \right]. \tag{8.113}$$

If the core is initially saturated by phase 2 to the immobile residual saturation of phase 1, S_{1r}, and then only phase 1 is injected, then

$$Q^* = Q_1^* \tag{8.114}$$

and

$$\rho_1 S_1 \big|_{t=0} = \rho_1 \big|_{t=0} S_{1r}. \tag{8.115}$$

Thus, Eqs. (8.108)–(8.110) become

$$\frac{d(\bar{\rho}_1 \bar{S}_1)}{dQ^*} = \frac{1}{A\phi L} \frac{d}{dQ^*} \left[\rho_1^* - \rho_1 \frac{dQ_1}{dQ^*} \right], \tag{8.116}$$

$$\bar{\rho}_1 \bar{S}_1 = \rho_1 \big|_{t=0} S_{1r} + \frac{1}{A\phi L} \left[\rho_1^* Q^* - \int_0^{Q^*} \rho_1 \frac{dQ_1}{dQ^*} dQ^* \right], \tag{8.117}$$

and

$$\frac{d^2(\bar{\rho}_1 \bar{S}_1)}{dQ^{*2}} = -\frac{1}{A\phi L} \frac{d}{dQ^*} \left[\rho_1 \frac{dQ_1}{dQ^*} \right]. \tag{8.118}$$

Again, assuming an incompressible fluid, Eqs. (8.116)–(8.118) become

$$\frac{d\bar{S}_1}{dQ^*} = \frac{1}{A\phi L} \left[1 - \frac{dQ_1}{dQ^*} \right] = \frac{1}{A\phi L}(1 - f_1), \tag{8.119}$$

$$\bar{S}_1 = S_{1r} + \frac{1}{A\phi L}(Q^* - Q_1) = S_{1r} + \frac{Q_2}{A\phi L}, \tag{8.120}$$

and

$$\frac{d^2 \bar{S}_1}{dQ^{*2}} = -\frac{1}{A\phi L} \frac{d^2 Q_1}{dQ^{*2}} = -\frac{1}{A\phi L} \frac{df_1}{dQ^*}. \tag{8.121}$$

Evaluation of p_1 and $\partial p_1/\partial x$ Consider a saturation-weighted two-phase fluid pressure defined by (Lewis and Schrefler, 1987)

$$p = S_1 p_1 + S_2 p_2. \tag{8.122}$$

The equation is based upon a volumetric average, although a more appropriate average for pressure would be an average over the total solid contact surface (Civan and Evans, 1993). Because the measurement of the fraction of the total area contacted by each fluid phase is difficult, the saturation weighting given in Eq. (8.122) represents an approximation.

We assume p can be measured along the core. By means of Eqs. (8.27) and (8.34), Eq. (8.122) can be written as

$$p_1 = p - (1 - S_1) p_c, \tag{8.123}$$

from which we obtain $\partial p_1/\partial x$ directly by a numerical method. Alternatively, by differentiating Eq. (8.123) with respect to x and rearranging, we can obtain

$$\frac{\partial p_1}{\partial x} = \frac{\partial p}{\partial x} + \left[p_c - (1 - S_1) \frac{\partial p_c}{\partial S_1} \right] \frac{\partial S_1}{\partial x}. \tag{8.124}$$

Evaluation of Dynamic Capillary Pressure p_c and $\partial p_c/\partial S_1$ can be calculated as follows. Consider an alternative definition of average saturation according to

$$\overline{S}_1 = \frac{\displaystyle\int_{p_c^*}^{p_c} \rho_1 S_1 dp_c}{\displaystyle\int_{p_c^*}^{p_c} \rho_1 dp_c} \tag{8.125}$$

and define

$$\overline{\rho}_1 = \frac{\displaystyle\int_{p_c^*}^{p_c} \rho_1 dp_c}{p_c - p_c^*}. \tag{8.126}$$

We assume Eqs. (8.125) and (8.126) yield the same averages obtained by Eqs. (8.97) and (8.98). Hence, Eqs. (8.125) and (8.126) can be combined to obtain

$$\overline{\rho}_1 \overline{S}_1 = \frac{\displaystyle\int_{p_c^*}^{p_c} \rho_1 S_1 dp_c}{p_c - p_c^*}. \tag{8.127}$$

p_c^* is the capillary pressure at the inlet conditions of the core. Since immediately after the injection begins the immobile residual saturation of phase 2 is reached, therefore, p_c^* is the capillary pressure at

$$S_1 = 1 - S_{2r}, x = 0, t > 0. \tag{8.128}$$

Differentiating Eq. (8.127) yields

$$\rho_1 S_1 = \overline{\rho}_1 \overline{S}_1 + \left(p_c - p_c^*\right) \frac{d\left(\overline{\rho}_1 \overline{S}_1\right)}{dp_c}. \tag{8.129}$$

Comparing Eqs. (8.102) and (8.129) yields

$$\frac{dp_c}{p_c - p_c^*} = -\frac{dQ^*}{Q^*}. \tag{8.130}$$

Integrating Eq. (8.130) results in the following expression for the dynamic capillary pressure:

$$p_c = \frac{C}{Q^*} + p_c^*. \tag{8.131}$$

C is an integration constant, which can be determined using a known condition. From Eq. (8.131),

$$\frac{\partial p_c}{\partial S_1} = -\frac{C}{Q^{*2}} \frac{\partial Q^*}{\partial S_1}. \tag{8.132}$$

Evaluation of the Static Capillary Pressure   Kalaydjian (1992) proposed the following expression for the correction of static capillary pressure to obtain the dynamic capillary pressure:

$$p_c = p_{c_{\text{static}}} + \frac{\partial p_c}{\partial S_1} \frac{\partial S_1}{\partial t}. \tag{8.133}$$

q is the total flow rate given by Eq. (8.43). An alternative similar expression including the variation of porosity is given by Pavone (1990). By means of Eqs. (8.38) and (8.43), Eq. (8.133) can be written as

$$p_c = p_{c_{\text{static}}} + \frac{\partial p_c}{\partial S_1} \frac{\partial S_1}{\partial Q^*} \tag{8.134}$$

or simply

$$p_c = p_{c_{\text{static}}} + \frac{dp_c}{dQ^*}. \tag{8.135}$$

Substituting Eq. (8.131) into Eq. (8.135) gives

$$p_{c_{\text{static}}} = C\left(\frac{1}{Q^*} + \frac{1}{Q^{*2}}\right) + p_c^*. \tag{8.136}$$

8.6.1.5 Applications   The method developed in the preceding section is applied with a typical nitrogen/brine displacement data for illustration. For this purpose, a vertical core is saturated fully with brine and then the displacement process is initiated by injecting nitrogen gas from the upper end of the core. The data of Ouenes et al. (1992) are described as the following. Core data: $L = 37.8\,\text{cm}$, $A = 11.7\,\text{cm}^2$, $\theta = 90°$, $\phi = 0.301$, $K = 0.311\,\text{darcy}$; fluid data: index $1 = $ nitrogen, index $2 = $ brine, $\rho_1 = 0.00875\,\text{g/cm}^3$, $\rho_2 = 1.003\,\text{g/cm}^3$, $\mu_1 = 0.018\,\text{cP}$, $\mu_2 = 1.02\,\text{cP}$, $S_{1r} = 0$, $S_{2r} = 0.633$, $q = 0.0417\,\text{cm}^3/\text{s}$; and assumed values: $\sigma = 72$ dyne/cm, $p_{c_{\text{max}}} = 0.003\,\text{atm}$, $C = 4$, and $W = 55$. The displacement data by Ouenes et al. (1992) are presented in Table 8.1. By using these data with Eqs. (8.31)–(8.33), (8.40), (8.44), (8.52), (8.53), (8.104), (8.105), (8.119)–(8.121), (8.124), (8.131), (8.132), and (8.136), the relative permeability and capillary pressure data for the nitrogen/brine system are calculated. Typical numerical results are presented in Table 8.1. Figures 8.2 and 8.3 show the plot of the results (Civan and Evans, 1993).

8.6.2 Tóth et al. Formulae for the Direct Determination of Relative Permeability from Unsteady-State Fluid Displacements

Tóth et al. (2002) derived convenient formulae for the construction of two-phase relative permeability curves from immiscible fluid displacement data obtained by laboratory core flow tests. Their method allows for the direct calculation of relative permeability from constant rate and constant pressure displacement tests. The tests need to be conducted at sufficiently high flow rates so that the capillary end effects

TABLE 8.1 Immiscible Displacement Experimental Data (Modified after Civan and Evans, © 1993 SPE, with Permission from the Society of Petroleum Engineers)

Data of Ouenes et al. (1992)			Results by Civan and Evans (1993)			
Q_1^* (cm³)	Q_2 (cm³)	ΔP (atm)	S_1	k_{r1}	k_{r2}	P_c (atm)
9	9	0.30	0	0	1.33	0.497
16	16	0.38	0	0	1.047	0.264
20	20	0.44	0.026	0.003	0.757	0.213
24	22	0.39	0.076	0.01	0.517	0.174
26	23	0.38	0.095	0.013	0.429	0.158
29	24	0.35	0.106	0.015	0.401	0.142
35	26	0.29	0.118	0.019	0.404	0.115
48	29	0.21	0.142	0.03	0.386	0.081
68	33	0.16	0.176	0.042	0.312	0.056
91	35	0.14	0.212	0.051	0.185	0.041
123	36	0.13	0.233	0.055	0.119	0.03
173	38	0.12	0.247	0.059	0.088	0.02
231	40	0.11	0.255	0.063	0.076	0.014
580	45	0.08	0.289	0.077	0.047	0.004
890	47	0.07	0.319	0.078	0.026	0.001
1150	48	0.06	0.34	0.08	0.015	0
1450	49	0.05	0.367	0.076	0	0

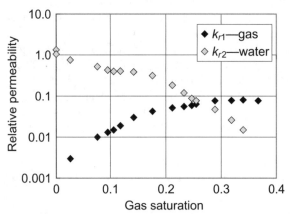

Figure 8.2   Relative permeability curves (modified after Civan and Evans, 1993; © 1993 SPE, with permission of the Society of Petroleum Engineers).

can be neglected. Their method can be applied only for processing the after-breakthrough displacement test data obtained by one-dimensional immiscible two-phase fluid displacements in horizontal core plugs so that the effect of gravity can be neglected. Further, core and fluid properties are assumed constant and homogeneous. This method utilizes special functions of fluid mobility and characteristic parameters of immiscible displacement.

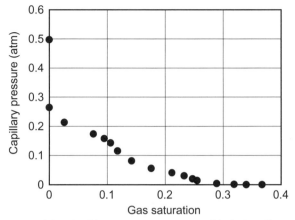

Figure 8.3   Capillary pressure curve (modified after Civan and Evans, 1993; © 1993 SPE, with permission of the Society of Petroleum Engineers).

8.6.2.1 Determination of Relative Permeability under Variable Pressure and Rate Conditions

Tóth et al. (2002) first derived the formulae for the calculation of relative permeability under variable pressure difference Δp and injection flow rate q_i test conditions as described in this section. Here, L, K, and A denote the length, permeability, and cross-sectional area of core plugs. P, μ, and k_r are the fluid pressure, viscosity, and relative permeability. The subscripts i, d, and k indicate the injected, displacing, and displaced fluid phases.

The gravity effect and the capillary pressure gradient (dp_c / dx) are neglected. Darcy's law is applied for flow of the displacing (d) and displaced (k) fluid phases, respectively, as

$$q_d = -KA \frac{k_{rd}}{\mu_d} \frac{\partial p}{\partial x} \qquad (8.137)$$

and

$$q_k = -KA \frac{k_{rk}}{\mu_k} \frac{\partial p}{\partial x}, \qquad (8.138)$$

where x denotes the distance measured from the core inlet.

For incompressible fluids, the volumetric balance of fluids reads as

$$q_i = q_d + q_k. \qquad (8.139)$$

Combining Eqs. (8.137)–(8.139) yields

$$-\frac{dp}{dx} = \frac{q_i}{KA} \frac{1}{Y(S_d)}, \qquad (8.140)$$

where S_d is the saturation of the displacing phase and $Y(S_d)$ is the total mobility function, given by

$$Y(S_d) = \frac{k_{rd}}{\mu_d} + \frac{k_{rk}}{\mu_k}. \qquad (8.141)$$

The core inlet- and outlet-end pressures are specified, respectively, as

$$p = p_1(t), \, x = 0 \tag{8.142}$$

and

$$p = p_2(t), \, x = L, \tag{8.143}$$

where $p_1 > p_2$. Integrating Eq. (8.140) over the core length and applying the boundary conditions given by Eqs. (8.142) and (8.143) yields

$$p_1 - p_2 = \Delta p = \frac{q_i}{KA} \int_0^L \frac{1}{Y(S_d)} dx. \tag{8.144}$$

Note that the Buckley and Leverett (1942) equation is given by

$$x(S_d) = \frac{V_i(t)}{A\varphi} \frac{df_d}{dS_d}. \tag{8.145}$$

The cumulative volume of the injected or displacing fluid phase is given by

$$V_i(t) = \int_0^t q_i dt. \tag{8.146}$$

The displacing phase is injected at the core inlet. Thus,

$$S_d = 1 - S_{kr}, f_d = 1, \frac{df_d}{dS_d} = 0, \frac{\partial}{\partial x}\left(\frac{df_d}{dS_d}\right) = 0, \text{ and } x = 0. \tag{8.147}$$

The displaced and displacing fluid phases are produced together at the core outlet after breakthrough. Thus,

$$S_d = S_{d_2}, f_d = f_{d_2}, \frac{df_d}{dS_d} = \left(\frac{df_d}{dS_d}\right)_2, \text{ and } x = L. \tag{8.148}$$

Combining Eqs. (8.144), (8.145), (8.147), and (8.148) yields

$$\Delta p(t) = \frac{q_i(t)V_i(t)}{KA^2\varphi} G, \tag{8.149}$$

where the G-function is given by

$$G = \int_0^{f_{d2}} \frac{1}{Y(S_d)} d\left(\frac{df_d}{dS_d}\right). \tag{8.150}$$

The derivative of Eqs. (8.149) and (8.150) with respect to time yields, respectively,

$$\frac{d[\Delta p(t)]}{dt} = \frac{1}{KA^2\varphi}\left[q_i(t)V_i(t)\frac{dG}{dt} + G\frac{d}{dt}[q_i(t)V_i(t)]\right] \tag{8.151}$$

and

$$\frac{dG}{dt} = -\frac{1}{Y(S_{d_2})} \frac{AL\varphi}{V_i^2(t)} q_i(t).$$ (8.152)

Substituting Eqs. (8.149) and (8.152) into Eq. (8.151) reproduces the equation of Marle (1981):

$$\frac{d[\Delta p(t)]}{dt} = \frac{\Delta p(t)}{q_i(t)V_i(t)} \frac{d}{dt}[q_i(t)V_i(t)] - \frac{q_i^2(t)L}{KAY(S_{d_2})V_i(t)}.$$ (8.153)

An empirical linear equation is used to correlate the after-breakthrough cumulative injected fluid volume (V_i) and the produced displacing and displaced fluid volumes (V_d, V_k) according to (Welge, 1952; Tóth, 1995; Tóth et al., 1998)

$$\frac{V_i(t)}{V_k} = a + b\frac{V_i(t)}{V_p}, (t \geq t_a).$$ (8.154)

Here, t_a is the breakthrough time. The parameters a and b are given as the following:

$$a = f_{kf} = 1 - f_{df}, 0 < a < 1$$ (8.155)

and

$$b = 1/(S_{d\,max} - S_{di}), b > 1,$$ (8.156)

where f_{kf} and f_{df} are the fractional flows of the displaced and displacing fluids at breakthrough, S_{dmax} is the maximum displacing fluid saturation obtained after an infinite throughput, and S_{di} is the initial displacing fluid saturation in the core.

The volumetric fluid flow rates at the core outlet are given by

$$q_k = \frac{dV_k}{dt}$$ (8.157)

and

$$q_d = \frac{dV_d}{dt}.$$ (8.158)

Eqs. (8.154) and (8.158) provide

$$f_d = \frac{q_d}{q_i} = 1 - f_k$$ (8.159)

and

$$f_k = \frac{q_k}{q_i} = \frac{a}{\left[a + b\frac{V_i(t)}{V_p}\right]^2}.$$ (8.160)

Combining Eqs. (8.137), (8.138), (8.159), and (8.160) yields an expression for the mobility ratio of the fluids as

$$M_{d2} = \frac{q_d}{q_k} = \frac{\dfrac{k_{rd}}{\mu_d}}{\dfrac{k_{rk}}{\mu_k}} = \frac{f_d}{f_k} = \frac{1}{f_k} - 1 = \frac{\left[a + b\dfrac{V_i(t)}{V_p}\right]^2}{a} - 1. \tag{8.161}$$

A solution of Eqs. (8.141) and (8.161) yields the following equations for the relative permeability:

$$k_{rd} = \mu_d \frac{M_{d2}Y(S_{d2})}{M_{d2} + 1} = \mu_d f_d Y(S_{d2}) \tag{8.162}$$

and

$$k_{rk} = \mu_k \frac{Y(S_{d2})}{M_{d2} + 1} = \mu_k f_k Y(S_{d2}). \tag{8.163}$$

The core average and outlet face saturations are given as functions of the injected cumulative fluid volume, respectively, by

$$\overline{S_d} - S_{di} = \frac{\dfrac{V_i(t)}{V_p}}{a + b\dfrac{V_i(t)}{V_p}} \tag{8.164}$$

and

$$S_{d2} = S_{di} + b(\overline{S_d} - S_{di})^2 = S_{di} + b\left[\frac{\dfrac{V_i(t)}{V_p}}{a + b\dfrac{V_i(t)}{V_p}}\right]^2. \tag{8.165}$$

8.6.2.2 Determination of Relative Permeability under Constant Pressure Conditions
When Δp is maintained constant, the injection flow rate varies according to

$$q_i = \frac{dV_i(t)}{dt}. \tag{8.166}$$

Eq. (8.153) becomes

$$Y(S_{d_2}) = \frac{L}{\Delta p K A} \frac{q_i^3(t)}{\dfrac{d}{dt}[q_i(t)V_i(t)]}. \tag{8.167}$$

Alternatively, substituting Eq. (8.166) into Eq. (8.167) yields

$$Y(S_{d_2}) = \frac{L}{\Delta p K A} \frac{\left[\dfrac{dV_i(t)}{dt}\right]^3}{\left[\dfrac{dV_i(t)}{dt}\right]^2 + V_i(t)\dfrac{d^2V_i(t)}{dt^2}}. \tag{8.168}$$

The cumulative injected fluid volume is correlated by an empirical power-law equation, given by (Tóth et al., 1998)

$$V_i(t) = a_2 t^{b_2}, b_2 \geq 1. \tag{8.169}$$

Eqs. (8.168) and (8.169) are combined to obtain

$$Y(S_{d2}) = \frac{La_2 b_2^2 \left(\dfrac{V_p}{a_2}\right)^{(1-1/b_2)}}{\Delta p K A (2b_2 - 1)} \left[\frac{V_i(t)}{V_p}\right]^{(1-1/b_2)}, b_2 \geq 1. \tag{8.170}$$

8.6.2.3 Determination of Relative Permeability under Constant Rate Conditions
When q_i is maintained constant, Δp varies and Eq. (8.146) becomes

$$V_i(t) = q_i t. \tag{8.171}$$

Then, combining Eqs. (8.153) and (8.171) reproduces the equation of Marle (1981):

$$Y(S_{d_2}) = \frac{q_i L}{K A} \frac{1}{\left[\Delta p(t) - t \dfrac{d\Delta p(t)}{dt}\right]}. \tag{8.172}$$

The after-breakthrough pressure difference can be correlated by an empirical power-law equation, given by (Toth et al., 1998)

$$\Delta p(t) = a_1 \left(\frac{q_i t}{V_p}\right)^{b_1} = a_1 \left[\frac{V_i(t)}{V_p}\right]^{b_1}, b_1 \leq 0, t \geq t_a. \tag{8.173}$$

Combining Eqs. (8.172) and (8.173) yields

$$Y(S_{d2}) = \frac{q_i L}{K A a_1 (1 - b_1) \left[\dfrac{V_i(t)}{V_p}\right]^{b_1}}, b_1 \leq 0. \tag{8.174}$$

8.6.2.4 Applications for Data Analysis
Tóth et al. (2002) calculated the relative permeability data by applying the above-given formulae using the imbibition experimental data of Jones and Roszelle (1978): core length, 12.7 cm; core cross-sectional area, 11.04 cm^2; core porosity, 0.215 fraction; core pore volume, 31.13 cm^3; core permeability, 0.0354 μm^2; core irreducible water saturation, 0.35 fraction; oil viscosity, 10.45 cP; and water viscosity, 0.97 cP. Figures 8.4–8.6 show the straight-line plots of Eqs. (8.154), (8.169), and (8.173), respectively, obtained by the least squares linear regression of the experimental data using the after-breakthrough displacement data. The best estimate values of parameters a and b of Eq. (8.154), a_1 and b_1 of Eq. (8.173) for constant rate displacement, and a_2 and b_2 of Eq. (8.169) for constant pressure displacement have been determined by the least squares method. The parameter values obtained for displacement under $\Delta p = 6.900$-bar conditions are $a = 0.392$, $b = 2.99$, $a_2 = 0.00708$, and $b_2 = 1.16$. The parameter

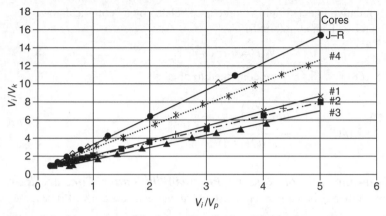

Figure 8.4 Linear plots of the injected-to-displaced fluid volume ratio versus the injected fluid-to-pore volume ratio for various core fluid systems (after Tóth et al., 2002; with permission from Elsevier).

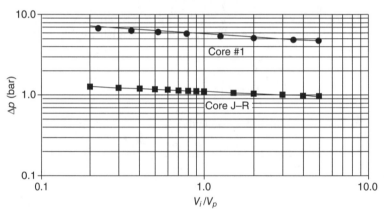

Figure 8.5 Linear plots of the pressure differential across the core versus the injected fluid-to-pore volume ratio for various core fluid systems (after Tóth et al., 2002; with permission from Elsevier).

values obtained for displacement under $q_i = 0.0222\,\text{cm}^3/\text{s}$ conditions are $a = 0.404$, $b = 2.99$, $a_1 = 12.5$, and $b_1 = 0.110$. The total mobility functions were calculated using Eqs. (8.170) and (8.174) for constant pressure and constant rate displacements, respectively, as a function of the saturation using Eqs. (8.164) and (8.165). The displacing fluid-phase fractional flow, f_d, versus S_{d2} was calculated using Eqs. (8.159) and (8.160). As can be seen in Figure 8.7, the imbibition relative permeability curves obtained under constant pressure and constant rate water injections match each other closely.

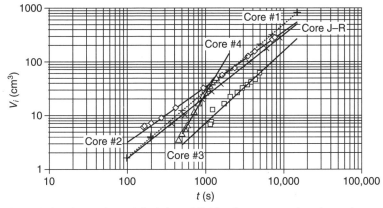

Figure 8.6   Linear plots of the injected fluid volume versus time for various core fluid systems (after Tóth et al., 2002; with permission from Elsevier).

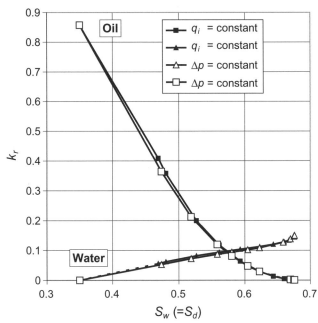

Figure 8.7   Comparison of Jones and Roszelle (1978) Test# J-R/a and b imbibition relative permeability curves obtained by constant pressure and constant rate water injections (after Tóth et al., 2002; with permission from Elsevier).

8.7 INDIRECT METHODS FOR THE DETERMINATION OF POROUS MEDIA FLOW FUNCTIONS AND PARAMETERS

Indirect methods for analysis of steady-state and unsteady-state core flow tests are presented according to Civan and Evans (1991, 1993) and Ucan et al. (1997). A piecewise functional representation of relative permeability and capillary pressure data from laboratory core fluid displacement data can be determined uniquely provided that the transient-state internal saturation profiles and the overall pressure differentials are used simultaneously for history matching. Global functional representations of relative permeability and capillary pressure data do not satisfactorily describe the flow functions. A discrete representation of the flow functions results in nonsmooth functions. A unique representation also requires that the number of estimated model parameters is less than or equal to the number of observable parameters and simulated annealing is a convenient method to achieve a global optimization for determining the best estimates of relative permeability and capillary pressure from laboratory core fluid displacement data.

8.7.1 Indirect Method for Interpretation of the Steady-State Core Tests

In the steady-state method, $\partial/\partial t = 0$ and Eqs. (8.175) and (8.176) lead to the following prescribed mass flux conditions for the fluid phases 1 and 2 according to Civan and Evans (1991, 1993):

$$\rho_1 u_1 = (\rho_1 u_1)_{x=0} = \text{ct.} \tag{8.175}$$

$$\rho_2 u_2 = (\rho_2 u_2)_{x=0} = \text{ct.} \tag{8.176}$$

The saturation and pressure profiles along the core plug are obtained by solving simultaneously

$$\frac{dS_1}{dx} = -\left(\frac{dp_c}{dS_1}\right)^{-1} \left[\begin{array}{c} \dfrac{\mu_2}{Kk_{r_2}} u_2 - \dfrac{\mu_1}{Kk_{r_1}} u_1 - (\rho_2 - \rho_1) g \sin\theta \\[3mm] + \dfrac{\rho_2}{\eta\eta_{r_2}} u_2^2 - \dfrac{\rho_1}{\eta\eta_{r_1}} u_1^2 - \dfrac{F_{12}}{S_1 S_2} \end{array} \right] \tag{8.177}$$

and

$$\frac{dp_1}{dx} = -\left[\frac{\mu_1}{Kk_{r_1}} u_1 - \rho_1 g \sin\theta + \frac{\rho_1}{\eta\eta_{r_1}} u_1^2 + \frac{F_{12}}{S_1} \right]. \tag{8.178}$$

In addition,

$$S_1 + S_2 = 1.0 \tag{8.179}$$

and

$$p_c = p_2 - p_1. \tag{8.180}$$

The conditions necessary for the solution of Eqs. (8.177) and (8.178) can be derived as the following. At the outlet face of the core, the fluids are at the same pressure as soon as they leave the core. Therefore,

$$p_c = 0 \quad \text{and} \quad p_1 = p_2 = p_{\text{out}}, \, x = L \tag{8.181}$$

for an intermediately wet system. The saturation value at this condition is determined for zero capillary pressure as

$$S_1 = (S_1)_{p_c=0}, \, x = L. \tag{8.182}$$

Only for a strongly fluid 1-wet system,

$$S_1 = 1 - S_{2i}, x = L. \tag{8.183}$$

Because the conditions at the core outlet face are specified for saturation and pressure, Eqs. (8.177) and (8.178) must be solved backward starting with the values prescribed by Eqs. (8.181) and (8.182). A numerical solution of Eqs. (8.177) and (8.178) subject to the conditions given by Eqs. (8.181) and (8.182) can be readily obtained by an appropriate numerical solution method, such as a Runge–Kutta method for ordinary differential equations. However, to obtain a numerical solution, the relative permeability and capillary pressure data are required. These unknown data are assumed by a trial-and-error method until the numerical solution satisfactorily matches the measured data, such as internal saturation and pressure profiles if they can be measured, or the core length average saturation and pressure difference across the core for a range of prescribed injection fluid mass fluxes. The initial guess of the permeability and capillary pressure data is assumed linear over the normalized saturation range. The best estimates of the permeability and capillary pressure curves are obtained iteratively until the numerical solution matches the measured saturation and/or pressure profiles. See exercise problem 1, for example.

8.7.2 Unsteady-State Core Test History Matching Method for the Unique and Simultaneous Determination of Relative Permeability and Capillary Pressure

Ucan et al. (1997) investigated the uniqueness and the simultaneous predictability of the flow functions by history matching as described in the following. For this purpose, (1) the flow functions are chosen as the only model parameters to be estimated; (2) the number of observable parameters is enlarged by using both external and internal data; (3) the flow functions are represented by global empirical functions, discrete values, and piecewise continuous local functions; (4) a finite difference solution of the model is used to describe the multiphase flow in a laboratory core; and (5) the simulated annealing method is used as a nonlinear global optimization technique because it does not require the evaluation of the first or second gradients of the objective function and it usually converges to the global optimum without increasing the computational effort significantly.

8.7.2.1 Formulation of a Two-Phase Flow in Porous Media   Ucan et al. (1997) describe the flow of oil/water systems assuming (1) immiscible fluids, (2) no adsorption, (3) isothermal flow, (4) Newtonian fluids, (5) Darcy's law is applicable, and (6) one-dimensional flow.

The conservation equations for water and oil phases are given, respectively, as

$$-\frac{\partial}{\partial x}\left(\frac{u_w}{B_w}\right) = \frac{\partial}{\partial t}\left(\frac{S_w}{B_w}\phi\right) + q_w \tag{8.184}$$

and

$$-\frac{\partial}{\partial x}\left(\frac{u_o}{B_o}\right) = \frac{\partial}{\partial t}\left(\frac{S_o}{B_o}\phi\right) + q_o, \tag{8.185}$$

where we substituted the following definitions of the formation volume factors:

$$B_w = \rho_s^w/\rho_w, \ B_o = \rho_s^o/\rho_o. \tag{8.186}$$

where density with subscript s indicates a value at the standard condition.

The saturation constraint is

$$S_w + S_o = 1. \tag{8.187}$$

The capillary pressure is given by

$$P_c = P_o - P_w. \tag{8.188}$$

The Darcy equations for the water and oil phases are given, respectively, by

$$u_w = -\frac{kk_{rw}}{\mu_w}\left(\frac{\partial P_w}{\partial x} - \gamma_w \sin\theta\right) \tag{8.189}$$

and

$$u_o = -\frac{kk_{ro}}{\mu_o}\left(\frac{\partial P_o}{\partial x} - \gamma_o \sin\theta\right), \tag{8.190}$$

where $\gamma_w = \rho_w g$ and $\gamma_o = \rho_o g$; g denotes the gravitational acceleration.

Substituting the Darcy equations into the mass conservation equations for each phase yields the following pressure saturation equations:

$$\frac{\partial}{\partial x}\left[K_w\left(\frac{\partial P_w}{\partial x} - \gamma_w \sin\theta\right)\right] = \frac{\partial}{\partial t}\left(\frac{S_w}{B_w}\phi\right) + q_w \tag{8.191}$$

and

$$\frac{\partial}{\partial x}\left[K_o\left(\frac{\partial P_o}{\partial x} - \gamma_o \sin\theta\right)\right] = \frac{\partial}{\partial t}\left(\frac{S_o}{B_o}\phi\right) + q_o, \tag{8.192}$$

where K_w and K_o are the conductivities of the oil and water phases given, respectively, by

$$K_o = \frac{kk_{ro}}{B_o\mu_o} \tag{8.193}$$

and

$$K_w = \frac{kk_{rw}}{B_w\mu_w}. \tag{8.194}$$

The initial condition for the pressure and saturation equations are given by

$$P_w = P_{initial} \ \ 0 \le x \le L, t = 0 \tag{8.195}$$

and

$$S_w = S_{initial} \ \ 0 \le x \le L, t = 0. \tag{8.196}$$

The boundary conditions can be specified as constant rate or pressure. The source and sink terms, q_w and q_o, are taken as zero everywhere except at the inlet and outlet. Hence, the flow is introduced as a source/sink. A no-flow boundary condition is used at the inlet, given by

$$\frac{\partial P_w}{\partial x} = 0. \; x = 0, t \geq 0. \tag{8.197}$$

The boundary condition at the outlet for pressure before breakthrough is given by

$$\frac{\partial P_w}{\partial x} = 0. \; X = L, t \geq 0, \tag{8.198}$$

and after breakthrough, the pressure is specified as outlet pressure:

$$P_w = P_{\text{specified}} \; x = L, t \geq 0. \tag{8.199}$$

The length of the core is divided into N equal blocks with the grid points located in the centers of these blocks. Thus, there are no points at the core inlet- and outlet-end boundaries. An implicit pressure and explicit saturation (IMPES) method is used for the numerical solution (Aziz and Settari, 1979).

8.7.2.2 Representation of Flow Functions

The flow functions can be described by discrete point values or by global or local functions. For a discrete description, the range of the initial to the residual saturation is divided into a number (N) of discrete points. Following Civan and Evans (1991), the initial guesses for the flow functions are first assumed to vary linearly with saturation. The values of the flow functions at discrete points are determined by history matching.

For oil and water flow, the empirical relative permeability and the capillary pressure expressions are given, respectively, by the following power-law functions of saturation (Brooks and Corey, 1966):

$$k_{rw} = k_{rw}^o (S_D)^{n_1}, \tag{8.200}$$
$$k_{ro} = k_{ro}^0 (1 - S_D)^{n_2}, \tag{8.201}$$

and

$$P_c = P_{c,\max} (S_D)^{n_3}, \tag{8.202}$$

where the normalized saturation is given by

$$S_D = \frac{S_w - S_{wc}}{1 - S_{or} - S_{wc}}. \tag{8.203}$$

The values of the n_1, n_2, and n_3 parameters are determined for a global functional representation by history matching. End-point values of relative permeability, k_{rw}^o and k_{ro}^o, are assumed to be known by other means, such as from transient well tests, injection, and production data. The maximum value of the capillary pressure, $P_{c,\max}$, was also assumed to be known.

Eqs. (8.200)–(8.202) are fit to the experimental measurements. However, the determination of a single general function is impractical for large data sets and strongly nonlinear behavior. Instead, it is more convenient to describe data locally in a piecewise continuous manner by fitting a series of empirical functions over a series of interdata segments using cubic splines.

For example, the relative permeability to water is described by the following expression:

$$k_{rwi} = a_i S_D^3 + b_i S_D^2 + c_i S_D + d_i, \tag{8.204}$$

where a_i, b_i, c_i, and d_i are fitting coefficients, which assume different values over various segments, and S_D is defined by Eq. (8.203). For N data points, there are $(N - 1)$ intervals. Consequently, there are $4(N - 1)$ unknown coefficients to be determined, that is, $a_1, a_2, \ldots, a_{N-2}, a_{N-1}; b_1, b_2, \ldots, b_{N-2}, b_{N-1}; c_1, c_2, \ldots, c_{N-2}, c_{N-1};$ and $d_1, d_2, \ldots, d_{N-2}, d_{N-1}$.

The continuity condition requires that the functional values at the interior points should be equal:

$$k_{rw_i}^- = k_{rw_i}^+ \quad i = 1, 2, 3, \ldots, N - 2. \tag{8.205}$$

The compatibility condition requires that the first and second derivatives at the interior points should be equal:

$$\left(\frac{\partial k_{rw_i}}{\partial S_D} \right)^- = \left(\frac{\partial k_{rw_i}}{\partial S_D} \right)^+ \quad i = 1, 2, 3, \ldots, N - 2 \tag{8.206}$$

and

$$\left(\frac{\partial^2 k_{rw_i}}{\partial S_D^2} \right)^- = \left(\frac{\partial^2 k_{rw_i}}{\partial S_D^2} \right)^+ \quad i = 1, 2, 3, \ldots, N - 2. \tag{8.207}$$

Further, the end-point conditions are given by

$$\frac{\partial k_{rw1}}{\partial S_D} = C_1 \; S_D = 0 \tag{8.208}$$

$$k_{rw1} = 0 \; S_D = 0, \tag{8.209}$$

and

$$k_{rw,N-1} = k_{rw}^0 \; S_D = 1, \tag{8.210}$$

where k_{rw}^0 is the end-point relative permeability, which is assumed to be known, and C_1 is the slope of relative permeability at the connate water and residual oil saturations depending on the degree of the wettability. This value can be also estimated as an additional parameter. For a strongly wet system, C_1 is very close to zero. In this study, it is assumed to be zero. Thus, applying Eqs. (8.208)–(8.210), Eqs. (8.205)–(8.207) provide

$$C_1 = 0, \tag{8.211}$$

$$d_1 = 0, \tag{8.212}$$

and

$$b_{N-2} = k_{rw}^0 - a_{N-2} - c_{N-2} - d_{N-2}. \tag{8.213}$$

The following equations are obtained using Eqs. (8.205)–(8.207) at the first interior point:

$$(b_1 - b_2)S_D^2 + (c_1 - c_2)S_D + d_1 - d_2 = a_2 S_D^3 - a_1 S_D^3, \tag{8.214}$$

$$(2b_1 - 2b_2)S_D + c_1 - c_2 = 3a_2 S_D^2 - 3a_1 S_D^2, \tag{8.215}$$

and

$$2b_1 - 2b_2 = 6a_2 S_D - 6a_1 S_D. \tag{8.216}$$

Eqs. (8.214)–(8.216) yield

$$d_2 = (a_1 - a_2)S_D^2 + d_1 \tag{8.217}$$

and

$$c_2 = 3(a_2 - a_1)S_D^2 + c_1. \tag{8.218}$$

Consequently, only $a_1, a_2, \ldots, a_{N-2}, a_{N-1}$ need to be estimated initially to start the optimal search method. Note that d_1 and c_1 are already calculated from the end-point conditions (Eqs. 8.211 and 8.212) and that d_2 and c_2 can be calculated from Eqs. (8.217) and (8.218). The remaining coefficients are determined by the recurrence relationship for the interior points according to Eqs. (8.217) and (8.218):

$$d_i = (a_{i-1} - a_i)S_D + d_{i-1} \quad i = 3, \ldots N-1 \tag{8.219}$$

and

$$c_i = 3(a_i - a_{i-1})S_D + c_{i-1} \quad i = 3, \ldots N-1. \tag{8.220}$$

The coefficients of $b_{N-1}, \ldots, b_1$ are obtained by using Eqs. (8.213) and (8.216) to obtain

$$b_i = 3(a_{i+1} - a_i)S_D + b_{i+1} \quad i = N-2, \ldots 1. \tag{8.221}$$

A similar approach is taken for the nonwetting-phase relative permeability and the capillary pressure.

8.7.2.3 Parameter Estimation Using the Simulated Annealing Method
The simulated annealing method (SAM) (Metropolis et al., 1953; Kirkpatric et al., 1983; Farmer, 1989) is a global optimization method. The best representation of the discrete values or global or local functions is accomplished to minimize the difference between the predicted and measured core fluid displacement data.

For this purpose, the objective function is expressed in terms of the saturation profiles and the production and pressure drop history (Chardaire et al., 1989):

$$J(a) = W_Q \sum_k (Q_k^c - Q_k^m)^2 + W_p \sum_k (\Delta P_k^c - \Delta P_k^m)^2$$
$$+ W_s \sum_k (S_{k,i}^c - S_{k,i}^m)^2. \tag{8.222}$$

To increase the rate of convergence, the terms defined in the objective function, such as the cumulative production rate and the pressure drop, are scaled to the same order of magnitude by appropriate weighting coefficients W_Q, W_p, and W_S or by using dimensionless forms of the immiscible displacement flow equations. The effect of core heterogeneity on the relative permeability and capillary pressure curves can be considered by using a permeability and porosity distribution along the core. The fitting parameters, such as $a_1, a_2, \ldots, a_n$ of the cubic splines (Eq. 8.204) or n_1, n_2, and n_3 of the power-law functions (Eqs. 8.200–8.202), are determined by simulated annealing to minimize Eq. (8.222).

The search is started by making arbitrary initial guesses for the model parameters. The objective function (J_i), Eq. (8.222), is then calculated by means of the numerical solution of the flow model (the forward problem). The value of the new objective function is sequentially reduced toward a global optimum according to the following procedure:

1. The flow functions and the objective function value, (J_{i+1}), are calculated by reducing or increasing the values of the parameter.

2. The new perturbed flow function is accepted unconditionally as the new flow function when the value of the new objective function decreases, $J_{i+1} < J_i$.

3. The new perturbed flow functions may still be accepted when the value of the new objective function increases, $J_{i+1} > J_i$, if the Metropolis acceptance rule described next is satisfied. Otherwise, the nonimproved new value is rejected and a new flow function is generated.

The acceptance of the nonimproved move, $J_{i+1} > J_i$, is stochastic and depends on the probability based on the value of $\exp(-\Delta J/T)$, where $\Delta J = J_{i+1} - J_i$ and T is the value of the control parameter to be decreased. The value of $\exp(-\Delta J/T)$ will always be between $(0, 1)$ because ΔJ is always positive for a nonimproved move and T is also always positive. A random number, generated from a uniform distribution $(0, 1)$, is compared with the exponential quantity, $\exp(-\Delta J/T)$. The nonimproved move, $J_{i+1} > J_i$, is accepted if the random number is less than the exponential number. If both left and right neighbor searches lead to a nonimproved solution, $(\Delta J > 0)$, then the decision to accept either nonimproved moves is based on the magnitude of the potential move. For large values of T, a large perturbation of parameters will be accepted. As the value of T approaches zero, no perturbation of parameters will be accepted at all. In order to jump out of a local minimum and to continue searching for a better flow function, the initial T should be sufficiently high. Therefore, most of the alterations are accepted at the beginning of the simulation.

Simulated annealing algorithms also require a decrement function that specifies the lowering of the value of the control parameter. After a complete pass through all the flow functions at each discrete points, T is replaced by the old T multiplied by a constant. The constant is chosen to be less than one. Different asymptotic decrement functions can also be used, such as the logarithmic schedule, which is suggested by Aarts and Korst (1989) and by Eglese (1990). However, Ucan et al. (1997) used the geometric decrement schedule because of its simplicity. For a given specific problem, a decrement function can be developed also by a trial-and-error method (Kirkpatric et al., 1983).

The data sets of drainage and imbibition tests were used in the following sections for the simultaneous determination of the relative permeability and capillary pressure curves.

8.7.2.4 Applications for Drainage Tests In the drainage case, displacement of oil by water in an oil-wet core, using the saturation history profiles and pressure differential data for global functional and piecewise local functional representation of flow functions, is considered.

First, Ucan et al. (1997) tested the computer code for the forward problem by using experimental data obtained from Richmond and Watson (1990) with the rock properties and operating conditions given as $k = 9.6\,\text{md}$, $\phi = 0.262$, $L = 7.13\,\text{cm}$, $A = 11.3\,\text{cm}^2$, $S_{wi} = 0.10$, $Q = 2.0\,\text{cm}^3/\text{min}$, μ_w, $= 0.262\,\text{cP}$, and $\mu_o = 0.725\,\text{cP}$. The relative permeability and capillary pressure curves (Fig. 8.8) are representative of an oil-wet system. Ucan et al., (1993) checked the core flood simulator to determine whether it could handle the end effect properly for drainage experiments, and the results were compared with Richmond and Watson's (1990) data. The pressure drop history and the cumulative production history were matched successfully.

As a second exercise, the uniqueness and the predictability of the flow functions were investigated. For this purpose, the global empirical functions were first attempted for a history match. As the initial estimate, flow function exponents were assumed to be $n_1 = 2$, $n_2 = 2$, and $n_3 = 2$ for the drainage experiment. Applying

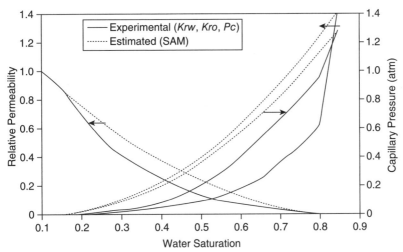

Figure 8.8 Comparison of the experimental (forward) and assumed global functional representation of flow function ($n_1 = 2$, $n_2 = 2$, $n_3 = 2$) (after Ucan et al., 1997; © 1997 SPE, with permission from the Society of Petroleum Engineers). SAM, simulated annealing method.

simulated annealing, the global optimal values of the exponent were determined to be $n_1 = 5.1444$, $n_2 = 2.8326$, and $n_3 = 2.4857$. A comparison of the experimental and the newly estimated flow functions is given in Figure 8.9. The relative permeability of water at high saturations is not as good as it is at low saturations. The pressure drop history was matched well, but the cumulative production history (Fig. 8.10) is

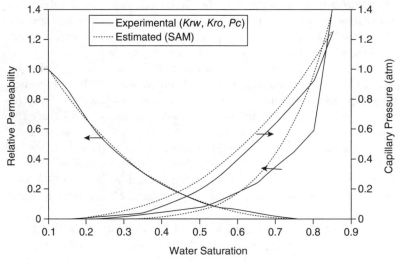

Figure 8.9   Comparison of experimental (forward) and estimated (inverse) representation of flow function by simulated annealing method ($n_1 = 5.1444$, $n_2 = 2.8326$, and $n_3 = 2.4857$) (after Ucan et al., 1997; © 1997 SPE, with permission from the Society of Petroleum Engineers).

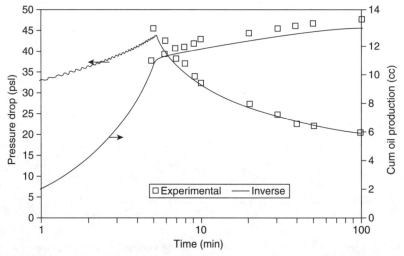

Figure 8.10   Pressure drop and cumulative production history for the drainage experiment by using global functional representation ($n_1 = 5.1444$, $n_2 = 2.8326$, and $n_3 = 2.4857$) (after Ucan et al., 1997; © 1997 SPE, with permission from the Society of Petroleum Engineers).

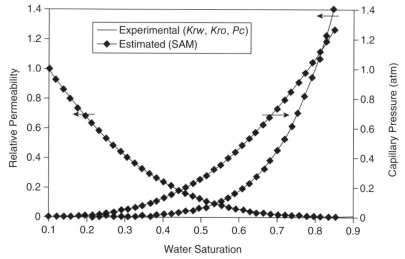

Figure 8.11   Comparison of experimental (forward) and estimated (inverse) global functional representations of flow function (after Ucan et al., 1997; © 1997 SPE, with permission from the Society of Petroleum Engineers).

still not satisfactory and flow functions are not retrieved accurately. This confirms the limitation of using an empirical global functional representation of the flow functions.

For the inverse numerical code, first, the newly estimated flow functions were used as the forward problem; then, the new observable parameters were generated. By using the generated observable parameters, the flow functions were recalculated. As shown in Figure 8.11, the global functional representations of the flow functions were recovered exactly by the simulated annealing method.

In order to improve the history matching and to retrieve the initial experimental flow function data, the saturation range was divided into five segments from the initial to the residual saturation and a piecewise functional representation was used. Saturation history profiles and the pressure drop history were used as constraints to generate the flow functions. Figure 8.12 shows that the initial flow functions were retrieved successfully within a reasonable range of accuracy. With these two constraints, the pressure drop history, the cumulative production history (Fig. 8.13), and the saturation history profiles (Fig. 8.14) were also matched well. Thus, a local piecewise continuous functional representation is a better approach for the determination of the flow functions.

8.7.2.5   *Applications for Imbibition Tests*

For imbibition processes, the capillary pressure reaches zero when the wetting-phase pressure equals the nonwetting-phase pressure and then the production of the injected phase begins. Consider the displacement of oil by water in a water-wet core, using (1) the recovery curves and pressure differential data and (2) the saturation history profiles and pressure

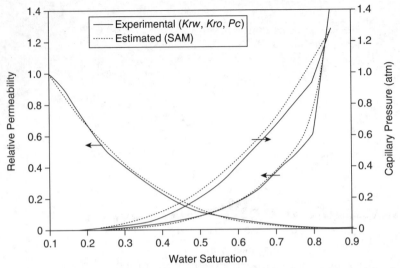

Figure 8.12 Comparison of experimental (forward) and estimated (inverse) local functional representation of flow function by the simulated annealing method (after Ucan et al., 1997; © 1997 SPE, with permission from the Society of Petroleum Engineers).

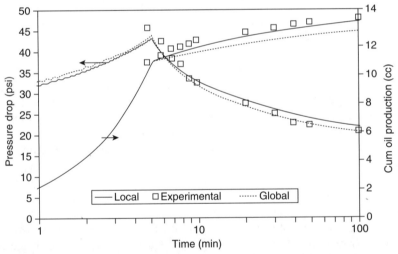

Figure 8.13 Pressure drop and cumulative production history for the drainage experiment by using global and local functional representations (after Ucan et al., 1997; © 1997 SPE, with permission from the Society of Petroleum Engineers).

differential data for discrete and piecewise local functional representations of the flow functions. Ucan et al. (1997) tested the numerical code using the simulated experimental data obtained from Richmond and Watson (1990). The rock properties and operating conditions are given as $k = 1270$ md, $\phi = 0.249$, $L = 7.62$ cm, $A = 5.06$ cm^2, $S_{wi} = 0.2882$, $Q = 0.36$ cm^3/min, $\mu_w = 1.0$ cP, and $\mu_o = 10.0$ cP. The

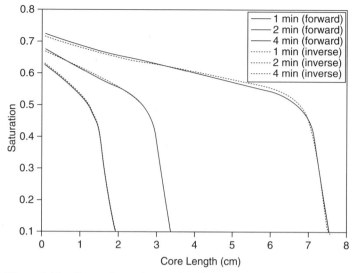

Figure 8.14   Comparison of saturation history profiles using the estimated and initial flow functions (after Ucan et al., 1997; © 1997 SPE, with permission from the Society of Petroleum Engineers).

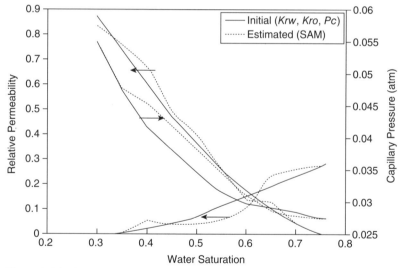

Figure 8.15   Comparison of initial and estimated discrete representations of flow function by the simulated annealing method (after Ucan et al., 1993, 1997; © 1993, 1997 SPE, with permission from the Society of Petroleum Engineers).

relative permeability and capillary pressure curves in Figure 8.15 are representative of a strong water-wet system. Results matched successfully the pressure drop history and the cumulative production histories.

First, the flow functions by discrete representation, Figure 8.15, were estimated by using only the recovery curves and pressure differential data. Second,

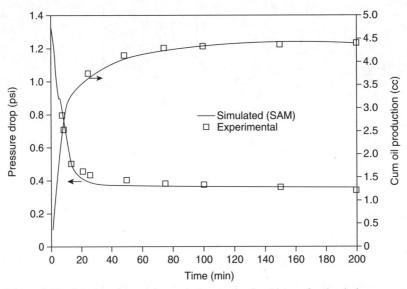

Figure 8.16   Pressure drop and cumulative production history for the drainage experiment by using discrete representation of flow functions (after Ucan et al., 1997; © 1997 SPE, with permission from the Society of Petroleum Engineers).

the pressure drop and cumulative production of oil and water were recalculated by using the estimated relative permeability and capillary pressure. Finally, the pressure drop history, the cumulative production history (Fig. 8.16), and the saturation history profiles (Fig. 8.17) calculated from the estimated flow properties were compared to the initial history performance. As shown in Figure 8.16, a good history match was obtained for the pressure drop and cumulative production history with the estimated flow function properties. However, the saturation history profiles (Fig. 8.17) did not match well. This indicates that matching saturation history profiles guarantees matching of the cumulative production history, but the reverse is not true. A similar procedure was applied for a piecewise functional representation by using only the external data. As seen from Figure 8.18, the relative permeability of oil does not match the initial curve, but the recovery of water relative permeability and capillary pressure curves are within an acceptable range. The history matching of the pressure drop and the cumulative production, and the saturation profiles are given in Figures 8.19 and 8.20, respectively. By using the piecewise functional representation, the history matching of the saturation history profiles (Fig. 8.20) was improved as compared to the discrete representation (Fig. 8.17).

Next, the saturation history profiles and pressure drop history were used as constraints to generate the flow functions by a piecewise functional representation (Fig. 8.21). The initial flow functions were retrieved successfully. With these two constraints, the pressure drop history, the cumulative production history (Fig. 8.22), and the saturation history profiles (Fig. 8.23) were matched well.

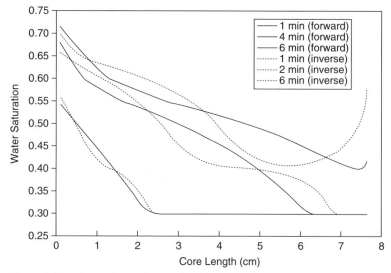

Figure 8.17   Comparison of saturation history profiles for the imbibition experiment by using experimental and discrete representations of flow functions (after Ucan et al., 1993, 1997; © 1993, 1997 SPE, with permission from the Society of Petroleum Engineers).

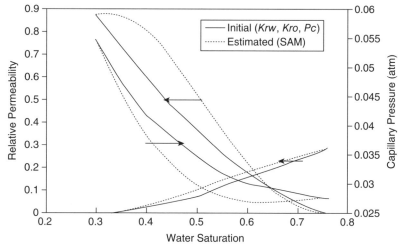

Figure 8.18   Comparison of initial and estimated local functional representations of flow functions by using external core data (after Ucan et al., 1997; © 1997 SPE, with permission from the Society of Petroleum Engineers).

As the last exercise, only the before-breakthrough information was used to generate the flow functions since conventional unsteady-state methods only provide information after breakthrough. The average water saturation prior to break-through is 0.63. The flow functions obtained from the unsteady-state method will be limited to the saturation range (0.63–0.7625). By using saturation history profiles

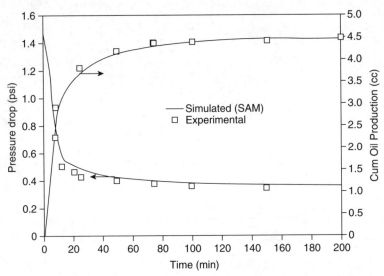

Figure 8.19 Pressure drop and cumulative production history for the drainage experiment by using local functional representation (after Ucan et al., 1997; © 1997 SPE, with permission from the Society of Petroleum Engineers).

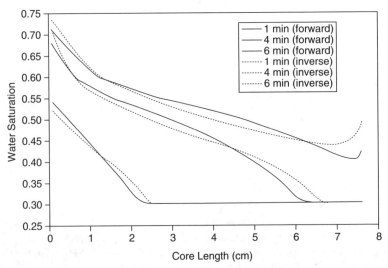

Figure 8.20 Comparison of saturation history profiles using the estimated and initial flow functions (after Ucan et al., 1997; © 1997 SPE, with permission from the Society of Petroleum Engineers).

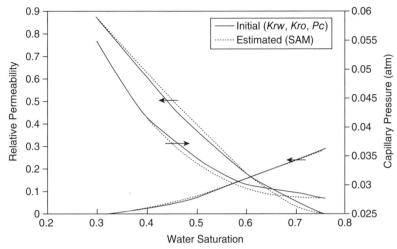

Figure 8.21   Comparison of initial and estimated local functional representations of flow functions by using internal and external core data (after Ucan et al., 1997; © 1997 SPE, with permission from the Society of Petroleum Engineers).

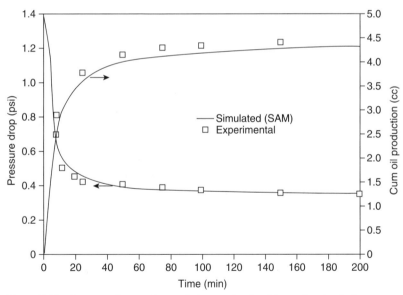

Figure 8.22   Pressure drop and cumulative production history for the drainage experiment by using local functional representation (after Ucan et al., 1997; © 1997 SPE, with permission from the Society of Petroleum Engineers).

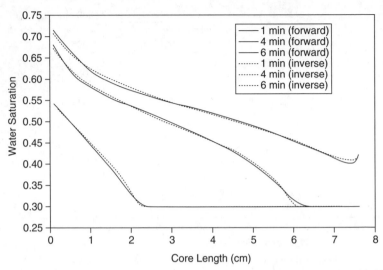

Figure 8.23   Comparison of saturation history profiles using estimated and initial flow functions (after Ucan et al., 1997; © 1997 SPE, with permission from the Society of Petroleum Engineers).

and pressure drop information prior to breakthrough, similar flow functions (Fig. 8.21) were obtained. From these three different runs, it can be concluded that using the recovery curves with the pressure drop history is not sufficient to determine the shape of the flow functions accurately. The estimated flow and the calculated saturation history profiles based on the discrete representation are not smooth monotonic functions like the saturation profiles and the flow functions obtained from the piecewise functional representation. By using the saturation history profiles, the pressure history drop along with a piecewise functional representation, the flow functions can be retrieved accurately. Removing the saturation history profiles as a constraint on the problem leads to nonunique flow functions.

8.8   EXERCISES

1. Calculate and plot the saturation profiles over the core length for various phase 1 and phase 2 mass flow rates by applying the indirect method for interpretation of the steady-state core flow tests given in the following for plug number 6686 (Arastoopour and Semrau, 1989): horizontal core, $L = 3.88\,cm$; $A = 5.07\,cm^2$; porosity is $\phi = 0.1367$; immobile wetting fluid saturation is $S_{1i} = 0.4065$; immobile gas saturation is $S_{2i} = 0.0$; wetting fluid is water ($l = 1$); nonwetting fluid is nitrogen ($l = 2$); temperature is $T = 306.5\,K$; base conditions are $p_b = 1\,atm$ and $T_b = 306.5\,K$; water density at base condition is $0.9947\,g/cm^3$ and water viscosity at base condition is $0.744\,cP$; gas viscosity at base condition is $1.11 \times 10^{-3}\,cP$; the effective permeability are

TABLE 8.2   Conditions of Various Experimental Data (Modified after Civan, 2004; © 2004 SPWLA, with the Permission of the Society of Petrophysicists and Well Log Analysts)

Source of Experimental Data	Porous Material	Fluid System	Property
Novak (1975)	Granular glass	Air/water	Work of immiscible
Nimmo and Miller (1986)	Glass beads	Air/water	displacement
Nimmo and Miller (1986)	Plainfield sand	Air/water	
Constantz (1991)	Oakley sand	Air/water	
Unpublished data in Nakano and Brown (1971)	Point Barrow silt	Water/ice	Unfrozen water content
Norinaga et al. (1999)	Yallourn coal Beulah–Zap coal Illinois #6 coal	Water/ice	
Madden and Strycker (1989)	Berea sandstone	Mineral oil/water New London without polars/water New London without asphatenes and polars/water	Wettability index
Li and Horne (2001)	Berea sandstone	Steam/water	Fluid saturation

given by $K_{e1} = 10.02 \times 10^{-6}[(S_1 - S_{1i})/(1 - S_{1i})]^4$ and $K_{e2} = 31.9 \times 10^{-6}(1.0 + 2.92/p_2)$ $(-0.0995 - 0.472 \ln S_1)$; and the capillary pressure is $p_c = 0.701 + 3.02/S_1^2 - p_b$. Make reasonable assumptions for any missing data.

2. Develop a numerical solution for the unsteady-state method.

3. Develop a numerical solution for the Ucan et al. (1997) method.

4. Develop a numerical solution for the Tóth et al. (2002) method.

5. The equations of Tóth et al. (2002) described in this chapter are applicable for linear flow, such as on laboratory core plugs used for fluid flow. Transform these equations so that they can be used for radial flow as described by Tóth et al. (2010)

6. Correlate the data of various wettability-related properties of porous media with temperature using the Arrhenius (1889) equation. For this purpose, consider the data described in Table 8.2.

7. Applying the Tóth et al. (2006) method, construct the relative permeability curves for the oil and water phases using the production data given by Odeh and Dotson (1985) for the case of oil displacing water in a water-saturated core summarized in Table 8.3 by Civan and Donaldson (1987). Note that $K = 0.025$ darcy, porosity = 0.227, core diameter $D = 2.54$ cm, core length $L = 7.62$ cm, water viscosity $\mu_w = 0.93$ cP, oil viscosity $\mu_o = 3.02$ cP, connate water saturation $S_{wc} = 0.42$, residual oil saturation $S_{ro} = 0.24$, and oil injection rate $q_i = 0.0084$ cm^3/s.

8. Applying the Toth et al. (2006) method, construct the relative permeability curves for the oil and water phases using the production data given in Table 8.4 for water injection

TABLE 8.3   Unsteady-State Core Flooding Test Data Involving Oil Displacing Water in a Water-Saturated Core (Modified after Odeh and Dotson, 1985; Civan and Donaldson, 1987)

Cumulative Volume of Oil Injected, V_i (cm^3)	Cumulative Volume of Water Produced, V_k (cm^3)	Pressure Difference, Δp (atm)
1.6E+00	1.4E+00	1.3E+01
2.4E+00	1.6E+00	1.2E+01
4.0E+00	1.7E+00	1.0E+01
6.0E+00	1.8E+00	9.0E+00
8.5E+00	1.9E+00	8.1E+00
1.3E+01	2.0E+00	7.2E+00
1.8E+01	2.1E+00	6.4E+00
3.0E+01	2.3E+00	5.0E+00
5.0E+01	2.4E+00	4.3E+00

TABLE 8.4   Water and Oil Production Data (Modified after Tóth et al., 2006)

t(d)	q_w(m^3/d)	q_o(m^3/d)
0.0E+00	0.0E+00	5.0E+02
8.9E+01	2.9E+00	5.0E+02
1.7E+02	9.9E+01	4.0E+02
2.2E+02	1.5E+02	3.5E+02
2.5E+02	1.9E+02	3.1E+02
2.6E+02	2.3E+02	2.7E+02
2.8E+02	2.8E+02	2.2E+02
3.0E+02	3.1E+02	1.9E+02
3.5E+02	3.3E+02	1.7E+02
4.3E+02	3.5E+02	1.5E+02
5.2E+02	3.9E+02	1.1E+02
8.0E+02	4.3E+02	7.2E+01
1.1E+03	4.5E+02	4.6E+01
1.3E+03	4.6E+02	3.6E+01
1.5E+03	4.6E+02	3.7E+01
1.7E+03	4.8E+02	2.5E+01
1.8E+03	4.8E+02	2.2E+01
2.0E+03	4.8E+02	1.5E+01
2.7E+03	4.9E+02	1.3E+01

under constant rate of $q_{wi} = 500\,\text{m}^3/\text{day}$. Wellbore radius $r_w = 0.1\,\text{m}$, external radius of well influence $r_e = 155\,\text{m}$, pay zone thickness $h = 29\,\text{m}$, porosity $\phi = 0.22$, permeability $K = 0.18\,\mu\text{m}^2$, skin factor $s = 0$, oil formation volume factor $B_o = 1.23$, water formation volume factor $B_w = 1.0$, oil viscosity $\mu_o = 1.32\,\text{mPa} \cdot \text{s}$, water viscosity $\mu_w = 1.0\,\text{mPa} \cdot \text{s}$, connate water saturation $S_w = 0.23$, maximum water saturation $S_{w\text{max}} = 0.73$, $a = 0.67$, $b = 2.0$, $a_2 = 2.3 \times 10^6$, and $b_2 = -0.104$.

CHAPTER 9

MASS, MOMENTUM, AND ENERGY TRANSPORT IN POROUS MEDIA

9.1 INTRODUCTION

This chapter demonstrates the applications of coupled mass, momentum, and energy conservation equations for various problems.* The transport of species through porous media by different mechanisms is described. Dispersivity and dispersion in heterogeneous and anisotropic porous media issues are reviewed. Formulation of compositional multiphase flow through porous media is presented in the following categories: general multiphase fully compositional nonisothermal mixture model, isothermal black oil model of nonvolatile oil systems, isothermal limited compositional model of volatile oil systems, and shape-averaged models. Formulation of source/sink terms in conservation equations is discussed. Analyses and formulations of problems involving phase change and transport in porous media, such as gas condensation, freezing/thawing of moist soil, and production of natural gas from hydrate-bearing formations, are presented. Typical applications and their results are provided for illustration of the theoretical treatise.

* Parts of this chapter have been reproduced with modifications from the following:

Civan, F. 2000d. Unfrozen water content in freezing and thawing soils—Kinetics and correlation. Journal of Cold Regions Engineering, 14(3), pp. 146–156, with permission from the American Society of Civil Engineers;

Civan, F. and Sliepcevich, C.M. 1984. Efficient numerical solution for enthalpy formulation of conduction heat transfer with phase change. International Journal of Heat Mass Transfer, 27(8), pp. 1428–1430, with permission from Elsevier;

Civan, F. and Sliepcevich, C.M. 1985b. Comparison of the thermal regimes for freezing and thawing of moist soils. Water Resources Research, 21(3), pp. 407–410, with permission from the American Geophysical Union; and

Civan, F. and Sliepcevich, C.M. 1987. Limitation in the apparent heat capacity formulation for heat transfer with phase change. Proceedings of the Oklahoma Academy of Science, 67, pp. 83–88, with permission from the Oklahoma Academy of Science.

9.2   DISPERSIVE TRANSPORT OF SPECIES IN HETEROGENEOUS AND ANISOTROPIC POROUS MEDIA

Spontaneous irreversible dispersive transport of species in fluid media may occur under the effect of various driving forces (Bird et al., 1960), including the presence of the gradients (difference over distance) of species concentration, total pressure, potential energy or body force, and temperature (the Soret effect), and by hydrodynamic spreading and mixing (Bear, 1972; Civan, 2002e, 2010b). However, the dominant mechanism for spontaneous species transport in fluids present in porous media is the hydrodynamic mixing or dispersion phenomenon caused during flow through irregular hydraulic flow paths formed inside the porous structure. Under special circumstances, other factors may also cause spontaneous species transport. For example, live organisms, such as bacteria, have the natural tendency to move from scarce to abundant nutrient-containing locations. This phenomenon is referred to as chemotaxis (Chang et al., 1992).

The total flux of spontaneous transport of species i in phase j is expressed as the sum of the fluxes of spontaneous transport of species by various factors:

$$\mathbf{J}_{ij} = \mathbf{J}_{ij}^M + \mathbf{J}_{ij}^D + \mathbf{J}_{ij}^P + \mathbf{J}_{ij}^B + \mathbf{J}_{ij}^T + \sum_O \mathbf{J}_{ij}^O, \qquad (9.1)$$

where the superscripts M, D, P, B, T, and O refer to spontaneous species transport owing to molecular diffusion, convective or mechanical dispersion, pressure, body force, temperature, and other factors, respectively. Table 9.1 presents the fundamental expressions of flux. Bear (1972) defines the sum of the molecular diffusion and convective mixing effects as hydrodynamic dispersion. Therefore, the spontaneous hydrodynamic dispersion flux can be defined as

$$\mathbf{J}_{ij}^H = \mathbf{J}_{ij}^M + \mathbf{J}_{ij}^D, \qquad (9.2)$$

where the superscript H refers to hydrodynamic dispersion.

Frequently, the flux of species is expressed by the gradient model, expressing the flux as being proportional to the gradient of the variable causing the spontaneous transport, associated with a usually empirically determined proportionality coefficient, referred to as the diffusion coefficient. In the following, commonly used gradient models for expressing species flux by various driving mechanisms are reviewed.

TABLE 9.1   Fundamental Expressions of Flux

Driving Force	Gradient Law	Species Flux
Flow potential	$\mathbf{u}_j = -\dfrac{k_{rj}}{\mu_j}\mathbf{K}\cdot\nabla\psi_j$	$\mathbf{J}_{ij}^B = w_{ij}\rho_j\mathbf{u}_j = w_{ij}\mathbf{m}_j$
Hydraulic dispersion or convective mixing effect	$\mathbf{m}_j^H = -\mathbf{D}_{jb}\cdot\nabla(\varepsilon_j\rho_j)$	$\mathbf{J}_{ij}^H = w_{ij}\mathbf{m}_j^H$
Molecular diffusion	$\mathbf{J}_j^M = -\mathbf{D}_{ij}\cdot\nabla w_{ij}$	$\mathbf{J}_{ij}^M = w_{ij}\mathbf{J}_j^M$
Soret effect (temperature gradient)	$\mathbf{J}_j^T = -\mathbf{D}_j^T\cdot\nabla T_j$	$\mathbf{J}_{ij}^T = w_{ij}\mathbf{J}_j^T$

9.2.1 Molecular Diffusion

The diffusive species flux is given by

$$\mathbf{J}_{ij}^{M} = -\varepsilon_j \rho_j \mathbf{D}_{ij}^{M} \cdot \nabla w_{ij}, \tag{9.3}$$

in which the molecular diffusion coefficient tensor, considering the effect of tortuous flow paths in porous media, is given by

$$\mathbf{D}_{ij}^{M} = D_{ij} \mathbf{T}^{-1} \cdot \mathbf{I}, \tag{9.4}$$

where D_{ij} is the molecular diffusion coefficient of species i present in phase j, $\mathbf{T}$ is the tortuosity tensor of porous media, and $\mathbf{I}$ is the unit diagonal tensor. $D_{ij}\mathbf{T}^{-1}$ represents the extended molecular diffusion coefficient of species i traveling through tortuous flow paths, generally longer than the porous media.

The molecular diffusion coefficient is usually estimated by empirical correlations, such as the Wilke and Chang correlation given by (Wilke, 1950; Perkins and Geankopolis, 1969)

$$D_i = \frac{7.4 \times 10^{-8} T M^{0.5}}{\mu v_i^{0.6}}, \tag{9.5}$$

where D_i denotes the molecular diffusivity of species i in the square centimeter per second units, and M and μ represent the molecular weight (gram per mole) and viscosity (centipoises) of the fluid phase in which species i diffuses; hence, M and μ do not include species i. v_i denotes the partial molar volume of species i (cubic centimeter per mole), and T is the absolute temperature (Kelvin).

9.2.2 Hydrodynamic Dispersion

The convective or mechanical mixing and dispersion of fluids is a dominant factor causing spontaneous spreading of species in porous media. Convective dispersion and mechanical mixing occur owing to the inhomogeneity of intricately complicated interconnected pore structure in porous media (Bear, 1972). Simultaneously, the molecular diffusion phenomenon takes place owing to species concentration difference.

Hydrodynamic dispersion can be primarily explained by means of the stream-splitting and mixing, and mixing-cell theories, although other mechanisms may also promote additional mixing affects (Skjaeveland and Kleppe, 1992). The stream-splitting and merging theory is based on the phenomenon of consecutive splitting and merging of flow as the fluid encountering the porous media grains is forced to flow around them. As a result, a lateral spreading of fluid and transverse dispersion of species contained in the spreading fluid occurs. The mixing-cell theory considers the mechanical mixing as a result of consecutive expansion and compression of fluid as it flows through pore bodies and pore throats, causing repetitive fluid accelerations and decelerations, respectively, and therefore a longitudinal dispersion of species. Insufficient pore connectivity, local fluid recirculation, flow constriction, dead-end pores, and retardation by sorption processes also affect convective mixing (Skjaeveland and Kleppe, 1992).

The dispersive species flux due to convective mixing is usually expressed by (Lake at al., 1984, Bear and Bachmat, 1990)

$$\mathbf{J}_{ij}^{D} = -\varepsilon_{j}\rho_{j}\mathbf{D}_{ijb}^{D}\cdot\nabla w_{ij},\qquad(9.6)$$

where $\mathbf{D}_{ij}^{D}$ is the hydrodynamic dispersion coefficient tensor in the length squared over time units (L^2/T). Although used frequently, this particular form or its variations may have some limitations. For example, when the species mass fraction is uniform, that is, $w_{ij} = $ ct., Eq. (9.6) implies a vanishing dispersive flux of species. However, dispersive flux of species occurs as long as convective mixing prevails. Therefore, extending the formulation of Civan (2002e, 2010b), and by intuition, a more appropriate gradient model can be proposed as (see Chapter 3)

$$\begin{aligned}\mathbf{J}_{ij}^{D} &= -\mathbf{D}_{ijb}^{D}\cdot\nabla\left\langle\rho_{ij}\right\rangle_{b} = -\mathbf{D}_{ijb}^{D}\cdot\nabla\left(\varepsilon_{j}\left\langle\rho_{ij}\right\rangle_{j}\right) = -\mathbf{D}_{ijb}^{D}\cdot\nabla\left(\varepsilon_{j}\left\langle\rho_{j}w_{ij}\right\rangle_{j}\right)\\ &= -\mathbf{D}_{ijb}^{D}\cdot\nabla\left[\varepsilon_{j}\left(\left\langle\rho_{j}\right\rangle_{j}\left\langle w_{ij}\right\rangle_{j} + \left\langle\rho_{j}'w_{ij}'\right\rangle_{j}\right)\right].\end{aligned}\qquad(9.7)$$

If the average of the product of deviations is neglected and the convention for averaging is dropped for convenience, Eq. (9.7) can be simplified as

$$\mathbf{J}_{ij}^{D} = -\mathbf{D}_{ijb}^{D}\cdot\nabla\left(\varepsilon_{j}\rho_{j}w_{ij}\right).\qquad(9.8)$$

Eq. (9.8) simplifies to Eq. (9.6) only when the volume fraction and density of the j-phase are uniform in porous media. Eq. (9.8) implies that dispersive species flux can occur so long as the volume fraction, density, and species concentration in the j-phase are not uniform in porous media.

The predictability of the hydrodynamic dispersion coefficient has occupied many researchers, including Bear and Bachmat (1990) and Liu and Masliyah (1996). However, further work is necessary for improved means of estimating the hydrodynamic dispersion coefficient in terms of the relevant parameters of porous media and pore fluids. Under certain simplifying assumptions, such as isochoric flow, uniform density, conservative extensive quantity, and impervious interface for diffusion, Bear and Bachmat (1990) derived an expression for the hydrodynamic dispersion coefficient. This equation will be adapted in the following form for the components of the hydrodynamic dispersion coefficient tensor:

$$D_{nm} = a_{nklm}\frac{v_{k}v_{l}}{|\mathbf{v}|}f(Pe_{i}, r),\qquad(9.9)$$

where n, $m = x$-, y-, and z-Cartesian directions, and a_{nklm} denote the components of the isotropic dispersivity, given by

$$a_{nklm} = a_{T}\delta_{nm}\delta_{kl} + \frac{a_{L} - a_{T}}{2}(\delta_{nk}\delta_{lm} + \delta_{nl}\delta_{mk}),\qquad(9.10)$$

in which a_{L} and a_{T} are referred to as the longitudinal and transverse dispersivity coefficients, respectively, in the length dimensions (L) and relative to the primary flow direction. δ_{pq} is the Kronecker delta, whose value is zero when $p \neq q$ and unity when $p = q$. The function $f(Pe_{i}, r)$ is given by

$$f(Pe_i, r) = \frac{Pe_i}{Pe_i + 1 + r}, \tag{9.11}$$

where Pe_i is the species i Peclet number and r is the ratio of the characteristic length along the flow direction to the characteristic length normal to the flow direction inside a pore space. Hence, Eq. (9.11) indicates that

$$\text{Limit } f(Pe_i, r) = 1 \tag{9.12}$$
$$Pe_i \to \infty.$$

Bear and Bachmat (1990) justify that $r = O\ (1)$ and therefore, $f(Pe_i, r) \approx 1$ when $Pe_i \gg 1$, and $f(Pe_i, r) = O\ (Pe_i)$ when $Pe_i \ll 1$.

Substituting Eq. (9.10) into Eq. (9.9) yields

$$D_{nm} = \left[a_T |\mathbf{v}| \delta_{nm} + (a_L - a_T) \frac{v_n v_m}{|\mathbf{v}|} \right] f(Pe_i, r), \tag{9.13}$$

where n, m = x-, y-, and z-Cartesian directions, and v_x, v_y, and v_z denote the components of the fluid velocity in the respective directions. Thus, the velocity vector is given by

$$\mathbf{v} = v_x \mathbf{i} + v_y \mathbf{j} + v_z \mathbf{k}, \tag{9.14}$$

in which $\mathbf{i}$, $\mathbf{j}$, and $\mathbf{k}$ denote the unit vectors in the x-, y-, and z-Cartesian directions. The magnitude of the velocity vector is given by

$$v \equiv |\mathbf{v}| = \sqrt{v_x^2 + v_y^2 + v_z^2}. \tag{9.15}$$

For applications of Eq. (9.13), see Exercises 1–3.

For convenience in developing empirical correlations, the convective dispersion coefficient can be scaled by the molecular diffusion coefficient and correlated by a power-law function of the Peclet number. For example, Eq. (9.7) can be scaled in the longitudinal or transverse directions as follows, defining the Peclet dimensionless number:

$$\mathbf{N}_{Pe} = \frac{\mathbf{D}_{ij\delta}^D}{D_{ij}^M} = \frac{\alpha_{\delta j} u_{\delta j}}{D_{ij}^M}, \delta = l \text{ or } t. \tag{9.16}$$

The right of Eq. (9.16) can be correlated as a power-law function of the porous media particle Peclet number.

9.2.3 Advective/Convective Flux of Species

The advective flux of species by potential gradient is given by

$$\mathbf{J}_{ij}^A = w_{ij} \rho_j \mathbf{u}_j = w_{ij} \mathbf{m}_j, \tag{9.17}$$

where, according to Darcy's law, for example,

$$\mathbf{u}_j = -\frac{k_{rj}}{\mu_j} \mathbf{K} \cdot \nabla \psi_j, \tag{9.18}$$

in which the flow potential gradient is defined by

$$\nabla \psi_j = \nabla p_j + \rho_j g \nabla z. \qquad (9.19)$$

Darcy's equation given above considers laminar flow due to viscous forces.

9.2.4 Correlation of Dispersivity and Dispersion

Auset and Keller (2004) observed, by means of experiments conducted using micromodels, that the magnitude of particle dispersion depends primarily on their preferential paths and velocities. Smaller particles tend to move along longer complicated tortuous paths and therefore are slower, and larger particles tend to move along shorter, straighter paths and therefore are faster. Hence, the dispersivity and dispersion coefficient of particles in porous media increase with decreasing particle size and vice versa, depending on the pore structure and the pore channel-to-particle size ratio.

Let u denote the volumetric flux, $k_B = 1.38\text{E}{-}23\,\text{m}^2\,\text{kg/s}^2/\text{K}$ is the Boltzmann constant, μ is viscosity, T is the absolute temperature, D_g is the porous media mean grain diameter, D_p is the mean diameter of diffusing particle (or molecule), and D_∞ denotes the bulk molecular diffusivity.

The Peclet number N_{Pe} based on grain diameter D_g is given by

$$N_{Pe} = \frac{uD_g}{D_\infty}. \qquad (9.20)$$

The Stokes–Einstein equation of bulk molecular diffusivity D_∞ of particles (or molecules) is given by

$$D_\infty = \frac{k_B T}{3\pi\mu D_p}. \qquad (9.21)$$

Many attempts have been made at developing empirical correlations for the estimation of the dispersion coefficients. In the following, two of the frequently used correlations are presented.

The correlation of Hiby (1962) for the dispersion coefficient is given by

$$\frac{D}{D_\infty} = 0.67 + \frac{0.65 N_{Pe}}{1 + 6.7 N_{Pe}^{-1/2}}, 0.01 \le N_{Pe} \le 100. \qquad (9.22)$$

The correlation of Blackwell et al. (1959) for the dispersion coefficient is given by

$$\frac{D}{D_\infty} = 8.8 N_{Pe}^{1.17}, N_{Pe} > 0.5. \qquad (9.23)$$

Let v denote the interstitial pore fluid velocity, given by (Dupuit, 1863)

$$v = \frac{\tau u}{\phi}. \qquad (9.24)$$

Thus, substituting Eq. (9.20) for the Peclet number into Eq. (9.23) and then applying Eq. (9.24) yields (Civan, 2010c)

$$\frac{D}{D_\infty} = 8.8\left(\frac{uD_g}{D_\infty}\right)^{1.17} = 8.8\left(\frac{v\phi D_g}{\tau D_\infty}\right)^{1.17}, N_{Pe} > 0.5. \tag{9.25}$$

We can obtain the following approximation in terms of the superficial velocity by rounding the exponent to 1.0:

$$D \cong \alpha_1 u, N_{Pe} > 0.5, \alpha_1 = 8.8D_g. \tag{9.26}$$

Alternatively, we can obtain the following approximation in terms of the interstitial velocity:

$$D \cong \alpha v, N_{Pe} > 0.5, \alpha = \frac{8.8\phi D_g}{\tau}. \tag{9.27}$$

Figure 9.1 shows a comparison of the correlation developed by the author and the Unice and Logan (2000) correlation of the dispersivity α (meter) data of Gelhar et al. (1992). Unice and Logan (2000) correlated the data of Gelhar et al. (1992), where x (meter) denotes the scale, by

$$\alpha = 10[1 - \exp(-x/120)]. \tag{9.28}$$

Unice and Logan (2000) also provided a more consevative correlation of the data of Gelhar et al. (1992) by

$$\alpha = 1 - \exp(-x/36). \tag{9.29}$$

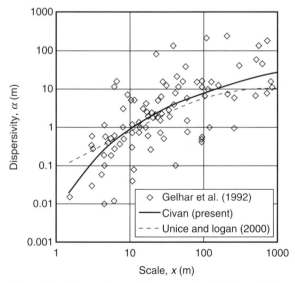

Figure 9.1 Comparison of the present and Unice and Logan (2000) correlation of the dispersivity data of Gelhar et al. (1992) (prepared by the author).

The present author correlated the same data more accurately by the following:

$$\log \alpha = 3.159 - \frac{7.663}{\log x + 1.4}, \quad R^2 = 0.476. \tag{9.30}$$

9.3 GENERAL MULTIPHASE FULLY COMPOSITIONAL NONISOTHERMAL MIXTURE MODEL

The multiphase mixture model describes a multiphase system as a pseudosingle phase. Hence, the pseudoproperties of the multiphase system are defined in a manner to formulate the conservation equations similar to those of the single-phase systems. The formulation of such a model by Cheng and Wang (1996), Wang and Cheng (1996), and Wang (1998) is presented in this section with modifications for consistency with the rest of the materials presented in this book.

The pseudoproperties represent the average properties over the multiphase system, defined as a volume fraction-weighted average of the properties of the various phases present in the multiphase system. Thus, the porosity ϕ and the mobile multiphase fluid volume fraction ε_k, pseudodensity ρ, pore velocity $\mathbf{v}$, species weight fraction w^α, and enthalpy h are expressed, respectively, as the following, where the subscript k indicates a phase k:

$$\phi = \sum_k \varepsilon_k, \tag{9.31}$$

$$\phi \rho = \sum_k \rho_k \varepsilon_k, \tag{9.32}$$

$$\phi \rho \mathbf{v} = \sum_k \rho_k \mathbf{v}_k \varepsilon_k \text{ or } \rho \mathbf{u} = \sum_k \rho_k \mathbf{u}_k, \tag{9.33}$$

$$\phi \rho w^\alpha = \sum_k \rho_k w_k^\alpha \varepsilon_k, \sum_\alpha w_k^\alpha = 1 \tag{9.34}$$

and

$$\phi \rho h = \sum_k \rho_k h_k \varepsilon_k. \tag{9.35}$$

It is apparent, based on the above-given expressions, that we consider the primary properties as the density ρ, the mass flux $\rho \mathbf{v}$, the species mass per unit volume ρw^α, where w^α denotes the weight fraction of species α, and the enthalpy per unit volume ρh. Note that ϕ denotes the total volume fraction of the bulk porous media occupied by all the mobile fluid phases. Hence, it excludes the volume occupied by the immobile fluid phases in the pore space.

A pseudokinematic viscosity for the multiphase fluid system is defined by

$$v = \frac{1}{\sum_k \dfrac{k_{rk}}{v_k}}, \tag{9.36}$$

where k_{rk} and v_k denote the phase k relative permeability and kinematic viscosity, respectively. The relative mobility of phase k is given by

$$\lambda_k = \frac{k_{rk}}{v_k} v, \quad \sum_k \lambda_k = 1. \tag{9.37}$$

The pseudopressure differential is defined as

$$dp = \sum_k \lambda_k dp_k. \tag{9.38}$$

Parker (1989) expressed the capillary pressure between any phase j and phase k in a multiphase fluid system as a function of porosity ϕ, interfacial tension σ_{jk} between these two phases, and the saturations $S_l : l = 1, 2, \ldots, N$ of various phases, where N denotes the total number of fluid phases, according to

$$p_{cjk} = p_j - p_k = f\left(\phi, \sigma_{jk}, S_l : l = 1, 2, \ldots, N\right). \tag{9.39}$$

Hence, the pseudopressure of a multiphase system can be expressed by

$$p = p_k + \sum_i \int_0^{S_i} Ca_{ik} dS_i + \sum_\alpha \int_0^{w^\alpha} Cs_{\alpha k} dw^\alpha + \int_0^T Ct_k dT. \tag{9.40}$$

Note, we can write

$$\sum_i \int_0^{S_i} Ca_{ik} dS_i = \frac{1}{\phi} \sum_i \int_0^{\varepsilon_i} Ca_{ik} d\varepsilon_i = \sum_i \int_0^{\varepsilon_i} \bar{C}a_{ik} d\varepsilon_i. \tag{9.41}$$

Here are Ca_{ik}, $Cs_{\alpha k}$, and Ct_k referred to as the capillary, solutal-capillary, and thermal factors, defined respectively, by

$$\bar{C}a_{ik} \equiv \frac{Ca_{ik}}{\phi} = \sum_j \lambda_j \frac{\partial p_{cjk}}{\partial \varepsilon_i}, \tag{9.42}$$

$$Cs_{\alpha k} = \sum_j \lambda_j \frac{\partial p_{cjk}}{\partial \sigma_{jk}} \frac{\partial \sigma_{jk}}{\partial w^\alpha}, \tag{9.43}$$

and

$$Ct_k = \sum_j \lambda_j \frac{\partial p_{cjk}}{\partial \sigma_{jk}} \frac{\partial \sigma_{jk}}{\partial T}. \tag{9.44}$$

Having described the above fundamental definitions, the pseudo-single-phase differential mass, momentum, species, and energy balance equations can now be derived from the individual-phase balance equations as demonstrated in the following.

The mass balance equation of the phase k is given by

$$\frac{\partial}{\partial t}\left(\varepsilon_k \rho_k\right) + \nabla \cdot \left(\rho_k \mathbf{u}_k\right) = \dot{m}_k. \tag{9.45}$$

Adding equations similar to Eq. (9.45) for all phases and then applying Eqs. (9.31)–(9.33) yields the following pseudo-single-phase equation of mass balance:

$$\frac{\partial}{\partial t}(\phi\rho)+\nabla\cdot(\rho\mathbf{u})=\dot{m}. \tag{9.46}$$

Consider the momentum balance equation expressed by Darcy's law as the following for phase k:

$$\rho_k\mathbf{u}_k=-\frac{\lambda_k}{\upsilon}\mathbf{K}\cdot(\nabla p_k-\rho_k\mathbf{g}). \tag{9.47}$$

Adding equations similar to Eq. (9.47) for all phases and then applying Eqs. (9.33) and (9.38) yields the following pseudo-single-phase Darcy law:

$$\rho\mathbf{u}=-\frac{1}{\upsilon}\mathbf{K}\cdot(\nabla p-\gamma_\rho\rho\mathbf{g}), \tag{9.48}$$

where γ_ρ is a density correction factor, given by

$$\gamma_\rho=\frac{\sum_k\rho_k\lambda_k}{\sum_k\rho_kS_k}=\frac{\phi\sum_k\rho_k\lambda_k}{\sum_k\rho_k\varepsilon_k}=\frac{1}{\rho}\sum_k\rho_k\lambda_k. \tag{9.49}$$

The diffusive mass flux of phase k can be defined as

$$\mathbf{j}_k=\rho_k\mathbf{u}_k-\lambda_k\rho\mathbf{u},\ \sum_k\mathbf{j}_k=0. \tag{9.50}$$

It can be shown that Eq. (9.50) can be expressed as the following:

$$\mathbf{j}_k=\sum_i\left[-\rho_k\mathbf{D}_{cik}\cdot\nabla\varepsilon_i+\frac{\lambda_k\lambda_i(\rho_k-\rho_i)}{\upsilon}\mathbf{K}\cdot\mathbf{g}\right]$$
$$+\sum_\alpha(-\rho_k\mathbf{D}_{s\alpha k}\cdot\nabla w^\alpha)+(-\rho_k\mathbf{D}_{tk}\cdot\nabla T). \tag{9.51}$$

The following expressions define the capillary diffusion coefficient D_{cik}, the solutal-capillary diffusion coefficient $D_{s\alpha k}$, and the thermocapillary diffusion coefficient D_{tk}, respectively:

$$\mathbf{D}_{cik}=\mathbf{K}\frac{\lambda_k}{\rho_k\upsilon}\sum_j\lambda_j\left(-\frac{\partial p_{cjk}}{\partial\varepsilon_i}\right), \tag{9.52}$$

$$\mathbf{D}_{s\alpha k}=\mathbf{K}\frac{\lambda_k}{\rho_k\upsilon}\sum_j\lambda_j\left(-\frac{\partial p_{cjk}}{\partial\sigma_{jk}}\frac{\partial\sigma_{jk}}{\partial w^\alpha}\right), \tag{9.53}$$

and

$$\mathbf{D}_{tk}=\mathbf{K}\frac{\lambda_k}{\rho_k\upsilon}\sum_j\lambda_j\left(-\frac{\partial p_{cjk}}{\partial\sigma_{jk}}\frac{\partial\sigma_{jk}}{\partial T}\right). \tag{9.54}$$

The species α mass balance equation of phase k is given by

$$\frac{\partial}{\partial t}\left(\varepsilon_k \rho_k w_k^\alpha\right) + \nabla \cdot \left(\rho_k \mathbf{u}_k w_k^\alpha\right) = \nabla \cdot \left(\varepsilon_k \rho \mathbf{D}_k^\alpha \cdot \nabla w_k^\alpha\right) + \dot{m}_k^\alpha. \tag{9.55}$$

Adding equations similar to Eq. (9.55) for all phases and then applying Eqs. (9.31)–(9.35) yields the following pseudo-single-phase equation of species α mass balance:

$$\frac{\partial}{\partial t}\left(\phi \rho w^\alpha\right) + \nabla \cdot \left(\gamma_\alpha \rho \mathbf{u} w^\alpha\right) = \nabla \cdot \left(\phi \rho \mathbf{D}^\alpha \cdot \nabla w^\alpha\right)$$

$$+ \nabla \cdot \left\{\sum_k \left[\varepsilon_k \rho_k \mathbf{D}_k^\alpha \cdot \left(\nabla w_k^\alpha - \nabla w^\alpha\right)\right]\right\} - \nabla \cdot \left(\sum_k w_k^\alpha \mathbf{j}_k\right). \tag{9.56}$$

The species α advection correction factor γ_α is given by

$$\gamma_\alpha = \frac{1}{w^\alpha}\sum_k \lambda_k w_k^\alpha. \tag{9.57}$$

The effective diffusion coefficient $\mathbf{D}^\alpha$ is defined by

$$\phi \rho \mathbf{D}^\alpha = \sum_k \varepsilon_k \rho_k \mathbf{D}_k^\alpha. \tag{9.58}$$

Assuming thermal equilibrium, the energy balance equation of the phase k is given by

$$\frac{\partial}{\partial t}\left(\varepsilon_k \rho_k h_k\right) + \nabla \cdot \left(\rho_k \mathbf{u}_k h_k\right) = \nabla \cdot \left(\frac{1}{\phi}\varepsilon_k \mathbf{k}_k \cdot \nabla T\right) + \dot{q}_k. \tag{9.59}$$

Adding equations similar to Eq. (9.59) for all phases and then applying Eqs. (9.31)–(9.35) yields the following pseudo-single-phase equation of energy:

$$\frac{\partial}{\partial t}\left(\phi \rho h\right) + \nabla \cdot \left(\gamma_h \rho \mathbf{u} h\right) = \nabla \cdot \left(\frac{1}{\phi}\mathbf{k}_e \cdot \nabla T\right) + \nabla \cdot \left(\sum_k h_k \mathbf{j}_k\right) + \dot{q}. \tag{9.60}$$

The energy advection correction factor γ_h is given by

$$\gamma_h = \frac{1}{h}\sum_k \lambda_k h_k. \tag{9.61}$$

The effective thermal conductivity $\mathbf{k}_e$ is defined by

$$\frac{1}{\phi}\mathbf{k}_e = \frac{1}{\phi}\sum_k \varepsilon_k \mathbf{k}_k. \tag{9.62}$$

Note that Eqs. (9.59)–(9.62) were derived here and are therefore somewhat different from those given by Wang and Cheng (1996) and Wang (1998).

If a stationary porous solid matrix at thermal equilibrium with the pore fluids is considered, then the overall energy balance equation can be written as

$$\frac{\partial}{\partial t}[(1-\phi)\rho_s h_s + \phi\rho h] + \nabla \cdot (\gamma_h \rho \mathbf{u} h) = \nabla \cdot \left(\frac{1}{\phi}\overline{\mathbf{k}}_e \cdot \nabla T\right) + \nabla \cdot \left(\sum_k h_k \mathbf{j}_k\right) + \dot{q}, \tag{9.63}$$

where $\overline{\mathbf{k}}_e$ denotes the effective thermal conductivity of the solid fluid system.
Consider the identity, given by

$$\nabla T = \frac{1}{c_j}\nabla h_j = \frac{1}{c_j}\nabla(h_j \pm h), \tag{9.64}$$

where c_j is the specific heat capacity. Then, Eq. (9.63) can be written as

$$\frac{\partial}{\partial t}[(1-\phi)\rho_s h_s + \phi\rho h] + \nabla \cdot (\gamma_h \rho \mathbf{u} h) = \nabla \cdot \left(\frac{1}{\phi c_j}\overline{\mathbf{k}}_e \cdot \nabla h\right)$$

$$+ \nabla \cdot \left(\frac{1}{\phi c_j}\overline{\mathbf{k}}_e \cdot \nabla(h_j - h)\right) + \nabla \cdot \left(\sum_k h_k \mathbf{j}_k\right) + \dot{q}. \tag{9.65}$$

The solution of the preceding equations requires the following auxiliary relationships. The local species α partitioning or equilibrium coefficient K_k^α is given by

$$K_k^\alpha = \frac{w_k^\alpha}{w^\alpha}. \tag{9.66}$$

The partitioning of species α between pairs of l- and m-phases is given by (Nernst law)

$$K_{l-m}^\alpha = \frac{w_l^\alpha}{w_m^\alpha}. \tag{9.67}$$

The interface mass transfer rate under nonequilibrium conditions is given by

$$\dot{m}_{l-m}^\alpha = k(w_l^\alpha - w_m^\alpha), \tag{9.68}$$

where k is the film mass transfer coefficient.

9.4 FORMULATION OF SOURCE/SINK TERMS IN CONSERVATION EQUATIONS

Formulation of transport processes involving various physical and chemical reactions in the presence of the various phases in porous media can be handled through the expression of the source and sink terms. Our objective is the characterization of the source/sink terms involved in the various balance equations. Therefore, consider a fluid phase present in a multiphase system of various fluid phases and a solid phase forming the porous matrix containing the fluid phases in its pore space.

The nomenclature used here essentially follows Slattery (1990) but with some modifications required for the purpose of discussion presented here. Let V denote the volume of the phase, A the surface area of the interface separating the phase from the other phases per unit bulk volume of porous media, and C the lines through which various interface surfaces intersect with each other. Further, ε denotes the volume fraction or the volume of a phase contained per unit bulk volume of porous media, ρ is the phase volume average mass density of the phase, ρ^A is the mass density of a quantity per interface surface area, m is the amount of a quantity generated per unit mass of the phase, m^E is the amount of a quantity added externally from the surrounding per unit mass of the phase present in porous media, m^A is the amount of a quantity generated per unit mass of the quantity present per unit area of the interface separating the phase from the other phases, $\mathbf{v}$ denotes the mass-averaged velocity of the phase, $\mathbf{v}^A$ is the mass-averaged surface velocity, $\mathbf{u}$ is the velocity at which a point on the interface moves, $\mathbf{T}$ denotes the stress tensor, $\mathbf{T}^A$ is the surface stress tensor, $\mathbf{b}$ is the body force vector such as owing to gravity, $\mathbf{b}^A$ is the body force vector per unit mass applying at the interface, U is the internal energy, $\mathbf{q}$ is the energy flux vector, $\mathbf{q}^A$ is the surface energy flux vector, Q^A is the surface energy flux per unit mass, $\mathbf{j}$ is the diffusive mass flux of species in the phase relative to phase mass-averaged phase velocity $\mathbf{v}$, $\mathbf{j}^A$ is the diffusive mass flux of species in the phase relative to the interface surface mass-averaged phase velocity $\mathbf{v}^A$, r^A is the rate of species mass generation per unit area of the interface surface, w is the mass fraction of species in the phase, w^A is the mass fraction of species at the interface surface, and $\mathbf{n}$ is the outward unit normal vector.

The substantial derivative is defined as

$$\frac{d_s}{dt} = \frac{\partial}{\partial t} + \mathbf{v} \cdot \nabla. \tag{9.69}$$

The change across an interface surface is defined by

$$\Delta[\ ] \equiv (\ )^+ - (\ )^-, \tag{9.70}$$

where the superscripts $+$ and $-$ indicate the phase side and the other side, respectively, of the interface surface.

There are essentially three constituents of the source/sink terms for a given phase as described in the following.

- **Internal or bulk volume transport processes (homogeneous processes).** Physical or chemical processes occurring inside the bulk volume of a fluid phase are considered as the bulk volume processes. Examples include the separation of solid matter by chemical or physical precipitation or crystallization and the reverse phenomena occurring within the fluid phase.

- **Interface transport processes (heterogeneous processes).** Physical and chemical processes occurring over the contact surfaces separating various phases are of the heterogeneous-type processes. Examples include adsorption/desorption, chemical reactions, crystallization, and interface transfer.

- **External interaction processes (heterogeneous processes).** Interaction of the phases present inside the bulk porous media with the surrounding systems in the forms of addition or removal of a quantity is considered heterogeneous processes of the external type.

Thus, in general, the amount of a quantity added per unit bulk volume of porous media can be expressed as the sum of these three types of the quantity, namely, the internal, interface, and external, respectively, as:

$$m_t = \varepsilon \rho m + \sum_A A \rho^A m^A + \varepsilon \rho m^E. \tag{9.71}$$

Thus, the source term of a quantity can be expressed as

$$r = \dot{m} \equiv \frac{\partial m_t}{\partial t}. \tag{9.72}$$

Slattery (1990) provides rigorous formulation of the interfacial transport for mass, species mass, momentum, and energy in terms of the jump balance conditions. These are summarized in the following.

The jump mass balance equation for expressing the mass conservation at an interface surface is given by

$$\frac{d_s \rho^A}{dt} + \rho^A \nabla \cdot \mathbf{v}^A + \Delta[\rho(\mathbf{v} - \mathbf{u}) \cdot \mathbf{n}] = 0. \tag{9.73}$$

The jump species mass balance equation for expressing the species mass conservation at an interface surface is given by

$$\rho^A \frac{d_s w^A}{dt} + \nabla \cdot \mathbf{j}^A - r^A + \Delta[\mathbf{j} \cdot \mathbf{n} + \rho(w - w^A)(\mathbf{v} - \mathbf{u}) \cdot \mathbf{n}] = 0. \tag{9.74}$$

The jump momentum balance equation for expressing the momentum conservation at an interface surface is given by

$$\rho^A \frac{d_s \mathbf{v}^A}{dt} - \nabla \cdot \mathbf{T}^A - \rho^A \mathbf{b}^A + \Delta[\rho\{(\mathbf{v} - \mathbf{u}) \cdot \mathbf{n}\}^2 \mathbf{n} - \mathbf{T} \cdot \mathbf{n}] = \mathbf{0}. \tag{9.75}$$

The jump energy balance equation for expressing the energy conservation at an interface surface is given by

$$\rho^A \frac{d_s}{dt}\left(U^A + \frac{1}{2}\mathbf{v}^A \cdot \mathbf{v}^A\right) - \nabla \cdot (\mathbf{T}^A \cdot \mathbf{v}^A)$$
$$- \sum \rho^A \mathbf{v}^A \cdot \mathbf{b}^A + \nabla \cdot \mathbf{q}^A - \rho^A Q^A \tag{9.76}$$
$$+ \Delta\left[\rho\left(U + \frac{1}{2}\mathbf{v} \cdot \mathbf{v} - U^A - \frac{1}{2}\mathbf{v}^A \cdot \mathbf{v}^A\right)(\mathbf{v} - \mathbf{u}) \cdot \mathbf{n} - \mathbf{v} \cdot \mathbf{T} \cdot \mathbf{n} + \mathbf{q} \cdot \mathbf{n}\right] = \mathbf{0}.$$

Nevertheless, implementation of the jump balance conditions depends on the prescribed conditions of specific problems.

There are many different types of natural and engineered processes. The rates of homogeneous and heterogeneous reactions can be described by means of properly stated constitutive kinetic equations (Quintard and Whitaker, 2005). Derivation of constitutive kinetic equations for specific applications involving certain chemical and physical processes can be found in many literatures and therefore is beyond the scope of this book. However, we will provide a few examples here for instructional purposes.

For example, as pointed out by Quintard and Whitaker (2005), the interfacial flux constitutive equation for adsorption/desorption processes occurring under non-equilibrium conditions is given by (Langmuir, 1916, 1917)

$$r^A = \dot{m}^A = k_a c - k_d c^A, \qquad (9.77)$$

where k_a and k_d represent the adsorption and desorption rate coefficients, respectively, and c and c^A denote the phase volume-averaged concentration and interface surface area-averaged concentration of the species undergoing adsorption/desorption, respectively.

Under equilibrium conditions, Eq. (9.77) yields the following frequently used equilibrium ratio relationship:

$$\dot{m}^A = 0, \; K = \frac{k_a}{k_d} = \frac{c^A}{c}. \qquad (9.78)$$

9.5 ISOTHERMAL BLACK OIL MODEL OF A NONVOLATILE OIL SYSTEM

The black oil model is a simplified two-species (dead-oil and dissolved-gas) compositional model, intended primarily for the description of flow of low-volatile oil, which is essentially made up of heavy hydrocarbons and some dissolved methane gas in petroleum reservoirs (Peaceman, 1977; Odeh, 1982; Thomas, 1982; Lake et al., 1984; Odeh and Heinemann, 1988). This model considers two hydrocarbon pseudospecies, namely, the dead-oil and dissolved-gas and the water pseudospecies (Peaceman, 1977).

A pseudospecies is defined as a system of some prescribed composition of various species, which remains unchanged during flow. Obviously, this is a simplifying assumption. The data required for describing the hydrocarbon equilibrium is obtained by means of differential vaporization tests conducted in a PVT cell using oil samples taken from petroleum reservoirs.

However, the black oil model has additional simplifying assumptions as schematically described in Figure 9.2 by Civan (2006). It considers a three-phase fluid system consisting of the gas, oil, and water phases, denoted by the lowercase letters, g, o, and w, respectively. The gas, oil, and water pseudocomponents are denoted by the uppercase letters, G, O, and W, respectively. The subscript s denotes the standard conditions, usually taken as 14.7 psia (1 atm) and 60°F (15.5°C) in the petroleum industry.

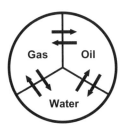

Figure 9.2   Interaction of various phases (prepared by the author).

The pseudospecies are defined from a practical point of view. When the gas/oil/water system in a petroleum reservoir is produced at the well head and processed at the surface, it separates into the gas, oil, and water products. The produced gas/oil/water system has a different composition than the reservoir fluid system because the dissolved gases present in the reservoir oil and water are liberated when the pressure is reduced. Here, the reservoir oil, referred to as the live oil, separates into the dead-oil and dissolved-gas pseudospecies at the surface. The same can be said of the water phase (Civan, 2006c).

For computational convenience and sufficient simplified description of the compositional three-phase fluid flow in petroleum reservoirs, the species mass conservation equation and Darcy's law are combined in the following manner:

$$\frac{\partial}{\partial t}\left(\varepsilon_g \rho_g w_g^\alpha + \varepsilon_o \rho_o w_o^\alpha + \varepsilon_w \rho_w w_w^\alpha\right)$$

$$= \nabla \cdot \left[\begin{array}{l} \dfrac{\rho_g w_g^\alpha}{\mu_g}\mathbf{K}_g \cdot (\nabla p_g - \rho_g \mathbf{g}\cdot\nabla z) + \dfrac{\rho_o w_o^\alpha}{\mu_o}\mathbf{K}_o \cdot (\nabla p_o - \rho_o \mathbf{g}\cdot\nabla z) \\ + \dfrac{\rho_w w_w^\alpha}{\mu_w}\mathbf{K}_w \cdot (\nabla p_w - \rho_w \mathbf{g}\cdot\nabla z) \end{array}\right] \tag{9.79}$$

$$+ \varepsilon_g q_g^\alpha + \varepsilon_o q_o^\alpha + \varepsilon_w q_w^\alpha, \alpha \in (G, O, W),$$

where the effective permeability tensors of the various fluid phases are given by

$$\mathbf{K}_l = k_{rl}\mathbf{K}, l \in (g, o, w), \tag{9.80}$$

where $\mathbf{K}$ denotes the absolute permeability and k_{rl} denotes the relative permeability of phase l. A comparison of this formulation with the general one presented in the previous section can reveal its inherent simplifying assumptions, such as neglecting transport by dispersion.

However, in the black oil model, the water phase is assumed to contain only the water pseudospecies, and therefore the oil and gas pseudocomponents do not exist in the water phase. Consequently, the weight fractions of the gas, oil, and water pseudocomponents in the water phase are expressed as $w_w^G = 0, w_w^O = 0, w_w^W = 1$, respectively. We consider that the gas pseudocomponent G can dissolve in the oil phase, but the water pseudocomponent W cannot. Therefore, $w_o^G \neq 0, w_o^O \neq 0, w_o^W = 0$. We consider that oil and water cannot vaporize into the gas phase and the gas phase contains only the gas pseudospecies. Therefore, $w_g^G = 1, w_g^O = 0, w_g^W = 0$. Under these conditions, Eq. (9.79) reduces to the following expressions for the gas, oil, and water pseudocomponents G, O, and W, respectively:

$$\frac{\partial}{\partial t}\left(\varepsilon_g \rho_g + \varepsilon_o \rho_o w_o^G\right)$$

$$= \nabla \cdot \left[\begin{array}{l} \dfrac{\rho_g}{\mu_g}\mathbf{K}_g \cdot (\nabla p_g - \rho_g \mathbf{g}\cdot\nabla z) \\ + \dfrac{\rho_o w_o^G}{\mu_o}\mathbf{K}_o \cdot (\nabla p_o - \rho_o \mathbf{g}\cdot\nabla z) \end{array}\right] + \varepsilon_g q_g^G + \varepsilon_o q_o^G, \tag{9.81}$$

$$\frac{\partial}{\partial t}\left(\varepsilon_o \rho_o w_o^O\right) = \nabla \cdot \left[\frac{\rho_o w_o^O}{\mu_o} \mathbf{K}_o \cdot \left(\nabla p_o - \rho_o \mathbf{g} \cdot \nabla z\right)\right] + \varepsilon_o q_o^O, \tag{9.82}$$

and

$$\frac{\partial}{\partial t}\left(\varepsilon_w \rho_w\right) = \nabla \cdot \left[\frac{\rho_w}{\mu_w} \mathbf{K}_w \cdot \left(\nabla p_w - \rho_w \mathbf{g} \cdot \nabla z\right)\right] + \varepsilon_w q_w^W. \tag{9.83}$$

It is customary to express the above-given equations in terms of the properties of fluids for which experimental data can be obtained by tests involving differential vaporization of reservoir oil samples in a PVT cell. For this purpose, consider the mass of the gas and dead-oil pseudospecies denoted by m^G and m^O, respectively. Let ρ_g and ρ_o denote the density of the gas and oil phases at the reservoir fluid conditions. Let ρ_s^G and ρ_s^O denote the density of the gas and dead-oil species at the standard conditions. Then, the following expressions of practical importance can be derived.

The pseudogas component G solubility in the oil phase o is defined by

$$R_{So}^{G/O} = \frac{V_s^G}{V_s^O} = \frac{m^G / \rho_s^G}{m^O / \rho_s^O} = \frac{m^G \rho_s^O}{m^O \rho_s^G}, \tag{9.84}$$

where V_s^G is the volume of dissolved-gas pseudospecies and V_s^O is the volume of dead-oil pseudospecies both measured at standard pressure and temperature conditions.

The gas-phase formation volume factor at reservoir conditions is given by

$$B_g = \frac{V_g}{V_s^G} = \frac{m^G / \rho_g}{m^G / \rho_s^G} = \frac{\rho_s^G}{\rho_g}, \tag{9.85}$$

where V_g is the volume of the gas phase measured at reservoir pressure and temperature conditions.

The oil-phase formation volume factor accounting for volume change by swelling of oil as a result of gas dissolution in the oil at reservoir conditions is given by

$$B_o = \frac{V_o}{V_s^O} = \frac{\left(m^G + m^O\right)/\rho_o}{m^O / \rho_s^O} = \frac{\rho_s^O}{\rho_o}\left(1 + \frac{m^G}{m^O}\right), \tag{9.86}$$

where V_o is the volume of the oil phase (dead-oil plus dissolved-gas pseudospecies) measured at reservoir pressure and temperature conditions.

The water-phase formation volume factor at reservoir conditions is given by

$$B_w = \frac{V_w}{V_s^W} = \frac{m^W / \rho_w}{m^W / \rho_s^W} = \frac{\rho_s^W}{\rho_w}, \tag{9.87}$$

where V_w is the volume of the water phase measured at reservoir pressure and temperature conditions.

The mass fraction of the gas pseudospecies in the oil phase at the reservoir conditions is given by

$$w_o^G = \frac{m^G}{m^G + m^O} = \frac{\dfrac{m^G}{m^O}}{1 + \dfrac{m^G}{m^O}}. \tag{9.88}$$

Combining Eqs. (9.84)–(9.88) yields

$$w_o^G = \frac{\rho_s^G R_{So}^{G/O}}{\rho_o B_o}. \tag{9.89}$$

The mass fraction of the dead-oil pseudospecies in the oil phase at the reservoir conditions is given by

$$w_o^O = \frac{m^O}{m^G + m^O} = \frac{1}{1 + \dfrac{m^G}{m^O}} = \frac{\rho_s^O}{\rho_o B_o}. \tag{9.90}$$

Note that the oil-phase density can be estimated by

$$w_o^G + w_o^O = \frac{\rho_s^G R_{So}^{G/O}}{\rho_o B_o} + \frac{\rho_s^O}{\rho_o B_o} = 1, \text{ and thus } \rho_o = \frac{\rho_s^G R_{So}^{G/O} + \rho_s^O}{B_o}. \tag{9.91}$$

Consequently, a substitution of Eqs. (9.84)–(9.91) and the expressions of the density of the various phases obtained earlier into Eqs. (9.81)–(9.83) result in the following expressions used in the development of the black oil simulators (modified after Peaceman, 1977):

$$\frac{\partial}{\partial t}\left(\frac{\varepsilon_g}{B_g} + \frac{\varepsilon_o R_{So}^{G/O}}{B_o}\right) = \nabla \cdot \left\{ \mathbf{K} \cdot \left[\begin{array}{l} \dfrac{k_{rg}}{B_g \mu_g}(\nabla p_g - \rho_g \mathbf{g} \cdot \nabla z) \\[2mm] + \dfrac{R_{So}^{G/O} k_{ro}}{B_o \mu_o}(\nabla p_o - \rho_o \mathbf{g} \cdot \nabla z) \end{array} \right] \right\} \\ + \frac{\varepsilon_g q_g^G + \varepsilon_o q_o^G}{\rho_s^G}, \tag{9.92}$$

$$\frac{\partial}{\partial t}\left(\frac{\varepsilon_o}{B_o}\right) = \nabla \cdot \left\{ \mathbf{K} \cdot \left[\frac{k_{ro}}{B_o \mu_o}(\nabla p_o - \rho_o \mathbf{g} \cdot \nabla z) \right] \right\} + \frac{\varepsilon_o q_o^O}{\rho_s^O}, \tag{9.93}$$

and

$$\frac{\partial}{\partial t}\left(\frac{\varepsilon_w}{B_w}\right) = \nabla \cdot \left\{ \mathbf{K} \cdot \left[\frac{k_{rw}}{B_w \mu_w}(\nabla p_w - \rho_w \mathbf{g} \cdot \nabla z) \right] \right\} + \frac{\varepsilon_w q_w^W}{\rho_s^W}, \quad B_w \cong 1. \tag{9.94}$$

9.6 ISOTHERMAL LIMITED COMPOSITIONAL MODEL OF A VOLATILE OIL SYSTEM

The applicability of the formulation of the black oil model derived in the previous section is limited to nonvolatile oils. For applications to volatile oils, it is necessary to allow oil vaporization into the gas phase. Therefore, we now consider the volatility of the oil in the gas phase denoted by $R_{Vg}^{O/G}$ and the solubility of the gas in the water phase $R_S^{G/W}$ in addition to the solubility of the gas in the oil phase $R_{So}^{G/O}$. Consequently, the mass fractions of the pseudospecies in the various phases are given as $w_g^G \neq 0, w_g^O \neq 0, w_g^W = 0, \quad w_o^G \neq 0, w_o^O \neq 0, w_o^W = 0,$ and $w_w^G \neq 0,$ $w_w^O = 0, w_w^W = 1.$ Then, the following equations are obtained (modified after Peaceman, 1977):

$$
\begin{aligned}
\frac{\partial}{\partial t} & \left(\frac{\varepsilon_g}{B_g} + \frac{\varepsilon_o R_{So}^{G/O}}{B_o} + \frac{\varepsilon_w R_{Sw}^{G/W}}{B_w} \right) \\
&= \nabla \cdot \left\{ \mathbf{K} \cdot \left[\begin{array}{l} \dfrac{k_{rg}}{B_g \mu_g}(\nabla p_g - \rho_g \mathbf{g} \cdot \nabla z) + \dfrac{R_{So}^{G/O} k_{ro}}{B_o \mu_o}(\nabla p_o - \rho_o \mathbf{g} \cdot \nabla z) \\[3mm] + \dfrac{R_{Sw}^{G/W} k_{rw}}{B_w \mu_w}(\nabla p_w - \rho_w \mathbf{g} \cdot \nabla z) \end{array} \right] \right\} \\
&\quad + \frac{\varepsilon_g q_g^G + \varepsilon_o q_o^G + \varepsilon_w q_w^G}{\rho_s^G},
\end{aligned}
\tag{9.95}
$$

$$
\frac{\partial}{\partial t}\left(\frac{\varepsilon_g R_{Vg}^{O/G}}{B_g} + \frac{\varepsilon_o}{B_o} \right) = \nabla \cdot \left\{ \mathbf{K} \cdot \left[\begin{array}{l} \dfrac{R_{Vg}^{O/G} k_{rg}}{B_g \mu_g}(\nabla p_g - \rho_g \mathbf{g} \cdot \nabla z) \\[3mm] + \dfrac{k_{ro}}{B_o \mu_o}(\nabla p_o - \rho_o \mathbf{g} \cdot \nabla z) \end{array} \right] \right\} + \frac{\varepsilon_g q_g^G + \varepsilon_o q_o^O}{\rho_s^O},
\tag{9.96}
$$

and

$$
\frac{\partial}{\partial t}\left(\frac{\varepsilon_w}{B_w} \right) = \nabla \cdot \left\{ \mathbf{K} \cdot \left[\frac{k_{rw}}{B_w \mu_w}(\nabla p_w - \rho_w \mathbf{g} \cdot \nabla z) \right] \right\} + \frac{\varepsilon_w q_w^W}{\rho_s^W}, \quad B_w \cong 1.
\tag{9.97}
$$

9.7 FLOW OF GAS AND VAPORIZING WATER PHASES IN THE NEAR-WELLBORE REGION

The modeling approach taken by Mahadevan et al. (2007) is presented here with modifications for consistency with the rest of the materials. The process is a radial immiscible displacement involving water evaporation into the gas phase during convection-dominated simultaneous flow of a gas/water system through a homogeneous porous media under isothermal and thermodynamic equilibrium conditions. While the water is treated as an incompressible fluid, the expanding gas behaves as a highly compressible fluid. Gas dissolution in the water phase, flow of the water phase, and the Joule–Thompson cooling and capillary pressure effects are neglected.

They assumed an ideal gas density, given by

$$
\rho_g = \frac{M_g p_g}{R_g T},
\tag{9.98}
$$

where M_g, p_g, and T denote the molecular weight, total pressure, and temperature of the gas phase, and R_g is the universal gas constant.

Applying Raoult's law, the mole fraction of the water vapor in the gas phase is given by

$$y_w = \frac{p_s}{p_g}, \tag{9.99}$$

where p_s denotes the saturation vapor pressure of the water in the gas phase.

The mass fraction of the water vapor in the gas phase is given by

$$w_g^W = \frac{M_w y_w}{M_g}, \tag{9.100}$$

where M_w is the molecular weight of water. The weight fraction of the gas component in the gas phase is

$$w_g^G = 1 - w_g^W. \tag{9.101}$$

The mass balance of the water component in the liquid water phase is given by

$$\frac{\partial}{\partial t}\left(\phi S_w \rho_w w_w^W\right) + \frac{1}{r}\frac{\partial}{\partial r}\left(r\rho_w u_w w_w^W\right) = 0. \tag{9.102}$$

Assuming all water $w_w^W = 1$, and constant properties $\phi = ct.$, $\rho_w = ct.$, Eq. (9.102) simplifies as

$$\phi\rho_w \frac{\partial S_w}{\partial t} + \rho_w \frac{1}{r}\frac{\partial(ru_w)}{\partial r} = 0. \tag{9.103}$$

The mass balance of the water component in the gas phase is given by

$$\phi\frac{\partial}{\partial t}\left(S_g \rho_g w_g^W\right) + \frac{1}{r}\frac{\partial}{\partial r}\left(r\rho_g u_g w_g^W\right) = 0. \tag{9.104}$$

Summing Eqs. (9.103) and (9.104) and substituting Eq. (9.98) yields a total water component mass balance equation as

$$\frac{\partial}{\partial t}\left[\alpha(1 - S_w) + \beta S_w\right] + \frac{1}{\phi}\frac{1}{r}\frac{\partial}{\partial r}\left[r(\alpha f_g + \beta f_w)u_T\right] = 0, \tag{9.105}$$

where

$$\alpha = \frac{y_w p_g}{R_g T}, \beta = \frac{\rho_w}{M_w}, u_T = u_w + u_g.$$

The mass balance of the gas component in the gas phase is given by

$$\phi\frac{\partial}{\partial t}\left(S_g \rho_g w_g^G\right) + \frac{1}{r}\frac{\partial}{\partial r}\left(r\rho_g u_g w_g^G\right) = 0. \tag{9.106}$$

Substituting Eq. (9.98) into Eq. (9.106) yields

$$\frac{\partial}{\partial t}\left[(1 - y_w)p_g S_g\right] + \frac{1}{\phi}\frac{1}{r}\frac{\partial}{\partial r}\left[r(1 - y_w)p_g f_g u_T\right] = 0. \tag{9.107}$$

Figure 9.3 shows the typical solutions obtained by Mahadevan et al. (2007) after some mathematical approximations and manipulations of the above-given

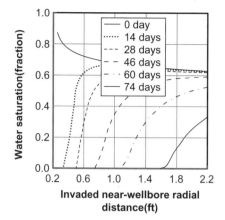

Figure 9.3 Time evolution of saturation profiles during the evaporative regime in a tight-gas sandstone (TG-3) at a pressure drop of 279 psi in an unfractured well (radial geometry). The saturation profiles show faster evaporation near the outlet end on the left side (modified after Mahadevan et al., 2007; © 2007 SPE, with permission from the Society of Petroleum Engineers).

equations. Typical water saturation profiles in the radial direction at various times after the initiation of flow are depicted.

9.8 FLOW OF CONDENSATE AND GAS PHASE CONTAINING NONCONDENSABLE GAS SPECIES IN THE NEAR-WELLBORE REGION

The modeling approach taken by Yu et al. (1996) is presented here with modifications for consistency.

The process considered here is a radial immiscible displacement involving the condensation of the condensable gas species while the noncondensable gas species remain in the gas phase during simultaneous flow of the gas/condensate system through porous media. They considered the fluid properties at the actual in situ fluid conditions.

Consider a gas phase in equilibrium with a liquid condensate phase. The gas phase contains condensable hydrocarbon components (C_{1-7}^+) and some noncondensable components (CO_2, N_2). The liquid phase contains only the condensable hydrocarbon components.

The liquid condensate and the gas phase mass fractions in the pore space are given by

$$x_c = \frac{\rho_o S_o}{\rho_g S_g + \rho_o S_o}, \, x_g = 1 - x_c. \tag{9.108}$$

Let the total condensable gas component weight fraction in the pore space or the maximum liquid dropout be denoted by x_{cm}. Then, the mass or weight fractions of the condensable and noncondensable gas components present in the gas phase at any reservoir temperature and pressure conditions are given by

$$w_{cg} = \frac{x_{cm} - x_c}{1 - x_c}, \, w_{ncg} = 1 - w_{cg}. \tag{9.109}$$

Darcy's law is given by the following for the gas and condensate (oil) phases:

$$u_g = -\frac{k_{rg}K}{\mu_g}\frac{\partial p_g}{\partial r} \tag{9.110}$$

and

$$u_o = -\frac{k_{ro}K}{\mu_o}\frac{\partial p_o}{\partial r}. \tag{9.111}$$

The mass balance of the noncondensable gas component in the gas phase is given by

$$\frac{\partial}{\partial t}\left(\phi S_g \rho_g w_{ncg}^G\right)+\frac{1}{r}\frac{\partial}{\partial r}\left(r\rho_g u_g w_{ncg}^G\right)=0. \tag{9.112}$$

Thus, substituting Eq. (9.110) into Eq. (9.112) yields

$$\frac{\partial}{\partial t}\left(\phi S_g \rho_g w_{ncg}^G\right)=\frac{1}{r}\frac{\partial}{\partial r}\left(r\frac{k_{rg}K}{\mu_g}\rho_g w_{ncg}^G\frac{\partial p_g}{\partial r}\right). \tag{9.113}$$

Similarly, the mass balance of the condensable gas component in the gas phase is given by

$$\frac{\partial}{\partial t}\left(\phi S_g \rho_g w_{cg}^G\right)=\frac{1}{r}\frac{\partial}{\partial r}\left(r\frac{k_{rg}K}{\mu_g}\rho_g w_{cg}^G\frac{\partial p_g}{\partial r}\right). \tag{9.114}$$

Further, the mass balance of the condensable gas component in the liquid condensate (oil) phase ($w_{co}^G = 1$) is given by

$$\frac{\partial}{\partial t}\left(\phi S_o \rho_o\right)=\frac{1}{r}\frac{\partial}{\partial r}\left(r\frac{k_{ro}K}{\mu_o}\rho_o\frac{\partial p_o}{\partial r}\right). \tag{9.115}$$

Neglecting the capillary pressure effect, they assume $p_g = p_o = p$. Thus, summing Eqs. (9.114) and (9.115) yields

$$\frac{\partial}{\partial t}\left(S_g \rho_g w_{cg}^G + S_o \rho_o\right)=\frac{K}{\phi}\frac{1}{r}\frac{\partial}{\partial r}\left[r\left(\frac{k_{rg}}{\mu_g}\rho_g w_{cg}^G+\frac{k_{ro}}{\mu_o}\rho_o\right)\frac{\partial p}{\partial r}\right]. \tag{9.116}$$

Similarly, summing Eqs. (9.113) and (9.116) yields

$$\frac{\partial}{\partial t}\left(S_g \rho_g + S_o \rho_o\right)=\frac{K}{\phi}\frac{1}{r}\frac{\partial}{\partial r}\left[r\left(\frac{k_{rg}}{\mu_g}\rho_g+\frac{k_{ro}}{\mu_o}\rho_o\right)\frac{\partial p}{\partial r}\right]. \tag{9.117}$$

Define a pseudopressure function as

$$\Psi(p)=\frac{\mu_r}{\rho_r}\int_{p_r}^{p}\left(\frac{k_{rg}}{\mu_g}\rho_g+\frac{k_{ro}}{\mu_o}\rho_o\right)dp, \tag{9.118}$$

where the subscript r represents the properties of the gas condensate system at some reference conditions. Consequently, Eq. (9.117) can be linearized as

$$\frac{\partial \Psi}{\partial t}=D_m\frac{1}{r}\frac{\partial}{\partial r}\left(r\frac{\partial \Psi}{\partial r}\right), \tag{9.119}$$

where D_m is the hydraulic diffusivity, given by

$$D_m = \frac{\phi}{K} \frac{\dfrac{k_{rg}}{\mu_g}\rho_g + \dfrac{k_{ro}}{\mu_o}\rho_o}{c_t + (\rho_o - \rho_g)\dfrac{\partial S_o}{\partial p}}, \qquad (9.120)$$

where the isothermal total fluid compressibility c_t is given by

$$c_t = S_g\rho_g c_g + S_o\rho_o c_o, c_g = \frac{1}{\rho_g}\frac{\partial \rho_g}{\partial p}, c_o = \frac{1}{\rho_o}\frac{\partial \rho_o}{\partial p}. \qquad (9.121)$$

The conditions of the solution are given by

$$\Psi = \Psi_i, r_w \le r \le r_e, t = 0 \qquad (9.122)$$

and

$$r\frac{\partial \Psi}{\partial r} = \frac{m_t u_r}{2\pi K h \rho_r}, r = r_w, t > 0, \qquad (9.123)$$

where the isothermal total fluid mass flow rate m_t is given by

$$m_t = m_g + m_o, m_g = S_g\rho_g, m_o = S_o\rho_o \qquad (9.124)$$

and

$$\frac{\partial \Psi}{\partial r} = 0, r = r_e, t > 0. \qquad (9.125)$$

Typical near-wellbore pressure and gas-saturation profiles calculated in the radial direction at various times are presented in Figures 9.4 and 9.5 by Yu et al. (1996).

Note that the above-mentioned treatment assumed Darcy flow for both the gas and condensate phases. However, the gas phase may involve inertial flow effect as

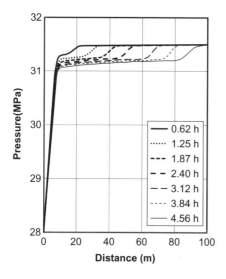

Figure 9.4   Pressure profiles varying with production time (modified after Yu et al., 1996; © 1996 SPE, with permission from the Society of Petroleum Engineers).

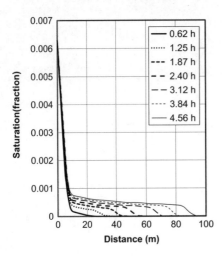

Figure 9.5   Gas-saturation profiles varying with production time (modified after Yu et al., 1996; © 1996 SPE, with permission from the Society of Petroleum Engineers).

a result of acceleration by gas expansion. Therefore, it is more appropriate to consider the following Forchheimer equation for the gas phase while retaining Darcy's law for the condensate (oil) phase. Thus,

$$\frac{\partial p_g}{\partial r} = \frac{\mu_g}{k_{rg} K} \frac{m_g}{2\pi r h \rho_g} + \beta \rho_g \left(\frac{m_g}{2\pi r h \rho_g} \right)^2 \tag{9.126}$$

and

$$\frac{\partial p_o}{\partial r} = \frac{\mu_o}{k_{ro} K} \frac{m_o}{2\pi r h \rho_o}. \tag{9.127}$$

Note that the negative signs were removed because we are now considering the production case for which the direction of flow is opposite to the radial direction.

By substituting Eqs. (9.126) and (9.127) into Eq. (9.118), assuming $p_g = p_o = p$, and by differentiating, we obtain

$$\begin{aligned}
\Psi(p) &= \frac{\mu_r}{\rho_r} \int_{p_r}^{p} \left[\frac{m_t}{2\pi r h K} + \beta \frac{k_{rg}}{\mu_g} \left(\frac{m_g}{2\pi r h} \right)^2 \right] dp \\
&= \frac{\mu_r}{\rho_r} \left[\frac{m_t}{2\pi h K} \ln\left(\frac{r_w}{r} \right) + \beta \frac{k_{rg}}{\mu_g} \left(\frac{m_g}{2\pi h} \right)^2 \left(\frac{1}{r_w} - \frac{1}{r} \right) \right].
\end{aligned} \tag{9.128}$$

Let us now define a dimensionless pseudopressure function as

$$\Psi_D(p) = \frac{\Psi(p)}{\dfrac{\mu_r}{\rho_r} \dfrac{m_t}{2\pi h K}} = \ln\left(\frac{r_w}{r} \right) + s_{ND}\left(1 - \frac{r_w}{r} \right) + s, \tag{9.129}$$

where we included a formation skin factor s for pressure loss by porous media alteration and also defined a non-Darcy skin factor s_{ND} as

$$s_{ND} = \frac{\beta K k_{rg} m_g^2}{2\pi h m_t \mu_g r_w}. \tag{9.130}$$

9.9 SHAPE-AVERAGED FORMULATIONS

Frequently, simplified equations are resorted by averaging the conservation equations over a region of the physical system. Several examples are provided in the following sections.

9.9.1 Thickness-Averaged Formulation

For example, we might seek a general thickness-averaged conservation equation derived for a control volume shown in Figure 9.6.

$$\frac{(\Delta x \Delta y HW)_{t+\Delta t} - (\Delta x \Delta y HW)_t}{\Delta t} = (n_x \Delta y H)_x - (n_x \Delta y H)_{x+\Delta x} \\ + (n_y \Delta x H)_y - (n_y \Delta x H)_{y+\Delta y} + \dot{m}, \tag{9.131}$$

where W is the quantity per unit volume, $H = H(x, y, t)$ is the thickness, n_x and n_y denote the fluxes in the x- and y-directions, and $\dot{m}$ is an external source.

Dividing by $(\Delta x \Delta y)$ and then taking the limit as $\Delta x, \Delta y, \Delta t \to 0$ yields

$$\frac{\partial (HW)}{\partial t} + \frac{\partial (n_x H)}{\partial x} + \frac{\partial (n_y H)}{\partial y} = \operatorname*{Limit}_{\Delta x, \Delta y \to 0} \frac{\dot{m}}{\Delta x \Delta y}. \tag{9.132}$$

For example, if we substitute $W = \phi \rho$, $n_x = \rho u_x$, $n_y = \rho u_y$, then we obtain the following thickness-averaged equation of continuity:

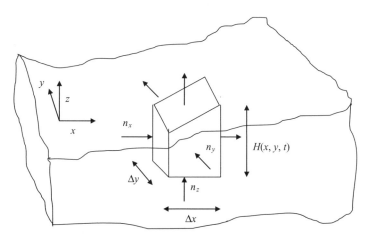

Figure 9.6   Thickness averaging (prepared by the author).

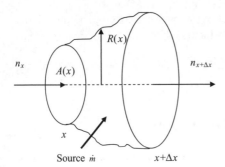

Figure 9.7 Cross-sectional averaging (prepared by the author).

$$\frac{\partial(\phi\rho H)}{\partial t} + \frac{\partial(\rho u_x H)}{\partial x} + \frac{\partial(\rho u_y H)}{\partial y} = \underset{\Delta x, \Delta y \to 0}{\text{Limit}} \frac{\dot{m}}{\Delta x \Delta y}. \tag{9.133}$$

Further, if $\dot{m} = (\Delta x \Delta y)\dot{M}$, where $\dot{M}$ denotes the mass added per unit surface area of the system and $H = $ ct., then Eq. (9.133) becomes

$$\frac{\partial(\phi\rho)}{\partial t} + \frac{\partial(\rho u_x)}{\partial x} + \frac{\partial(\rho u_y)}{\partial y} = \frac{\dot{M}}{H}. \tag{9.134}$$

9.9.2 Cross-Sectional Area-Averaged Formulation

For example, we might seek a general cross-sectional area $A(x, t)$ averaged conservation equation derived for a control volume shown in Figure 9.7:

$$\frac{(\Delta x A W)_{t+\Delta t} - (\Delta x A W)_t}{\Delta t} = (n_x A)_x - (n_x A)_{x+\Delta x} + \dot{m}. \tag{9.135}$$

Dividing by Δx and then taking the limit as $\Delta x, \Delta t \to 0$ yields

$$\frac{\partial(A W)}{\partial t} + \frac{\partial(n_x A)}{\partial x} = \underset{\Delta x \to 0}{\text{Limit}} \frac{\dot{m}}{\Delta x}. \tag{9.136}$$

For example, for illustration purposes, assume constant properties for the solid and fluid phases and equilibrium conditions. If we substitute

$$W = \phi\rho c T, n_x = \rho c u_x T - \phi k \frac{\partial T}{\partial x}, A = \pi R^2, \dot{m} = (2\pi R\phi\Delta x)U(T - T_R),$$

where R is the radius, T the temperature, ϕ the porosity, c the heat capacity, k the thermal conductivity, u_x the volumetric flux, T_R the surrounding temperature, and U the overall heat transfer coefficient, then we obtain the following cross-sectional averaged equation of continuity:

$$\frac{\partial(\pi R^2 \phi\rho c T)}{\partial t} + \frac{\partial}{\partial x}\left[\pi R^2\left(\rho c u_x T - \phi k \frac{\partial T}{\partial x}\right)\right] = (2\pi R\phi)U(T - T_R). \tag{9.137}$$

Further, if all physical properties are constant values and $R = $ ct., then Eq. (9.133) becomes

$$\phi\frac{\partial T}{\partial t} + u\frac{\partial T}{\partial x} = \phi\alpha\frac{\partial^2 T}{\partial x^2} + \beta(T - T_R), \alpha = \frac{k}{\rho c}, \beta = \frac{2U\phi}{R\rho c}. \tag{9.138}$$

The conditions of the solution may be selected as the following:

$$T = T_o, 0 \leq x \leq L, t = 0, \tag{9.139}$$

$$uT - \phi\alpha\frac{\partial T}{\partial x} = uT_i, x = 0, t > 0 \tag{9.140}$$

and

$$\frac{\partial T}{\partial x} = 0, x = L, t > 0, \tag{9.141}$$

where T_o and T_i denote the initial and injection fluid temperatures, and T_R is the external temperature.

Consider now the following transformation to a new function, f, given by (Civan and Sliepcevich, 1984)

$$f = T\exp(x/x_o), \tag{9.142}$$

where x_o is a characteristic length to be determined.

Substituting Eq. (9.142) into Eq. (9.138) and demanding that the coefficient of

$$\frac{\partial f}{\partial x}$$

vanishes so that $x_o = -2\phi\alpha/u$ is obtained, then Eqs. (9.138)–(9.142) become

$$\frac{\partial f}{\partial t} + \left(\frac{u^2}{4\phi^2\alpha} - \frac{\beta}{\phi}\right)f + \frac{\beta}{\phi}f_R = \alpha\frac{\partial^2 f}{\partial x^2}, \tag{9.143}$$

where

$$f_R = T_R\exp(x/x_o), \tag{9.144}$$

$$f = f_i, 0 \leq x \leq L, t = 0, \tag{9.145}$$

$$\frac{\partial f}{\partial x} = \frac{u}{2\phi\alpha}f - \frac{u}{\phi\alpha}f_i, x = 0, t > 0, \tag{9.146}$$

and

$$\frac{\partial f}{\partial x} = -\frac{u}{2\phi\alpha}f, x = L, t > 0. \tag{9.147}$$

9.10 CONDUCTIVE HEAT TRANSFER WITH PHASE CHANGE

Parts of this section are presented following Civan (2000d), who emphasized that engineering applications in cold regions frequently involve buried pipelines, cables, storage tanks; construction of buildings and roads; mining operations; and drilling and completion of wells in groundwater and petroleum reservoirs. Accurate prediction of the apparent thermal properties of seasonally frozen soil is essential for effective design and mitigation of such engineering projects. Modeling and simulation of systems involving seasonally freezing and thawing wet soils require correlation of the water and ice content as functions of the soil temperature. As pointed out by Civan and Sliepcevich (1985b),

> the thermal regimes of moist soils undergo freezing or thawing at a single temperature for coarse-grained gravel or sand or over a temperature interval for fine-grained silts or clays. The principle difference in the thermal regimes for these two types is due to the amount that the thermal physical properties of the soil are affected by the unfrozen water content.

> The unfrozen water content for a phase change at a single temperature can be represented mathematically by

$$w_u = w_o U(T) \qquad (9.148)$$

> in which U is a unit step function whose value is zero when $T < 0°C$ but equals one otherwise. For the phase change occurring over a finite temperature interval the amount of water that remains unfrozen can be estimated from an empirical correlation as a function of the surface area per unit mass of soil and the temperature below the normal freezing point of water according to Anderson and Tice (1972), or it can be measured directly.

Lunardini (1981, 1991) gives a comprehensive summary of various analytical and numerical techniques that have been used in solving heat conduction with phase change. Lukianov and Golovko (1957) pioneered the apparent heat capacity concept and formulation, and Hashemi and Sliepcevich (1967) introduced the technique of distributing the latent heat over a finite temperature interval in order to circumvent problems in numerical computation arising from the apparent heat capacity approaching infinity during phase change at a fixed temperature.

As described by Civan and Sliepcevich (1985b),

> two approaches have been used to formulate this class of problems. The first formulation is based on a moving interface or Stefan concept. Unfortunately, the moving interface formulation is not well suited to numerical solution techniques without resorting to complicated techniques such as deforming grids. Another deficiency in the moving interface formulation is the inherent limitation that the entire phase change occurs at a single temperature, whereas in reality for many systems—particularly the freezing of fine grained moist soils—the phase change occurs gradually over a temperature range of several degrees.

The formulation that is applicable for either instantaneous or gradual phase changes is a *second* type of formulation, referred to as the apparent heat capacity method (Hashemi and Sliepcevich, 1967; Nakano and Brown, 1971) or its integral, the enthalpy method (Wood et al., 1981; Voller and Cross, 1983), which includes the latent heat effect in the energy equation. A difficulty encountered in these techniques is the jump discontinuities in the thermophysical properties accompanying an instantaneous phase change. Accordingly, variable grid spacings are required to control the stability of the numerical solution (Goodrich, 1978).

The accuracy of the numerical computations can be improved by means of a group transformation, which reduces the number of variables. In general, this type of transformation is suited to initial value problems; it has limited application for initial-boundary value problems. Specifically, the similarity (Boltzmann) transformation is the only one that works for initial-boundary value problems in the semi-infinite or infinite domains. Although the Boltzmann transformation has been used extensively to obtain *analytical* solutions for special, simple cases of heat conduction with phase change, its value in facilitating numerical solutions of problems for which analytical solutions are not available appears to have been overlooked prior to the study by Anderson and Ford (1981). As pointed out by Crank (1957), the Boltzmann transformation does not work for a finite domain.

9.10.1 Unfrozen Water in Freezing and Thawing Soils: Kinetics and Correlation

The experimental data indicate that water in freezing or thawing wet soils undergoes a gradual phase change. Theoretical treatment of the gradual freezing/thawing process by Civan (2000d) provides insight into the mechanism of the gradual phase change of water in wet soils and the proper means of correlating experimental measurements. The analysis presented by Civan (2000d) demonstrates that the unfrozen water content can be adequately correlated by means of an exponential decay function shifted by the amount of the water adsorbed over the soil grains, which cannot freeze.

The reported studies mostly assumed instantaneous phase change for computational convenience. However, transition between water and ice in soil occurs by gradual phase change over a temperature range below the freezing temperature of pure water as indicated by the experimental data by Nakano and Brown (1971), Lunardini (1981), and Inaba (1983). Nakano and Brown (1971) and Civan and Sliepcevich (1985b) considered the gradual freezing/thawing effect in the numerical modeling of soil freezing/thawing.

The apparent thermal properties of wet soils depend, in a complicated manner, on various factors, including texture, fabric, mineral composition, and specific grain surface of the soil matrix; the frozen and unfrozen waters and gas (mostly air) filling the pore space; and thermal properties of the constituents of soil (Anderson et al., 1973).

Various empirical correlations of the unfrozen water content in wet soils have been reported in the literature. Anderson and Tice (1973) used power-law-type empirical correlations, which can be expressed as

$$w_u = \alpha \theta^\beta, \qquad (9.149)$$

where ϕ_u is the mass (or weight) of the unfrozen water per unit mass (or weight) of dry soil; θ is the soil temperature, T, below the freezing temperature of pure water, $T_t(0°\text{C at 1 atm})$, given by

$$\theta = T_t - T, \qquad (9.150)$$

and α and β are empirically determined characteristic parameters of soils given respectively, by the power-law functions of the specific surface of soil as

$$\alpha = 1.30 S^{0.552} \qquad (9.151)$$

and

$$\beta = -1.45S^{-0.264}, \tag{9.152}$$

where S is the specific surface of soil grains in cubic meter per gram dry soil. Their experimental data and other reported data indicate that a fraction of water is adsorbed over the soil grains and remains unfrozen. Therefore, other researchers including the present author preferred correlations that can be expressed in the following form:

$$w_u = w_a + (w_o - w_a)\exp(-k\theta) \tag{9.153}$$

and

$$\phi_u = w_u / w_o = \phi_a + (1 - \phi_a)\exp(-k\theta), \tag{9.154}$$

in which w_o and w_a denote the total and adsorbed water masses (or weights) per unit mass (or weight) of dry soil; ϕ_u and ϕ_a represent the mass fractions of the unfrozen and adsorbed water, respectively, expressed as mass (or weight) per total mass (or weight) of water present in soil; and k is a freezing rate coefficient.

Nakano and Brown (1971) correlated the unfrozen water content of the Point Barrow silt by

$$w_u = 0.0722 + 0.1278\exp(-0.763\theta), \tag{9.155}$$

where θ is in Celsius and w_u is in gram unfrozen water per gram dry soil. Thus, for $\theta = 0°C$, Eq. (9.155) indicates that the total water content of this soil is $w_o = 0.20$ g total water/g dry soil. Comparing Eq. (9.155) with Eq. (9.153), the adsorbed, or nonfreezing, water content is $w_a = 0.0722$ g water/g dry soil. Dividing Eq. (9.155) by the total water content, the unfrozen water content of this soil can be expressed in mass (or weight) fraction as

$$\phi_u = w_u / w_o = 0.361 + 0.639\exp(-0.763\theta). \tag{9.156}$$

Lunardini (1981) reports that the unfrozen water content of a gravel–sand mixture is given by

$$w_u = \exp(0.17\theta - 0.73), \tag{9.157}$$

where θ is in Fahrenheit and w_u is in pound unfrozen water per pound dry soil. For $\theta = 0°F$, Eq. (9.157) indicates that the total water content of the gravel–sand mixture is $w_o = \exp(-0.73) = 0.482$ lb total water/lb dry soil. Hence, expressing the temperature below the freezing temperature of pure water in °C and dividing Eq. (9.157) by the total water content, the unfrozen water content of the gravel–sand mixture can be expressed in mass fraction as

$$\phi_u = w_u / w_o = \exp(-0.306\theta). \tag{9.158}$$

A comparison of Eq. (9.158) with Eq. (9.154) indicates that the adsorbed water content is zero.

Often, the physical data of natural systems do not readily reveal insight into the mechanisms of the governing processes, unless a meaningful theory and formulation are developed. The correlations of the unfrozen water content, Eqs. (9.149)–(9.158), have been developed solely by empirical fitting of measured data without any theoretical bases. Civan (2000d) developed a theoretical model and derived rate equation for the freezing/thawing processes and demonstrated that the solution of this equation, under reasonable simplifying conditions, conforms to Eqs. (9.153) and (9.154), indicating that Eqs. (9.153) and (9.154) are adequate for correlation of the unfrozen water content in wet soils. However, a number of approximation functions, as described in this section, are also available.

9.10.2 Kinetics of Freezing/Thawing Phase Change and Correlation Method

Consider the reversible equation of phase change between water and ice given by

$$\text{Water} \rightleftarrows \text{ice} + Q, \tag{9.159}$$

where Q denotes the heat removed from water to form ice. It is assumed that water and ice are at thermal equilibrium.

Civan (2000d) expressed the rate of ice formation as the following in terms of two driving factors: (1) the heat efflux, q, and (2) the difference between the instantaneous unfrozen water, ϕ_u, and the nonfreezing adsorbed water, ϕ_a, retained on the soil grains:

$$-d\left(w_u - w_a\right)/dt = k_r q^\alpha \left(w_u - w_a\right)^\beta, \tag{9.160}$$

where t is the time, k_r is the rate coefficient, and α and β are empirically determined exponents. For many processes, the exponents or orders of kinetics are given by

$$\alpha = \beta = 1. \tag{9.161}$$

The rate (q) and cumulative (Q) of heat removed are related by

$$q = dQ/dt. \tag{9.162}$$

The rate, q, depends on the properties of wet soils in a complicated manner, as shown in the following.

The following energy balance equation of the lth phase in porous media can be applied to wet soil (Civan, 2000d):

$$\frac{\partial}{\partial t}\left(\varepsilon_l \rho_l h_l\right) + \nabla \cdot \left(u_l \rho_l h_l\right)$$

$$= \nabla \cdot \left(\varepsilon_l k_l \cdot \nabla T_l\right) + \frac{\partial}{\partial t}\left(\varepsilon_l p_l\right) + u_l \cdot \nabla p_l - \varepsilon_l \left(q_l \sum_{\substack{a=1 \\ a\neq l}} q_{l,a}\right), \tag{9.163}$$

where l denotes the various phases present in wet soil, such as the soil grains, s, unfrozen water, u, frozen water, f, and gas, g. ε denotes the fraction of the volume occupied in the bulk volume of the wet soil. Thus,

$$\sum_l \varepsilon_l = 1. \tag{9.164}$$

$\rho, h, k, T, p,$ and u are the density, enthalpy, thermal conductivity tensor, temperature, pressure, and volume flux vector, respectively. q_l and $q_{l,\alpha}$ denote the rates of external and internal (interface between any pairs of phases l and α) heat losses per unit volume of the lth phase present in soil. The total energy balance equation of the wet soil system can be obtained by summing Eq. (9.163) over all the phases as

$$\frac{\partial}{\partial t}\sum_l (\varepsilon_l \rho_l h_l) + \nabla \cdot \sum_l (u_l \rho_l h_l) = \nabla \cdot \sum_l (\varepsilon_l k_l \cdot \nabla T_l)$$

$$+ \frac{\partial}{\partial t}\sum_l (\varepsilon_l p_l) + \sum_l (u_l \cdot \nabla p_l) - \sum_l (\varepsilon_l q_l). \tag{9.165}$$

The interface heat transfer rates vanished in Eq. (9.165) because they are equal and opposite between the various phases:

$$\sum_l \varepsilon \sum_{\substack{\alpha=l \\ \alpha \neq l}} q_{l,\alpha} = 0. \tag{9.166}$$

When the heat transfer in soil occurs sufficiently gradual, then the constituents of soil can be considered at thermal equilibrium. Thus, an instantaneous soil temperature, T, can be defined, for all phases, as

$$T_l = T. \tag{9.167}$$

The dry soil density is given by

$$\rho_d = \varepsilon_s \rho_s. \tag{9.168}$$

The mass of the lth phase per unit mass of dry soil is given by

$$w_l = \frac{m_l}{m_s} = \frac{\varepsilon_l \rho_l}{\varepsilon_s \rho_s} = \frac{\varepsilon_l \rho_l}{\rho_d}, \tag{9.169}$$

where m_l and m_s denote the masses (or weights) of the lth phase and dry soil, respectively. Thus, the total volumetric (or apparent) enthalpy of the wet soil can be expressed by

$$\rho h = \sum_l (\varepsilon_l \rho_l h_l) = \rho_d \sum_l (w_l h_l). \tag{9.170}$$

Differentiation of Eq. (9.170) yields

$$\frac{d(\rho h)}{dT} = \frac{d\rho_d}{dT} \sum_l (w_l h_l) + \rho_d \sum_l \left(w_l \frac{dh_l}{dT} + h_l \frac{dw_l}{dT} \right). \tag{9.171}$$

The total volumetric (or apparent) heat capacity of wet soil and the intrinsic specific heat capacities of the various phases at constant pressure are defined, respectively, by

$$\rho c = \frac{d(\rho h)}{dT} \tag{9.172}$$

and

$$c_l = \frac{dh_l}{dT}.$$

(9.173)

Assume that the wet soil is fully saturated with water and therefore there is no gas, $w_g = 0$, and that the variation of the dry soil density by temperature is negligible, and therefore $\rho_d \cong$ constant. Further, note that by Eq. (9.169), $w_s = 1.0$ and the total water content, w_0, is the sum of the unfrozen and frozen water contents, as given by

$$w_o = w_u + w_f.$$

(9.174)

The density ρ_d and the specific heat capacity c_d of dry soil can be expressed in terms of the wet soil density ρ and specific heat capacity c, and the total water content w_o, respectively, by

$$\rho_d = \rho/(1 + w_o)$$

(9.175)

and

$$c_d = c(1 + w_o).$$

(9.176)

The latent heat of freezing of water is defined by

$$L_{u,f} = h_f - h_u.$$

(9.177)

Invoking Eqs. (9.172)–(9.177) into Eq. (9.171) leads to the apparent volumetric heat capacity of soil as

$$\rho c = \rho_d c_d = \rho_d (1 + w_o) c = \rho_d \left[c_s + c_f w_o + (c_u - c_f) w_u - L_{u,f} \frac{dw_u}{dT} \right].$$

(9.178)

Eq. (9.178) has been frequently used by previous studies, although it is subject to the inherent limitations of the above-stated assumptions. Eq. (9.178) can be used, with additional assumptions that c_s, c_f, and c_u remain approximately constant, to derive a simple analytical expression for the unfrozen water content. However, accurate numerical solutions can be generated using Eqs. (9.171)–(9.173) directly with the thermal property data of the constituents of wet soil. However, the analyses of reported experimental data presented later in the application section will demonstrate that the simplified analytical solution is sufficient for practical purposes.

Inferred by Eq. (9.165), the other volumetric average (or apparent) properties of wet soil can be defined as the following. The thermal conductivity is given by

$$k = \sum_l \varepsilon_l k_l.$$

(9.179)

Under the same conditions leading to Eq. (9.178), Eq. (9.179) can be expressed for the apparent thermal conductivity of soil as

$$k = \rho_d \left[\frac{k_s}{\rho_s} + \frac{k_f}{\rho_f} w_o + \left(\frac{k_u}{\rho_u} - \frac{k_f}{\rho_f} \right) w_u \right].$$

(9.180)

The volumetric average pressure is given by

$$p = \sum_l \varepsilon_l p_l. \tag{9.181}$$

However, assuming stress equilibrium, we can write, for all phases,

$$p_l = p. \tag{9.182}$$

The volumetric average density and mass and heat flux vectors are defined, respectively, by

$$\rho = \sum_l \varepsilon_l \rho_l, \tag{9.183}$$

$$\rho u = \sum_l \rho_l u_l, \tag{9.184}$$

and

$$\rho h u = \sum_l \rho_l h_l u_l. \tag{9.185}$$

Consequently, by invoking Eqs. (9.170), (9.172), and (9.179)–(9.185), Eq. (9.165) can be written in a compact form as

$$\rho c \frac{\partial T}{\partial t} + \nabla \cdot (u \rho h) = \nabla \cdot (k \cdot \nabla T) + \frac{\partial p}{\partial t} + u \cdot \nabla p - q. \tag{9.186}$$

Eq. (9.186) can be used to calculate the total rate of heat lost, q, by the wet soil. However, it will be simplified for conditions of typical laboratory soil freezing/thawing tests.

The data, considered in the applications presented later, have been obtained by subjecting small amounts of soil samples to slow cooling or thawing at a certain pressure, using an experimental system, such as that described by Inaba (1983). Therefore, there is no flow involved, $u = 0$, the pressure remains constant, $p = $ constant, and the temperature is uniform throughout the soil sample but varies with time; that is, $T = T(t)$. As a result, Eq. (9.186) can be simplified and the heat loss can be expressed in terms of the temperature decrease by

$$dQ = q dt = -\rho c dT. \tag{9.187}$$

Thus, substituting Eqs. (9.150), (9.161), (9.162), (9.178), and (9.187) into Eq. (9.160) and separating variables yields

$$\frac{[1 + C(w_u - w_a)] d(w_u - w_a)}{[A + B(w_u - w_a)](w_u - w_a)} = -d\theta, \tag{9.188}$$

subject to

$$w_u = w_o, \theta = 0, \tag{9.189}$$

where the coefficients are defined by

$$A = k_r \rho_d [c_s + c_f w_o + (c_u - c_f) w_a], \tag{9.190}$$

$$B = k_r \rho_d (c_u - c_f), \tag{9.191}$$

and

$$C = k_r \rho_d L_{u,f}. \tag{9.192}$$

Eqs. (9.188) and (9.189) should be integrated numerically when the parameters in Eqs. (9.190)–(9.192) vary appreciably with temperature. However, when the parameters are approximately constant, an analytical solution is obtained as

$$\left[\frac{w_u - w_a}{w_o - w_a} \right] \left[\frac{A + B(w_u - w_a)}{A + B(w_o - w_a)} \right]^{(-1 + AC/B)} = \exp(-A\theta). \tag{9.193}$$

If an average specific heat capacity, $\bar{c}$, over the freezing temperature range is substituted for the unfrozen and frozen water specific heat capacities, that is,

$$c_u = c_f = \bar{c}, \tag{9.194}$$

then Eqs. (9.190) and (9.191) lead to

$$k \equiv A = k_r \rho_d (c_s + \bar{c} w_o), \ B = 0 \tag{9.195}$$

and Eq. (9.193) simplifies as

$$\phi_u = w_u / w_o = \phi_a + (1 - \phi_a) \exp(-k\theta), \tag{9.196}$$

which is the same as the empirically derived Eq. (9.154).

Civan (2000d) demonstrated the validity of Eq. (9.196) by analyzing various empirical data.

The parameters of Eq. (9.196), that is, the adsorbed water content and the rate constant, have been determined by regression of Eq. (9.178), using Eq. (9.196), to the Dynatech soil apparent (sensible plus latent) heat capacity measurements, referred to as Data #3 (AGA, 1974). The unfrozen water content of the soil was correlated in Figure 9.8 as

$$\phi_u = w_u / 0.103 = 0.117 + 0.883 \exp(-0.30\theta), \tag{9.197}$$

where θ is in Celsius and w_u is in kilogram water per kilogram dry soil.

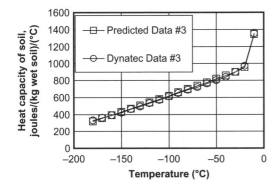

Figure 9.8 Correlation of the Dynatech Data #3 measurements (after Civan, 2000d; with permission from the American Society of Civil Engineers).

Figure 9.9 shows the straight-line plot of the Anderson (1969) data (unpublished data reported by Nakano and Brown, 1971) for the unfrozen water content of the Point Barrow silt according to Eq. (9.196), which yields the following correlation:

$$\phi_u = w_u/0.20 = 0.380 + 0.620\exp(-0.708\theta), \tag{9.198}$$

where θ is in Celsius and w_u is in gram water per gram dry soil. Note that the parameters of the present Eq. (9.198) are slightly different from those of Eq. (9.156), given by Nakano and Brown (1971), possibly due to the differences in the processing of the experimental data.

Figure 9.10 shows the effect of using an average specific heat capacity value to simplify Eq. (9.193) to Eq. (9.196). The solution obtained by the

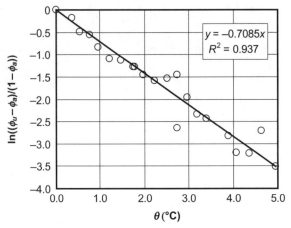

Figure 9.9 Correlation of the unfrozen water content of the Point Barrow silt (after Civan, 2000d; with permission from the American Society of Civil Engineers).

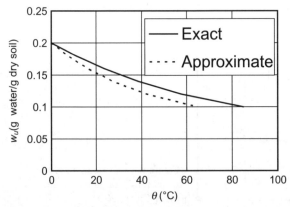

Figure 9.10 Comparison of the unfrozen water predictions by Eqs. (9.44) and (9.47) (after Civan, 2000d; with permission from the American Society of Civil Engineers).

simplified Eq. (9.196) matches reasonably well with that of Eq. (9.193), using $c_u =$ 4186 J/kg·K, $c_f = 2300$ J/kg·K, $\bar{c} = (4186 + 2300)/2 = 3243$ J/kg · K, $\rho_d = 1680$ kg/m^3, $k_r = 1.0 \times 10^{-8}$ K^{-1}, $L_{u,f} = 333{,}880$ J/kg, $w_o = 0.20$ kg H$_2$O/kg dry soil, and $w_a = 0.07$ kg H$_2$O/kg dry soil.

These exercises indicate that the variation of the unfrozen water content in freezing or thawing soils adequately follows an exponential decay function shifted by the amount of the water adsorbed over the soil grains, which cannot freeze. This confirms the validity of the approximate analytic solution derived for the unfrozen water content and justifies the theory presented by Civan (2000d). However, when accurate thermal property data are available, direct numerical solutions of the equations presented earlier may provide better results than the approximate analytic solution obtained by using average properties for the soil constituents. The theoretical treatment of the gradual freezing/thawing process given earlier provides an insight into the mechanism of the gradual phase change of water in wet soils and a proper means of correlating experimental measurements of the unfrozen water in soil.

9.10.3 Representation of the Unfrozen Water Content for Instantaneous Phase Change

Parts of this section are presented following Civan and Sliepcevich (1987), who stated

> Of the various approaches for modeling problems in heat conduction with phase change (the classical Stefan problem) the apparent heat capacity formulation still appears to be one of the preferred methods. The principal advantage of this approach is that temperature is the primary dependent variable that derives directly from the solution. However, it is well known that this formulation suffers from a singularity problem for a phase change that occurs at a fixed temperature. This difficulty can be circumvented by assuming that the phase change occurs over a small temperature interval (Hashemi and Sliepcevich, 1967).

The following function describes the volumetric fraction of phase 1 when the phase transition takes place instantaneously at a fixed temperature:

$$\phi_1 = U(T - T_t). \tag{9.199}$$

U is a step function whose value is zero when $T < T_t$ but equals one otherwise. The derivative of this function expresses the variation of the phase 1 fraction with temperature according to

$$d\phi_1/dT = \delta(T - T_t), \tag{9.200}$$

where $\delta(T - T_t)$ is the Dirac delta function whose value is infinity at the transition temperature T_t but is zero at all other temperatures. This singularity problem can be circumvented using the Dirac delta function approximation by a uniform distribution over a finite temperature interval so that Eq. (9.200) is approximated by

$$d\phi_1/dT = 1/(2\Delta T_t). \tag{9.201}$$

The same can be accomplished also using the normal distribution function, given by (Hashemi and Sliepcevich, 1967)

$$d\phi_1/dT = \left(\varepsilon\pi^{-1/2}\right)\exp\left[-\varepsilon^2\left(T-T_t\right)^2\right], \tag{9.202}$$

where ε is selected in a manner that $\text{erf}(\varepsilon\Delta T_t) = 1.0 - \lambda$, where ΔT_t is one-half of the assumed phase change interval and λ is a sufficiently small positive number. These remedial measures provide the linear and error function approximations for the integrals of Eqs. (9.201) and (9.202), respectively. Thus, the numerical instabilities arising from the jump in the values of the volumetric fraction of phase 1 from zero to one can be avoided conveniently in conventional numerical solution methods (finite differences and finite elements).

9.10.4 Apparent Heat Capacity Formulation for Heat Transfer with Phase Change

In this section, the two-phase Stefan problem for a one-dimensional system undergoing a phase transition at a fixed temperature is solved for the case of constant physical properties in each phase, according to Civan and Sliepcevich (1987). The results are compared with the analytical solution given by Neumann in Carslaw and Jaeger (1969). For this purpose, normalized variables, along with the Boltzmann similarity variable and coordinate mapping transformation, are used to eliminate computational difficulties and problems that frequently arise with conventional techniques, such as the finite difference method (Goodrich, 1978). For the present purpose, a typical example of a pure incompressible substance undergoing cooling and freezing (or conversely heating and melting) is solved. Civan and Sliepcevich (1987) state that the same methodology can be applied to other processes involving discontinuities, such as shock phenomena and interfacial mass transport.

The energy equation for a two-phase system of a pure substance is given by

$$(\partial/\partial t)(\rho_1 h_1 \phi_1 + \rho_2 h_2 \phi_2) + \nabla \cdot \left[(\rho_1 h_1 V_1 + \rho_2 h_2 V_2) + (q_1 \phi_1 + q_2 \phi_2)\right] - Dp/Dt = 0, \tag{9.203}$$

in which subscripts distinguish the phases, ρ and h are the density and enthalpy, and ϕ is the volume fraction (volume of one phase divided by the volume occupied by both phases), respectively. V and q are the volumetric and heat fluxes associated with each of the phases, respectively. D/Dt represents a substantive derivative, $(\partial/\partial t + V \cdot \nabla)$; p is pressure; and t is time. By definition,

$$\phi_1 + \phi_2 = 1. \tag{9.204}$$

Eliminating ϕ_2 by means of Eq. (9.204) and rearranging permits Eq. (9.203) to be written as

$$(\partial/\partial t)[\rho_2 h_2 + (\rho_1 h_1 - \rho_2 h_2)\phi_1] + \nabla \cdot \{(\rho_1 h_1 V_1 + \rho_2 h_2 V_2) \\ + [q_2 + (q_1 - q_2)\phi_1]\} - Dp/Dt = 0. \tag{9.205}$$

The phases are assumed at thermal equilibrium during the phase transition. On the assumption that all heat transferred is by conduction, then Fourier's equation applies:

$$q = -k \cdot \nabla T, \tag{9.206}$$

in which k is the thermal conductivity tensor and T is temperature. Eq. (9.205) can be rearranged as

$$\{\partial(\rho_2 h_2)/\partial T + [\partial(\rho_1 h_1)/\partial T - \partial(\rho_2 h_2)/\partial T]\phi_1 + (\rho_1 h_1 - \rho_2 h_2)(d\phi_1/dT)\}$$
$$\partial T/\partial t + \nabla \cdot \{(\rho_1 h_1 V_1 + \rho_2 h_2 V_2) - [k_2 + (k_1 - k_2)\phi_1] \cdot \nabla T\} - Dp/Dt = 0. \tag{9.207}$$

The energy equation in the form of Eq. (9.207) is called an apparent heat capacity formulation, in which the quantities enclosed in the braces preceding $\partial T/\partial t$ represent an "apparent" volumetric heat capacity. Because the volumetric latent heat effect of phase change, $(\rho_1 h_1 - \rho_2 h_2)$, is included in the apparent volumetric heat capacity definition, the apparent heat capacity formulation allows for a continuous treatment of a system involving phase transfer. In contrast, the Stefan formulation requires solutions of separate, single-phase equations for each of the phases; these solutions are later matched along the phase-change interface boundary for compatibility and consistency.

Consider an infinitely long one-dimensional medium (x-direction) containing a material in a thermodynamic state represented by phase 1, the temperature of which is uniform. A transient-state heat transfer process is initiated by contacting the leading surface of this material with a sink of infinite extent whose temperature is such that it results in the formation of phase 2. In other words, the material undergoes a transition from phase 1 to phase 2 with the interface progressing gradually with time from the leading surface to infinite distance.

Assume that the pressure is constant, the phases are isotropic and homogeneous, the phase densities are equal, and there is no motion in either of the phases. Accordingly, Eq. (9.207) reduces to

$$\rho c(\partial T/\partial t) = (\partial/\partial x)(k\partial T/\partial x), \tag{9.208}$$

in which

$$\rho = \rho_1 = \rho_2, \tag{9.209}$$
$$c = c_2 + (c_1 - c_2)\phi_1 + Ld\phi_1/dT, \tag{9.210}$$

and

$$k = k_2 + (k_1 - k_2)\phi_1. \tag{9.211}$$

In Eq. (9.210), c denotes the specific heat capacity and L the specific latent heat for change from phase 2 to phase 1, which are determined according to the specific enthalpies of phases 1 and 2, respectively,

$$h_1 = L + \int_{T_t}^{T} c_1 dT \text{ for } T > T_t \tag{9.212}$$

and

$$h_2 \int_{T_t}^{T} c_2 dT \text{ for } T > T_t. \tag{9.213}$$

For comparison with an analytical solution, consider the initial and boundary conditions as

$$T = T_i, t = 0, 0 \le x < \infty, \tag{9.214}$$

$$T = T_s, t > 0, x = 0, \tag{9.215}$$

and

$$T = T_i, t > 0, x \to \infty. \tag{9.216}$$

To achieve the optimum accuracy in the numerical solution of Eq. (9.212), apply the Boltzmann similarity transformation

$$y = x/t^{1/2} \tag{9.217}$$

and the transformation to map the semi-infinite domain to a unit size domain

$$z = 1 - \exp(-y/b) \tag{9.218}$$

as demonstrated by Civan and Sliepcevich (1984, 1985). In Eq. (9.218), b is an arbitrary scaling factor. Assume that

$$b = 2^{1/2}(k/\rho c)_0^{1/2}, \tag{9.219}$$

in which $(k/\rho c)_0$ is the maximum value of the heat diffusivity in the range (T_i, T_s). In addition, a normalized temperature is defined by

$$u = \int_{T_i}^{T} k\,dt \bigg/ \int_{T_i}^{T_s} k\,dt. \tag{9.220}$$

As a result, Eqs. (9.208) and (9.214)–(9.216) are transformed to

$$\ln(1-z)(du/dz) = \psi(d/dz)[(1-z)(du/dz)], \tag{9.221}$$

$$u = 1, z = 0, \tag{9.222}$$

and

$$u = 0, z = 1. \tag{9.223}$$

In Eq. (9.221), ψ is the diffusivity ratio given by

$$\psi = (k/\rho c)/(k/\rho c)_0. \tag{9.224}$$

The solution to Eq. (9.221) is accomplished by separating it into two first-order, ordinary differential equations:

$$du/dz = v/(1-z), \tag{9.225}$$

$$u = 1, z = 0, \tag{9.226}$$

and

$$u = 0, z = 1; \tag{9.227}$$

and

$$dv/dz = (1/\psi)\ln(1-z)(du/dz). \tag{9.228}$$

The numerical solution has been obtained by solving Eqs. (9.225) and (9.228) simultaneously by a variable-step Runge–Kutta-Fehlberg four (five) method (Fehlberg, 1969).

For numerical calculations, the data from Goodrich (1978) have been considered, in which the thermal conductivities of the frozen and unfrozen phases are given as 2.25 and 1.75 W/m·K, the volumetric heat capacities as 1.5×10^6 and 2.5×10^6 J/m³·K, respectively. The transition temperature is 0°C and the volumetric latent heat is 1×10^8 J/m³. Computations were carried out for cases in which the initial and leading surface temperatures, T_i and T_s, were considered to be 2 and −2°C, 2 and −10°C, and 10 and −10°C, respectively.

Calculations were made by approximating the Dirac delta function by a normal distribution and by a uniform distribution over the temperature intervals of ±0.01, 0.1, 0.5, 1.0, and 1.5°C. The results with the uniform distributions, Eq. (9.210), were about twice as greatly in error as with the normal distribution, Eq. (9.211). Only the results with the normal distribution are shown in Figures 9.11–9.13, in terms of the normalized variables in order to magnify the errors.

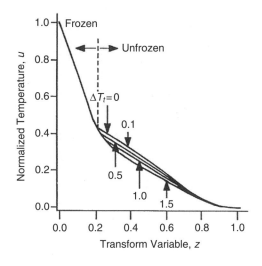

Figure 9.11 Temperature profiles for different ΔT_t (Celsius) values for a case of $T_i = 2°C$ and $T_s = -2°C$ (after Civan and Sliepcevich, 1987; reprinted by permission from Oklahoma Academy of Science).

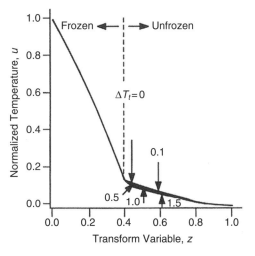

Figure 9.12 Temperature profiles for different ΔT_t (Celsius) values for a case of $T_i = 2°C$ and $T_s = -10°C$ (after Civan and Sliepcevich, 1987; reprinted by permission from Oklahoma Academy of Science).

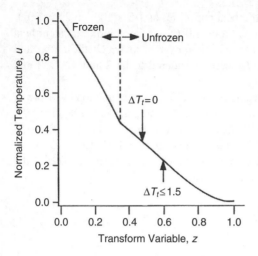

Figure 9.13 Temperature profiles for different ΔT_t (Celsius) values for a case of $T_i = 10°C$ and $T_s = -10°C$ (after Civan and Sliepcevich, 1987; reprinted by permission from Oklahoma Academy of Science).

It is evident that the error introduced by approximating a singularity (Dirac delta function) by a gradual change over a small temperature interval is more pronounced in the unfrozen region, which is effectively of infinite thickness, than in the thin, frozen region near the leading surface; the latter conforms closely to the exact solution. Another observation is that the effect of distributing the latent heat over a temperature interval diminishes as the ratio of the temperature interval to the overall temperature variation of the system becomes smaller.

It appears from these numerical solutions that the results obtained by approximating the phase change at a fixed temperature by a gradual change over a small temperature interval should be acceptable if

$$2\Delta T_t / |(T_i - T_s)| < 0.1. \tag{9.229}$$

9.10.5 Enthalpy Formulation of Conduction Heat Transfer with Phase Change at a Fixed Temperature

In the enthalpy method, the energy equation for transient heat conduction in Cartesian coordinates is given by

$$\frac{\partial H}{\partial t} = \sum_{i=1}^{N \leq 3} \frac{\partial}{\partial x_i}\left(\frac{\partial T}{\partial x_i}\right), \tag{9.230}$$

where t is time, x_i is the Cartesian distance, N is the number of space dimensions, k is the thermal conductivity, T is temperature, and H is enthalpy. The variation of enthalpy with respect to temperature is the volumetric heat capacity given by

$$\rho c = \partial H / \partial T. \tag{9.231}$$

The volumetric heat capacity and the thermal conductivity for a composite system containing a constituent, which is partially phase changed, can be predicted using, respectively

$$\rho c = (\rho c)_{tf} + \left[(\rho c)_{tu} - (\rho c)_{tf}\right](w_{au}/w_{ao}) + \rho_r L_a\, \partial w_{au}/\partial T \qquad (9.232)$$

and

$$k = k_{tf} + (k_{tu} - k_{tf})(w_{au}/w_{ao}), \qquad (9.233)$$

in which ρ_r is the density of a reference component in the composite, preferably an inactive component which does not undergo a phase change; w_{au} is the mass of active component, which has not yet undergone a phase change per unit mass of reference component; w_{ao} is the total mass of active component originally present per unit mass of reference component; $(\rho_c)_{tu}$ and k_{tu} are the volumetric sensible heat capacity and the thermal conductivity, respectively, of the composite (overall system) that has not undergone any phase change; $(\rho_c)_{tf}$ and k_{tf} are the volumetric sensible heat capacity and the thermal conductivity, respectively, of the composite in which the active component has undergone a complete phase change; and L_a is the latent heat of phase change for the active component.

The normalized temperature and enthalpy are defined, respectively, by

$$u = \int_{T_1}^{T} kdT \,\bigg/ \int_{T_1}^{T_2} kdT \qquad (9.234)$$

and

$$e = (H - H_1)/(H_2 - H_1), \qquad (9.235)$$

so that Eqs. (9.230) and (9.231) become

$$(H_2 - H_1)\frac{\partial e}{\partial t} = \sum_{i=1}^{N \leq 3} \frac{\partial}{\partial x_i}\left(\int_{T_1}^{T_2} kdT \cdot \frac{\partial u}{\partial x_i}\right) \qquad (9.236)$$

and

$$\gamma \cdot \rho c = \partial e/\partial u, \qquad (9.237)$$

in which

$$\gamma = \int_{T_1}^{T_2} kdT \big/ (H_2 - H_1), \qquad (9.238)$$

where T_1 and T_2 are the range of temperature variation for the system and H_1 and H_2 are the corresponding enthalpy values, respectively.

The dimensionless distance in a given Cartesian direction can be defined by either of the following forms depending on the nature of the problem.

Case I: A finite domain—then the dimensionless distance is

$$y_i = x_i/a_i, \qquad (9.239)$$

where a_i is a characteristic distance in the x_i dimension.

Case II: A semi-infinite domain—then the semi-infinite domain is converted to a unit size finite domain by

$$y_i = 1 - e^{-x_i/a_i}, \qquad (9.240)$$

in which a_i is a geometric scaling factor. This transformation is desirable to avoid uncertainties in representing infinity. For this purpose, Hashemi and Sliepcevich (1967) used the transformation $x_i/(1 + x_i)$. However, it is more appropriate to use the transformation given by Eq. (9.240) because an exponential function is more suited to scale variables extending to infinity.

Case III: A semi-infinite media whose temperature is initially uniform (constant)—then the number of independent variables can be reduced by one using the Boltzmann transformation (Crank, 1957):

$$y_i = x_i/t^{1/2}. \qquad (9.241)$$

Because the Boltzmann transformation still expresses the new variable y_i between zero and infinity, an additional transformation similar to Eq. (9.240) is needed to convert the semi-infinite domain to a finite unit domain. Thus,

$$z_i = 1 - e^{-y_i/a_i}, \qquad (9.242)$$

where a_i is some arbitrary scaling factor, which should be selected, such that better resolution is achieved in the region where the temperature varies more. Applying Eqs. (9.241) and (9.242) to Eq. (9.236) for a media extending to infinity in all Cartesian coordinates, for example, gives, respectively,

$$y_i \frac{\partial e}{\partial y_i} + \frac{2}{(H_2 - H_1)} \sum_{i=1}^{N \leq 3} \frac{\partial}{\partial y_i} \left(\int_{T_1}^{T_2} kdT \cdot \frac{\partial u}{\partial y_i} \right) = 0, \qquad (9.243)$$

in which the term $y_i \, \partial e/\partial y_i$ can be written in terms of $i = 1$, 2, or 3, and

$$-(1 - z_i)\ln(1 - z_i)\frac{\partial e}{\partial z_i} + \frac{2}{(H_2 - H_1)} \sum_{i=1}^{N \leq 3} (1/a_i^2)(1 - z_i)$$

$$\times \frac{\partial}{\partial z_i} \left[\int_{T_1}^{T_2} kdT \cdot (1 - z_i)\frac{\partial u}{\partial z_i} \right] = 0. \qquad (9.244)$$

As an example, the two-phase Stefan problem for a one-dimensional system undergoing an instantaneous phase transition will be solved for the case with constant physical properties as used by Goodrich (1978). The results are then compared with the analytical solution given by Carslaw and Jaeger (1959, 1969).

The heat transfer is represented by

$$\partial H/\partial t = k\partial^2 T/\partial x^2, \qquad (9.245)$$
$$T = T_1, t = 0, 0 \leq x < \infty, \qquad (9.246)$$
$$T = T_2, x = 0, t > 0, \qquad (9.247)$$

and

$$T = T_1, x \to \infty, t > 0. \qquad (9.248)$$

For the one-dimensional problem considered here, Eq. (9.244) is separated into two first-order ordinary differential equations. Thus, Eqs. (9.245)–(9.248) reduce to the following:

$$\frac{du}{dz} = \frac{v + \ln(1-z)e}{2\gamma(1-z)},$$ (9.249)

$$u = 1, z = 0,$$ (9.250)

and

$$u = 0, z = 1;$$ (9.251)

and

$$dv/dz = e/(1-z).$$ (9.252)

For a solution, the initial value of v is guessed until the final boundary value is obtained. The numerical solution is obtained by integrating the equations simultaneously using a variable-step Runge–Kutta–Fehlberg four (five) method.

The temperature values and the frost depths in terms of the Boltzmann variable obtained via the enthalpy method reproduced those obtained from the exact analytical solution with an average relative error of the order of 10^{-8}.

Goodrich (1978) solved this problem using the apparent heat capacity formulation assuming a freezing range of 0.5°C and employing a Crank–Nicolson method. He compares the frost depth progressing with time with that obtained from Neumann's exact analytical solution. The oscillatory and distorted numerical results reported by Goodrich are due to the use of large step sizes in the Crank–Nicolson method. In addition, in his solution, the arbitrary value assigned to represent infinite distance probably accounts for some of the accuracy loss in the numerical solution.

The formulation presented by Civan and Sliepcevich (1984) has certain advantages:

- The use of dimensionless variables and similarity and scaling transformations not only speeds up the computational process but also improves the accuracy of the results.

- Because of the reduction in the number of independent variables, the accuracy of the solution can be controlled more efficiently.

- The need for representing infinity by approximate numbers (as required in standard numerical methods) is eliminated by converting the semi-infinite domain to a finite domain.

- Even though the dimensionless Eq. (9.236) is general, the subsequent transformed Eqs. (9.243) and (9.244) are restricted to a semi-infinite domain whose initial temperature is uniform.

- In the existence of a convective boundary condition, the time variable in the boundary condition equation cannot be eliminated by the Boltzmann transformation. Accordingly, the transformed energy equation needs to be solved repetitively at prescribed time values of the convective boundary condition.

- Because the present method combines the space and time variables into one Boltzmann variable, the frost depth is defined by a unique value of the

Boltzmann variable at which value the temperature profile has a slope discontinuity.

9.10.6 Thermal Regimes for Freezing and Thawing of Moist Soils: Gradual versus Fixed Temperature Phase Change

As stated by Civan and Sliepcevich (1985b), "Once the unfrozen water content for a particular soil is known as a function of temperature, the thermal conductivity and the volumetric apparent heat capacity of moist soil can be estimated." Civan and Sliepcevich (1985b) considered the energy equation in the form of

$$\frac{\partial H}{\partial t} = \frac{\partial}{\partial x}\left(k \frac{\partial T}{\partial x} \right), \tag{9.253}$$

subject to the conditions

$$T = T_i \quad t = 0 \quad 0 \le x < \infty, \tag{9.254}$$
$$T = T_s \quad x = 0 \quad t > 0, \tag{9.255}$$

and

$$T = T_i \quad x \to \infty \quad t > 0. \tag{9.256}$$

The following steps are taken to improve the accuracy of the numerical solutions:

1. Express Eqs. (9.253)–(9.256) in terms of normalized volumetric enthalpy:

$$h \equiv (H - H_i)/(H_s - H_i), \tag{9.257}$$

in which $H_i = H(T_i)$ and $H_s = H(T_s)$ and normalized temperature

$$u \equiv (T - T_i)/(T_s - T_i) \tag{9.258}$$

to obtain

$$\frac{\partial h}{\partial t} = \beta \frac{\partial}{\partial x}\left(k \frac{\partial u}{\partial x} \right), \tag{9.259}$$

where

$$\beta = (T_s - T_i)/(H_s - H_i) \tag{9.260}$$

and

$$u = 0 \quad t = 0 \quad 0 \le x \le \infty, \tag{9.261}$$
$$u = 1 \quad x = 0 \quad t > 0, \tag{9.262}$$

and

$$u = 0 \quad x \to \infty \quad t > 0. \tag{9.263}$$

2. Use the Boltzmann similarity transformation

$$y = ax/t^{1/2}, \tag{9.264}$$

where $a = 1/2(k/\rho c)_0^{1/2}$ and $(k/\rho c)_0$ is a characteristic diffusivity. This transformation is permissible, even though Eq. (9.259) contains a temperature-dependent thermal conductivity and two different functions, h and u, which are related by Eq. (9.231), because it is applied only to the independent variables, x and t, in order to reduce the number of independent variables by one. In this study, $(k/\rho c)_0$ was selected to be its maximum value in the temperature range of interest. Eq. (9.259) then becomes

$$-y\frac{dh}{dy} = 2a^2\beta\frac{d}{dy}\left(k\frac{du}{dy}\right).\tag{9.265}$$

Eq. (9.265) is a second-order ordinary differential equation, which can be solved by separating it into two first-order differential equations. However, a direct separation on Eq. (9.265) will lead to an equation containing the derivative dh/du, which is equal to $\beta\rho c$, as can be seen from Eqs. (9.231), (9.257), (9.258), and (9.260). Because c approaches infinity as the temperature interval for the phase change narrows, a singularity problem arises. Fortunately, this problem can be circumvented by a rearrangement similar to the procedure used by Anderson and Ford (1981). Thus, Eq. (9.255) can be written as

$$\frac{d}{dy}\left(yh + 2a^2\beta k\frac{du}{dy}\right) = h,\tag{9.266}$$

and when Eq. (9.264) is applied, Eqs. (9.261)–(9.263) become

$$u = 1 \quad y = 0 \tag{9.267}$$

and

$$u = 0 \quad y \to \infty.\tag{9.268}$$

3. Introduce another transformation prior to making a separation even though Eq. (9.266) can be separated into two equations, neither of which contain the derivative of h and its attendant problems. The purpose of this transformation is to convert the semi-infinite domain to a finite unit size domain, thereby circumventing the need for approximating infinity. Thus, the semi-infinite domain y can be reduced to a finite domain z using

$$z = 1 - e^{-y/b},\tag{9.269}$$

in which b is a scaling factor. Thus, Eqs. (9.266)–(9.268) become

$$(1-z)\frac{d}{dz}\left[-\ln(1-z)h + 2(a/b)^2\beta k(1-z)\frac{du}{dz}\right] = h,\tag{9.270}$$

$$u = 1 \quad z = 0, \tag{9.271}$$

and

$$u = 0 \quad z = 1.\tag{9.272}$$

The solution of Eq. (9.270) can now be accomplished by separating it into two first-order, ordinary differential equations:

$$\frac{du}{dz} = [v + \ln(1-z)h]/[k(1-z)],\tag{9.273}$$

$$u = 1 \quad z = 0, \tag{9.274}$$

and

$$u = 0 \quad z = 1; \tag{9.275}$$

and

$$\frac{dv}{dz} = h/(1 - z). \tag{9.276}$$

Note that the scaling factor is chosen as $b = a(2\beta)^{1/2}$ for convenience. Consequently, the constants a, b, and β do not appear in Eq. (9.273). Had the value of b been selected differently, such that the constants a, b, and β appeared in Eq. (9.273), the solution curves would nevertheless remain the same.

For a numerical solution, the value of $v(\equiv k \, du/dz$ from Eq. 9.273) is guessed at $z = 0$ until the solution satisfies the boundary conditions given by Eqs. (9.274) and (9.275). The value of the normalized volumetric enthalpy h, appearing on the right of Eqs. (9.273) and (9.276), is calculated by means of Eqs. (9.231), (9.257), and (9.258) during the solution.

For application, Civan and Sliepcevich (1985b) used the example problem of Nakano and Brown (1971) of the Point Barrow silt. The properties of the Point Barrow silt are the following: bulk density of dry soil, $1680 \, \text{kg/m}^3$; moisture content, 20% dry weight; unfrozen water content (kilogram water per kilogram dry soil), $w_u = 0.1278 \exp \ (0.7630T) + 0.0722$; thermal conductivity for frozen soil, $3.46 \, \text{W/m/K}$, and thawed soil, $2.42 \, \text{W/m/K}$; volumetric heat capacity for frozen soil, $2.00 \times 10^6 \, \text{J/m}^3\text{/K}$, and thawed soil, $2.64 \times 10^6 \, \text{J/m}^3\text{/K}$; latent heat of freezing water, $0.335 \times 10^6 \, \text{J/kg}$; initial temperature, $-4°\text{C}$ (thawing) and $+4°\text{C}$ (freezing); and leading surface temperature, $+4°\text{C}$ (thawing) and $-4°\text{C}$ (freezing).

Numerical solutions for Eqs. (9.273)–(9.276) were obtained using a Runge–Kutta–Fehlberg four (five) method described by Fehlberg (1969). The solution for the phase change at a single temperature for both cases gave results identical to the values obtained from Neumann's exact solution given by Carslaw and Jaeger (1959). The solution for the phase change over a temperature interval was obtained using the expression for the unfrozen water content given earlier. The temperature field for thawing and freezing cases is presented in Figure 9.14 in terms of the Boltzmann similarity variable. The deviation in the temperature profiles between the phase change over a temperature interval and for a single temperature is much greater for thawing than it is for freezing.

9.11 SIMULTANEOUS PHASE TRANSITION AND TRANSPORT IN POROUS MEDIA CONTAINING GAS HYDRATES

Selim and Sloan (1989) describe that "gas hydrates are crystalline ice-like solids that form when water and sufficient quantities of certain gases, relatively small in molecular size, are combined under the right conditions of temperature and pressure." Kamath et al. (1991) point out that in situ natural gas hydrates may be found in sediments below arctic onshore permafrost layers and arctic offshore waters, and

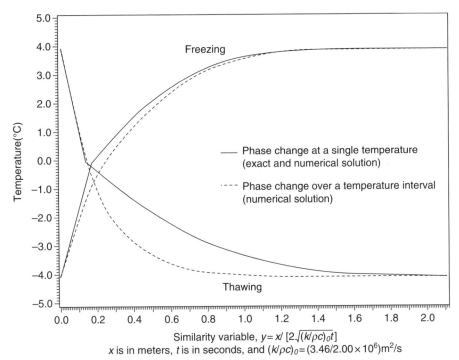

Figure 9.14   Temperature profile for Point Barrow silt (after Civan and Sliepcevich, 1985b; American Geophysical Union).

in offshore marine sediments present deep below the tropical regions. The production of the natural gas from these reserves is of continuing interest. Among the proposed techniques of extraction of gas from in situ natural gas hydrates are pressure depletion, heating, and facilitating hydrate inhibitors (Kamath et al., 1991). Among other factors, the development of optimal strategies and effective techniques for the production of gas from hydrates heavily depends on the understanding of the kinetics of the dissociation of gas hydrates, and the dynamics of the heterogeneous system of the gas hydrate, gas, water, and ice under the natural and/or induced conditions of the applied techniques (Bishnoi et al., 1994). In this section, the description of the hydrate dissociation process and the flow of the gas and liquid phases in hydrate-bearing formations is presented. This is accomplished by combining the various approaches, including those by Selim and Sloan (1989), Tsypkin (1998), Yousif and Sloan (1991), and Yousif et al. (1991), in a manner consistent with the rest of the material.

Selim and Sloan (1989) and Tsypkin (1998) considered and derived analytical solutions for simplified cases of hydrate dissociation in a one-dimensional, semi-infinite horizontal porous media. Tsypkin's (1998) approximate analytical solutions for two cases are presented as an example of modeling the hydrate dissociation in hydrate-bearing formations.

Figure 9.15 depicts a typical gas hydrate phase diagram. The *ABC* line represents the equilibrium hydrate dissociation/formation conditions; the *GBH* line

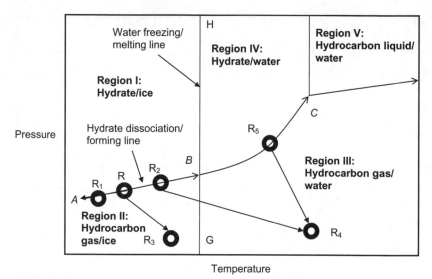

Figure 9.15   Schematic water/hydrocarbon mixture phase diagram (modified after McCain, 1990; prepared by the author).

represents the equilibrium water freezing/melting conditions. The quadruple point *B* represents the coexistence of hydrate, ice, water, and gas at equilibrium. In region I, hydrate and ice coexist. In region II, ice and gas coexist. In region III, gas and water coexist. In region IV, hydrate and water coexist. In region V, liquified gas and water coexist.

As described by Tsypkin (1998), in a process of hydrate decomposition by pressure reduction beginning with the equilibrium thermodynamic conditions represented, for example, by a point R in Figure 9.15, first, a transition occurs from point R to a point R_1 as temperature decreases due to pressure reduction. This induces a heat flow toward the system from the surrounding formation, causing a slight temperature rise, leading to a condition represented by point R_3 at the production well. Hence, the R_2R_3 path is followed. Tsypkin (1998) points out that the hydrate decomposition by simultaneous heating and pressure depletion may follow two different paths; that is, the temperature increase by heating may or may not cause a pressure increase beyond the ice melting line of GH. Thus, the conditions represented by points R_2 or R_5 may be attained. Then, the reduction of pressure causes the system to reach a condition represented by point R_4 at the production well. Consequently, these two possible paths are RR_2R_4 and $RR_2R_5R_4$.

For the analysis and formulation, consider a heterogeneous system of the gas hydrate, gas, water, and ice phases present at thermodynamic equilibrium in hydrate-bearing porous formation. In the following formulation, the hydrate, gas, water, ice, and porous matrix are denoted by the subscripts h, g, w, i, and s, respectively. T and P are the temperature and pressure of the heterogeneous system. S denotes the phase saturation in porous media. ϕ and K denote the porosity and permeability of porous media.

Considering the hydrate and ice a part of the stationary solid phases, the effective porosity and permeability of the porous media vary with time because of the variation of their saturations during phase change and hence the pore-size distribution of the medium (Yousif et al., 1991).

The saturations of the various phases occupying porous media add up to one:

$$S_h + S_g + S_w + S_i = 1.0. \tag{9.277}$$

Let ρ_{ih} and ρ_{gh} denote the masses of the ice and gas components per unit volume of the hydrate, that is, the mass concentrations of the ice and gas components in the hydrate. Let M_g and M_w denote the molecular weights of the gas and water, respectively, and let N_h represent the hydrate number expressing the number of water occupancy per methane in the hydrate. If, for example, $N_h = 6$ water molecules per methane molecule, the weight fractions of the methane gas and ice components present in the hydrate are calculated as

$$w_{gh} = \frac{M_g}{M_g + N_h M_w} = \frac{16}{16 + 6(18)} = 0.129 \frac{\text{kg CH}_4}{\text{kg hydrate}} \tag{9.278}$$

and

$$w_{ih} = 1 - w_{gh} = 0.871 \frac{\text{kg H}_2\text{O}}{\text{kg hydrate}}. \tag{9.279}$$

The mass concentrations of the gas and ice in the hydrate can be expressed, respectively, by

$$\rho_{gh} = \rho_h w_{gh} \tag{9.280}$$

and

$$\rho_{ih} = \rho_h w_{ih}. \tag{9.281}$$

Thus, if the hydrate density is $\rho = 900 \, \text{kg/m}^3$, the mass concentrations of the gas and ice are estimated as $\rho_{gh} = 116.13 \, \text{kg CH}_4/\text{m}^3$ hydrate and $\rho_{ih} = 783.87 \, \text{kg H}_2\text{O/m}^3$ hydrate (Tsypkin, 1998). Let $\dot{m}_w$, $\dot{m}_g$, and $\dot{m}_i$ represent the mass rates of water (liquid), gas, and ice generated per unit volume. Inferred by Yousif et al. (1991), these quantities are related by

$$\dot{m}_h = \dot{m}_g + \dot{m}_w + \dot{m}_i, \tag{9.282}$$
$$\dot{m}_g = \dot{m}_h w_{gh}, \tag{9.283}$$

and

$$\dot{m}_w + \dot{m}_i = \dot{m}_h w_{ih}. \tag{9.284}$$

Yousif et al. (1991) applied the Kim–Bishnoi model to estimate the rate of gas generation by hydrate dissociation:

$$\dot{m}_g = k_d A_s (p_e - p), \tag{9.285}$$

where p_e is the equilibrium pressure and p is the gas pressure. A_s is the specific surface area of the hydrate. Denoting the pore volume occupied by the gas and water per unit bulk volume by ϕ_{wg}, and the permeability by K, Yousif et al. (1991) approximated the specific surface area according to the parallel cylinder model (Kozeny, 1927):

$$A_s = \sqrt{\frac{\phi_{wg}^3}{2K}}.$$ (9.286)

k_d is the dissociation rate constant, which follows the Arrhenius equation of temperature dependency (Kim et al., 1987):

$$k_d = k_d^o \exp\left(-\frac{\Delta E}{RT}\right),$$ (9.287)

where k_d^o is the infinite temperature dissociation rate constant, ΔE denotes the activation energy required for hydrate dissociation, T is the absolute temperature, and R is the universal gas constant. Typically, $\Delta E/R = 9,400\ K^{-1}$ (Kim et al., 1987).

Selim and Sloan (1989) and Tsypkin (1998) used the empirical form of the Clasius–Clapeyron equation to represent the conditions of thermodynamic equilibrium, known as the Antoine equation:

$$\ln\left(\frac{p}{p_a}\right) = A - \frac{B}{T}.$$ (9.288)

This equation expresses the equilibrium relationship between the hydrate dissociation temperature and the gas pressure at the hydrate–gas contact surface (Selim and Sloan, 1989).

Tsypkin (1998) considered the dissociation of hydrate by thermal stimulation and/or pressure reduction at the hydrate-bearing formation face, located at $x = 0$ in Figure 9.16. Therefore, in general, Tsypkin (1998) considered three distinct zones for the analysis as indicated in Figure 9.16. Zone I is the hydrate-bearing region, containing the hydrate and free gas at equilibrium. Zone II contains the ice and gas phases produced by hydrate decomposition. Zone III is a region of fully decomposed hydrate products of water and gas, which exists when the temperature applied at the formation face is above the ice melting temperature at the prevailing pressure conditions. Zones I and II are separated by a moving hydrate dissociation front. Zones II and III are separated by a moving ice melting front. Neglecting the compressibility effects, the solid matrix, gas hydrate, and ice phases are assumed immobile. The water and gas are the mobile phases. The governing equations of these phases are presented in the following, inferred by Tsypkin (1998), Selim and Sloan (1989), and Yousif et al. (1991).

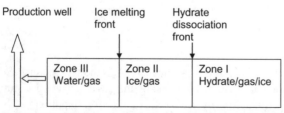

Figure 9.16   Hydrate dissociation process (prepared by the author).

The mobile water and gas-phase mass balance equations are given, respectively, by

$$\frac{\partial(\phi\rho_w S_w)}{\partial t} + \nabla\cdot(\rho_w u_w) = \dot{m}_w \tag{9.289}$$

and

$$\frac{\partial(\phi\rho_g S_g)}{\partial t} + \nabla\cdot(\rho_g u_g) = \dot{m}_g. \tag{9.290}$$

The stationary ice mass balance equation is given by

$$\frac{\partial(\phi\rho_i S_i)}{\partial t} = \dot{m}_i. \tag{9.291}$$

The mass balance equations for the stationary ice and gas present in the gas hydrate are given, respectively, by

$$\frac{\partial(\phi\rho_{ih} S_h)}{\partial t} = -\dot{m}_i \tag{9.292}$$

and

$$\frac{\partial(\phi\rho_{gh} S_h)}{\partial t} = -\dot{m}_g. \tag{9.293}$$

The horizontal flow of the mobile gas and water phases is represented by Darcy's law, respectively, as

$$u_g = -\frac{k_{rg}}{\mu_g} K\cdot\nabla p_g \tag{9.294}$$

and

$$u_w = -\frac{k_{rw}}{\mu_w} K\cdot\nabla p_w. \tag{9.295}$$

The pressures of the gas and water phases are related by the capillary pressure:

$$p_c(S_w) = p_g - p_w. \tag{9.296}$$

The following energy balance equation can be written, assuming thermal equilibrium between the various phases and considering only the water and gas phases are mobile:

$$\frac{\partial}{\partial t}(\rho e) + \nabla(\rho_g h_g u_g + \rho_w h_w u_w) = \nabla(\lambda\cdot\nabla T) + q. \tag{9.297}$$

The internal energy e and the enthalpy h are related by

$$e = h - p/\rho. \tag{9.298}$$

The specific heat capacities at constant volume and pressure are defined, respectively, by

$$c_v = de/dT \tag{9.299}$$

and

$$c_p = dh/dT. \tag{9.300}$$

The total thermal conductivity of the heterogeneous system at thermal equilibrium is given by

$$\lambda = \phi\left(S_h\lambda_h + S_i\lambda_i + S_w\lambda_w + S_g\lambda_g\right) + (1-\phi)\lambda_s. \tag{9.301}$$

The total internal energy of the heterogeneous system is given by

$$\rho e = \phi\left(S_h\rho_h e_h + S_i\rho_i e_i + S_w\rho_w e_w + S_g\rho_g e_g\right) + (1-\phi)\rho_s e_s. \tag{9.302}$$

Prior to hydrate decomposition, the initial hydrate, ice, and gas saturations and the temperature and pressure are assumed known. Thus, the initial conditions are prescribed as

$$S_h = S_{ho}, S_i = S_{io}, S_g = S_{go}, T = T_o, P = P_o, t = 0. \tag{9.303}$$

The conditions at the semi-infinite distance ($x \to \infty$) are assumed the same as the initial values.

The temperature and pressure values at the formation face are denoted by T^o and p^o, respectively:

$$T = T^o, p = p^o, x = 0, t > 0. \tag{9.304}$$

In addition, several interface boundary conditions on the phase-transition moving fronts are required.

The position and velocity of a moving front are represented by

$$x = X(t), V_n = dX(t)/dt, t > 0 \tag{9.305}$$

starting with

$$x = 0, V_n = 0, t = 0. \tag{9.306}$$

Tsypkin (1998) considers the temperature and pressure continuous, and the velocity and saturations of the various phases discontinuous across a moving front. Therefore, the compatibility and consistency relationships across a front, or the jump conditions, given by Tsypkin (1998) can be written as the following with some modifications for generality. The asterisk represents the values of the relevant quantities at the moving front. n indicates the normal to the front. The superscripts plus and minus denote the values of the relevant quantities on the fore and post sides of the moving front.

The temperature and pressure continuity conditions are expressed by

$$T^- = T^+ = T_*. \tag{9.307}$$

and

$$p^- = p^+ = p_*. \tag{9.308}$$

The Antoine equation is used to represent the equilibrium thermodynamic conditions:

$$\ln\left(\frac{p_*}{p_a}\right) = A - \frac{B}{T_*}.$$ (9.309)

The jump mass balance of the ice across the front is given by

$$\rho_i S_i^+ + \rho_{ih} S_h^+ = \rho_i S_i^-.$$ (9.310)

The jump mass balance of the gas across the front is given by

$$\phi\left[S_h^+ \rho_{gh} + S_g^+ \rho_{g*}\right]\left(V_n - u_{gn}^+\right) = \phi S_g^- \rho_{g*}\left(V_n - u_{gn}^-\right).$$ (9.311)

Similarly, for liquid water,

$$\rho_i S_i^+ = \rho_i S_i^- + \rho_w S_w^-.$$ (9.312)

The jump energy balance of the ice, gas, and water system is given by

$$\phi\left(S_h^+ \rho_h h_h + S_i^+ \rho_i h_i\right)V_n + \phi S_g^+ \rho_g h_g\left(V_n - u_{gn}^+\right) + \phi S_w^+ \rho_w h_w\left(V_n - u_{wn}^+\right) + \lambda^+\left(\nabla T\right)_n^+$$
$$= \phi S_i^- \rho_i h_i V_n + \phi S_g^- \rho_g h_g\left(V_n - u_{gn}^-\right) + \phi S_w^- \rho_w h_w\left(V_n - u_{wn}^-\right) + \lambda^-\left(\nabla T\right)_n^-.$$ (9.313)

Tsypkin (1998) developed two separate models. The first model considers hydrate decomposition by pressure reduction at the formation face, while the temperature remains below the ice melting temperature. The second model considers thermal stimulation by heat application at the formation face sufficient to melt the ice, in addition to pressure reduction. The approximate self-similar analytical solutions to these problems were obtained in terms of the following variables:

$$X(t) = 2\gamma\sqrt{a_f t}$$ (9.314)

and

$$\zeta = \frac{x}{2\sqrt{a_f t}},$$ (9.315)

where the subscript f denotes the front. The thermal diffusivity is defined by

$$a = \frac{\lambda}{\rho c}.$$ (9.316)

Subscripts 1, 2, and 3 correspond to zones I, II, and III, respectively.

The data used by Tsypkin (1998) for the first problem are $A = 12.88$, $B = 2655.58$, $N_h = 6$, $\rho_{ih} = 783.87\,\mathrm{kg/m^3}$, $\rho_{gh} = 116.13\,\mathrm{kg/m^3}$, $\rho_s = 2.\times 10^3\,\mathrm{kg/m^3}$, $\rho_I = 0.91\times 10^3\,\mathrm{kg/m^3}$, $\rho_h = 0.9\times 10^3\,\mathrm{kg/m^3}$, $\lambda_s = 2\,\mathrm{W/m\text{-}K}$, $\lambda_s = 2\,\mathrm{W/m\text{-}K}$, $\lambda_I = 2.23\,\mathrm{W/m\text{-}K}$, $\lambda_h = 2.11\,\mathrm{W/m\text{-}K}$, $c_s = 1000\,\mathrm{J/kg\text{-}K}$, $c_i = 2090\,\mathrm{J/kg\text{-}K}$, $c_h = 2500\,\mathrm{J/kg\text{-}K}$, $R = 520\,\mathrm{J/kg\text{-}K}$, $\mu_g = 1.8\times 10^{-5}\,\mathrm{Pa\cdot s}$, $T_* = 263.41\,\mathrm{K}$, $p^* = 1.58\times 10^6\,\mathrm{Pa}$, $S_i^+ = 0.461$, $S_h^+ = 0.28$, $\gamma_z = 0.775$, $T_o = T^o = 270\,\mathrm{K}$, $p^o = 1.5\times 10^6\,\mathrm{Pa}$, $K = 0.5\times 10^{-14}\,\mathrm{m^2}$, $\phi = 0.20$, $S_{ho} = 0.7$, $S_{io} = 0.1$, and $k_{rg} = S_g$.

For the second problem, Tsypkin (1998) used these additional parameter values: $\rho_w = 1000\,\mathrm{kg/m^3}$, $\lambda_w = 0.58\,\mathrm{W/m\text{-}K}$, $c_w = 4200\,\mathrm{J/kg\text{-}K}$, $\mu_w = 1.8\times 10^{-3}\,\mathrm{Pa\cdot s}$, $q_i = 3.34\times 10^5\,\mathrm{J/kg}$. But $T_* = 273.15\,\mathrm{K}$, $T^o = 333\,\mathrm{K}$, and $K = 10^{-15}\,\mathrm{m^2}$ were used instead of the previous values used for the first problem.

The approximate analytical solutions of the first problem are as follows (Tsypkin, 1998):

Zone I (hydrate–ice–gas)

$$\frac{T}{T_o} = 1 + \left[\frac{T_*}{T_o} - 1\right] \frac{erfc(\zeta)}{erfc(\gamma)}$$

$$\frac{p}{p_o} = \frac{B}{T_o} \left[\frac{T_*}{T_o} - 1\right] + 1$$

$$\frac{S_h}{S_{ho}} = \sigma_1 \left[\frac{T_*}{T_o} - 1\right] + 1 \tag{9.317}$$

$$\frac{S_i}{S_{io}} = -\frac{\rho_{ih}\sigma_1 S_{ho}}{\rho_i S_{io}} \left[\frac{T_*}{T_o} - T_o\right] + 1$$

Zone II (ice–gas)

$$\frac{T}{T_o} = \frac{T^o}{T_o} + \left[\frac{T_*}{T_o} - \frac{T^o}{T_o}\right] \frac{erf(\zeta)}{erf(\gamma)}$$

$$\frac{p}{p_o} = \frac{p^o}{p_o} + \left[\frac{p_*}{p_o} - \frac{p^o}{p_o}\right] \frac{\zeta}{\gamma} \tag{9.318}$$

S_h and S_i are given by Eqs. (9.310) and (9.311). These solutions are depicted in Figure 9.17 by Tsypkin (1998).

Tsypkin's (1998) approximate analytic solutions for the second problem are given in the following:

Zone I (hydrate–ice–gas)

$$\frac{T}{T_o} = 1 + \left[\frac{T_{*1}}{T_o} - 1\right] \frac{erfc(\zeta)}{erfc(\gamma_1)}$$

$$\frac{p}{p_o} = \frac{B}{T_o} \left[\frac{T}{T_o} - 1\right] + 1$$

$$\frac{S_h}{S_{ho}} = \sigma_1 \left[\frac{T}{T_o} - 1\right] + 1 \tag{9.319}$$

$$\frac{S_i}{S_{io}} = -\frac{\rho_{ih}\sigma_1 S_{ho}}{\rho_i S_{io}} \left[\frac{T}{T_o} - T_o\right] + 1$$

Zone II (ice–gas)

$$\frac{T}{T_o} = \frac{T_{*2}}{T_o} + \left[\frac{T_{*1}}{T_o} - \frac{T_{*2}}{T_o}\right] \frac{erf\left(\zeta\sqrt{a_f/a_2}\right) - erf\left(\gamma_2\sqrt{a_f/a_2}\right)}{erf\left(\gamma_1\sqrt{a_f/a_2}\right) - erf\left(\gamma_2\sqrt{a_f/a_2}\right)}$$

$$\frac{p}{p_o} = \frac{p_{*2}}{p_o} + \left[\frac{p_{*1}}{p_o} - \frac{p_{*2}}{p_o}\right] \frac{\zeta - \gamma_2}{\gamma_1 - \gamma_2} \tag{9.320}$$

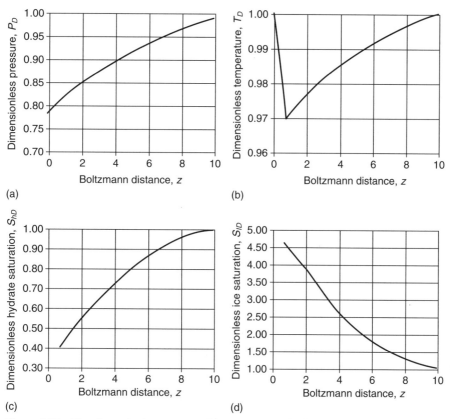

Figure 9.17   The dimensionless pressure (a), temperature (b), hydrate saturation (c), and ice saturation (d) profiles for hydrate decomposition in a negative temperature region (prepared by the author with modifications after Tsypkin, 1998; with permission from Springer).

Zone III (water–gas)

$$\frac{T}{T_o} = \frac{T^o}{T_o} + \left[\frac{T_{2*}}{T_o} - \frac{T^o}{T_o}\right]\frac{erf(\zeta)}{erf(\gamma_2)}$$

$$\frac{p}{p_o} = \frac{p^o}{p_o} + \left[\frac{p_{2*}}{p_o} - \frac{p^o}{p_o}\right]\frac{\zeta}{\gamma_2}$$

(9.321)

These solutions are depicted in Figure 9.18 by Tsypkin (1998) in terms of the dimensionless variables defined by

$$p_D = p/p_o,$$ (9.322)

$$T_D = T/T_o,$$ (9.323)

$$S_{hD} = S_h/S_{ho},$$ (9.324)

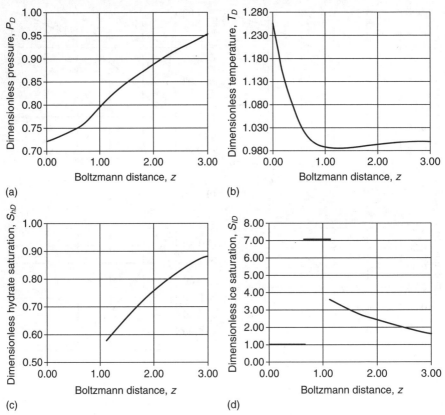

(a)

(b)

(c)

(d)

Figure 9.18 The dimensionless pressure pressure (a), temperature (b), hydrate saturation, (c) and water–ice (d) saturation profiles for hydrate decomposition by depression (prepared by the author with modifications after Tsypkin, 1998; with permission from Springer).

$$S_{iD} = S_i / S_{io}, \qquad (9.325)$$

$$z = \zeta \sqrt{a_f / a}, \qquad (9.326)$$

and

$$\gamma_z = \gamma \sqrt{a_f / a}. \qquad (9.327)$$

9.12 MODELING NONISOTHERMAL HYDROCARBON FLUID FLOW CONSIDERING EXPANSION/COMPRESSION AND JOULE–THOMSON EFFECTS

The formulation presented by App (2008, 2009) is presented in the following with some modifications for consistency with the materials presented in the rest of the book.

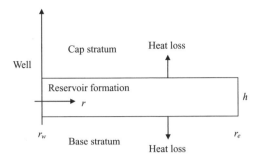

Figure 9.19   Schematic of cross section of a radial reservoir formation (prepared by the author).

9.12.1   Model Considerations and Assumptions

App (2008, 2009) modeled the nonisothermal transient-state one-dimensional radial single-phase hydrocarbon fluid flow through a horizontal uniform thick isotropic porous media. A formation thickness-averaged formulation is pursued in the following according to the schematic of a cylindrical-shaped reservoir region of well influence presented in Figure 9.19. The porous formation contains a single-phase mobile hydrocarbon phase (gas or oil), an immobile connate water phase, and a stationary porous media (rock) bounded between two parallel and impermeable base and cap strata. The effect of formation deformation on permeability is neglected. The fluid density and viscosity vary with temperature and pressure. The porous formation material density is constant. The isothermal compressibility of the hydrocarbon fluid (gas or oil) is dependent on pressure. The isothermal compressibility of water and porous formation are assumed constant. The coefficients of thermal expansion are assumed constant for the hydrocarbon, water, and porous formation. Further, assume that the porosity varies with temperature and pressure. The capillary pressure between the hydrocarbon and water is neglected. The Joule–Thomson coefficient was taken constant for oil and variable with pressure for gas. The hydrocarbon, water, and porous formation are assumed to attain an instantaneous thermal equilibrium.

9.12.2   Temperature and Pressure Dependency of Properties

Note that density is dependent on temperature and pressure:

$$\rho = \rho(p, T). \tag{9.328}$$

Therefore,

$$d\rho = \left(\frac{\partial \rho}{\partial p}\right)_T dp + \left(\frac{\partial \rho}{\partial T}\right)_p dT. \tag{9.329}$$

Define the isothermal fluid compressibility coefficient by

$$c = \frac{1}{\rho}\left(\frac{\partial \rho}{\partial p}\right)_T. \tag{9.330}$$

Define the thermal fluid expansion coefficient by

$$\beta = -\frac{1}{\rho}\left(\frac{\partial \rho}{\partial T}\right)_p. \tag{9.331}$$

Thus, Eq. (9.329) can be written as

$$\frac{\partial \rho}{\partial t} = \rho c \frac{\partial p}{\partial t} - \rho \beta \frac{\partial T}{\partial t}. \tag{9.332}$$

Note that the density can be expressed in terms of the formation volume factor B as

$$\rho = \frac{\rho^s}{B}, \tag{9.333}$$

where ρ^s is the density expressed at standard conditions (273.15°K and 101,325 Pa). Similarly, note that

$$\phi = \phi(p, T). \tag{9.334}$$

Therefore,

$$d\phi = \left(\frac{\partial \phi}{\partial p}\right)_T dp + \left(\frac{\partial \phi}{\partial T}\right)_p dT. \tag{9.335}$$

Define the isothermal rock compressibility coefficient by

$$c_r = \frac{1}{\phi}\left(\frac{\partial \phi}{\partial p}\right)_T. \tag{9.336}$$

Define the thermal rock expansion coefficient by

$$\beta_r = -\frac{1}{\phi}\left(\frac{\partial \phi}{\partial T}\right)_p. \tag{9.337}$$

Thus, Eq. (9.335) can be written as

$$\frac{\partial \phi}{\partial t} = \phi c_r \frac{\partial p}{\partial t} - \phi \beta_r \frac{\partial T}{\partial t}. \tag{9.338}$$

Hence,

$$\frac{\partial(\phi p)}{\partial t} = p\frac{\partial \phi}{\partial t} + \phi\frac{\partial p}{\partial t} = \phi(1 + pc_r)\frac{\partial p}{\partial t} - p\phi\beta_r \frac{\partial T}{\partial t}. \tag{9.339}$$

Next, consider that

$$H = H(p, T). \tag{9.340}$$

Therefore,

$$dH = \left(\frac{\partial H}{\partial p}\right)_T dp + \left(\frac{\partial H}{\partial T}\right)_p dT. \tag{9.341}$$

Define the Joule–Thomson throttling coefficient by

$$\sigma = \left(\frac{\partial H}{\partial p} \right)_T. \tag{9.342}$$

Define the specific heat capacity at constant pressure by

$$c_p = \left(\frac{\partial H}{\partial T} \right)_p. \tag{9.343}$$

Thus, Eq. (9.341) can be written as

$$dH = \sigma dp + c_p dT. \tag{9.344}$$

Consequently, the Joule–Thomson effect of compression/expansion, which is the change in temperature because of a change in pressure involving an isenthalpic thermodynamic process, is usually expressed by the Joule–Thomson coefficient given by (Moore, 1972)

$$\mu_{J-T} = \left(\frac{\partial T}{\partial p} \right)_H = -\frac{\left(\dfrac{\partial H}{\partial p} \right)_T}{\left(\dfrac{\partial H}{\partial T} \right)_p} = -\frac{\sigma}{c_p}. \tag{9.345}$$

9.12.3 Mixture Properties

Now consider the following expressions for properties of a mixture of m (gas or oil), w (water), and r (rock):

$$1 = S_m + S_w, \tag{9.346}$$
$$\rho = \phi(\rho_m S_m + \rho_w S_w) + (1-\phi)\rho_r, \tag{9.347}$$
$$\rho H = \phi(\rho_m H_m S_m + \rho_w H_w S_w) + (1-\phi)\rho_r H_r, \tag{9.348}$$
$$\rho H \mathbf{u} = \rho_m H_m \mathbf{u}_m + \rho_w H_w \mathbf{u}_w, \tag{9.349}$$
$$\lambda \equiv \lambda_t = \phi(\lambda_m S_m + \lambda_w S_w) + (1-\phi)\lambda_r, \tag{9.350}$$

and

$$p \equiv p_t = \phi(p_m S_m + p_w S_w) + (1-\phi)p_r. \tag{9.351}$$

The pressure of the fluid system is given by

$$p_f = p_m S_m + p_w S_w. \tag{9.352}$$

However, neglecting the capillary pressure effect,

$$p_f = p_m = p_w. \tag{9.353}$$

However, applying Biot's theorem, Civan (2010b) showed that

$$p_b = [\alpha + (1-\alpha)\phi]p_f, \tag{9.354}$$

where $0 \le \alpha \le 1$ denotes the effective stress coefficient, ordinarily called Biot's coefficient. $\alpha = 1$ represents the special case involving porous media made of incompressible grains according to Terzaghi (1923).

The total heat source is defined by

$$q_c = \phi S_m q_{cm} + \phi S_w q_{cw} + (1-\phi) q_{cr}. \tag{9.355}$$

The total exchange of heat between the formation and the base and cap strata per unit contact surface is estimated by the following expression (Zolotukhin, 1979; Satman et al., 1984):

$$q_c = -\frac{2U(T - T_i)}{h}, \tag{9.356}$$

where T_i and T denote the initial and instantaneous fluid-saturated formation temperatures, h is the formation thickness, and U is the overall heat transfer coefficient, which varies with time according to (Zolotukhin, 1979)

$$U = 2\sqrt{\frac{(\lambda \rho c_p)_r}{\pi t}}, \tag{9.357}$$

where λ, ρ, and c_p denote the thermal conductivity, density, and specific heat capacity at constant pressure, respectively; the subscript r denotes the rock formation; and t is time.

9.12.4 Equations of Conservations

App (2008, 2009) assumed the water phase consisting of the connate water only is immobile ($\mathbf{u}_w = 0$) and the hydrocarbon phase is the only mobile phase. The flow of the mobile hydrocarbon phase is described by Darcy's law as

$$\mathbf{u}_m = -\frac{K_m}{\mu_m} \nabla p. \tag{9.358}$$

The equation for conservation of mass (microscopic) is given by

$$\frac{\partial \rho}{\partial t} + \nabla \cdot (\rho \mathbf{u}) = \dot{m}, \tag{9.359}$$

where $\dot{m}$ denotes the mass added per unit volume per unit time.

Applying the equation for conservation of mass (microscopic) given by Eq. (9.359) for the hydrocarbon and immobile water phases yields the following macroscopic conservation of mass, respectively:

$$\frac{\partial (\phi \rho_m S_m)}{\partial t} + \nabla \cdot (\rho_m \mathbf{u}_m) = \dot{m}_m \tag{9.360}$$

and

$$\frac{\partial (\phi \rho_w S_w)}{\partial t} = \dot{m}_w. \tag{9.361}$$

Then, adding Eqs. (9.360) and (9.361) results in

$$\frac{\partial [\phi (\rho_m S_m + \rho_w S_w)]}{\partial t} + \nabla \cdot (\rho_m \mathbf{u}_m) = \dot{m}_m + \dot{m}_w. \tag{9.362}$$

Expanding, substituting Eqs. (9.332) and (9.338), and considering that S_m and S_w are constant yields

$$(\rho_m S_m + \rho_w S_w)\frac{\partial \phi}{\partial t} + \phi\left(S_m \frac{\partial \rho_m}{\partial t} + S_w \frac{\partial \rho_w}{\partial t}\right) + \nabla \cdot (\rho_m \mathbf{u}_m) = \dot{m}_m + \dot{m}_w. \qquad (9.363)$$

Thus,

$$-\phi[(\rho_m S_m + \rho_w S_w)\beta_r + \rho_m S_m \beta_m + \rho_w S_w \beta_w]\frac{\partial T}{\partial t}$$

$$+\phi[(\rho_m S_m + \rho_w S_w)c_r + \rho_m S_m c_m + \rho_w S_w c_w]\frac{\partial p}{\partial t} \qquad (9.364)$$

$$= \nabla \cdot \left(\frac{\rho_m K_m}{\mu_m}\nabla p\right) + \dot{m}_m + \dot{m}_w.$$

Let

$$c_t = (\rho_m S_m + \rho_w S_w)c_r + \rho_m S_m c_m + \rho_w S_w c_w \qquad (9.365)$$

and

$$\beta_t = (\rho_m S_m + \rho_w S_w)\beta_r + \rho_m S_m \beta_m + \rho_w S_w \beta_w. \qquad (9.366)$$

Hence,

$$-\phi\beta_t \frac{\partial T}{\partial t} + \phi c_t \frac{\partial p}{\partial t} = \nabla \cdot \left(\frac{\rho_m K_m}{\mu_m}\nabla p\right) + \dot{m}_m + \dot{m}_w. \qquad (9.367)$$

In radial coordinates,

$$-\phi\beta_t \frac{\partial T}{\partial t} + \phi c_t \frac{\partial p}{\partial t} = \frac{1}{r}\frac{\partial}{\partial r}\left(\frac{\rho_m K_m}{\mu_m}r\frac{\partial p}{\partial r}\right) + \dot{m}_m + \dot{m}_w. \qquad (9.368)$$

The equation for conservation of energy (microscopic) is given by

$$\frac{\partial(\rho H)}{\partial t} + \nabla \cdot (\rho H \mathbf{u}) = \nabla \cdot (\lambda \nabla T) + (\boldsymbol{\tau}:\nabla\mathbf{u}) + \frac{Dp}{Dt} + q_c, \qquad (9.369)$$

where q_c is the energy added per unit volume and unit time, H is enthalpy, and $\boldsymbol{\tau}$ is the shear stress tensor. The substantial derivative is defined by

$$\frac{Dp}{Dt} \equiv \frac{\partial p}{\partial t} + \mathbf{u}\cdot\nabla p. \qquad (9.370)$$

Incorporating Eqs. (9.370) and (9.359) into Eq. (9.369) yields

$$\rho\frac{\partial H}{\partial t} + \rho\mathbf{u}\cdot\nabla H + \dot{m}H = \nabla \cdot (\lambda\nabla T) + (\boldsymbol{\tau}:\nabla\mathbf{u}) + \frac{\partial p}{\partial t} + \mathbf{u}\cdot\nabla p + q_c. \qquad (9.371)$$

For fluids, the term $(\boldsymbol{\tau}:\nabla\mathbf{u})$ involved in the conservation of energy equation denotes the viscous dissipation of energy by frictional forces. Following Al-Hadhrami et al. (2003), the viscous dissipation term for low permeability ($K \to 0$) can be approximated as (see Chapter 5)

$$(\boldsymbol{\tau}:\nabla\mathbf{u}) \cong \mu\mathbf{u}\cdot\mathbf{K}^{-1}\cdot\mathbf{u} = -\mathbf{u}\cdot\nabla\mathbf{p}. \tag{9.372}$$

Hence, applying Eq. (9.372) to Eq. (9.371) yields

$$\rho\frac{\partial H}{\partial t} + \rho\mathbf{u}\cdot\nabla H + \dot{m}H = \nabla\cdot(\lambda\nabla T) + \frac{\partial p}{\partial t} + q_c. \tag{9.373}$$

Note that the density is variable in Eq. (9.373).

If the density ρ varies, then an expression similar to Eq. (9.373) can be written also for a solid phase. However, if the density is assumed to remain constant, then the following microscopic equation of energy conservation can be written for the solid phase by simplifying Eq. (9.369):

$$\frac{\partial(\rho H)}{\partial t} = \nabla\cdot(\lambda\nabla T) + q_c. \tag{9.374}$$

The following macroscopic equations of conservation can be written based on Eqs. (9.373) and (9.347)–(9.354):

$$\begin{aligned}
\phi\rho_m S_m \frac{\partial H_m}{\partial t} &+ \rho_m\mathbf{u}_m\cdot\nabla H_m + \dot{m}_m H_m \\
&= \nabla\cdot(\phi S_m\lambda_m\nabla T) + \frac{\partial(\phi\rho_m S_m)}{\partial t} + \phi S_m q_{cm}
\end{aligned} \tag{9.375}$$

and

$$\phi\rho_w S_w \frac{\partial H_w}{\partial t} + \rho_w\mathbf{u}_w\cdot\nabla H_w + \dot{m}_w H_w = \nabla\cdot(\phi S_w\lambda_w\nabla T) + \frac{\partial(\phi\rho_w S_w)}{\partial t} + \phi S_w q_{cw}. \tag{9.376}$$

Assuming ρ_r is constant and applying Eq. (9.374) yields

$$\begin{aligned}
\frac{\partial[(1-\phi)\rho_r H_r]}{\partial t} &= (1-\phi)\rho_r\frac{\partial H_r}{\partial t} + \rho_r H_r\frac{\partial(1-\phi)}{\partial t} \\
&= \nabla\cdot((1-\phi)\lambda_r\nabla T) + (1-\phi)q_{cr}.
\end{aligned} \tag{9.377}$$

Then, adding Eqs. (9.375)–(9.377) and assuming the water phase consisting of the connate water only is immobile ($\mathbf{u}_w = 0$) and the hydrocarbon phase is the only mobile phase yield

$$\begin{aligned}
\phi\rho_m S_m\frac{\partial H_m}{\partial t} &+ \phi\rho_w S_w\frac{\partial H_w}{\partial t} + (1-\phi)\rho_r\frac{\partial H_r}{\partial t} \\
&+ \rho_m\mathbf{u}_m\cdot\nabla H_m + \dot{m}_m H_m + \dot{m}_w H_w \\
&= \nabla\cdot[(\phi S_m\lambda_m + \phi S_w\lambda_w + (1-\phi)\lambda_r)\nabla T] \\
&+ (\rho_r H_r + p)\frac{\partial\phi}{\partial t} + \phi\frac{\partial p}{\partial t} + \phi S_m q_{cm} + \phi S_w q_{cw} + (1-\phi)q_{cr}.
\end{aligned} \tag{9.378}$$

Therefore, applying Eqs. (9.358) and (9.344) to Eq. (9.378) yields (note $\sigma_r = 0$ for solid)

$$\phi\left[\rho_m S_m c_{pm} + \rho_w S_w c_{pw} + \left(\frac{1-\phi}{\phi}\right)\rho_r c_{pr} + (\rho_r H_r + p)\beta_r\right]\frac{\partial T}{\partial t}$$

$$+ \phi\left[\rho_m S_m \sigma_m + \rho_w S_w \sigma_w - (\rho_r H_r + p)c_r - 1\right]\frac{\partial p}{\partial t} \qquad (9.379)$$

$$+ \dot{m}_m H_m + \dot{m}_w H_w + \dot{m}_r H_r$$

$$= \frac{\rho_m K_m}{\mu_m}\left[c_{pm}\nabla p \cdot \nabla T + \sigma_m\left(\nabla p\right)^2\right] + \nabla \cdot \left(\lambda_t \nabla T\right) + q_c.$$

In radial coordinates, applying Eqs. (9.355)–(9.357) yields

$$\phi\left[\rho_m S_m c_{pm} + \rho_w S_w c_{pw} + \left(\frac{1-\phi}{\phi}\right)\rho_r c_{pr} + (\rho_r H_r + p)\beta_r\right]\frac{\partial T}{\partial t}$$

$$+ \phi\left[\rho_m S_m \sigma_m + \rho_w S_w \sigma_w - (\rho_r H_r + p)c_r - 1\right]\frac{\partial p}{\partial t} + \dot{m}_m H_m + \dot{m}_w H_w + \dot{m}_r H_r$$

$$= \frac{\rho_m K_m}{\mu_m}\left[c_{pm}\frac{\partial p}{\partial r}\frac{\partial T}{\partial r} + \sigma_m\left(\frac{\partial p}{\partial r}\right)^2\right] + \frac{1}{r}\frac{\partial}{\partial r}\left(\lambda_t r \frac{\partial T}{\partial r}\right) - \frac{4U(T-T_i)}{h}\sqrt{\frac{(\lambda \rho c_p)_r}{\pi t}}.$$

$$(9.380)$$

9.12.5 Applications

The equations of the conservations of mass and energy, namely, Eqs. (9.368) and (9.380), can be solved simultaneously for pressure and temperature distributions along the radial distance at different times using an appropriate numerical method. For example, Figures 9.20 and 9.21 given by App (2008) present typical

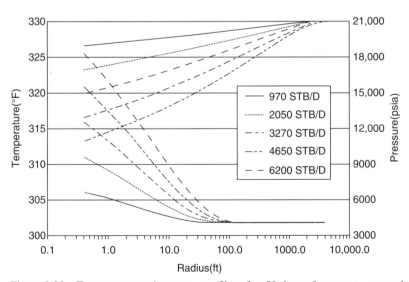

Figure 9.20 Temperature and pressure profiles after 50 days of constant rate production (after App, 2008; © 2008 SPE, with permission from the Society of Petroleum Engineers). STB/D, stock-tank barrels per day.

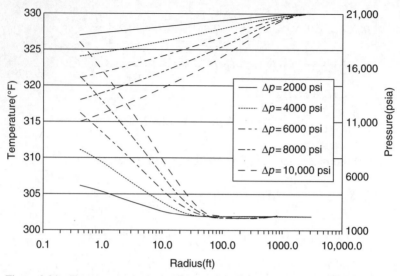

Figure 9.21 Temperature and pressure profiles after 50 days of production based on constant drawdown constraint (after App, 2008; © 2008 SPE, with permission from the Society of Petroleum Engineers).

results obtained under constant rate and constant pressure drawdown conditions, respectively.

9.13 EXERCISES

1. Beginning with Eq. (9.13) and assuming $f(Pe_i, r) \approx 1$, derive the following expressions of the hydrodynamic dispersion coefficients in the x-, y-, and z-Cartesian coordinates given by Bear and Bachmat (1990)

$$D_{xx} = \frac{a_T \left(v_y^2 + v_z^2 \right) + a_L v_x^2}{v}, \tag{9.381}$$

$$D_{xy} = D_{yx} = \frac{(a_L - a_T) v_x v_y}{v}, \tag{9.382}$$

$$D_{xz} = D_{zx} = \frac{(a_L - a_T) v_x v_z}{v}, \tag{9.383}$$

$$D_{yy} = \frac{a_T \left(v_x^2 + v_z^2 \right) + a_L v_y^2}{v}, \tag{9.384}$$

$$D_{yz} = D_{zy} = \frac{(a_L - a_T) v_y v_z}{v}, \tag{9.385}$$

and

$$D_{zz} = \frac{a_T \left(v_x^2 + v_y^2 \right) + a_L v_z^2}{v}, \tag{9.386}$$

where

$$v \equiv |\mathbf{v}| = \sqrt{v_x^2 + v_y^2 + v_z^2}$$

2. Beginning with Eq. (9.13), derive the convective dispersion coefficient tensor in a two-dimensional porous media and single-phase fluid, given by Bruneau et al. (1995)

$$\mathbf{D}_i^D = \frac{a_L}{v}\begin{bmatrix} v_x^2 & v_x v_y \\ v_x v_y & v_y^2 \end{bmatrix} + \frac{a_T}{v}\begin{bmatrix} v_y^2 & -v_x v_y \\ -v_x v_y & v_y^2 \end{bmatrix}, \tag{9.387}$$

where

$$v \equiv |\mathbf{v}| = \sqrt{v_x^2 + v_y^2}$$

3. Show that Eq. (9.387) can also be written as (Yang, 1998)

$$\mathbf{D}_i^D = \begin{bmatrix} \left(a_L\dfrac{v_x^2}{v} + a_T\dfrac{v_y^2}{v}\right) & (a_L - a_T)\dfrac{v_x v_y}{v} \\ (a_L - a_T)\dfrac{v_x v_y}{v} & \left(a_L\dfrac{v_y^2}{v} + a_T\dfrac{v_x^2}{v}\right) \end{bmatrix}. \tag{9.388}$$

4. Derive Eqs. (9.40)–(9.44).
5. Derive Eq. (9.46).
6. Repeat all derivations presented in this chapter if flux is expressed by Eq. (9.8) according to Civan (2010b).
7. Derive Eqs. (9.51)–(9.54).
8. Derive Eqs. (9.56)–(9.57).
9. Derive Eqs. (9.61), (9.63), and (9.65).
10. Derive mass, momentum, species, and energy equations for a gas/liquid two-phase system.
11. Reformulate Eqs. (9.92)–(9.94) by applying the definition of the flow potential given by

$$\Psi = \int_{p_o}^{} \frac{dp}{\rho} + g(z - z_o). \tag{9.389}$$

The subscript o denotes an arbitrarily or conveniently selected reference, datum, or base level. z is the upward vertical distance in the gravity direction.
12. Derive Eqs. (9.95)–(9.97) of the limited compositional model presented in this chapter.
13. Modify the limited compositional model developed in this chapter if also the water is allowed to vaporize into the gas phase.
14. Eqs. (9.138)–(9.141) describe the cross-sectional average temperature variation for the fluid flowing along a cylindirical porous core sample, assuming thermal equilibrium between the fluid and porous matrix. Derive the corresponding modified formulation when the fluid and porous matrix are not at thermal equilibrium.
15. Obtain a second-order accurate numerical solution for Eqs. (9.138)–(9.141) by means of the finite difference method using the following data: The core length is 10 cm and diameter is 1 cm; the porosity is 20% and permeability is 100 md; the fluid density is 1000 kg/m^3; thermal conductivity is 2.5 W/m-K, and specific heat capacity is 2.6×10^6 J/

m^3-K; and the overall heat transfer coefficient is $2.0\,kJ/h\text{-}m^2$-K. The initial tremperature is $50°C$; the injection fluid temperature is $80°C$; and the core outlet surface is maintained at $25°C$ temperature. The fluid inlet and outlet pressures are 5 and 1 atm, respectively. Apply Darcy's law to estimate the fluid flux across the core sample.

16. Repeat the problem described in the previous exercise using Eqs. (9.143)–(9.147) and compare the numerical solution obtained with the direct solution obtained in the previous problem. What do you conclude about the quality of solutions generated using Eqs. (9.138)–(9.141) and Eqs. (9.143)–(9.147)?

17. Consider the equations of the conservations of mass and energy, given by Eqs. (9.368) and (9.380), that can be solved simultaneously for pressure and temperature distributions along the radial distance at different times using an appropriate numerical method. Obtain a fully implicit second-order accurate finite difference numerical solution by considering a horizontal layer containing 20 grid cells, which are logarithmically expanded from the wellbore to the external boundary of the porous formation according to App (2009). Use the following data provided by App (2009): The wellbore radius is 0.41 ft; the external radius is 4000 ft; the formation porosity is 18%; the permeability is 20 md; and the thickness is 100 ft. The initial fluid pressure is 19,100 psia and the initial temperature is 231°F. The immobile (connate) water saturation is 15%. The isothermal compressibility of the formation rock is $3.0 \times 10^{-6}\,psi^{-1}$, and the water is $3.0 \times 10^{-6}\,psi^{-1}$. The bubble point pressure of the oil is below 10,000 psia. The density of the oil, water, and rock are 51.9, 62.43, and $165.0\,lb_m/ft^3$; the thermal expansion coefficients are $4.0 \times 10^{-4}, 5.0 \times 10^{-4}, 5.0 \times 10^{-5}°F^{-1}$, and the specific heat capacity at constant pressure are 0.52, 1.0, and 0.23, respectively. The Joule–Thomson coefficients for the oil and water are -0.0051 and $-0.0024°F/psi$, respectively. The total heat conductivity of the fluid-saturated formation rock is $1.8\,Btu/hr\text{-}ft\text{-}°F$. Make reasonable assumptions for any missing data.

18. Manoranjan and Stauffer (1996) obtained an analytical solution for contaminant transport undergoing a sorption process according to the Langmuir nonequilibrium kinetics, described by the following equations. Let ϕ (volume fraction) denote the porosity of porous media; ρ_b (kilogram per cubic meter) is the bulk mass density of porous media; v (meter per second) is the pore fluid velocity; t (second) is time; x (meter) is distance; c (kilogram per cubic meter) is the mass concentration of species in the flowing fluid; q (kilogram per kilogram) is the mass of species adsorbed per unit mass of porous media; q_o (kilogram per kilogram) is the maximum value of q; k_a (cubic meter per kilogram-second) and k_d (per second) denote the adsorption and desorption coefficients, respectively; and D (square meter per second) is the diffusion coefficient.

The transient-state advection–dispersion equation is given by

$$\frac{\partial c}{\partial t} + v\frac{\partial c}{\partial x} = D\frac{\partial^2 c}{\partial x^2} + \dot{m}, -\infty < x < \infty, t > 0. \tag{9.390}$$

The constitutive equation for the nonequilibrium sorption process is given by

$$\frac{\partial q}{\partial t} = k_a c(q_o - q) - k_d q. \tag{9.391}$$

(a) Show that the source term $\dot{m}$ in Eq. (9.390) is expressed as

$$\dot{m} = -\frac{\rho_b}{\phi}\frac{\partial q}{\partial t}. \tag{9.392}$$

(b) Check and confirm if the exact solution given by Manoranjan and Stauffer (1996) as the following satisfies Eq. (9.390):

$$c(x,t) = \frac{1}{-(a_2 / a_1) + A\exp[-a_1(x - \alpha vt)]},$$ (9.393)

where

$$a_1 = \frac{k_d}{\alpha v} - \frac{(1-\alpha)}{D}, a_2 = \frac{k_a}{2\alpha v}.$$ (9.394)

$A > 0$ is an integration constant that needs to be specified for a given problem. Following Manoranjan and Stauffer (1996), determine the value of A for $c = (-1/2)$ (a_1/a_2) when $x - \alpha vt = 0$. The value of the wave speed α is obtained by solving

$$2\alpha^3 - 4\alpha^2 + \left(2 + \frac{Dk_d}{v} - \frac{Dk_a\rho_b q_o}{\phi v}\right)\alpha - \frac{Dk_d}{v} = 0.$$ (9.395)

(c) Prepare a plot of the analytical solution obtained at various times using the following data given by Manoranjan and Stauffer (1996) but presented in the consistent International System of Units (SI): $\phi = 0.41$, $\rho_b = 1640\,kg/m^3$, $v = 1.1 \times 10^{-4}\,m/s$, $q_o = 7.85 \times 10^{-3}\,kg/kg$, $k_a = 2.8 \times 10^{-4}\,m^3/kg\cdot s$, $k_d = 1.8 \times 10^{-4}\,s^{-1}$, and $D = 1.0 \times 10^{-6}\,m^2/s$.

19. Derive the macroscopic mass balance equations for the single- and multiphase systems flowing through porous media by applying the control volume approach.

20. The molar concentration of the methane gas is $6.25 \times 10^{-5}\,mol/cc$ in a natural gas of density $0.005\,g/cc$. Calculate the weight fraction of methane in the natural gas.

21. One liter of a mixture of sand and water contains 50-g sand. The density of water is $1.0\,g/cm^3$ and that of sand is $2.5\,g/cm^3$. Calculate the following:

   (a) Mass fraction of sand in the mixture

   (b) Volume fraction of sand in the mixture

22. The composition of a mixture of methane and propane gases at 1-atm pressure and 25°C temperature condition is given as 80 mol % methane and 20 mol % propane.

   (a) What is the mass concentration of methane in the units of kilogram per cubic meter?

   (b) What is the mole concentration of propane in the units of kilomole per cubic meter?

23. What is the geometric average permeability for an anisotropic porous medium having 100-, 300-, and 700-md permeability values in the x-, y-, and z-Cartesian directions?

24. Consider the hydraulic fracture completed in a two-dimensional unit-thick porous medium as shown in Figure 9.22. The fracture is symmetrical with respect to the x-axis. The fracture medium is packed with sand so that it is a porous medium itself. The permeability and porosity of the fracture medium are functions of distance and time, and therefore $K = K(x, t)$ and $\phi = \phi(\mathbf{x}, \mathbf{t})$. The width of the fracture varies with the distance in the x-Cartesian direction and therefore $w = w(x)$. As the fluid flows through the fracture, some fluid is lost to the surrounding reservoir formation through the fracture face as shown. The leak-off rate (mass per unit fracture surface per unit time) is denoted by q_y on both sides of the fracture as shown. Write a material balance for the fluid present in the control volume shown in Figure 9.22 and derive a differential

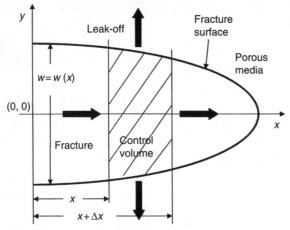

Figure 9.22 Control volume in a hydraulic fracture completed in a two-dimensional unit-thick porous medium (prepared by the author).

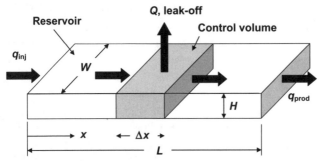

Figure 9.23 Control volume in a two-dimensional unit-thick porous medium (prepared by the author).

material balance equation specially formulated for fluid flowing through the hydraulic fracture.

25. Consider Figure 9.23 of a porous medium having a length L, thickness H, and width W. Assume that the reservoir contains a single-component gas, such as methane, in the pore space, a real gas behavior, and the gas is leaking into the upper layer at a mass rate of Q per unit surface area of the cap rock (kilogram per second per square meter).

  (a) Derive a differential equation based on the transient-state mass balance equation written over the control volume for one-dimensional flow occurring in the opposite x-direction as shown in Figure 9.23.

  (b) Write down the expression of the boundary condition for pressure if the injection rate into the porous medium is constant q_{inj} at the injection boundary of the reservoir as shown.

(c) Write down the expression of the boundary condition for pressure if the production rate from the porous medium is constant q_{prod} at the production boundary of the porous medium as shown in Figure 9.23.

26. Consider a gas and condensate two-phase fluid system having 60% gas saturation by volume. The density of the gas is $125 \, kg/m^3$ and that of the liquid condensate is $750 \, kg/m^3$. What is the mass fraction of the liquid condensate in this system?

CHAPTER 10

SUSPENDED PARTICULATE TRANSPORT IN POROUS MEDIA

10.1 INTRODUCTION

Migration and retention of particulate matter during flow of particulate suspensions in porous media, such as encountered in petroleum, geothermal, and groundwater reservoirs, are influenced strongly in a complicated manner by various transport and retention mechanisms, and temperature variation during injection of fluids at temperatures different from those of porous media, such as for purposes of secondary oil recovery, acid stimulation, and carbon sequestration (Civan, 2010c).* Description of particulate behavior by phenomenological modeling including dispersive transport and nonisothermal conditions is required for generalized handling of various applications of practical importance.

However, whether the particle dispersion phenomenon is of practical importance depends on particular applications and appears to be a matter of continuing debate (Unice and Logan, 2000; Altoe et al., 2006; Lominé and Oger, 2009). Logan et al. (1997) point out that particle dispersion is different from chemical dispersion. They indicate, for example, that the bromide tracer dispersion is a chemical dispersion issue and the bacteria dispersion is a particle dispersion issue. Logan et al. (1997) explain that particle dispersion involves other factors such as the sticking phenomenon, slow desorption, and heterogeneity of particles and porous media grains. Lominé and Oger (2009) state, "Dispersion of particles results from collisions with the porous matrix and with other moving particles." Lominé and Oger (2009) determined that the value of the dispersion coefficient increases with the particle concentration.

Auset and Keller (2004) observed by means of experiments conducted using micromodels that the magnitude of particle dispersion depends primarily on their

---

* Parts of this chapter have been reproduced with modifications from Civan, F. 2010c. Non-isothermal permeability impairment by fines migration and deposition in porous media including dispersive transport. Transport in Porous Media, 85(1), pp. 233–258, with permission from Springer.

Porous Media Transport Phenomena, First Edition. Faruk Civan.
© 2011 John Wiley & Sons, Inc. Published 2011 by John Wiley & Sons, Inc.

preferential paths and velocities. Smaller particles tend to move along longer complicated tortuous paths and therefore are slower, and larger particles tend to move along shorter straighter paths and therefore are faster. Hence, the dispersivity and dispersion coefficient of particles in porous media increase with decreasing particle size and vice versa, depending on the pore structure and the pore channel-to-particle size ratio.

Temperature variation affects the governing particulate transport and rate processes in porous media in very complicated ways. The particulate and pore surface interactions are strongly affected by temperature (Schembre and Kovscek, 2005; Civan, 2007a,b, 2008a,c, 2010c). Particles tend to deposit more preferentially over the pore surface at lower temperatures than higher temperatures. Hence, colder temperature conditions are favorable for more pore surface attachment and retention of fine particles (Civan, 2007b). Conversely, at sufficiently high temperatures, pore surface conditions become more suitable for particle detachment, and therefore fine particles are less likely to deposit over the pore surface but rather migrate toward the pore throats and form particle bridges nevertheless only under favorable conditions (Pandya et al., 1998; Civan, 2007a, and Tran et al., 2009).

The balance between the effects of the pore surface retention and pore-throat plugging conditions varies with the prevailing temperature, fine particle suspension, and porous matrix conditions, and determines the severity of permeability impairment. In general, pore-throat plugging renders more severe permeability damage than pore surface retention. Such conditions are implicitly included in the value of the filter coefficient estimated using the correlations developed by Tufenkji and Elimelech (2004) and Chang et al. (2009). Such effects are important in various applications, such as water injection into hot subsurface formations for extraction of geothermal energy and injection of steam with or without mixing with light hydrocarbons into heavy oil reservoirs for enhanced oil recovery.

Particulate processes occurring in porous media should be expressed in terms of the interstitial fluid velocity rather than the pore velocity, volumetric flux, or flow rate because the fluid forces acting upon particles in porous media are determined by the actual fluid velocity in tortuous preferential fluid flow paths, referred to as the interstitial fluid velocity (Dupuit, 1863; Civan, 2007b). Fine particles present in tortuous flow paths are drifted more effectively along with the flowing fluid because the fluid moves faster in tortuous paths than in straight paths (Civan, 2007b, 2008d).

The consequences of injecting particulate suspensions into porous media, such as porous media clogging, permeability reduction, and effluent solution particle concentration, depend on the essential characteristics of the particles, carrier fluids, and porous media, and the prevailing conditions, such as temperature, pressure, wettability, and electrolytic and potentiometric activity. Information on values of the parameters involved in various dimensionless groups are necessary, for example, for the calculation of the filter coefficient using the empirical correlation given by Tufenkji and Elimelech (2004) and Chang et al. (2009).

In this chapter, first, a phenomenological model is presented for deep-bed filtration by considering temperature variation and particle transport by advection and dispersion according to Civan (2010c). Migration and retention of particles in porous media are formulated by theoretically modeling the relevant processes under both isothermal and nonisothermal conditions and with and without inclusion of the dispersive transport. The solutions generated with and without the inclusion of the dispersion effect under isothermal and nonisothermal conditions indicate the effect of dispersion and temperature on particulate processes occurring in porous media. Obviously, the transverse dispersion can play an important role in multidimensional cases depending on the ratio of the transverse-to-longitudinal dispersivity. Thus, the dispersion in various directions is taken into account by means of the directional elements of the dispersion tensor. Significant effects of temperature variation and particulate transport by advection and dispersion on particle concentration in the flowing suspension of particles, particle retention, and permeability variation during flow are demonstrated by examples under isothermal and nonisothermal conditions.

Second, a formulation of compressible filter cake formation undergoing small particle packing in between large particles is presented according to Tien et al. (1997). Filter cake is treated as a special porous media that forms over a filter used for the separation of particles from a suspension of particles or slurry. Larger particles are assumed to construct the essential porous structure of the filter cake. Smaller particles are assumed to be trapped and packed inside the pore space created by the larger particles.

The particle retention rate and the variation of porous matrix properties, such as porosity and permeability, are formulated by considering the temperature and particle retention effects. The nonisothermal and dispersion effects are considered, which may be important under certain conditions for the accurate description of transport and retention of colloidal and fine particles during flow through porous formations. It is demonstrated that temperature variation has a significant effect on particulate transport through porous media because it affects the filter coefficient, porous matrix thermal deformation, and pore-throat constriction. Comparison of the results obtained with and without the dispersion effect included indicates that transport by dispersion has some effect on suspended fines migration and retention, and on the resulting permeability variation in porous media. Dispersion causes the spreading of the effect of the particulate phenomenon on permeability reduction over a long range from the injection port. However, such effect is more pronounced over the region near the injection side of porous media when the dispersion mechanism of particle transport is ignored.

10.2 DEEP-BED FILTRATION UNDER NONISOTHERMAL CONDITIONS

The essential constituents of the overall model, some of which have been adopted with and without modifications after Civan (2008a, 2010c), are described in the following.

Consider a porous medium whose initial and instantaneous effective (interconnected) porosities are denoted by ϕ_i and ϕ. Let u and ρ represent the volumetric flux and density of the flowing fluid containing fine particles. ρ_p is the particle material mass density. w and σ denote the mass and volume fractions of particles present in the flowing fluid medium. ε is the volume fraction of particles deposited in bulk porous media. t and x denote the time and distance from the injection point.

10.2.1 Concentration of Fine Particles Migrating within the Carrier Fluid

The mass fraction of fine particles w present in a carrier fluid containing such suspended particles can be expressed in terms of the volume fraction of particles σ as the following:

$$w = \frac{\rho_p}{\rho} \sigma. \tag{10.1}$$

For sufficiently low concentrations of fine particles, the density of the flowing fluid ρ (carrier fluid, i.e., water, plus suspended particles) may be assumed to be the same as the carrier fluid density ρ_w, that is, when $\sigma \ll 1.0$, then $\rho \cong \rho_w$.

The retention of suspended particles reduces the instantaneous porosity of porous media according to (see Civan, 2007a for a discussion about particle packing efficiency)

$$\phi = \phi_i - \varepsilon. \tag{10.2}$$

The momentum balance of fluid flowing through porous media is given by Darcy's law as

$$\mathbf{u} = -\frac{1}{\mu} \mathbf{K} \cdot (\nabla p + \rho \mathbf{g} \cdot \nabla z), \tag{10.3}$$

where $\mathbf{K}$ is the permeability tensor, $\mathbf{g}$ is the gravitational acceleration vector, z is the Cartesian distance covered in the gravitational acceleration direction, and p and μ are the pressure and dynamic viscosity, respectively, of the flowing suspension of particles.

The total mass balance equation for the flowing fluid (carrier fluid plus suspended particles) is given by

$$\frac{\partial(\rho\phi)}{\partial t} + \nabla \cdot (\rho\mathbf{u}) = -\dot{m}. \tag{10.4}$$

The suspended particle mass balance equation is given by

$$\frac{\partial(\rho\phi w)}{\partial t} + \nabla \cdot (\rho\mathbf{u}w) + \nabla \cdot \mathbf{j} = -\dot{m}. \tag{10.5}$$

Combining Eqs. (10.4) and (10.5) yields the suspended particle mass balance equation for the flowing fluid as

$$\rho\left(\phi\frac{\partial w}{\partial t} + \mathbf{u} \cdot \nabla w \right) + \nabla \cdot \mathbf{j} = -(1-w)\dot{m}, \tag{10.6}$$

where $\dot{m}$ is the mass rate of particles deposited per unit bulk volume of porous media from the flowing phase and $\mathbf{j}$ denotes the dispersive mass flux vector of suspended particles in the flowing phase. Note that Eq. (10.6) is applicable regardless of whether the density and porosity are variable or constant.

The dispersive mass flux of particles is given by

$$\mathbf{j} = -\phi\rho\mathbf{D}\cdot\nabla w, \tag{10.7}$$

where $\mathbf{D}$ is the coefficient of dispersion of suspended particles migrating in the flowing phase assumed to be a linear function of the interstitial pore fluid velocity according to

$$\mathbf{D} = \alpha\cdot\mathbf{v}, \tag{10.8}$$

where α is the longitudinal dispersivity and $\mathbf{v}$ is the interstitial pore fluid velocity, given by (Dupuit, 1863)

$$\mathbf{v} = \frac{\tau\cdot\mathbf{u}}{\phi}, \tag{10.9}$$

where τ denotes the tortuosity of the preferential flow paths in porous media. Thus, the formulation presented here allows for the modeling of fines migration and retention effects also without including the dispersion effect simply by setting $\alpha = 0$. Note that Eq. (10.7) is more rigorous than the analogous equation given by Altoe et al. (2006) because here, the effect of tortuous flow paths is accounted for.

The particle capture (retention) rate $\dot{m}$ is assumed proportional to the total suspended particle flux modifying the equation given by Iwasaki (1937) as the following:

$$\frac{\partial(\rho_p\varepsilon)}{\partial t} = \dot{m} \equiv \frac{\partial m}{\partial t} = k\mathbf{J}\cdot\mathbf{n}, \tag{10.10}$$

where k is referred to as the filtration rate coefficient or filter coefficient discussed later in the following, $\mathbf{n}$ denotes the normal unit vector, and $\mathbf{J}$ is the total mass flux of suspended particles due to the combined effects of bulk flow (advection) and dispersion, given by

$$\mathbf{J} = \mathbf{u}\rho w + \mathbf{j}. \tag{10.11}$$

The preceding equations, Eqs. (10.6)–(10.11), can be combined readily to obtain the following equation of convenience:

$$\frac{\partial w}{\partial t} + \left[\frac{\tau\cdot\mathbf{v}(1 - k_1\alpha\cdot\tau)}{\tau^2} - \frac{1}{\rho\phi}\nabla\cdot(\mathbf{v}\rho\phi\cdot\alpha)\right]\cdot\nabla w + \frac{\tau\cdot\mathbf{v}k_1}{\tau^2}w = \mathbf{v}\cdot\alpha\cdot\nabla^2 w, \tag{10.12}$$

where, considering Eq. (10.1), a new parameter k_1 is defined as

$$k_1 = (1 - w)k = \left(1 - \frac{\rho_p}{\rho}\sigma\right)k. \tag{10.13}$$

For convenience in the numerical applications presented later, consider constant density and flow for the flowing fluid and neglect the effect of the small amount of fine particle retention on the variation of porosity (but not the porosity itself) ($\varepsilon \cong 0$ and $\phi = \phi_i - \varepsilon$, $\partial\phi \cong 0$). Thus, Eq. (10.12) simplifies as the following:

$$\frac{\partial w}{\partial t} + \frac{\boldsymbol{\tau} \cdot \mathbf{v}\left(1 - k_1 \boldsymbol{\alpha} \cdot \boldsymbol{\tau}\right)}{\tau^2} \cdot \nabla w + \frac{\boldsymbol{\tau} \cdot \mathbf{v} k_1}{\tau^2} w = \mathbf{v} \cdot \boldsymbol{\alpha} \cdot \nabla^2 w. \tag{10.14}$$

The variation of density and porosity with time and space by gradual thermal expansion is negligible relative to their absolute values. But the variation of viscosity with temperature and its effect on the filter coefficient value are more pronounced and therefore are considered in the present formulation.

The conditions of solution are described by the following. The initial condition throughout the porous media is given by

$$w = w^o, t = 0. \tag{10.15}$$

The boundary condition at the injection side is given by

$$\left(\mathbf{u}\rho w\right)_{\text{in}} = \mathbf{J} \text{ or } w_{\text{in}} = w - \boldsymbol{\alpha} \cdot \boldsymbol{\tau} \cdot \nabla w \cdot \mathbf{n}, t > 0. \tag{10.16}$$

The boundary condition at the outlet boundary side is given by

$$\nabla w \cdot \mathbf{n} = 0, t > 0, \tag{10.17}$$

where L denotes the length of the spatial region, the subscript "in" represents the inlet condition, and the superscript o represents the initial condition.

Note that Eqs. (10.14)–(10.17) can be expressed in volume fraction of particles in the flowing fluid, by means of Eq. (10.1), as

$$\frac{\partial \sigma}{\partial t} + \mathbf{a} \cdot \nabla \sigma + \mathbf{b}\sigma = \mathbf{c} \cdot \nabla^2 \sigma. \tag{10.18}$$

The conditions of solution are described by the following. The initial condition throughout the porous media is given by

$$\sigma = \sigma^o, t = 0. \tag{10.19}$$

The boundary condition at the injection side is given by

$$\sigma_{in} = \sigma - \mathbf{d} \cdot \nabla \sigma \cdot \mathbf{n}, t > 0. \tag{10.20}$$

The boundary condition at the outlet boundary side is given by

$$\nabla \sigma \cdot \mathbf{n} = 0, t > 0, \tag{10.21}$$

where

$$\mathbf{a} = \frac{\boldsymbol{\tau} \cdot \mathbf{v}\left(1 - k_1 \boldsymbol{\alpha} \cdot \boldsymbol{\tau}\right)}{\tau^2}, \mathbf{b} = \frac{\boldsymbol{\tau} \cdot \mathbf{v} k_1}{\tau^2}, \mathbf{c} = \boldsymbol{\alpha} \cdot \mathbf{v}, \mathbf{d} = \boldsymbol{\alpha} \cdot \boldsymbol{\tau}. \tag{10.22}$$

Obviously, the values of the parameters a and b vary as a result of variation of the value of the parameter k_1 according to Eq. (10.13) by the variation of the filter coefficient k and the volume fraction of particles σ in the flowing fluid with changing local conditions. Therefore, the numerical solution is obtained iteratively at a time attained after each finite time increment until convergence starting with a value of σ known at the previous time, which is the initial value prescribed at the beginning (see Exercise 4).

10.2.2 Concentration of Fine Particles Deposited inside the Pores of the Porous Matrix

The volume fraction of particles deposited in porous media can be calculated by combining Eqs. (10.7)–(10.11) as the following:

$$\rho_p \frac{\partial \varepsilon}{\partial t} = \frac{\boldsymbol{\tau} \cdot \mathbf{v} k \rho \phi}{\tau^2} (w - \boldsymbol{\alpha} \cdot \boldsymbol{\tau} \cdot \nabla w). \tag{10.23}$$

Eq. (10.23) can be expressed in volume fraction of suspended particles in the flowing fluid, by means of Eq. (10.1), as

$$\frac{\partial \varepsilon}{\partial t} = \frac{\boldsymbol{\tau} \cdot \mathbf{v} k \phi}{\tau^2} (\sigma - \boldsymbol{\alpha} \cdot \boldsymbol{\tau} \cdot \nabla \sigma), \tag{10.24}$$

where the values of the porosity ϕ and the filter coefficient k vary with changing local conditions.

The following initial condition is imposed throughout the porous media:

$$\varepsilon = \varepsilon^o, t = 0, \tag{10.25}$$

where the superscript o represents the initial condition.

10.2.3 Variation of Temperature in the System of Porous Matrix and Flowing Fluid

Variation of temperature as a result of injecting a fluid at a temperature different from the porous media temperature will affect the physical properties of materials involved, the filtration rate coefficient, permeability, and porosity, and hence the particle retention rate.

For purposes of the present application, a simplified form of the energy equation for the fluid flowing through porous media is used here. Similar to the exercise demonstrated earlier, first, we simplify the general equation of energy balance by means of the total mass balance equation (the equation of continuity given by Eq. 10.4). Then, we neglect the heat removed from the flowing fluid medium by the small amount of deposited particles, the kinetic energy and the viscous effects of the rather slow moving fluid, and the substantial derivatives of the slowly varying pressure and density. Finally, we drop the gravity term for horizontal flow.

Thus, the temperature T of the flowing fluid is described with reasonable accuracy by

$$\rho c \left(\phi \frac{\partial T}{\partial t} + \mathbf{u} \cdot \nabla T \right) = \nabla \cdot (\phi \boldsymbol{\kappa} \cdot \nabla T) + \dot{q}. \tag{10.26}$$

The conditions of solution are described by the following. The initial condition throughout the porous media is given by

$$T = T^o, t = 0. \tag{10.27}$$

The boundary condition at the injection side is given by

$$(\rho c \mathbf{u} T)_{\text{in}} \cdot \mathbf{n} = (\rho c \mathbf{u} T - \phi \kappa \cdot \nabla T) \cdot \mathbf{n}, \, t > 0. \tag{10.28}$$

The boundary condition at the effluent side is given by

$$\nabla T \cdot \mathbf{n} = 0, \, t > 0, \tag{10.29}$$

where $\dot{q}$ denotes the amount of heat added to the fluid, the subscript "in" represents the inlet condition, and the superscript o represents the initial condition.

The thermal expansion of the porous matrix is neglected here only for the rate of matrix displacement ($u_m = 0$) but is considered for reduction of permeability by pore-throat constriction by thermal expansion of porous media grains. Hence, the temperature T_m of the porous matrix is described by the following equation:

$$\rho_m c_m \left[(1-\phi) \frac{\partial T_m}{\partial t} + \mathbf{u}_m \cdot \nabla T_m \right] = \nabla \cdot [(1-\phi) \kappa_m \cdot \nabla T_m] + \dot{q}_m. \tag{10.30}$$

The subscript m indicates properties for the porous matrix and $\dot{q}_m$ denotes the amount of heat added to the porous matrix.

The conditions of solution are described by the following. The initial condition throughout the porous media is given by

$$T_m = T_m^o, \, t = 0. \tag{10.31}$$

The boundary condition at the injection side is given by

$$\nabla T_m \cdot \mathbf{n} = 0, \, t > 0. \tag{10.32}$$

The boundary condition at the outlet boundary side is given by

$$\nabla T_m \cdot \mathbf{n} = 0, \, t > 0. \tag{10.33}$$

The interface heat exchange rate between the flowing fluid and the porous matrix across the pore surface is given by

$$\dot{q}_m = -\dot{q} = A_m h (T - T_m), \tag{10.34}$$

where A_m is the matrix pore surface available per unit bulk volume and h is the film heat transfer coefficient.

If fluid flows sufficiently slowly through porous media, such as typically encountered in subsurface reservoirs, then it is reasonable to assume a thermal equilibrium between the flowing fluid system and the porous matrix. Thus, Eqs. (10.26) and (10.30) can be added together and then rearranged as

$$\frac{\partial T}{\partial t} + \left(\frac{\mathbf{u} \rho c - \nabla \cdot [\phi \kappa + (1-\phi) \kappa_m]}{[\phi \rho c + (1-\phi) \rho_m c_m]} \right) \cdot \nabla T = \frac{[\phi \kappa + (1-\phi) \kappa_m]}{[\phi \rho c + (1-\phi) \rho_m c_m]} \cdot \nabla^2 T. \tag{10.35}$$

The conditions of solution are described by the following. The initial condition throughout the porous media is given by

$$T = T^o, \, t = 0. \tag{10.36}$$

The boundary condition at the injection side is given by

$$(\rho c \mathbf{u} T)_{\text{in}} \cdot \mathbf{n} = (\rho c \mathbf{u} T - \phi \boldsymbol{\kappa} \cdot \nabla T) \cdot \mathbf{n}, t > 0. \tag{10.37}$$

The boundary condition at the outlet boundary side is given by

$$\nabla T \cdot \mathbf{n} = 0, t > 0. \tag{10.38}$$

If all the physical properties are assumed constant, neglecting the small effect of gradual temperature variation in liquid and solid media, then Eq. (10.35) can be written as

$$\frac{\partial T}{\partial t} + \mathbf{a}_1 \cdot \nabla T = \mathbf{b}_1 \cdot \nabla^2 T. \tag{10.39}$$

The conditions of solution are described by the following. The initial condition throughout the porous media is given by

$$T = T^o, t = 0. \tag{10.40}$$

The boundary condition at the injection side is given by

$$T_{\text{in}} = T - \mathbf{c}_1 \cdot \nabla T \cdot \mathbf{n}, t > 0. \tag{10.41}$$

The boundary condition at the outlet boundary side is given by

$$\nabla T \cdot \mathbf{n} = 0, t > 0, \tag{10.42}$$

where

$$\mathbf{a}_1 = \frac{\mathbf{u}\rho c}{\phi \rho c + (1 - \phi)\rho_m c_m}, \quad \mathbf{b}_1 = \frac{\phi \boldsymbol{\kappa} + (1 - \phi)\boldsymbol{\kappa}_m}{\phi \rho c + (1 - \phi)\rho_m c_m}, \quad \mathbf{c}_1 = \frac{\phi \mathbf{u} \cdot \boldsymbol{\kappa}}{u^2 \rho c}. \tag{10.43}$$

10.2.4   Initial Filter Coefficient

The initial filter coefficient k_o is defined as the value of the filter coefficient determined prior to particle retention, given by (Tien, 1989; Elimelech et al., 1995)

$$k_o = \frac{3}{2} \frac{(1 - \phi)}{D_g} \beta_1 \eta_o, \tag{10.44}$$

where η_o denotes the single-collector contact efficiency (a physical effect) representing the rate of collisions between porous media grains and suspended fine particles migrating within the flowing pore fluid, and the sticking coefficient β_1 denotes the particle attachment efficiency (a chemical effect) representing the fraction of particle collisions leading to successful particle attachment to porous media grains (Tufenkji and Elimelech, 2005).

The value of η_o is denoted by η_{oS} in the absence of electrostatic repulsive force (case of high ionic strength colloidal suspension) and is given by the following correlation (Tufenkji and Elimelech, 2004):

$$\eta_{oS} = 2.4 A_s^{1/3} N_R^{-0.081} N_{Pe}^{-0.715} N_{vdW}^{0.052} + 0.55 A_s N_R^{1.675} N_A^{0.125} + 0.22 N_R^{-0.24} N_G^{1.11} N_{vdW}^{0.053}. \tag{10.45}$$

The correction factor α_{CORR} required in the presence of electrostatic repulsive force is given by (Chang et al., 2009)

$$\alpha_{CORR} = \frac{\eta_o}{\eta_{oS}} = \exp\left[(\ln\alpha)_{CORR}\right] = \exp\left[\frac{1}{2}(\ln\alpha_{C-C} + \ln\alpha_{B-T})\right] \quad (10.46)$$

The correlation of α_{C-C} is given by Chang and Chan (2008) for submicroparticles:

$$\alpha_{C-C} = 0.024N_{DL}^{0.969}N_{E1}^{-0.423}N_{E2}^{2.88}N_{Lo}^{1.5} + 3.176A_s^{1/3}N_R^{-0.081}N_{Pe}^{-0.715}N_{Lo}^{2.687}$$
$$+ 0.222A_s N_R^{3.041}N_{Pe}^{-0.514}N_{Lo}^{0.125} + N_R^{-0.24}N_G^{1.11}N_{Lo}. \quad (10.47)$$

The correlation of α_{B-T} is given by Bai and Tien (1999) for small and large particles:

$$\alpha_{B-T} = 0.002527N_{DL}^{1.352}N_{E1}^{-0.3121}N_{E2}^{3.5111}N_{Lo}^{0.7031}. \quad (10.48)$$

The expressions of the dimensionless numbers facilitated in these correlations are given in the following. These include the following parameters: H denotes the Hamaker constant, κ_E denotes the reciprocal electric double-layer thickness, v_o is the dielectric constant, ξ_p is the zeta potential of suspended particles, ξ_g is the zeta potential of porous media grains, k_B is the Boltzmann constant, ρ_p and ρ_w denote the density of the particles and the carrier fluid (water for example), respectively, g is the gravitational acceleration coefficient, and D_∞ denotes the bulk diffusivity. The interstitial velocity v is given by Eq. (10.9). R_p and D_p denote the mean radius and diameter, respectively, of the suspended particles. D_g is the mean diameter of the porous media grains. Other variables were defined in the preceding sections.

The aspect ratio (particle-to-grain diameter ratio) N_R is given by

$$N_R = \frac{D_p}{D_g}. \quad (10.49)$$

The attraction number N_A is given by

$$N_A = \frac{H}{12\pi\mu R_p^2 u}, \quad (10.50)$$

where u denotes the magnitude of the volumetric flux vector $\mathbf{u}$.

The electric double-layer force parameter N_{DL} is given by

$$N_{DL} = \kappa_E R_p. \quad (10.51)$$

The first electrokinetic parameter N_{E1} is given by

$$N_{E1} = \frac{v_o R_p\left(\xi_p^2 + \xi_g^2\right)}{4k_B T}. \quad (10.52)$$

The second electrokinetic parameter N_{E2} is given by

$$N_{E2} = \frac{2\left(\xi_p/\xi_g\right)}{1+\left(\xi_p/\xi_g\right)^2}. \quad (10.53)$$

The gravity number N_G is given by

$$N_G = \frac{2R_p^2\left(\rho_p - \rho_w\right)g}{9\mu u}. \quad (10.54)$$

The London force parameter N_{Lo} is given by

$$N_{Lo} = \frac{H}{6k_B T}.$$ (10.55)

The Peclet number N_{Pe} is given by

$$N_{Pe} = \frac{uD_g}{D_\infty}.$$ (10.56)

The van der Waals number N_{vdW} is given by

$$N_{vdW} = \frac{H}{k_B T}.$$ (10.57)

The Stokes–Einstein equation of bulk molecular diffusivity D_∞ is given by

$$D_\infty = \frac{k_B T}{3\pi\mu D_p}.$$ (10.58)

The parameter A_s is given by

$$A_s = \frac{2(1-\gamma^5)}{2 - 3\gamma + 3\gamma^5 - 2\gamma^6}, \gamma = (1-\phi)^{1/3}.$$ (10.59)

10.2.5 Filter Coefficient Dependence on Particle Retention Mechanisms and Temperature Variation

Particles can deposit over the pore surface and/or behind the pore throats restricted for particle transport by means of the particle bridges formed across the pore throats. When particles form bridges across the pore throats but allow some flow of the carrier fluid through the gaps present between the bridging particles, then the pore-bridging particles act as a filter for the approaching particles leading to the capture and retention of particles behind the bridged pore throats. Then, the filter coefficient assumes a value depending on the degree of plugging restriction. When all the pore throats are clogged and sealed completely, then the value of the filter coefficient becomes zero because the completely plugged pore throats do not allow the flow of the carrier fluid and the flow paths involving such pore throats are left out of service.

The dependency of the filtration rate coefficient k_o on temperature T alone, that is, when $\varepsilon = 0$, can be expressed by a Vogel–Tammann–Fulcher (VTF)-type equation, given by (Civan, 2008b)

$$\frac{k_o(T)}{k_c} = \exp\left(\frac{A_k}{T - T_{ck}}\right), A_k \equiv -\frac{E_k}{R},$$ (10.60)

where T is the absolute temperature, T_{ck} is a characteristic-limit absolute temperature, k_c is a pre-exponential coefficient, R is the universal gas constant, and E_k is the activation energy. Note that the temperature difference $(T - T_{ck})$ has the same value regardless of whether the ordinary (°C) or absolute (K) temperatures are used. The pre-exponential rate coefficient k_c represents the high-temperature limit value of the filtration rate coefficient.

Thus, the temperature dependency factor $G(T)$ of the filter coefficient can be derived from Eq. (10.60) as

$$\frac{k_o(T)}{k_o(T_{ik})} = G(T) \cong \exp\left[A_k\left(\frac{1}{T-T_{ck}} - \frac{1}{T_{ik}-T_{ck}}\right)\right],$$ (10.61)

where the subscript i denotes a reference condition and T_{ik} is a reference absolute temperature.

The effect of temperature on the filter coefficient is profound because particle detachment force is higher at higher temperatures and therefore particle retention over the pore surface is less probable (Civan, 2007b). Consequently, particles are more likely to move toward and to deposit behind the pore throats when the conditions are favorable for pore-throat bridging. Hence, rapid particle retention and accumulation occurs behind the bridged pore throats, rendering a pore-filling phenomenon.

On the other hand, the particle retention dependency factor $F(\varepsilon)$ of the filter coefficient can be expressed as (Ives, 1967)

$$\frac{k(\varepsilon, T)}{k_o(T)} = F(\varepsilon) \cong \left(1 - \frac{\varepsilon}{\varepsilon_M}\right)^{m_1}\left(1 + \frac{b\varepsilon}{\phi_o}\right)^{m_2}\left(1 - \frac{\varepsilon}{\phi_o}\right)^{m_3},$$ (10.62)

$$\varepsilon_M \leq \phi_o, \quad k_o(T) \equiv k(\varepsilon = 0, T),$$

where ε_M is the maximum amount of retention at which limit condition the filter coefficient becomes zero, and therefore the particle retention phenomenon ceases; b is an empirical constant; and m_1, m_2, and m_3 are empirically determined exponents of intensity. In the following, the exponent values of $m_2 = m_3 = 0$ are assumed (Ives, 1987).

Thus, the dependency of the filtration coefficient on temperature and fines retention can be expressed based on the method of separation of variables as

$$\frac{k(\varepsilon, T)}{k_i} = \left[\frac{k(\varepsilon, T)}{k_o(T)}\right]\left[\frac{k_o(T)}{k_o(T_{ik})}\right] = F(\varepsilon)G(T)$$

$$= \left(1 - \frac{\varepsilon}{\varepsilon_M}\right)^{m_k}\exp\left[A_k\left(\frac{1}{T-T_{ck}} - \frac{1}{T_{ik}-T_{ck}}\right)\right], \quad k_i \equiv k_o(T_{ik}).$$ (10.63)

Note in general that $\varepsilon_M = \varepsilon_M(T)$ and $\phi_o = \phi_o(T)$. For example, expressing the grain volume V_g by $V_g = (1-\phi)V_b$ and assuming that the bulk volume V_b is constant for confined porous material, the porous media grain volume expansion coefficient c_g is given by (Civan, 2008c)

$$c_g = \frac{1}{V_g}\frac{\partial V_g}{\partial T} = \frac{1}{(1-\phi)}\frac{\partial(1-\phi)}{\partial T}.$$ (10.64)

Thus, the variation of porosity with temperature can be expressed by (Gupta and Civan, 1994b)

$$\phi \cong 1 - (1 - \phi_o) \exp[c_g(T - T_o)]. \tag{10.65}$$

The thermal expansion coefficient of minerals is rather small. For example, the coefficient of thermal expansion for quartz and calcite minerals is only in the order of 10^{-5} K^{-1} (Gupta and Civan, 1994b). It is reasonable to neglect the effect of temperature on porosity because the thermal expansion coefficient is very small and porosity reduction occurs by pore volume reduction. However, as described in the next section, the effect of temperature on permeability cannot be neglected because permeability reduction occurs primarily by pore-throat constriction. Even a small increase in temperature can cause sufficient grain expansion to choke the pore-throat openings and to reduce the permeability substantially.

10.2.6 Permeability Alteration by Particle Retention and Thermal Deformation

When externally confined porous materials are heated, the pore throats are constricted by thermal expansion of porous matrix grains acting like valves to reduce the flow through the preferential hydraulic flow paths. Applying the method of separation of variables as demonstrated in the previous section, the total permeability variation can be expressed by the product of the permeability variations resulting from particle retention and temperature variation.

The dependency of permeability K_o on temperature T alone, that is, when $\varepsilon = 0$, can be expressed by a VTF-type equation, given by (Civan, 2008c)

$$\frac{K_o(T)}{K_c} = \exp\left(\frac{A_K}{T - T_{cK}}\right), A_K \equiv -\frac{E_K}{R}, \tag{10.66}$$

where T is the absolute temperature, T_{cK} is the absolute characteristic temperature, K_o is the permeability of porous media, R is the universal gas constant, and E_K is the activation energy. The pre-exponential permeability coefficient K_c represents the high-temperature limit value of permeability. Civan (2008c) has demonstrated the validity of this equation by successfully correlating a number of experimental data.

Thus, the temperature dependency factor can be derived from Eq. (10.66) as

$$\frac{K_o(T)}{K_o(T_{iK})} = \exp\left[A_K\left(\frac{1}{T - T_{cK}} - \frac{1}{T_{iK} - T_{cK}}\right)\right]. \tag{10.67}$$

The subscript i denotes the reference condition and T_{iK} is a reference absolute temperature.

To describe the reduction of permeability by particle retention, consider Civan's power-law flow unit equation of permeability based on a bundle of leaky-tube model of tortuous preferential flow paths involving cross-flow between them depending on the pore interconnectivity in porous media (Civan, 2001):

$$\sqrt{\frac{K}{\phi}} = \Gamma\left(\frac{\phi}{1 - \phi}\right)^{\beta}, 0 \leq \beta < \infty, \tag{10.68}$$

where the pore interconnectivity parameter Γ is given by

$$\Gamma = \phi^\upsilon + e, \upsilon < 0, e > 0, \tag{10.69}$$

where the parameter e represents the effect of the threshold condition, corresponding to the minimum porosity below which permeability vanishes, and β and υ are empirically determined parameters, which can be related to the fractal parameters of pore structure. Thus, using Eqs. (10.68) and (10.69), the particle retention dependency factor can be formulated as

$$\frac{K(\phi,T)}{K_o(T)} = \frac{\phi}{\phi_o}\left(\frac{\phi^\upsilon + e}{\phi_o^\upsilon + e}\right)^2\left[\frac{\phi}{\phi_o}\left(\frac{1-\phi_o}{1-\phi}\right)\right]^{2\beta}, \tag{10.70}$$

$$\phi = \phi_o - \varepsilon, 0 \le \beta < \infty, \upsilon < 0, e > 0, K_o(T) \equiv K(\phi_o, T).$$

For convenience, Eq. (10.70) can be approximated reasonably upon substitution of Eq. (10.2) as

$$\frac{K(\varepsilon, T)}{K_o(T)} \cong \left(1 - \frac{\varepsilon}{\phi_o}\right)^{m_K}, m_K = 1 + 2(\beta + \upsilon), K_o(T) \equiv K(\varepsilon = 0, T). \tag{10.71}$$

Thus, the dependency of permeability on temperature and fines retention can be expressed based on the method of separation of variables as

$$\frac{K(\varepsilon, T)}{K_i} = \left(1 - \frac{\varepsilon}{\phi_o}\right)^{m_K}\exp\left[A_K\left(\frac{1}{T - T_{cK}} - \frac{1}{T_{iK} - T_{cK}}\right)\right], K_i(T) \equiv K_o(T_{iK}). \tag{10.72}$$

10.2.7 Applications

The numerical solution of the above-given differential equations was obtained for one-dimensional particulate transport in a porous core plug using a finite difference scheme according to Civan (2009, 2010c). The details of this approach are described in Exercise 4.

The representative parameter values employed in the following numerical simulation studies are $R_p = 1.5\text{E-06}\,\text{m}$, $D_g = 4.6\text{E-04}\,\text{m}$, $\rho_p = 1050\,\text{kg/m}^3$, $\rho = 1020\,\text{kg/m}^3$, $\rho_m = 1600\,\text{kg/m}^3$, $c = 4186\,\text{J/(kg-K)}$, $c_m = 820\,\text{J/(kg-K)}$, $\kappa = 0.65\,\text{W/(m-K)}$, $\kappa_m = 2.0\,\text{W/(m-K)}$, $\xi_p = -1.0\text{E} - 03\,\text{V}$, $\xi_g = -3.0\text{E-03 V}$, $\kappa_E = 2.2\text{E} + 08\,\text{1/m}$, $\upsilon_o = 8.9\text{E-11}$, $\phi = 0.2$, $\tau = 1.41$, $L = 0.1\,\text{m}$, $H = 1.0\text{E-20 J}$, $k_B = 1.38\text{E-23 m}^2\text{kg/s}^2\text{/K}$, $T_{c\mu} = -100°\text{C}$, $A_\mu = 370\,\text{K}$, $\mu_c = 0.045\,\text{cp}$, $T_{ik} = 25°\text{C}$, $T_{ck} = -200°\text{C}, A_k = -251\,\text{K}, k_c = 41.2\,\text{m}^{-1}, m_k = 2, \beta_1 = 1.0, T_{iK} = 50°\text{C}, T_{cK} = -101°\text{C}$, $A_K = 500\,\text{K}$, $m_K = 5$, $\varepsilon_M = 0.17\,\text{m}^3\text{/m}^3$, $u = 1.0\text{E-04 m/s}$, $T^o = 50°\text{C}$, $\sigma^o = 0$, $\varepsilon^o = 0$, $T_{in} = 80°\text{C}$, and $\sigma_{in} = 5.0\text{E-3 m}^3\text{/m}^3$.

The effect of temperature on the carrier fluid (water) viscosity was correlated using the water viscosity data of Bett and Cappi (1965) at 1 atm by the VTF equation as (Civan, 2008b)

$$\frac{\mu(T)}{\mu_c} = \exp\left(\frac{A_\mu}{T - T_{c\mu}}\right), A_\mu \equiv -\frac{E_\mu}{R}. \tag{10.73}$$

The best estimate values of $T_{c\mu}$, A_μ, and μ_c reported earlier were determined by the least squares linear regression of their data.

The temperature dependency of the filter coefficient was obtained also by using a VTF-type equation (Eq. 10.63) by correlating the filter coefficient values predicted by means of the correlations given by Tufenkji and Elimelech (2004) and Chang et al. (2009) as described previously using the best estimate parameter values as reported earlier. Eq. (10.73) was used for the calculation of viscosity in predicting the filter coefficient using these correlations.

The numerical calculations of the variation of the temperature, suspended particle volume fraction, deposited particle volume fraction, and permeability reduction profiles along the porous medium were carried out by Civan (2010c) with and without the dispersion effect included under isothermal ($T = 50°C$) and nonisothermal ($50°C \le T \le 80°C$) conditions by means of the numerical solution scheme described in Exercise 4 using $\Delta x = 0.001$-m grid-point spacing and $\Delta t = 5$-s time increments. The predictions of the above-given phenomenological model were calculated at various times after starting the injection of the fluid containing suspended fine particles.

Figure 10.1 shows the typical nonisothermal temperature profiles obtained at various times (minutes). As can be seen, the temperature effect propagates with time (minutes) from the injection port to the effluent port. A comparison of the results presented in Figures 10.2–10.4 indicate that the thermal and dispersion factors affect the particle volume fraction in the flowing suspension, volume fraction of the bulk porous media occupied by the deposited particles, and permeability impairment in porous media.

Figure 10.2 shows the suspended particle volume fraction profiles at various times (minutes) with and without the dispersion effect included under isothermal and elevated nonisothermal conditions. When the dispersion effect is ignored, the particle migration occurs with sharp progressing fronts as indicated in Figure 10.2a,c.

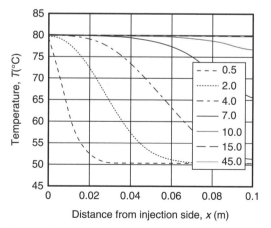

Figure 10.1   Nonisothermal temperature profiles at various times (minutes) (after Civan, 2010c; reprinted with permission from Springer).

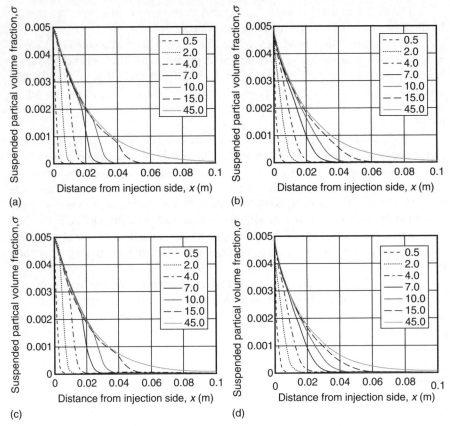

Figure 10.2   Isothermal suspended particle volume fraction profiles at various times (minutes) (a) without and (b) with the dispersion effect included. Nonisothermal suspended particle volume fraction profiles at various times (minutes) (c) without and (d) with the dispersion effect included (after Civan, 2010c; reprinted with permission from Springer).

When the dispersion effect is considered, the particle migration spreads over a long range as indicated in Figure 10.2b,d. Figure 10.2a,b obtained under isothermal conditions indicate higher suspended particle concentrations than those indicated by the corresponding Figure 10.2c,d obtained under elevated nonisothermal conditions.

Figure 10.3 shows the deposited particle volume fraction profiles at various times (minutes) with and without the dispersion effect included under isothermal and elevated nonisothermal conditions. When the dispersion effect is ignored, the particle retention is more pronounced over the region near the injection port as indicated in Figure 10.3a,c. When the dispersion effect is considered, the particle retention spreads over a long range as indicated in Figure 10.3b,d. But the effect is small because of low suspended particle concentration in the injected fluid. Figure 10.3a,b obtained under isothermal conditions indicate lower deposited particle concentrations than those indicated by the corresponding Figure 10.3c,d obtained under

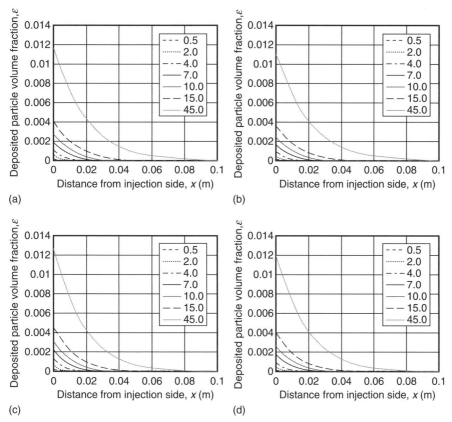

Figure 10.3 Isothermal deposited particle volume fraction profiles at various times (minutes) (a) without and (b) with the dispersion effect included. Nonisothermal deposited particle volume fraction profiles at various times (minutes) (c) without and (d) with the dispersion effect included (after Civan, 2010c; reprinted with permission from Springer).

elevated nonisothermal conditions. More retention occurs behind constricted and bridged pore throats at elevated temperatures.

Figure 10.4 shows the permeability reduction profiles at various times (minutes) with and without the dispersion effect included under isothermal and elevated nonisothermal conditions. When the dispersion effect is ignored, the permeability impairment occurs more severely over the region near the injection port as indicated in Figure 10.4a,c. When the dispersion effect is considered, the permeability impairment spreads over a long range as indicated in Figure 10.4b,d. But the effect is small because of low suspended particle concentration in the injected fluid. Figure 10.4a,b obtained under isothermal conditions indicate less permeability impairment compared with those indicated by the corresponding Figure 10.4c,d obtained under elevated nonisothermal conditions.

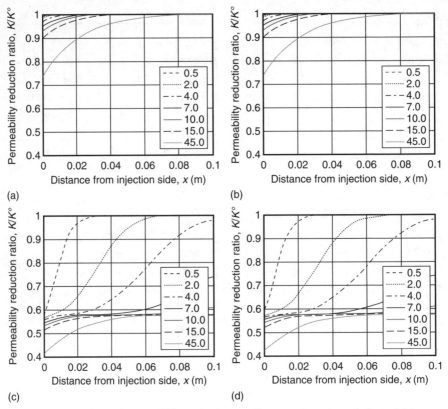

Figure 10.4 Isothermal permeability reduction profiles at various times (minutes) (a) without and (b) with the dispersion effect included. Nonisothermal permeability reduction profiles at various times (minutes) (c) without and (d) with the dispersion effect included (after Civan, 2010c; reprinted with permission from Springer).

10.3 CAKE FILTRATION OVER AN EFFECTIVE FILTER

Filter cake is a special porous media formed over a filter facilitated for separation of particles from a suspension of particles or slurry. Smaller particles are deposited in between the larger particles, forming a compressible and porous filter cake.

Detailed formulations of incompressible and compressible cake filtration processes are presented elsewhere by Civan (1998a,b). However, the Tien et al. (1997) model for one-dimensional linear flow is reviewed here for instructional purposes. Their formulation is expanded to explain the steps leading to model construction. The formulation of this model is presented in a manner consistent with the symbolism used in this chapter. A schematic description of the problem is depicted in Figure 10.5. A slurry of particles of various sizes is fed into a filter. The filter is assumed impermeable for particle invasion and, therefore, the filtrate is a particle-free carrier fluid.

Tien et al. (1997) classified the various particles in the slurry fed into the filter into two groups as large and small particles, denoted by indices 1 and 2, respectively.

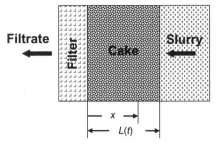

Figure 10.5   Cake filtration and the coordinate system used in the analysis (prepared by the author with modifications after Tien et al., 1997; © 1997 AIChE, reprinted by permission from the American Institute of Chemical Engineers).

They based this classification on Tien's (1989) particle retention criteria, which requires the ratio of the small to large particle diameters to be $d_{p2}/d_{p1} < 0.1$. In the following presentation, the filter cake and the flowing suspension of particles are denoted by the indices s and f. The carrier phase (liquid) is denoted by l and the particles by p. It is considered that large particles cannot enter the filter cake and, therefore, are deposited over the filter cake, whereas the small particles can penetrate the filter cake and deposit within the porous cake matrix.

The volume fractions (or concentrations) of the large and small particles are denoted by σ_{p1} and σ_{p2}. Thus, the total particle volume fraction in the slurry is given by

$$\sigma_p = \sigma_{p1} + \sigma_{p2}. \tag{10.74}$$

The volume fraction of all the particles (small plus large) forming the filter cake can be expressed in terms of the cake porosity, $\varepsilon_f \equiv \phi_c$, as

$$\varepsilon_s = 1 - \varepsilon_f. \tag{10.75}$$

Because the filter cake is formed from the large particles plus the deposited small particles, it is also true that

$$\varepsilon_s = \varepsilon_{s1} + \varepsilon_{s2}, \tag{10.76}$$

where ε_{s1} and ε_{s2} denote the volume fractions of the large and small particles in the filter cake, respectively.

The superficial (or macroscopic) velocities of the flowing suspension of particles and the particles forming the cake are indicated by u_f and u_s, respectively. The fluid pressure and the compressive stress of the particles of the filter cake generated by the fluid drag forces are represented by p and p_s, respectively. Therefore, neglecting the inertial effects, the effective pressure at the cake–slurry interface can be expressed as (Tiller and Crump, 1985)

$$p_o = p + p_s. \tag{10.77}$$

Next consider the general mass balance equation given in terms of the mass concentration, c_{ij}, of a species i in a phase j, as

$$\frac{\partial(\varepsilon_j c_{ij})}{\partial t} + \frac{\partial(c_{ij}u_j)}{\partial x} = \frac{\partial}{\partial x}\left[\varepsilon_j \rho_j D_{ij}\frac{\partial}{\partial x}\left(\frac{c_{ij}}{\rho_j}\right)\right] + R_{ij}, \qquad (10.78)$$

where ε_j is the volume fraction of phase j in the bulk of the filter cake, $\vec{u}_j$ is its superficial velocity, ρ_j is its density, $\vec{D}_{ij}$ is the coefficient of dispersion of species i in phase j, and R_{ij} is the mass rate of addition of species i to phase j.

The mass concentration, c_{ij}, can be expressed in terms of the volume fraction (or concentration), σ_{ij}, as

$$c_{ij} = \rho_i \sigma_{ij}. \qquad (10.79)$$

Thus, invoking Eq. (10.79), the mass balance equation given by Eq. (10.78) can be expressed in terms of the volume fraction as

$$\frac{\partial(\varepsilon_j \rho_i \sigma_{ij})}{\partial t} + \frac{\partial(\rho_i \sigma_{ij}u_j)}{\partial x} = \frac{\partial}{\partial x}\left[\varepsilon_j \rho_j D_{ij}\frac{\partial}{\partial x}\left(\frac{\rho_i \sigma_{ij}}{\rho_j}\right)\right] + R_{ij}. \qquad (10.80)$$

Eqs. (10.78) and (10.80) were adopted from Civan (2007a) with modifications for the present one-dimensional case in the x-direction.

When the system is assumed incompressible and the dispersion term is neglected following Tien et al. (1997), Eq. (10.80) simplifies as

$$\frac{\partial(\varepsilon_j \sigma_{ij})}{\partial t} + \frac{\partial u_{ij}}{\partial x} = N_{ij}, \qquad (10.81)$$

where

$$u_{ij} = \sigma_{ij}u_j \qquad (10.82)$$

is the volumetric flux of species i in phase j and

$$N_{ij} = R_{ij}/\rho_i \qquad (10.83)$$

is the volumetric rate of generation of species i in phase j.

Thus, the volumetric balance equation of the carrier fluid in the suspension can be expressed according to Eq. (10.81) for $i \equiv l$ and $j \equiv f$. For a diluted suspension of small particles, the carrier fluid volume fraction is nearly $\sigma_{lf} \cong 1.0$, and $\varepsilon_f = 1 - \varepsilon_s$ and $N_{lf} = 0$. Then, Eq. (10.81) can be written in one dimension, according to Tien et al. (1997), as

$$\frac{\partial \varepsilon_s}{\partial t} = \frac{\partial u_{lf}}{\partial x}. \qquad (10.84)$$

For the particles of the cake, $i \equiv p$, $j \equiv s$, and $\sigma_{ps} = 1.0$ (thus, $u_{ps} = u_s$). Thus, Eq. (10.81) becomes

$$\frac{\partial \varepsilon_s}{\partial t} + \frac{\partial u_s}{\partial x} = N_{ps} \equiv N, \qquad (10.85)$$

or elimination of $\partial \varepsilon_s/\partial t$ between Eqs. (10.84) and (10.85) leads to

$$\frac{\partial u_{lf}}{\partial x} + \frac{\partial u_s}{\partial x} = N_{ps} \equiv N. \tag{10.86}$$

For small particles contained in the filter cake, Eq. (10.81) becomes

$$\frac{\partial}{\partial t}\left(\varepsilon_s \sigma_{p2s}\right) + \frac{\partial u_{p2s}}{\partial x} = N_{p2s} \equiv N. \tag{10.87}$$

The volume fraction of small particles in the cake is given by

$$\varepsilon_{s2} = \varepsilon_s \sigma_{p2s} \tag{10.88}$$

and the volume flux is given by

$$u_{p2s} = u_s \sigma_{p2s} = u_s \frac{\varepsilon_{s2}}{\varepsilon_s}. \tag{10.89}$$

Therefore, substituting Eq. (10.89) into Eq. (10.87) yields

$$\frac{\partial \varepsilon_{s2}}{\partial t} + \frac{\partial}{\partial x}\left(u_s \frac{\varepsilon_{s2}}{\varepsilon_2}\right) = N_{p2s} \equiv N. \tag{10.90}$$

For small particles of the suspension flowing through the filter cake, substituting $\varepsilon_f = 1 - \varepsilon_s$ into Eq. (10.81) yields

$$\frac{\partial}{\partial t}\left[(1-\varepsilon_s)\sigma_{p2f}\right] + \frac{\partial u_{p2f}}{\partial x} = N_{p2f} = -N, \tag{10.91}$$

or eliminating N between Eqs. (10.86) and (10.91) yields

$$\frac{\partial}{\partial t}\left[(1-\varepsilon_s)\sigma_{p2f}\right] + \frac{\partial u_{p2f}}{\partial x} + \frac{\partial u_{lf}}{\partial x} + \frac{\partial u_s}{\partial x} = 0. \tag{10.92}$$

The superficial velocity, u_{fr}, of the flowing suspension relative to that of the compressing filter cake, u_s, is given by Darcy's law as

$$u_{fr} = u_f - \frac{1-\varepsilon_s}{\varepsilon_s}u_s = -\frac{k}{\mu}\frac{\partial p}{\partial x}, \tag{10.93}$$

where k is the permeability of the cake and μ is the viscosity of the flowing suspension.

The rate of small particle retention within the filter cake is assumed proportional to the small particle volume flux through the cake as

$$N = \lambda |u_{fr}| \sigma_{p2f}, \tag{10.94}$$

where λ is the retention rate constant.

Next, Tien et al. (1997) algebraically manipulate the preceding equations to derive a set of workable equations suitable for numerical solution.

Tien et al. (1997) integrate Eq. (10.86) over a distance, x, from the filter surface in the filter cake as

$$\left(u_{lf}+u_s\right)_x = \left(u_{lf}+u_s\right)_{x=0} + \int_0^x Ndx. \tag{10.95}$$

Because the solids are not permitted to enter the filter, $u_s|_{x=0}=0$ and Eq. (10.95) simplifies as

$$\left(u_{lf}+u_s\right)_x = u_{lf}|_{x=0} + \int_0^x Ndx, \tag{10.96}$$

in which $u_{lf}|_{x=0}$ is the superficial velocity of the filtrate (carrier fluid) entering the filter, expressed by Darcy's law as

$$u_{lf}|_{x=0} = \frac{k_m}{\mu}\frac{p|_{x=0}-p|_{x=L_m}}{L_m}, \tag{10.97}$$

where k_m and L_m denote the permeability and thickness of the filter.

Thus, substituting Eq. (10.77) and $p|_{x=L_m}=0$ gage pressure into Eq. (10.97) yields an equation for the filtrate invasion volumetric flux into the filter as

$$u_{lf}|_{x=0} = \frac{k_m}{\mu}\frac{p_o-p_s|_{x=0}}{L_m}. \tag{10.98}$$

Eliminating the filter cake superficial velocity, u_s, between Eqs. (10.93) and (10.96) and substituting $u_f \cong u_{lf}$ for a dilute suspension leads to

$$u_{lf} = -\varepsilon_s \frac{k}{\mu}\frac{\partial p}{\partial x} + (1-\varepsilon_s)u_{lf}|_{x=0} + (1-\varepsilon_s)\int_0^x Ndx. \tag{10.99}$$

Then, invoking Eq. (10.99) into Eq. (10.96) results in

$$u_s = \varepsilon_s \frac{k}{\mu}\frac{\partial p}{\partial x} + \varepsilon_s u_{lf}|_{x=0} + \varepsilon_s \int_0^x Ndx. \tag{10.100}$$

A substitution of Eq. (10.100) into Eq. (10.85) yields the following differential equation for the volume fraction of the particles forming the cake as

$$\frac{\partial\varepsilon_s}{\partial t} + \left[u_{lf}|_{x=0} + \int_0^x Ndx\right]\frac{\partial\varepsilon_s}{\partial x} + \frac{\partial}{\partial x}\left[\varepsilon_s \frac{k}{\mu}\frac{\partial p}{\partial x}\right] = (1-\varepsilon_s)N. \tag{10.101}$$

The boundary conditions at the slurry–cake interface are given by

$$p_s = 0, \ \varepsilon_s = \varepsilon_s^o, \ x = L(t). \tag{10.102}$$

Based on Shirato et al. (1987) and Tien (1989), Tien et al. (1997) facilitated the following power-law constitutive relationships for the variation of the volume fraction of particles in a compressible cake by pressure by

$$\frac{\varepsilon_s}{\varepsilon_s^o} = \left(1+\frac{p_s}{\lambda}\right)^{\beta} \tag{10.103}$$

and for the flow efficiency factor due to the retention of small particles by

$$\gamma = \frac{1}{1 + \alpha_1 \varepsilon_{s2}^{\alpha 2}}. \tag{10.104}$$

Therefore, the permeability variation is empirically expressed as

$$\frac{k}{k^o} = \left(1 + \frac{p_s}{\lambda}\right)^{-\delta} \left(1 + \alpha_1 \varepsilon_{s2}^{\alpha 2}\right)^{-1}, \tag{10.105}$$

or substituting Eq. (10.103) into Eq. (10.105),

$$\frac{k}{k^o} = \left(\frac{\varepsilon_s}{\varepsilon_s^o}\right)^{-\delta/\beta} \left(1 + \alpha_1 \varepsilon_{s2}^{\alpha 2}\right)^{-1}. \tag{10.106}$$

Eq. (10.103) can be rearranged as

$$p_s = \lambda \left[\left(\frac{\varepsilon_s}{\varepsilon_s^o}\right)^{1/\beta} - 1\right]. \tag{10.107}$$

Next, Tien et al. (1997) derive an equation to determine the filter cake thickness. For this purpose, consider the jump mass balance of species i in a phase l at the interface of the slurry and the cake given by

$$c_{ij}^- \left(u_j^- - u_j^{\sigma-}\right) = c_{ij}^+ \left(u_j^+ - u_j^{\sigma+}\right) + R_{ij}^\sigma, \tag{10.108}$$

where $-$ and $+$ indicate the slurry and cake sides of the interface, c_{ij} is the mass concentration of species i in the jth phase, R_{ij}^σ is the amount of the species i deposited over the cake surface from the jth phase, and the superficial velocities of the interface for phase j at the slurry and cake sides are given, respectively, by

$$u_j^{\sigma-} = \varepsilon_j^- dL/dt \tag{10.109}$$

and

$$u_j^{\sigma+} = \varepsilon_j^+ dL/dt, \tag{10.110}$$

where ε_j denotes the volume fraction of the jth phase.

For convenience, Eq. (10.108) can also be written in an alternative form as

$$m_{ij}^- - m_{ij}^+ = c_{ij}^- u_j^{\sigma-} - c_{ij}^+ u_j^{\sigma+} + R_{ij}^\sigma, \tag{10.111}$$

where

$$m_{ij} = c_{ij} u_j \tag{10.112}$$

is the mass flux of species i in the jth phase.

Dividing Eq. (10.108) by the density of species i leads to the following expression:

$$\sigma_{ij}^- u_j^- - \sigma_{ij}^+ u_j^+ = \sigma_{ij}^- u_j^{\sigma-} - \sigma_{ij}^+ u_j^{\sigma+} + N_{ij}^\sigma, \tag{10.113}$$

where σ_{ij} is the volume concentration (or fraction) of species i in the jth phase and

$$N_{ij}^\sigma \equiv R_{ij}^\sigma / \rho_i \tag{10.114}$$

denotes the volume of species i deposited over the cake surface from the jth phase. Again, for convenience, Eq. (10.113) can be written as

$$u_{ij}^- - u_{ij}^+ = \sigma_{ij}^- u_j^{\sigma-} - \sigma_{ij}^+ u_j^{\sigma+} + N_{ij}^\sigma, \tag{10.115}$$

where

$$u_{ij} = \sigma_{ij} u_j \tag{10.116}$$

denotes the volume flux of species i in the jth phase.

The application of Eqs. (10.109), (10.110), and (10.115) to the liquid in the flowing suspension for $i \equiv l, j \equiv f, N_{lf}^\sigma = 0, \varepsilon_f^- = 1.0, \varepsilon_f^+ = 1 - \varepsilon_s^+$, where ε_s^+ is the volume fraction of the solids in the cake; $\sigma_{lf}^- = 1 - \sigma_p^-$, and $\sigma_{lf}^+ = 1 - \sigma_{p2}^+$ (because only the small particles can penetrate the cake) yields

$$\frac{dL}{dt} = \frac{u_{lf}^- - u_{lf}^+}{\left(1 - \sigma_p^-\right) - \left(1 - \sigma_{p2}^+\right)\left(1 - \varepsilon_s^+\right)}. \tag{10.117}$$

The application of Eqs. (10.109), (10.110), and (10.115) to the particles (small plus large) for $i \equiv p, j \equiv f, \varepsilon_f^- = 1.0, \varepsilon_f^+ = 1 - \varepsilon_s^+, \sigma_{pf}^- = \sigma_p^-$ and $\sigma_{pf}^+ = \sigma_{p2}^+$ (because only the small particles can penetrate the cake), and $N_{pf}^\sigma \equiv N_{p1f}^\sigma$ (because only the large particles deposit over the cake surface) yields

$$\frac{dL}{dt} = \frac{u_{pf}^- - u_{pf}^+ - N_{p1f}^\sigma}{\sigma_p^- - \sigma_{p2}^+\left(1 - \varepsilon_s^+\right)}. \tag{10.118}$$

Similarly, for the large particles of the cake for $i \equiv p_1, j \equiv s, \varepsilon_s^- = 0$, $\varepsilon_s^+ = \varepsilon_s^+, \sigma_{p1s}^- = 0$, and $u_{p1s}^- = 0, \sigma_{p1s}^+ = 1.0$,

$$\frac{dL}{dt} = \frac{u_{p1s}^+ + N_{p1s}^\sigma}{\varepsilon_s^+}. \tag{10.119}$$

Assuming there are less small particles than the large particles, Tien et al. (1997) neglected the small particles in the slurry, that is, $\sigma_{p2}^- \cong 0, \sigma_p^- \cong \sigma_{p1}^- \equiv \sigma^o, \varepsilon_s^+ \equiv \varepsilon_s^o, u_{lf}^- \equiv u_{lo}$, and neglected the effect of the small particles in the suspension flowing through the filter cake, that is, $\sigma_{p2}^+ \cong 0$. Consequently, Eq. (10.117) simplifies as

$$\frac{dL}{dt} \cong \frac{u_{lf}^+ - u_{lo}}{\sigma^o - \varepsilon_s^o}, \tag{10.120}$$

and the following expression can be obtained by the elimination of $N_{p1f}^\sigma = N_{p1s}^\sigma \equiv N$ between Eqs. (10.118) and (10.119), and substituting $u_{p1s} \cong u_s$ (neglecting small particles) and $u_{p1f} \equiv u_{so}$ yields

$$\frac{dL}{dt} = \frac{u_s^+ - u_{so}}{\varepsilon_s^o - \sigma^o}. \tag{10.121}$$

Therefore, equating Eqs. (10.120) and (10.121) and considering Eq. (10.96) yields

$$u_{lf}^+ + u_s^+ = u_{lf}\big|_{x=0} + \int_o^L N dx. \tag{10.122}$$

Next, Tien et al. (1997) apply Darcy's law given by Eq. (10.93) at the filter surface $(x = 0)$ (with $u_s\big|_{x=0} = 0$) and at the interface of the slurry and the cake $(x = L)$ (with $u_{lf}^+ \equiv u_{lf}\big|_{L^+}$ and $u_{lf} = \sigma_{lf} u_f$, where $\sigma_{lf} \cong 1.0$ in the suspension flowing through the cake) to obtain the following respective expressions:

$$u_{lf}\big|_{x=0} = -\left(\frac{k}{\mu}\frac{\partial p}{\partial x}\right)_{x=0} \tag{10.123}$$

and

$$u_s^+ \equiv u_s\big|_{x=L^+} = \frac{\varepsilon_s^o}{1-\varepsilon_s^o}\left[u_{lf}\big|_{x=L^+} + \left(\frac{k}{\mu}\frac{\partial p}{\partial x}\right)_{x=L^+}\right]. \tag{10.124}$$

Thus, substituting Eqs. (10.123) and (10.124) into Eq. (10.122) and rearranging gives

$$u_{lf}\big|_{x=L} = -\varepsilon_s^o\left(\frac{k}{\mu}\frac{\partial p}{\partial x}\right)_{L^+} - \left(1-\varepsilon_s^o\right)\left(\frac{k}{\mu}\frac{\partial p}{\partial x}\right)_0 + \left(1-\varepsilon_s^o\right)\int_o^L N dx. \tag{10.125}$$

Assuming that the particles and the carrier liquid move at the same velocity in the slurry, Tien et al. (1997) write

$$u_{p1f}^- \cong u_{sf}^- = \frac{\sigma^o}{1-\sigma^o}u_{lf}^- \tag{10.126}$$

Thus, substituting Eqs. (10.123) and (10.126) into Eq. (10.122) yields:

$$u_{lo} \equiv u_{lf}^- = -\left(1-\sigma^o\right)\left(\frac{k}{\mu}\frac{\partial p}{\partial x}\right)_0 + \left(1-c^o\right)\int_o^L N dx \tag{10.127}$$

Then, substituting Eqs. (10.125) and (10.127) into Eq. (10.120) results in the following cake growth expression:

$$\frac{dL}{dt} = \frac{\varepsilon_s^o}{\varepsilon_s^o - \sigma^o}\left(\frac{k}{\mu}\frac{\partial p}{\partial x}\right)_{L^+} - \left(\frac{k}{\mu}\frac{\partial p}{\partial x}\right)_O + \int_o^L N dx, \tag{10.128}$$

subject to the condition that the cake thickness is initially zero:

$$L = 0, t = 0. \tag{10.129}$$

The filter cake problem is simulated by numerically solving Eqs. (10.77), (10.90), (10.91), (10.93), (10.94), (10.98)–(10.101), (10.106), (10.107), and (10.128) simultaneously.

For illustration purposes, Tien et al. (1997) obtained the numerical solutions of the above-mentioned model for typical constant rate and constant pressure filtration cases by using the parameter values given as the following: $\varepsilon_s^o = 0.27$, $\delta = 0.49$, $\beta = 0.09$, $p_a = 1200\,\text{Pa}$, $k^o = 3.5 \times 10^{-15}\,\text{m}^2$, $\mu = 0.001\,\text{Pa·s}$, $p_o = 9.0 \times 10^5\,\text{Pa}$,

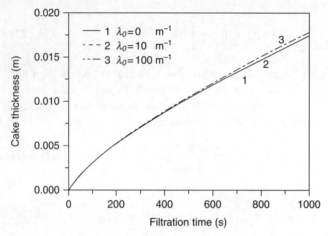

Figure 10.6 Predicted cake thickness versus time. Constant rate filtration with $n_2 = 0.05$ and three different values of λ_o (after Tien et al., 1997; © 1997 AIChE, reprinted by permission from the American Institute of Chemical Engineers).

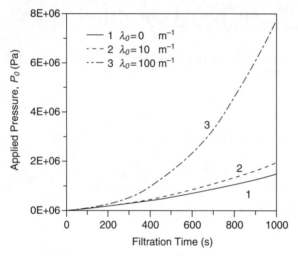

Figure 10.7 Predicted pressure requirement versus time. Constant rate filtration with $n_2 = 0.05$ and three different values of λ_o (after Tien et al., 1997; © 1997 AIChE, reprinted by permission from the American Institute of Chemical Engineers).

$q_{lm} = 2.0 \times 10^{-5}\,\text{m}^3/\text{m}^2\cdot\text{s}$, $c^o = 0.2$, $t_o = 0.1\,\text{s}$, $R = 100\,\text{m}^{-1}$, $\beta' = 5.0$, $\alpha_1 = 30$, $\alpha_2 = 1.0$, $n_{2o} = 0.05$, and $\lambda_o = 0$, 10, and 100. The predicted cake thickness and applied slurry pressure for constant rate filtration are shown in Figures 10.6 and 10.7, respectively. As can be seen, the small particle retention rate constant significantly affects the applied slurry pressure, while the filter cake thickness is not appreciable affected. The predicted cake thickness and cumulative filtrate volume are given in Figures 10.8 and 10.9, respectively, for constant pressure filtration. It is observed that the small particle retention rate constant significantly affects both the cake thickness and the cumulative filtrate volume.

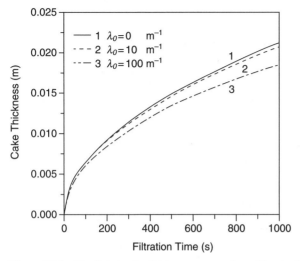

Figure 10.8   Predicted cake thickness versus time. Constant pressure filtration with $n_2 = 0.05$ and three different values of λ_o (after Tien et al., 1997; © 1997 AIChE, reprinted by permission from the American Institute of Chemical Engineers).

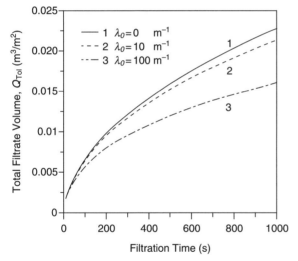

Figure 10.9   Total filtration volume versus. time. Constant pressure filtration with $n_2 = 0.05$ and three different values of λ_o (after Tien et al., 1997; © 1997 AIChE, reprinted by permission of the American Institute of Chemical Engineers).

10.4   EXERCISES

1. Derive the equations for the one-dimensional case involving laboratory core tests by expressing the equations of the nonisothermal deep-bed filtration model presented in this chapter according to Civan (2010c).

2. Applying the numerical solution scheme presented in the following according to Civan (2010c), carry out parametric sensitivity studies by varying the values of the parameters of the one-dimensional nonisothermal deep-bed filtration model in a core plug. The special numerical scheme facilitated here accomplishes the solution by means of the second-order accurate discretization of both the differential equations and their associated boundary conditions. Civan (2010c) obtained the numerical solution of the differential equations presented here by using $\Delta x = 0.001$-m grid-point spacing and $\Delta t = 5$-s time increments. Carry out numerical solutions to investigate the effect of the values of Δx and Δt on the quality of numerical results.

Consider the following general transport equation of the transient-state, advection, dispersion, and source/sink type:

$$\frac{\partial f}{\partial t} + a\frac{\partial f}{\partial x} + bf = c\frac{\partial^2 f}{\partial x^2}. \tag{10.130}$$

The initial and boundary conditions are specified as the following:

$$f = f^o, 0 \le x \le L, t = 0, \tag{10.131}$$

$$f_{in} = f - d\frac{\partial f}{\partial x}, x = 0, t > 0, \tag{10.132}$$

and

$$\frac{\partial f}{\partial x} = 0, x = L, t > 0. \tag{10.133}$$

A fully implicit solution of the above partial differential equation, first-order accurate in time and second-order accurate in space, is obtained by applying the finite difference method for discretization of Eqs. (10.130)–(10.133) as the following (Civan, 2008a, 2010c):

$$\frac{f_i^n - f_i^{n-1}}{\Delta t} + a_i\frac{f_{i+1}^n - f_{i-1}^n}{2\Delta x} + b_i f_i^n = c_i\frac{f_{i+1}^n - 2f_i^n + f_{i-1}^n}{\Delta x^2}, i = 1, 2, 3, \ldots, N, n > 0. \tag{10.134}$$

The initial and boundary conditions are discretized as the following:

$$f_i^n = f_i^o, i = 1, 2, 3, \ldots, n = 0, \tag{10.135}$$

$$f_{in} = f_i^n - d_i\frac{f_{i+1}^n - f_{i-1}^n}{2\Delta x}, i = 1, n > 0, \tag{10.136}$$

and

$$\frac{f_{i+1}^n - f_{i-1}^n}{2\Delta x} = 0, i = N, n > 0. \tag{10.137}$$

Rearranging Eqs. (10.134)–(10.137) yields the following algebraic analog equations, respectively.

The partial differential equation is represented by the following expression for the interior points:

$$-A1_i f_{i-1}^n + f_i^n - A2_i f_{i+1}^n = A3_i f_i^{n-1}, i = 2, 3, \ldots, N-1, n > 0, \tag{10.138}$$

where

$$A1_i = \frac{1}{e_i}\left(\frac{c_i}{\Delta x} + \frac{a_i}{2}\right), \quad A2_i = \frac{1}{e_i}\left(\frac{c_i}{\Delta x} - \frac{a_i}{2}\right),$$

$$A3_i = \frac{\Delta x}{\Delta t e_i}, \quad e_i = \frac{\Delta x}{\Delta t} + \Delta x b_i + \frac{2c_i}{\Delta x}.$$

(10.139)

The initial discrete point values are given by

$$f_i^n = f_i^o, i = 1, 2, 3, \ldots, N, n = 0.$$

(10.140)

The fictitious point values are expressed by the following equations:

$$f_{i-1}^n = f_{i+1}^n - \frac{2\Delta x}{d_i}\left(f_i^n - f_{in}\right), i = 1, n > 0$$

(10.141)

and

$$f_{i+1}^n = f_{i-1}^n, i = N, n > 0.$$

(10.142)

Applying Eq. (10.141) to Eq. (10.138), the partial differential equation is represented by the following expression for the inlet boundary point:

$$f_i^n - B1_i f_{i+1}^n = B2_i f_i^{n-1} + B3_i, i = 1, n > 0$$

$$E1_i = 1 + \frac{2\Delta x A1_i}{d_i}, \quad B1_i = \frac{A1_i + A2_i}{E1_i}, \quad B2_i = \frac{A3_i}{E1_i}, \quad B3_i = \frac{2\Delta x A1_i f_{in}}{d_i E1_i}.$$

(10.143)

Applying Eq. (10.142) to Eq. (10.138), the partial differential equation is represented by the following expression for the outlet boundary point:

$$-C1_i f_{i-1}^n + f_i^n = A3_i f_i^{n-1}, i = N, n > 0$$

$$C1_i = A1_i + A2_i.$$

(10.144)

Eq. (10.24) is discretized as the following and is solved explicitly:

$$\frac{\varepsilon_i^n - \varepsilon_i^{n-1}}{\Delta t} = \frac{vk\phi}{\tau}\left(\sigma_i^n - \frac{\alpha\tau}{2\Delta x}\Delta\sigma_i^n\right), i = 1, 2, 3, \ldots, N \text{ and } n > 0,$$

(10.145)

where

$$\Delta\sigma_i^n = -3\sigma_i^n + 4\sigma_{i+1}^n - \sigma_{i+2}^n, i = 1 \text{ and } n > 0,$$

(10.146)

$$\Delta\sigma_i^n = \sigma_{i+1}^n - \sigma_{i-1}^n, i = 2, 3, \ldots, N-1 \text{ and } n > 0,$$

(10.147)

and

$$\Delta\sigma_i^n = 3\sigma_i^n - 4\sigma_{i-1}^n + \sigma_{i-2}^n, i = N \text{ and } n > 0.$$

(10.148)

When the parameters a, b, c, and d are dependent upon the function value f, then the above-described numerical solution procedure is iterated until convergence starting with their values evaluated at the previous time, which is the initial time at the beginning.

3. Show that the application of Eqs. (10.109), (10.110), and (10.115) to the large particles depositing over the cake surface for $i \equiv p_1$, $j \equiv f$, $\varepsilon_f^- = 1.0$, $\varepsilon_f^+ = 1 - \varepsilon_s^+$, $\sigma_{p1f}^- = \sigma_{p1}^-$, and $\sigma_{p1f}^+ = 0$, and $q_{p1f}^+ = 0$) yields the following expression:

$$\frac{dL}{dt} = \frac{u_{p1f}^- - N_{p1f}^\sigma}{\sigma_{p1}^-}.$$

(10.149)

4. Show that the application of Eqs. (10.109), (10.110), and (10.115) leads to the following expression for the small particles for $i \equiv p_2$, $j \equiv f$, $\varepsilon_f^- = 1.0$, $\varepsilon_f^+ = 1 - \varepsilon_s^+$, $\sigma_{p2f}^- = \sigma_{p2}^-$, $\sigma_{p2f}^+ = \sigma_{p2}^+$, and $N_{p2f}^\sigma = 0$:

$$\frac{dL}{dt} = \frac{u_{p2f}^- - u_{p2f}^+}{\sigma_{p2}^- - \sigma_{p2}^+ \left(1 - \varepsilon_s^+\right)}. \tag{10.150}$$

5. Tien et al. (1997) used the applied pressure versus time given in Figure 10.7 and the cumulative filtrate volume versus time given in Figure 10.9 as a substitute for experimental data in order to illustrate the method of estimating the model parameters from experimental data. They determined the parameters of the model to minimize the difference between the measured filter cake thickness data and those predicted by the above-mentioned model. For this purpose, Tien et al. (1997) define an objective function as

$$J = \sum_{i}^{N} \left(f_i^m - f_i^p\right)^2, \tag{10.151}$$

in which $i = 1, 2, \ldots, N$ denote the data points, and f_i^m and f_i^p are the measured and predicted values, respectively, of a measurable quantity, such as the slurry application pressure necessary to maintain a constant filtration rate or the cumulative filtrate volume for constant pressure slurry applications, used here. The best estimates of the model parameters obtained by an optimization method to minimize the objective function given by Eq. (10.151) were determined to be very close to the assumed parameter values given earlier, which were used to generate the numerical solutions, substituted for experimental data. Carry out a similar exercise to determine the best estimate values of the parameters.

CHAPTER 11

TRANSPORT IN HETEROGENEOUS POROUS MEDIA

11.1 INTRODUCTION

Naturally fissured or fractured porous media are frequently encountered in subsurface geological formations, including groundwater, geothermal, and petroleum reservoirs (Moench, 1984; Chang, 1993).* Most fractured formations are anisotropic and heterogeneous systems, characterized by a network of intersecting fractures partitioning the porous matrix into various regions, as depicted in Figure 11.1 (Civan and Rasmussen, 2005). For convenience in describing transport through such complex porous media, Barenblatt et al. (1960) introduced the double-porosity realization. In this approach, the fractures and porous matrix are envisioned as two separate but overlapping continua, interacting through the matrix–fracture interface (Moench, 1984; Zimmerman et al., 1993). Usually, the porous matrix, referred to as the primary continuum, is of a low-permeability and high-pore volume region, while the fracture, referred to as the secondary continuum, is of a high-permeability and

---

* Parts of this chapter have been reproduced with modifications from the following:

Civan, F. 1998c. Quadrature solution for waterflooding of naturally fractured reservoirs. SPE Reservoir Evaluation & Engineering Journal, 1(2), p. 141–147, © 1998 SPE, with permission from the Society of Petroleum Engineers;

Civan, F. and Rasmussen, M.L. 2001. Asymptotic analytical solutions for imbibition water floods in fractured reservoirs. SPE Journal, 6(2), pp. 171–181, © 2001, with permission from the Society of Petroleum Engineers;

Civan, F., Wang, W., and Gupta, A. 1999. Effect of wettability and matrix-to-fracture transfer on the waterflooding in fractured reservoirs. Paper SPE 52197, 1999 SPE Mid-Continent Operations Symposium (March 28–31, 1999), Oklahoma City, OK, © 1999 SPE, with permission from the Society of Petroleum Engineers;

Rasmussen, M.L. and Civan, F. 1998. Analytical solutions for water floods in fractured reservoirs obtained by an asymptotic approximation. SPE Journal, 3(3), 249–252, © 2001 SPE, with permission from the Society of Petroleum Engineers; and

Rasmussen, M.L. and Civan, F. 2003. Full, short, and long-time analytical solutions for hindered matrix-fracture transfer models of naturally fractured petroleum reservoirs. Paper SPE 80892, SPE Mid-Continent Operations Symposium (March 22–25, 2003), Oklahoma City, OK, © 2003 SPE, with permission from the Society of Petroleum Engineers.

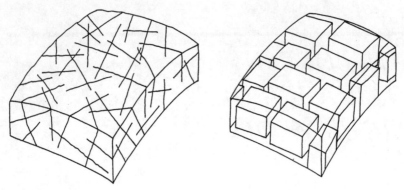

Figure 11.1 Representation of naturally fractured porous media (Civan and Rasmussen, 2005; © 2005 SPE, reproduced by permission of the Society of Petroleum Engineers).

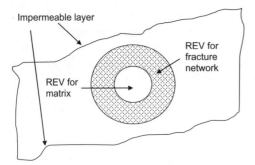

Figure 11.2 Schematic naturally fractured porous media (prepared by the author with modifications after Moench, 1984).

low-pore volume region in most fractured porous media (Moench, 1984; Chang, 1993; Zimmerman et al., 1993). The macroscopic description of transport in these two different continua requires different orders of magnitude representative elementary volumes (REVs). As depicted in Figure 11.2, the REV for the fracture network should be significantly larger than that for the porous matrix (Moench, 1984). However, the majority of the double-porosity models consider the larger of the REVs to define the volume averaged, that is, the effective properties of the various constituents and properties in both continua. For convenience in the mathematical modeling, Warren and Root (1963) resorted to a representation of naturally fractured porous media by an idealized and simplified geometrical model, referred to as the sugar cube model, as depicted in Figure 11.1.

Panfilov (2000) distinguished two possible types of elementary flow patterns in a cell, referred to as the translation or flow through matrix and source flow or flow around matrix, respectively, as depicted in Figure 11.3a,b. The translation flow occurs when the fracture and matrix media permeability are comparable with each other. Then, flows through both media are considered. Consequently, this approach requires two sets of similar equations for the naturally fractured porous media. The source flow occurs when the fracture medium permeability is significantly greater

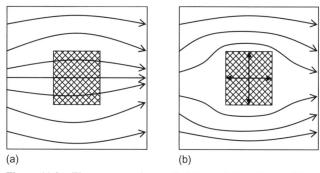

(a)   (b)

Figure 11.3   Flow patterns in a cell: (a) translation flow and (b) source flow (modified after Panfilov, 2000).

than the matrix medium permeability. Then, the fracture medium forms the preferential flow path, while the matrix medium acts as a source/sink for the fracture medium (Civan, 1998c; Rasmussen and Civan, 1998; Civan and Rasmussen, 2001). As a result, equations for the fracture medium are sufficient to describe the flow through naturally fractured porous media. Because the latter case prevails in most fractured reservoirs, the fracture–flow and matrix–source/sink formulation approach received the most attention. Ordinarily, the numerical solution of both the translation and source flow approaches requires the spatial discretization of both the fracture and matrix media, as demonstrated by Pruess and Narasimhan (1985). However, for practical reasons, the source/sink flow approach based on the sugar cube model has been frequently solved by a semianalytical method. This is based on prescribing the interchange or transport between the matrix and fracture media by a suitable analytical expression to be used as a source/sink function for the fracture flow equation, as described in the following sections.

In the following sections, a number of important topics are discussed, including transport units and transport in heterogeneous porous media, models for transport in fissured/fractured porous media, species transport in fractured porous media, immiscible displacement in naturally fractured porous media, and numerical solutions by the methods of weighted sums (quadrature) and finite difference.

11.2   TRANSPORT UNITS AND TRANSPORT IN HETEROGENEOUS POROUS MEDIA

11.2.1   Transport Units

The properties of heterogeneous porous media are spatially different. For convenient realizations and mathematical description of transport processes, heterogeneous porous media can be partitioned into various transport units, each involving transport of different types and orders of magnitude rates. Such models have been variably referred to by different names. For example, the frequently used models of macropores–micropores (Bai et al., 1995), double or dual porosity (Coats, 1989; Bai and Civan, 1998a,b), matrix–fracture (Warren and Root, 1963; Rasmussen and

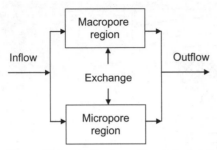

Figure 11.4 Realization of heterogeneous porous media by interacting micropore and macropore regions (prepared by the author).

Civan, 1998c; Civan and Rasmussen, 2001), and plugging–nonplugging paths (Gruesbeck and Collins, 1982; Civan and Nguyen, 2005) are based on a realization of heterogeneous porous media in two distinct scales of transport units. Different transport units are treated as mutually interacting continua (Valliappan et al., 1998) as schematically depicted in Figure 11.4 for a two-transport unit partitioning of heterogeneous porous media.

The definition of proper transport units in prescribed heterogeneous porous media depends on the specific applications and nature of porous media. For example, naturally fractured porous media can be analyzed by considering porous matrix interacting with fractures. Granular porous media usually display a bimodal pore size distribution and therefore, such media can be viewed as having the micropore and macropore regions interacting with each other by the exchange process. However, applying the Kozeny equation, the transport or flow units in heterogeneous porous media can be more adequately distinguished in terms of a quality index (Amaefule et al., 1993). The quality index is best defined by the mean hydraulic flow path diameter D_h in porous media (Civan, 2007a):

$$QI \equiv D_h = 4\sqrt{2\tau_h}\sqrt{K/\phi}, \qquad (11.1)$$

where τ_h is the tortuosity of preferential flow paths, K is permeability, and ϕ is porosity.

The volume fractions of the various transport units, denoted by f_j: $j = 1, 2, ...,$ N, are a characteristic of heterogeneous porous media. By definition of fractions,

$$\sum_{j=1}^{N-\text{units}} f_j = 1.0. \qquad (11.2)$$

11.2.2 Sugar Cube Model of Naturally Fractured Porous Media

The sugar cube model is one of the most frequently used heterogeneous porous media models introduced by Barenblatt et al. (1960) and Warren and Root (1963) for the realization of naturally fractured porous media. As depicted in Figure 11.1, the naturally fractured porous media are characterized in terms of a network of intersect-

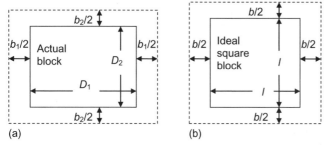

(a)   (b)

Figure 11.5   Representing (a) a rectangular block by (b) an idealized square block (prepared by the author).

ing fractures, separating the porous media into a number of matrix blocks (Warren and Root, 1963; Chang, 1993).

Consider the representative elemental volume shown in Figure 11.5a. D_1 and D_2 represent the fracture spacing. b_1 and b_2 denote the aperture widths for the fractures, and Γ_1 and Γ_2 denote the exchange rates between the matrix block and fracture in the x- and y-directions, respectively. Therefore, the following expressions for the representative elemental volume shown in Figure 11.5a are given.

The cross-sectional area A and perimeter P of the matrix block are given, respectively, by

$$A = l^2 = D_1 D_2 \tag{11.3}$$

and

$$P/2 = l + l = D_1 + D_2. \tag{11.4}$$

Hence, the following expressions can be derived using Eqs. (11.3) and (11.4):

$$l = \frac{D_1 D_2}{l} = \frac{D_1 + D_2}{2} \tag{11.5}$$

Thus,

$$l = \frac{2 D_1 D_2}{D_1 + D_2}. \tag{11.6}$$

The pore and bulk volumes are given, respectively, by

$$V_{PV} = (D_2 + b_2) b_1 + D_1 b_2 \tag{11.7}$$

and

$$V_{BV} = (D_1 + b_1)(D_2 + b_2). \tag{11.8}$$

Hence, using Eqs. (11.7) and (11.8), the fracture porosity is given by

$$\phi_f = \frac{V_{PV}}{V_{BV}} = \frac{(D_2 + b_2) b_1 + D_1 b_2}{(D_1 + b_1)(D_2 + b_2)}. \tag{11.9}$$

When $b_1 \ll D_1$ and $b_2 \ll D_2$, Eq. (11.9) simplifies as

$$\phi_f \cong \frac{D_2 b_1 + D_1 b_2}{D_1 D_2} = \frac{b_1}{D_1} + \frac{b_2}{D_2}. \tag{11.10}$$

If the same conditions are applied for $b_1 = b_2 = b$ (Fig. 11.5b), the result becomes (Valliappan et al., 1998):

$$\phi_f \cong \left(\frac{D_2 + D_1}{D_1 D_2} \right) b = \left(\frac{1}{D_1} + \frac{1}{D_2} \right) b. \tag{11.11}$$

11.3 MODELS FOR TRANSPORT IN FISSURED/FRACTURED POROUS MEDIA

The commonly used models are reviewed in the following.

11.3.1 Analytical Matrix–Fracture Interchange Transfer Functions

Derivation of the matrix–fracture interchange transfer functions has occupied many researchers, including Warren and Root (1963), Kazemi et al. (1976), Moench (1984), and Zimmerman et al. (1993). As pointed out by Moench (1984) and Zimmerman et al. (1993), the various approaches consider either pseudo-steady-state or transient-state conditions for describing the internal flow of the matrix medium. Moench (1984) claims that well test data support the presence of flows both at pseudo-steady-state or transient-state conditions. Most approaches, including those of Warren and Root (1963), Kazemi et al. (1976), and Lim and Aziz (1995), have utilized constant fracture fluid pressure along the matrix–fracture interface, referred to as the Dirichlet boundary condition. Moench (1984) considered the resistance to flow at the fracture–matrix interface owing to various reasons, including mineral deposition and alteration, and thus introduced the fracture skin concept. Consequently, Moench (1984) applied a Cauchy-type boundary condition. The mathematical fundamentals of these approaches are described in the following according to Moench (1984), Chang (1993), and Zimmerman et al. (1993).

Referring to Figure 11.6, describing a source flow, the flow inside a matrix block of a, b, and c dimensions in the x-, y-, and z-Cartesian coordinates is described by the equation of continuity and Darcy's law. Assuming a single-phase, constant viscosity and a slightly compressible fluid in a matrix block having constant porosity ϕ_m and anisotropic permeability K_{mx}, K_{my}, and K_{mz}, and a constant total compressibility c_m yields the following diffusion equation for the fluid pressure:

$$\phi_m c_m \frac{\partial p_m}{\partial t} = \nabla \cdot \left(\frac{1}{\mu} \mathbf{K}_m \cdot \nabla p_m \right), x, y, z \in V_m, t > 0, \tag{11.12}$$

where $\mathbf{K}_m$ and ϕ_m denote the permeability tensor and porosity of the matrix, and V_m denotes the volume of the matrix. Initially, consider that the fluid pressure throughout the matrix block is uniform:

$$p_m = p_i, x, y, z \in V_m, t = 0. \tag{11.13}$$

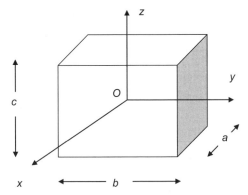

Figure 11.6   Rectangular parallelepiped shape matrix block (prepared by the author).

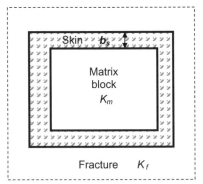

Figure 11.7   Matrix block with fracture skin (prepared by the author with modifications after Moench, 1984).

Along the outer surfaces of the matrix block, consider the following Cauchy-type boundary condition, accounting for the fracture skin as illustrated in Figure 11.7 (Moench, 1984):

$$u_n = -\frac{K_{mn}}{\mu}\frac{\partial p_m}{\partial n} = \frac{K_s}{\mu}\left(\frac{p_m - p_f}{b_s}\right), \quad A_m, t > 0, \tag{11.14}$$

where n denotes the outward normal direction at the matrix surface, K_s and b_s denote the permeability and thickness of the skin, p_f is the fracture fluid pressure assumed constant, and A_m denotes the outer surface area of the matrix.

The total flow rate through the matrix surfaces is given by (Duguid and Lee, 1977)

$$q = \frac{-1}{V_m}\int_{A_m}\frac{K_{mn}}{\mu}\frac{\partial p_m}{\partial n}\,dA, \quad t > 0. \tag{11.15}$$

As pointed out by Moench (1984), the boundary condition given by Eq. (11.14) adequately justifies the validity of neglecting the divergence of flow to drive the pseudo-steady-state model. $K_s \to \infty$ in the absence of the skin effect and, therefore, Eq. (11.14) simplifies to

$$p_m = p_f, A_m, t > 0. \tag{11.16}$$

In the following section, a number of special solutions to Eqs. (11.12)–(11.16) are presented, with modifications for the consistency with the rest of the presentation of this chapter.

11.3.2 Pseudo-Steady-State Condition and Constant Fracture Fluid Pressure over the Matrix Block: The Warren–Root Lump-Parameter Model

As described by Zimmerman et al. (1993), the Warren and Root (1963) model treats the matrix blocks in a lump-parameter form. Thus, integrating Eq. (11.12) over the matrix block yields

$$\phi_m c_m \frac{\partial}{\partial t} \int_{V_m} p_m dV = \int_{A_m} \frac{K_{mn}}{\mu} \frac{\partial p_m}{\partial n} dA, t > 0, t > 0. \tag{11.17}$$

Next, define the average fluid pressure within the matrix block as

$$\bar{p}_m = \frac{1}{V_m} \int_{V_m} p_m dV. \tag{11.18}$$

Hence, invoking Eqs. (11.15) and (11.18) into Eq. (11.16) results in the following equation:

$$\phi_m c_m \frac{d\bar{p}_m}{dt} = -q, \tag{11.19}$$

where q $[L^3/T]$ denotes the total volumetric flow rate of the fluid across the matrix surfaces. Eq. (11.19) is often referred to as the pseudo-steady-state assumption (Dake, 1978). A simplistic approach similar to Kazemi et al. (1976) to estimating this term is to apply the right of Eq. (11.14) in Eq. (11.15) using the average fluid pressure placed at the center of the matrix block, $p_m = \bar{p}_m$, substituting $K_s \equiv K_{mn}$ in the normal direction and taking the distance from the matrix center to the matrix surface for b_s. This leads to the following expression:

$$q = \frac{1}{V_m} \left[2\frac{K_{mx}A_x}{\mu(a/2)} + 2\frac{K_{my}A_y}{\mu(b/2)} + 2\frac{K_{mz}A_z}{\mu(c/2)} \right] (\bar{p}_m - p_f), \tag{11.20}$$

where A_x, A_y, and A_z represent the surface areas of the matrix block normal to the x-, y-, and z-Cartesian directions, given by

$$A_x = bc, A_y = ac, A_z = ab. \tag{11.21}$$

The volume of the block is

$$V_m = abc. \tag{11.22}$$

Also define a geometric average permeability for the anisotropic porous matrix as (Muskat, 1937)

$$\bar{K}_m = \left(K_{mx}K_{my}K_{mz}\right)^{1/3}. \tag{11.23}$$

As a result, substituting Eqs. (11.21)–(11.23) into Eq. (11.20) yields the following transfer function:

$$q = \sigma \frac{\bar{K}_m}{\mu}\left(\bar{p}_m - p_f\right), \tag{11.24}$$

where the parameter σ is called the shape factor, given by

$$\sigma = \frac{4}{\bar{K}_m}\left(\frac{K_{mx}}{a^2} + \frac{K_{my}}{b^2} + \frac{K_{mz}}{c^2}\right). \tag{11.25}$$

Eq. (11.25) simplifies to the shape factor of Kazemi et al. (1976) for an isotropic porous matrix. Eq. (11.25) implies that the shape factor σ has a reciprocal area dimension $[L^{-2}]$. Invoking Eq. (11.24) into Eq. (11.19) yields

$$\phi_m c_m \frac{d\bar{p}_m}{dt} = -\sigma \frac{\bar{K}_m}{\mu}\left(\bar{p}_m - p_f\right). \tag{11.26}$$

A solution of Eq. (11.26) subject to the initial condition given by Eq. (11.13) yields a decay function for the matrix average fluid pressure as

$$\frac{\bar{p}_m - p_i}{p_f - p_i} = 1 - \exp\left(-\frac{\sigma \bar{K}_m t}{\mu \phi_m c_m}\right). \tag{11.27}$$

Substituting Eq. (11.27) into Eq. (11.19) yields the following expression for the total volumetric flow rate of the fluid across the matrix block surfaces per unit matrix block volume:

$$q = -\phi_m c_m \frac{d\bar{p}_m}{dt} = -(p_f - p_i)\left(\frac{\sigma \bar{K}_m}{\mu}\right)\exp\left(-\frac{\sigma \bar{K}_m t}{\mu \phi_m c_m}\right). \tag{11.28}$$

11.3.3 Transient-State Condition and Constant Fracture Fluid Pressure over the Matrix Block

As demonstrated by Coats (1989), Chang (1993), and Lim and Aziz (1995), an analytical solution for Eq. (11.12), subject to Eqs. (11.13) and (11.16), can be derived by applying the separation of variables method for the transient-state pressure distribution for the fluid present in the matrix block. Then, integrating this analytical solution over the matrix block according to Eq. (11.18) yields (Lim and Aziz, 1995)

$$\frac{\bar{p}_m - p_i}{p_f - p_i} = 1 - \left(\frac{8}{\pi^2}\right)^3 \sum_{l=0}^{\infty}\sum_{m=0}^{\infty}\sum_{n=0}^{\infty} A\exp\left(-B\frac{\pi^2 t}{\mu \phi_m c_m}\right), \tag{11.29}$$

where

$$A = \frac{1}{(2l+1)^2 (2m+1)^2 (2n+1)^2} \qquad (11.30)$$

and

$$B = \frac{K_{mx}}{a^2}(2l+1)^2 + \frac{K_{my}}{b^2}(2m+1)^2 + \frac{K_{mz}}{c^2}(2n+1)^2. \qquad (11.31)$$

As a first-order approximation to estimating the shape factor, Lim and Aziz (1995) considered only the first terms in the summation series given in Eq. (11.29), leading to

$$\frac{\bar{p}_m - p_i}{p_f - p_i} \cong 1 - \left(\frac{8}{\pi^2}\right)^3 \exp\left[-\left(\frac{K_{mx}}{a^2} + \frac{K_{my}}{b^2} + \frac{K_{mz}}{c^2}\right)\frac{\pi^2 t}{\mu \phi_m c_m}\right]. \qquad (11.32)$$

Then, substituting Eq. (11.32) into Eq. (11.26) yields the shape factor expression of Lim and Aziz (1995) as

$$\sigma = \frac{\pi^2}{\bar{K}_m}\left(\frac{K_{mx}}{a^2} + \frac{K_{my}}{b^2} + \frac{K_{mz}}{c^2}\right). \qquad (11.33)$$

Consequently, the total volumetric flow across the matrix block surfaces can be obtained by substituting Eq. (11.33) into Eq. (11.28). However, a direct substitution of Eq. (11.29) into Eq. (11.26) yields a time-dependent expression for the shape factor as (Coats, 1989; Chang, 1993)

$$\sigma = \frac{\pi^2}{\bar{K}_m}\frac{\sum\limits_{l=0}^{\infty}\sum\limits_{m=0}^{\infty}\sum\limits_{n=0}^{\infty} AB\exp\left(-B\dfrac{\pi^2 t}{\mu \phi_m c_m}\right)}{\sum\limits_{l=0}^{\infty}\sum\limits_{m=0}^{\infty}\sum\limits_{n=0}^{\infty} A\exp\left(-B\dfrac{\pi^2 t}{\mu \phi_m c_m}\right)}. \qquad (11.34)$$

11.3.4 Single-Phase Transient Pressure Model of de Swaan for Naturally Fractured Reservoirs

In the following, the formulation and solution of the model by de Swaan (1990) are presented in a manner consistent with the rest of the chapter.

Consider an elementary cell volume of the sugar cube model of naturally fractured porous media given in Figure 11.1. Denote the representative elementary cell volume by V_e, and the volumes of the fracture and matrix block by V_f and V_m, respectively. The permeability of the fracture and matrix block is represented by K_f and K_m, the porosities by ϕ_f and ϕ_m, and the total compressibility by c_f and c_m, respectively.

The effective permeability of the fracture medium can be estimated by

$$K_f = K_f^*/\tau_f, \qquad (11.35)$$

where K_f^* is the intrinsic permeability and τ_f is the tortuosity of the fractures surrounding the matrix block in an elementary cell.

The elementary cell volume and its total pore volume are given, respectively, by

$$V_e = V_f + V_m \tag{11.36}$$

and

$$V_p = V_f \phi_f + V_m \phi_m. \tag{11.37}$$

Note that the volume fraction of the interconnected fractures can be reduced for various reasons, including the presence of deposits inside the fractures and isolation and/or connecting to dead-end vugs.

Considering the fracture–flow and matrix–source/sink approach for modeling, the fracture medium pressure diffusivity equation is given by

$$\phi_f c_f \frac{\partial p_f}{\partial t} + r = \frac{1}{\mu} K_f \nabla^2 p_f, t > 0, \tag{11.38}$$

where r denotes the mass rate of fluid lost from the fracture to matrix per unit fracture medium volume, given by

$$r = \frac{V_m \phi_m c_m}{V_f} q_{p_f = p_f(t)}. \tag{11.39}$$

Applying Duhamel's theorem, de Swaan (1990) corrected the matrix–fracture interface flow rate assuming constant fracture fluid pressure $p_f = $ ct. to account for the variable fracture fluid pressure $p_f = p_f(t)$ as

$$q_{p_f = p_f(t)} = \int_0^t q(t-\tau)_{p_f=\text{ct.}} \frac{\partial p_f}{\partial \tau} \partial \tau. \tag{11.40}$$

Thus, combining Eqs. (11.38)–(11.40) yields the following fracture fluid diffusivity equation:

$$\phi_f c_f \frac{\partial p_f}{\partial t} + \frac{V_m \phi_m c_m}{V_f} \int_0^t q(t-\tau)_{p_f=\text{ct.}} \frac{\partial p_f}{\partial \tau} \partial \tau = \frac{1}{\mu} K_f \nabla^2 p_f, t > 0. \tag{11.41}$$

The initial condition for radial flow around a well is given by

$$p_f = p_{fi}, 0 \leq r < \infty, t = 0. \tag{11.42}$$

The boundary condition at the wellbore is given by

$$q_w = \frac{\mu B}{A_w K_f} \frac{\partial p_f}{\partial r}, A_w = 2\pi r_w h, r = r_w, t > 0. \tag{11.43}$$

The boundary condition at a sufficiently long distance from the well, where the pressure remains undisturbed, is given by

$$p_f = p_{f\infty}, r \to \infty, t > 0. \tag{11.44}$$

As demonstrated by de Swaan (1990), an analytical solution for Eqs. (11.41)–(11.44) can be derived by using the matrix–fracture function given by Eq. (11.28) based on the model of Warren and Root (1963) and then by applying a Laplace transformation.

11.4 SPECIES TRANSPORT IN FRACTURED POROUS MEDIA

The description of flow and species transport in fractured porous media has been of practical importance for various purposes, including environmental contaminant pollution modeling (Valliappan et al., 1998). In this chapter, the formulation by Valliappan et al. (1998) is presented with modifications for consistency with the rest of the material presented in this book. This approach considers the fractured porous medium to comprise the fracture network and the porous matrix as two separate continua interacting through a matrix–fracture transfer term.

The total mass and species–mass balance equations are given, respectively, by

$$\frac{\partial}{\partial t}(\phi\rho) + \nabla \cdot (\rho u) = \phi\dot{m} \tag{11.45}$$

and

$$\rho\left(\phi\frac{\partial w_A}{\partial t} + \mathbf{u} \cdot \nabla w_A\right) = \nabla \cdot (\phi\rho\mathbf{D} \cdot \nabla w_A) + \phi(\dot{m}_A - w_A\dot{m}), \tag{11.46}$$

where $\dot{m}$ and $\dot{m}_A$ denote the mass rate of fluid and species A added per unit volume, respectively; $\mathbf{D}$ is the molecular diffusion coefficient tensor. Considering density variation by species concentration and pressure, and porosity variation by pressure, Eq. (11.45) can be expanded as

$$\phi\frac{\partial \rho}{\partial t} + \rho\frac{\partial \phi}{\partial t} + \nabla \cdot (\rho u) = \phi\dot{m}. \tag{11.47}$$

The volumetric flux of fluid is given by Darcy's law:

$$\mathbf{u} = -\frac{\rho}{\mu}\mathbf{K} \cdot \nabla\Psi, \tag{11.48}$$

where Ψ denotes the flow potential defined by

$$\Psi = \int_{P_o}^{P} \frac{dp}{\rho} + g(z - z_o). \tag{11.49}$$

Next, consider that the fluid density is a function of the pressure and concentration at isothermal conditions; that is,

$$\rho = \rho(p, w_A). \tag{11.50}$$

Hence, it follows that

$$dp = \frac{\partial \rho}{\partial p} dp + \frac{\partial \rho}{\partial w_A} dw_A. \tag{11.51}$$

A Taylor series expansion around a reference state, such as no species present and 1-atm pressure conditions, yields

$$\rho = \rho_o + \frac{\partial \rho}{\partial p}\bigg|_o (p - p_o) + \frac{\partial \rho}{\partial w_A}\bigg|_o (w_A - w_{Ao}) + \text{higher-order terms.} \tag{11.52}$$

In addition, assuming a slightly compressible fluid and porous matrix, the coefficients of compressibility can be defined by

$$c_p = \frac{1}{\rho} \frac{\partial \rho}{\partial p} \tag{11.53}$$

and

$$c_\phi = \frac{1}{\phi} \frac{\partial \phi}{\partial p}. \tag{11.54}$$

Also, define a coefficient of density variation by species concentration as

$$c_{w_A} = \frac{1}{\rho} \frac{\partial \rho}{\partial w_A}. \tag{11.55}$$

The total compressibility coefficient is given by

$$c_t = c_\phi + c_p. \tag{11.56}$$

Therefore, in view of Eqs. (11.48)–(11.56), Eqs. (11.47) and (11.46) can be expressed, respectively, as

$$\nabla \cdot \left(\frac{\rho}{\mu} \mathbf{K} \cdot \nabla \Psi \right) = \rho \phi c_t \frac{\partial p}{\partial t} + \rho \phi c_{w_A} \frac{\partial w_A}{\partial t} - \phi \dot{m} \tag{11.57}$$

and

$$\nabla \cdot (\phi \rho \mathbf{D} \cdot \nabla w_A) - \rho \mathbf{u} \cdot \nabla w_A = \rho \phi \frac{\partial w_A}{\partial t} - \phi \dot{m}_A + \phi w_A \dot{m}. \tag{11.58}$$

Eq. (11.58) assumes that all the species A present in the solution are available for transport. This assumption is valid if the porous material is inert and therefore does not react with species A in the solution. However, most porous materials tend to interact with the fluid media in various forms, including adsorption/desorption, absorption/desorption, deposition/dissolution, and chemical reaction. Such effects can be accounted for by a retardation factor (Neretnieks, 1980; Huyakorn et al., 1983), given by

$$R = 1 + \left(\frac{1 - \phi}{\phi} \right) \rho_s k_d, \tag{11.59}$$

where ρ_s denotes the density of the solid material of the porous media and k_d is a species distribution coefficient.

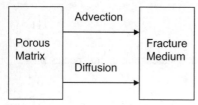

Figure 11.8 Matrix–fracture interaction by advection and diffusion (prepared by the author).

The fluid added to the fracture medium comes through the matrix–fracture interface from the matrix medium (Fig. 11.8). The mass rate of fluid Γ transferred from the matrix-to-fracture media can be expressed by

$$\Gamma = \phi_f \dot{m}_f = -\phi_m \dot{m}_m = \rho_m \sigma \frac{K_m}{\mu}(\Psi_m - \Psi_f), \qquad (11.60)$$

where σ denotes the shape factor. ρ_m is the fluid density.

The mass rate of species A Γ_A added to the fracture medium (Fig. 11.8) is the sum of the rate of mass diffusing from the matrix–fracture interface into the fracture medium, the species A carried by the fluid transfer from matrix to fracture through the matrix–fracture interface, and the species A lost by chemical reactions within the fracture medium, expressed by

$$\Gamma_A = \phi_f \dot{m}_{Af} = \bar{\alpha}\phi_m D_m(w_{Am} - w_{Af}) + \Gamma[\theta w_{Af} + (1-\theta)w_{Am}] - \lambda R_f \phi_f w_{Af}, \qquad (11.61)$$

where $\bar{\alpha}$ is an empirical coefficient. θ is a flow direction parameter, whose value is zero when the matrix–fracture interface fluid flow is from the matrix into the fracture media, and unity otherwise. D is the molecular diffusion coefficient. λ is a rate coefficient for the first-order decay reaction of species A.

Consequently, in view of the above-mentioned discussion, Eq. (11.58) should be modified as

$$\nabla \cdot (\phi\rho\mathbf{D} \cdot \nabla w_A) - \rho\mathbf{u} \cdot \nabla w_A = R\rho\phi \frac{\partial w_A}{\partial t} + \Gamma_A + w_A\Gamma. \qquad (11.62)$$

11.5 IMMISCIBLE DISPLACEMENT IN NATURALLY FRACTURED POROUS MEDIA

For example, an application of immiscible displacement in naturally fractured porous media is waterflooding. Incidentally, this is one of the economically viable techniques for recovery of additional oil following the primary recovery. However, the applications to naturally fractured reservoirs have certain challenges from the points of accurately describing the mechanism of oil recovery and flow of fluids in naturally fractured formations and the efficient numerical solution of the resulting equations. Parts of the following formulation of immiscible displacement in naturally fractured porous media are presented from Civan (1998c), Civan et al. (1999), Civan and Rasmussen (2001), and Rasmussen and Civan (1998).

In general, the multiple-porosity modeling approaches proposed for naturally fractured reservoirs are highly computationally intensive or impractical for large-scale applications. Therefore, in search for a more practical approach, Kazemi et al. (1992) have adopted the single porosity with source modeling approach by deSwaan (1978) and have demonstrated that this approach is advantageous over the frequently used multiple-porosity approaches because only the solution of the fracture flow problem is required. Basically, deSwaan described the immiscible displacement process in fractured porous media by adding a source term to the conventional Buckley–Leverett equation to represent the matrix-to-fracture oil transport by an empirical function given by Aronofsky et al. (1958). Kazemi et al. (1992) extended this function into a multiparameter empirical function. deSwaan's approach is applicable when the fracture permeability is much greater than the matrix permeability, so that the fractures provide the preferential paths for flow and the matrix becomes the source of oil for the fractures. The resulting transport equation is an integro-differential equation.

Most researchers including Kazemi et al. (1992) neglected the capillary pressure effect in the modeling of flow through fractures. However, Gilman (1983) and Pruess and Tsang (1989) considered that the capillary pressure effect should be included. Civan (1993) extended the modeling approach by deSwaan (1978) to include the gravity and capillary pressure effects. Civan (1993, 1994a, 1998c) and Gupta and Civan (1994a) derived two- and three-exponent matrix-to-fracture transfer functions based on a proposed mechanism of oil transfer from matrix to fracture to incorporate into the deSwaan (1978) model. The analytical and numerical solutions of the model were obtained by various researchers.

11.5.1   Correlation of the Matrix-to-Fracture Oil Transfer

The permeability of the fracture system in most naturally fractured reservoirs is much greater than that of the porous matrix, and therefore the fractures form the preferential flow paths while the matrix acts as the source of oil for fractures (Civan, 1998c). Then, modeling oil recovery by waterflooding of naturally fractured reservoirs by the fracture porosity and matrix source approach provides a computationally convenient method (Aronofsky et al., 1958; deSwaan 1978). The accuracy of this method, however, depends on the implementation of a properly defined matrix-to-fracture transfer function in the Buckley–Leverett formulation. For this purpose, Aronofsky et al. used a one-parameter empirical function obtained by correlation of the cumulative matrix-to-fracture oil transfer given by

$$V_f(t) = V_\infty \left(1 - e^{-\lambda t}\right), \tag{11.63}$$

where λ is an empirical constant and V_∞ is the volume of the movable oil initially available in the porous matrix, measured per unit bulk volume, although they recommended adding more exponential terms for better correlation of the experimental data. Hence, Kazemi et al. (1992) proposed an empirical function composed of an infinite series of exponential terms, given by

$$V_f(t) = V_\infty \left(1 - \sum_{j=1}^{\infty} a_j e^{-\lambda_j t} \right).$$ (11.64)

Civan (1993, 1998c) and then Gupta and Civan (1994a) perceived naturally fractured porous media as having primary fracture porosity and secondary matrix porosity. The porous structure of the matrix consists of interconnected and dead-end pores (see Fig. 11.9). Because the fractures have relatively larger permeability than the matrix, the flow is considered to occur essentially through the fracture network, and the matrix feeds oil into the fractures owing to the imbibition of water and thus acts as a source. The imbibition of water into the matrix causes oil to discharge into the fractures. The oil existing in the dead-end pores passes into the interconnected pores and then to the matrix–fracture interface to accumulate over the fracture face, where it is entrained and removed by the fluid system flowing through the fracture. Civan (1993, 1998c) and then Gupta and Civan (1994a) derived two- and then three-exponential matrix-to-fracture transfer functions, respectively, based on the principle that dynamic processes occur at rates proportional to the governing driving forces. The proportionality factors are called the rate constants. Thus, for the present case, the oil transfers at various points of naturally fractured formations were assumed to occur at rates proportional to the oil available at those sites. Thus, Gupta and Civan (1994a) theoretically derived and verified by experimental data that a maximum of three-exponential terms is sufficient for accurate phenomenological representation of the oil transfer from matrix to fracture. However, the contribution of the dead-end pores varies for different types of porous media as indicated in the following example.

It is perceived that the oil expelled from the matrix by water imbibition accumulates over the fracture surface as oil droplets attach to the fracture surface, and then these droplets are entrained by the fluid system flowing through the fracture medium (Fig. 11.9). Thus, the transfer of oil from matrix to fracture is considered to occur through three irreversible rate processes in series as (Civan, 1993; Gupta and Civan, 1994a)

$$\alpha V_d \rightarrow \beta V_n \rightarrow \gamma V_i \rightarrow V_f.$$ (11.65)

Gupta and Civan (1994a) included the effect of the dead-end pores by expanding Civan's (1993) rate equations as the following:

$$dV_d/dt = -\lambda_3 V_d^\alpha,$$ (11.66)
$$dV_n/dt = \lambda_3 V_d^\alpha - \lambda_1 V_n^\beta,$$ (11.67)
$$dV_i/dt = \lambda_1 V_n^\beta - \lambda_2 V_i^\gamma,$$ (11.68)

and

$$dV_f/dt = \lambda_2 V_i^\gamma,$$ (11.69)

where V_d, V_n, V_i, and V_f denote the cumulative volumes of oils remaining in the dead-end and interconnected pores of the matrix, accumulating over the fracture surface as oil droplets, and entrained by the fracture–medium fluid, respectively, expressed per unit bulk volume of porous media (see Fig. 11.10). λ_1, λ_2, and λ_3 are empirically

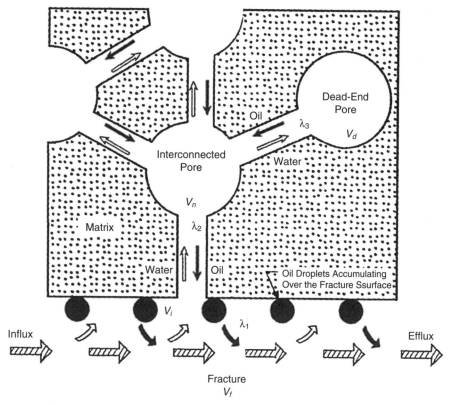

Figure 11.9 Imbibition-induced oil recovery from a porous media containing interconnected and dead-end pores and natural fractures (after Civan and Rasmussen, 2001; © 2001 SPE, reproduced with permission from the Society of Petroleum Engineers).

determined rate constants, and $\alpha, \beta,$ and γ represent the orders of the governing transfer processes.

The underlying physics for Eqs. (11.66)–(11.69) is that the rate at which a dynamic process occurs is proportional to the pertinent driving forces, and the proportionality factor is the rate constant. Eq. (11.66) expresses that the rate of oil depletion in the dead-end pores is proportional to the oil remaining in the dead-end pores. Eq. (11.69) expresses the rate of entrainment of the oil droplets from the fracture surface by the fluid system flowing through the fracture as being proportional to the oil droplets available over the fracture surface. Eq. (11.68) expresses the accumulation rate of oil droplets over the fracture surface as being equal to the difference between the rate of oil expulsion from the matrix and the rate of oil entrainment by the fluid flowing through the fracture. Eq. (11.67) can be interpreted similarly to Eq. (11.68).

A simultaneous solution of Eqs. (11.66)–(11.69), subject to the initial conditions, given by

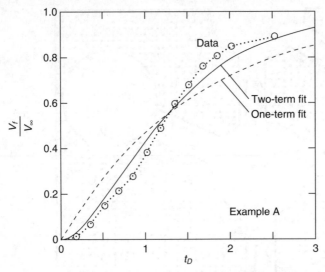

Figure 11.10 Representation of the Example A data by the one- and two-exponential transfer functions (after Civan and Rasmussen, 2001; © 2001 SPE, reproduced with permission from the Society of Petroleum Engineers).

$$V_d = f_d^\circ V_\infty, V_n = f_n^\circ V_\infty, V_i = 0, V_f = 0, t = 0, \tag{11.70}$$

yields the following expression when $\alpha = \beta = \gamma = 1$ (Gupta and Civan, 1994a):

$$V_f(t) = V_\infty \left(1 - \sum_{j=1}^{3} a_j e^{-\lambda_j t} \right). \tag{11.71}$$

The parameters a_j and λ_j denote the empirically determined pre-exponential coefficients and the rate constants associated with the oil entrainment by the fracture fluid, oil transfer from the interconnected pores to the fracture face, and discharge of oil from the dead-end pores to the interconnected pores, for $j = 1, 2, 3$, respectively (see the schematic shown in Figure 11.10). The expressions for the pre-exponential coefficients are given by Gupta and Civan (1994a) for the respective processes as

$$a_1 = \left[\frac{\lambda_2}{\lambda_2 - \lambda_1} \right] \left\{ f_n^\circ - \left[\frac{\lambda_3}{\lambda_1 - \lambda_3} \right] f_d^\circ \right\}, \tag{11.72}$$

$$a_2 = -\left[\frac{\lambda_1}{\lambda_2 - \lambda_1} \right] \left\{ f_n^\circ - \left[\frac{\lambda_3}{\lambda_2 - \lambda_3} \right] f_d^\circ \right\}, \tag{11.73}$$

and

$$a_3 = \left[\frac{\lambda_1}{\lambda_1 - \lambda_3} \right] \left[\frac{\lambda_2}{\lambda_2 - \lambda_3} \right] f_d^\circ, \tag{11.74}$$

where f_n^o and f_d^o denote the volume fractions of the initially present movable oil contained in the interconnected and dead-end pores of the matrix. It follows that

$$f_n^o + f_d^o = 1. \tag{11.75}$$

It also follows from Eqs. (11.72)–(11.74) that (Civan and Rasmussen, 2001)

$$a_1 + a_2 + a_3 = 1, \tag{11.76}$$

$$a_1\lambda_1 + a_2\lambda_2 + a_3\lambda_3 = 0, \tag{11.77}$$

$$a_1\lambda_1^2 + a_2\lambda_2^2 + a_3\lambda_3^2 = -\lambda_1\lambda_2 f_n^o, \tag{11.78}$$

and

$$a_1\lambda_1^3 + a_2\lambda_2^3 + a_3\lambda_3^3 = -\lambda_1\lambda_2(\lambda_1 + \lambda_2)f_n^o + \lambda_1\lambda_2\lambda_3 f_d^o. \tag{11.79}$$

These results will be used subsequently. Eqs. (11.76) and (11.77) are tantamount to the conditions $V_f(0) = 0$ and $\dot{V}_f(0) = 0$ applied to Eq. (11.64).

The three-exponential matrix-to-fracture transfer function given by Eq. (11.71) contains several parameters, which are the characteristics of given rock and fluid systems. These are the pre-exponential coefficients $a_1, a_2,$ and a_3 (defined by Eqs. 11.72–11.74) in terms of the rate constants $\lambda_1, \lambda_2,$ and λ_3; the movable oil initially present in the matrix V_∞; the volume fractions of the initially present movable oil contained in the interconnected and dead-end pores of the matrices f_n^o and f_d^o. Because there are nine unknown parameters related by four equations, namely, Eqs. (11.72)–(11.74), $9 - 4 = 5$ parameters must be estimated by means of experimental data, using a least squares regression of Eq. (11.71). However, because Eq. (11.71) is nonlinear, the parameter values cannot be determined uniquely. Therefore, some parameters should be directly measured. For example, the fraction of the dead-end and interconnected pores can be determined by a petrographical analysis of thin sections of porous rock. Uniqueness in the parameter values can also be achieved by enlarging the experimental database, similar to Ucan et al. (1997), who developed a method of unique and simultaneous determination of relative permeability and capillary pressure curves from displacement data. For this purpose, waterflood oil recovery data obtained by injecting water into fractured cores along with the imbibition oil recovery data obtained by exposing oil-saturated matrix blocks to water can be used together.

The matrix imbibition drive oil recovery experimental data by Guo et al. (1998) were considered by Civan and Rasmussen (2001). They used 1.5-in-diameter and 2-in-long core plugs exposed to water and measured the cumulative volume of oil recovered as a function of time, expressed in percentage of the bulk volume of the cores, representing the matrix in naturally fractured reservoirs. A least squares, nonlinear regression of these data yielded the values of $\lambda_1, \lambda_2, \lambda_3,$ and f_d^o. The remainder of the parameters was calculated by Eqs. (11.72)–(11.74) using these values. The movable oil initially present in the matrix was determined by extrapolation, as the limit of the Guo et al. (1998) measured data. The location of the inflection point of the plotted curves of the Guo et al. data was facilitated to aid in the regression process. The values of the parameters estimated using the Guo et al. (1998) data are presented in the following. The parameters of Example A are $\lambda_1 = 0.247$/day;

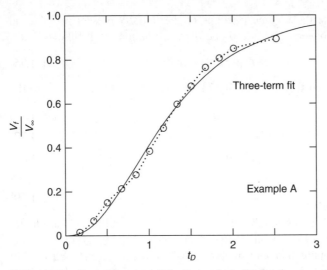

Figure 11.11 Representation of the Example A data by the one-, two-, and three-exponential transfer functions (after Civan and Rasmussen, 2001; © 2001 SPE, reproduced with permission from the Society of Petroleum Engineers).

$\lambda_2 = 0.414$/day; $\lambda_3 = 0.640$/day; a_1, dimensionless $= 3.886$; a_2, dimensionless $= -4.107$; a_3, dimensionless $= 1.221$; f_d^o, fraction $= 0.99$; V_∞, fraction $= 0.471$; ϕ_f, fraction $= 0.001$; $L = 1000$ ft; $v_i = 168$ ft/day; $\tau = 5.95$ days; S_{wc}, fraction $= 0$; and S_{or}, fraction $= 0$. The parameters of Example B are $\lambda_1 = 0.0993$/day; $\lambda_2 = 1.83$/day; $\lambda_3 = 0.0060$/day; a_1, dimensionless $= 0.471$; a_2, dimensionless $= -0.0175$; a_3, dimensionless $= 0.557$; f_d^o, fraction $= 0.522$; V_∞, fraction $= 0.375$; ϕ_f, fraction $= 0.001$; $L = 1000$ ft; $v_i = 168$ ft/day; $\tau = 5.95$ days; S_{wc}, fraction $= 0$; and S_{or}, fraction $= 0$.

Figures 11.10 and 11.11 show a comparison of the cumulative recovery predicted by Eq. (11.71), using one-, two-, and three-exponential terms, respectively, with the SP-33 data extracted from Guo et al. (1998). It is apparent that the three-exponent function fits the experimental data better than the others. These data are used in Example A by Civan and Rasmussen (2001). Figure 11.12 shows that the three-exponential function, given by Eq. (11.71), represents the SP-H9 data extracted from Guo et al. (1998) exactly because the number of parameters (five) is equal to the number of data points. These data are used in Example B.

11.5.2 Formulation of the Fracture Flow Equation

Consider the natural fracture network shown in Figure 11.9. Civan (1993, 1994d) has shown that a representative elemental volume averaging of the microscopic equation of continuity for a phase ℓ leads to the following macroscopic, porous media equation of continuity:

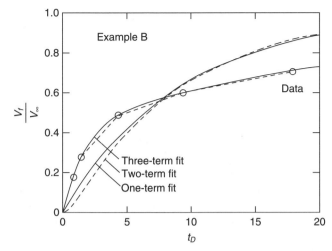

Figure 11.12   Exact representation of the Example B data by the one-, two-, and three-exponential transfer functions (after Civan and Rasmussen, 2001; © 2001 SPE, reproduced with the permission of the Society of Petroleum Engineers).

$$\frac{\partial}{\partial t}(\varepsilon_\ell \rho_\ell) + \nabla \cdot (\rho_\ell \mathbf{u}_\ell) - \nabla \cdot (\varepsilon_\ell \mathbf{D}_\ell \cdot \nabla \rho_\ell) + r_\ell = 0, \qquad (11.80)$$

in which $\mathbf{D}_\ell$ is an empirical hydraulic dispersion tensor. The third term can be interpreted as a hydraulic dispersion due to spatial variations in porous media. The mass rate of phase ℓ lost from the fracture to porous matrix is denoted by r_ℓ. The volume fraction, density, and volume flux of phase ℓ are denoted by $\varepsilon_\ell, \rho_\ell$, and $\mathbf{u}_\ell$, respectively.

For the purposes of the analysis presented here, assume incompressible and immiscible fluid phases in a constant porosity fracture medium and substitute

$$\varepsilon_\ell = \phi_f S_\ell. \qquad (11.81)$$

In this equation, ϕ_f is the fracture porosity and S_ℓ is the saturation of phase ℓ in the fracture. Also, define the fractional volume, f_ℓ, of the flowing phase ℓ according to Civan (1994b):

$$\mathbf{u}_\ell = f_\ell \mathbf{u}. \qquad (11.82)$$

In this equation, $\mathbf{u}$ is the total volumetric flux of the flowing phase given by

$$\mathbf{u} = \sum_\ell \mathbf{u}_\ell. \qquad (11.83)$$

The volume rate of phase ℓ lost from the fracture to porous matrix is given by

$$\dot{q} = r_\ell / \rho_\ell. \qquad (11.84)$$

By substituting Eqs. (11.81)–(11.83), Eq. (11.80) simplifies to

$$\phi_f \frac{\partial S_\ell}{\partial t} + \mathbf{u} \cdot \nabla f_\ell + \dot{q}_\ell = 0. \tag{11.85}$$

The volumetric rate of oil transfer into the fracture is given by

$$\dot{q}_o = -\dot{q}_w = \frac{dV_f}{dt}, 0 \le S_w \le 1.0. \tag{11.86}$$

Because the volume of water imbibed is equivalent to the volume of oil recovered, Eq. (11.86) also represents the rate of water imbibed by the matrix.

The derivation of Eq. (11.71) is based on the assumption that oil transferred into the fracture is rapidly carried away by the water flowing through the fracture so that the fracture surface is always exposed to 100% water. Because the water saturation S_w in a fracture varies, Eq. (11.71) should be used in Eq. (11.86) by a convolution according to deSwaan (1978) by applying Duhamel's theorem:

$$\dot{q} = \int_0^t \dot{q}_{S_{w=1}}(t-\tau) \frac{\partial S_w}{\partial \tau} \partial \tau. \tag{11.87}$$

11.5.3 Exact Analytical Solution Using the Unit End-Point Mobility Approximation

The fractional flow function is given by (Luan, 1995)

$$f_w = \frac{MS_{w,n}}{1+(M-1)S_{w,n}}, \tag{11.88}$$

where M is the end-point mobility ratio.

Analytic solutions of the one-dimensional Buckley–Leverett flow problem involving a one-parameter matrix-to-fracture transfer function have been presented by deSwaan (1978), Davis and Hill (1982), Kazemi et al. (1992), and Luan and Kleppe (1992). These analytical solutions have been possible after linearizing the governing integro-differential equation by invoking the unit end-point mobility ratio ($M = 1$) assumption, which allows for approximating the fractional flowing water f_w of the fracture medium by the normalized water saturation $S_{w,n}$ according to

$$f_w = S_{w,n} \equiv \frac{S_w - S_{wc}}{1 - S_{or} - S_{wc}}, \tag{11.89}$$

where S_{wc} and S_{or} are the fracture medium connate water and the residual oil saturations, respectively, and S_w is the fracture medium water saturation.

Considering Eq. (11.63) and substituting Eq. (11.87) into Eq. (11.85), deSwaan (1978) derived the following equation for the Buckley–Leverett immiscible displacement through a fracture network in porous media:

$$\frac{\partial S_w}{\partial t} + \frac{u}{\phi_f} \frac{\partial f_w}{\partial x} + \frac{V_\infty \lambda}{\phi_f} \int_0^t e^{-\lambda(t-\tau)} \frac{\partial S_w(t-\tau)}{\partial \tau} d\tau = 0, \tag{11.90}$$

where u is the water injection flux into the porous medium, subject to the initial and boundary conditions given by

$$S_w = S_{wc}, 0 \le x \le L, t = 0 \tag{11.91}$$

and

$$f_w = 1.0, x = 0, t > 0. \tag{11.92}$$

deSwaan (1978) applied the unit end-point mobility approximation given by Eq. (11.89) and obtained an analytical solution given by

$$S_{w,n}(x,t) = 0, t < \alpha \tag{11.93}$$

and

$$S_{w,n}(x,t) = 1 - e^{-\lambda(t-\alpha)} \int_0^\beta e^{-\tau} I_o \left[2\sqrt{\lambda(t-\alpha)\tau} \right] d\tau, t \ge \alpha. \tag{11.94}$$

Kazemi et al. (1989) derived an alternative form for Eq. (11.94) as

$$S_{w,n}(x,t) = e^{-\beta} \left\{ \begin{matrix} e^{-\lambda(t-\alpha)} I_o \left[2\sqrt{\beta\lambda(t-\alpha)} \right] \\ + \lambda \int_\alpha^t e^{-\lambda(\tau-\alpha)} I_o \left[2\sqrt{\beta\lambda(\tau-\alpha)} \right] d\tau \end{matrix} \right\}, t \ge \alpha, \tag{11.95}$$

where I_o is modified Bessel's function of the zero type, and

$$\alpha = x\phi_f / u, \beta = V_\infty \lambda\alpha / \phi_f. \tag{11.96}$$

Luan and Kleppe (1992) have proven that Eq. (11.94) can be transformed into Eq. (11.95), and therefore, they are identical.

11.5.4 Asymptotic Analytical Solutions Using the Unit End-Point Mobility Approximation

Civan and Rasmussen (2001) stated that the analytical solutions have resulted in complicated mathematical forms as demonstrated earlier, involving modified Bessel functions and quadratures, and thus require tedious and frequently inaccurate procedures to generate numerical values. Civan and Rasmussen (2001) presented a generalized methodology that obtains analytical solutions for imbibition waterfloods in naturally fractured oil reservoirs undergoing multistep matrix-to-fracture transfer processes. The phenomenological representation of the oil transfer from matrix to fracture is based on a three-exponential matrix-to-fracture transfer function, the necessity for which is seen by the examination of the experimental data. The resulting intego-differential equation is linearized by invoking the unit end-point mobility ratio assumption, converted to a fourth-order partial differential equation, and solved analytically by asymptotic means. It is shown that the asymptotic approximation approach significantly reduces the complexity of the solution process and yields adequate solutions for a long-time evaluation of waterfloods in naturally fractured reservoirs. The solution is not only computationally advantageous, but it also provides a physically meaningful interpretation of the propagation speed and diffusive spreading of the progressing wave front, which could not be readily obtained from the usual type of solution methods. Such analytical solutions are desired for convenient interpretation and correlation of laboratory core flood tests and verification of numerical solution schemes.

11.5.4.1 Formulation

For the one-dimensional horizontal flow in fractured porous media considered here, the Civan et al. (1999) volumetric balance of flowing water phase in fractures, obtained by combining Eqs. (11.64) and (11.85)–(11.87), can be written in partially dimensionless form as (Civan and Rasmussen, 2001)

$$\frac{\partial S_{w,n}}{\partial t_D} + \frac{v_i \tau}{L} \frac{\partial f_w}{\partial x_D} + \frac{V_\infty \tau}{\phi_f} \int_0^{t_D} \left[\sum_{j=1}^3 a_j \lambda_j e^{-\tau \lambda_j (t_D - t_D')} \right] \frac{\partial S_{w,n}}{\partial t_D'} dt_D' = 0, \qquad (11.97)$$

where v_i is the fracture fluid interstitial velocity given by

$$v_i \equiv \frac{u}{\phi_f (1 - S_{or} - S_{wc})}. \qquad (11.98)$$

u is the fracture–fluid volumetric flux, ϕ_f is the fracture porosity, L is the reservoir length, and V_∞ is the volume of the movable oil available in the matrix per unit bulk volume of porous media. Eq. (11.97) is an integro-differential equation.

The dimensionless distance and time are defined as

$$x_D \equiv \frac{x}{L} \qquad (11.99)$$

and

$$t_D \equiv \frac{t}{\tau}, \qquad (11.100)$$

where τ is an appropriately selected characteristic timescale.

As inferred by Eq. (11.97), the characteristic timescale can be defined in a variety of ways. The convection, matrix oil depletion, and rate process timescales can be defined alternatively as

$$\tau = \frac{L}{v_i} \qquad (11.101)$$

or

$$\tau = \frac{\phi_f}{a_j \lambda_j V_\infty}, \, j = 1, 2, \text{ or } 3 \qquad (11.102)$$

or

$$\tau = \frac{1}{\lambda_j}, \, j = 1, 2, \text{ or } 3. \qquad (11.103)$$

Because the flow in fractures is convection dominated, Eq. (11.101) is selected for the characteristic time; that is, $\tau = L / v_i$.

It is convenient to define several more nondimensional parameters:

$$\Lambda_j \equiv \frac{\lambda_j L}{v_i}, j = 1, 2, 3 \tag{11.104}$$

and

$$K_j \equiv \frac{V_\infty L a_j \lambda_j}{\phi_f v_i} = \frac{V_\infty a_j}{\phi_f} \Lambda_j, j = 1, 2, 3. \tag{11.105}$$

Further, define the following nondimensional functions of x_D and t_D:

$$C_j(x_D, t_D) = K_j e^{-\Lambda_j t_D} \int_0^{t_D} e^{+\Lambda_j t_D'} \frac{\partial S_{w,n}}{\partial t_D'} dt_D', j = 1, 2, 3 \tag{11.106}$$

When the above-mentioned nondimensional parameters and variables are used, then Eq. (11.97) can be written in the following simplified nondimensional form:

$$\frac{\partial S_{w,n}}{\partial t_D} + \frac{\partial f_w}{\partial x_D} + \sum_{j=1}^3 C_j(x_D, t_D) = 0. \tag{11.107}$$

The initial and boundary conditions for either Eq. (11.97) or (11.107) are

$$S_{w,n}(x_D \geq 0, t_D = 0) = 0 \tag{11.108}$$

and

$$f_w(x_D = 0, t_D > 0) = 1. \tag{11.109}$$

11.5.4.2 Small-Time Approximation
For small times, the approximate analytical solution is given by

$$S_{w,n}(x_D, t_D) \approx \left[1 - \frac{x_D \tilde{E}_1(\alpha - 1)}{\tilde{E}_2} \left(1 - e^{-\tilde{E}_2(t_D - x_D)} \right) \right] \tag{11.110}$$

$$H(t_D - x_D), t_D \to 0,$$

where $H(t_D - x_D)$ is the Heaviside unit step function ($H(y) = 0$ when $y < 0$, and $H(y) = 1$ when $y \geq 0$). The nondimensional diffusivity that governs the spreading out of the prevailing long-time bulk wave front is given by

$$v \equiv \frac{\tilde{E}_1(\beta - \alpha)}{\beta^3 \tilde{D}}, \tag{11.111}$$

where

$$\alpha = 1 + \frac{\Lambda_1 \Lambda_2 f_n^o}{(\Lambda_1 + \Lambda_2)\Lambda_3 + \Lambda_1 \Lambda_2} \left[\frac{V_\infty}{\phi_f} \right], \tag{11.112}$$

$$\beta = 1 + \frac{V_\infty}{\phi_f}, \tag{11.113}$$

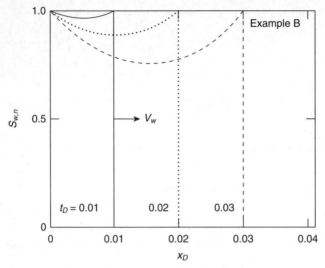

Figure 11.13   Short-time behavior of the saturation profiles at the dimensionless times $tD = 0.01$, 0.02, and 0.03 for Example B (after Civan and Rasmussen, 2001; © 2001 SPE, reproduced with permission from the Society of Petroleum Engineers).

$$\tilde{D} \equiv \Lambda_1 \Lambda_2 \Lambda_3, \tag{11.114}$$

$$\tilde{E}_2 \equiv \Lambda_1 + \Lambda_2 + \Lambda_3, \tag{11.115}$$

and

$$\tilde{E}_1 \equiv (\Lambda_1 + \Lambda_2)\Lambda_3 + \Lambda_1 \Lambda_2. \tag{11.116}$$

Thus, initially, a discontinuity wave propagates to the right with the nondimensional speed $v_w = 1$. The spatial variation between the injection face $x_D = 0$ and the front location at $x_D = t_D$ can be seen from a numerical example. The data used for the numerical examples, denoted as Examples A and B, are given in the above. These values will be explained in a later section. The small-time spatial variations with x_D are shown in Figure 11.13 for $t_D = 0.01, 0.02$, and 0.03, according to Example B and Eq. (11.110). For small times, $S_{w,n}$ decreases from unity at the injection face to a minimum value and then increases to unity at the outward propagating discontinuity surface. After a long enough time, near the injection face, $S_{w,n}$ begins to increase back toward unity, and the slope of the curve with respect to x_D approaches its long-time value of zero.

11.5.4.3   *Approximation for Large Time*   For large times, the approximate analytical solution is given as

$$S_{w,n}(x_D, t_D) = \frac{1}{2}\left\{ erfc\left[\frac{b(ax_D - t_D)}{\sqrt{t_D}}\right] + e^{4ab^2 x_D} erfc\left[\frac{b(ax_D + t_D)}{\sqrt{t_D}}\right]\right\}, \tag{11.117}$$

where $erfc(y)$ is the complementary error function with argument y. The parameters a and b are given by

$$a \equiv \beta \quad \text{and} \quad b \equiv \frac{1}{2\beta\sqrt{v}}, \tag{11.118}$$

where v is the effective diffusivity given by Eq. (11.111). Letting $\Lambda_3 \to \infty$ and then $\Lambda_2 \to \infty$ yields the single-term transfer function result of Rasmussen and Civan (1998).

When the argument of the second error function in Eq. (11.117) is very large, the value of the error function is exponentially small. Some computer software packages may set the error function $erfc(y)$ identically equal to zero when y is large enough. This may cause some error in numerical calculations when x_D is also large enough for the exponential multiplying term $\exp(4ab^2x_D)$ to compensate for the smallness of the error function. It is thus useful to use a two-term asymptotic expansion for $erfc(y)$ when y is large and to express the second term in Eq. (11.117) as

$$e^{4ab^2x_D} erfc\left[\frac{b(ax_D + t_D)}{\sqrt{t_D}}\right] \approx \frac{\sqrt{t_D}}{\sqrt{\pi}b(ax_D + t_D)}$$
$$\left[1 - \frac{t_D}{2b^2(ax_D + t_D)^2}\right]\exp\left[-\frac{b^2(ax_D - t_D)^2}{t_D}\right]. \tag{11.119}$$

Therefore, we obtain the following solution at the bulk wave front $x_D = t_D/a$ for large t_D:

$$S_{w,n}\left(\frac{t_D}{a}, t_D\right) = \frac{1}{2}\left[1 + \frac{1}{2b\sqrt{\pi t_D}}\left(1 - \frac{1}{8b^2 t_D}\right)\right]. \tag{11.120}$$

For a comparison of results, Rasmussen and Civan (1998) used the values used previously by Kazemi et al. (1992) and Civan (1993): $v_l = 168$ ft/day, $L = 1000$ ft, $\lambda = 0.1$/day, $\phi = 0.001$, $R_\infty = 0.08$, and $S_{wc} = S_{or} = 0$. These values yield $\tau = 5.95$ days, $\Lambda = 0.595$, $K = 47.6$, $a = 81$, and $b = 0.39$. The profiles for $S_{w,n}$ as a function of nondimensional distance $x_D \equiv x/L$ are shown in Figure 11.14 for the nondimensional times of $\tau_D \equiv t/\tau = 5$, 10, 15, and 20, which are relatively small. The approximate solution, according to Eq. (11.94), compares well with the exact solution as computed by the formula of Kazemi et al. (1992), and it improves as t_D increases. (Note, for example, that $t_D = 5$ represents a time t of about 30 days.) When $S_{w,n}$ is plotted for larger values of t_D on a correspondingly expanded scale of x_D, the difference between the approximate and exact solutions becomes harder to detect. The maximum error in $S_{w,n}$, for a given value of x_D near the central part of the wave, is about 3.5%. This error decreases as t_D increases. This error is quite acceptable for field applications, considering that many reservoir parameters cannot be reckoned within this accuracy.

The long-time behaviors for Examples A and B are shown in Figure 11.15 for $t_D = 100$. The dimensionless diffusivity for Example A is $v = 5.99 \times 10^{-6}$, whereas the value for Example B is much larger at $v = 1.15 \times 10^{-4}$. The dimensionless wave

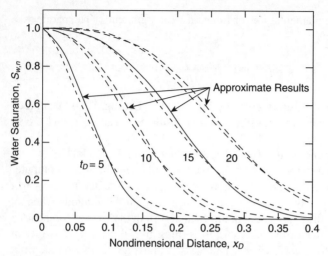

Figure 11.14 Comparison of exact and approximate normalized water saturation as a function of nondimensional distance at various nondimensional times (after Rasmussen and Civan, 1998; © 1998 SPE, reproduced with permission from the Society of Petroleum Engineers).

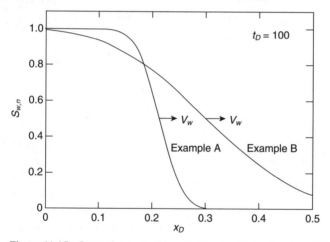

Figure 11.15 Long-time saturation profiles for Examples A and B for the dimensionless time of $t_D = 1000$ (after Rasmussen and Civan, 1998; © 1998 SPE, reproduced with permission from the Society of Petroleum Engineers).

speed of the bulk wave front, which is the propagation speed at approximately $S_{w,n} = 1/2$, is $v_w = 0.00212$ for Example A and $v_w = 0.00266$ for Example B.

11.6 METHOD OF WEIGHTED SUM (QUADRATURE) NUMERICAL SOLUTIONS

Civan (1993, 1994a,c,d,e, 1998c) presented an efficient solution for deSwaan's formulation of waterflooding in naturally fractured porous media by the quadrature method. Comparisons of the analytic, finite difference, and quadrature solutions of the unit end-point mobility ratio problem reveal that the quadrature solutions are more accurate and are much more easily obtained than the finite difference solutions. The computational requirements of the nonlinear problem with capillary pressure included are comparable to that of the unit end-point mobility ratio problem. Civan (1998c) has shown that the fracture medium capillary pressure has a small effect, while the oil viscosity has a large effect on the oil recovery from naturally fractured reservoirs.

11.6.1 Formulation

Civan (1998c) considered a one-dimensional immisicible displacement of oil by water with a constant-rate water injection and constant fluid and formation properties.

By substituting the interstitial fracture medium fluid velocity defined by Eq. (11.98), Eq. (11.85) can then be written for the fracture medium water phase as

$$
\frac{\partial S_w}{\partial t} + (1 - S_{wc} - S_{or})v\frac{\partial f_w}{\partial x} - \frac{V_\infty \lambda}{\phi_f}\int_0^t e^{-\lambda(t-\tau)}\frac{\partial S_w}{\partial \tau}\partial \tau \tag{11.121}
$$

$$
= 0, 0 \le x \le L, t > 0.
$$

The conditions of the solution are considered as the following:

$$
S_w = S_{wc}, 0 \le x \le L, t = 0 \tag{11.122}
$$

and

$$
f_w = 1, x = 0, t > 0. \tag{11.123}
$$

For computational convenience, Civan (1991, 1993, 1994a) also introduced a series of manipulations of Eq. (11.121): (1) multiplying by $e^{\lambda t}$, (2) differentiating with respect to time and dividing by $e^{\lambda t}$, and (3) integrating with respect to time and then applying the initial condition given by Eq. (11.122) to obtain

$$
\frac{\partial S_w}{\partial t} + (1 - S_{wc} - S_{or})\frac{v}{L}\left[\frac{\partial f_w}{\partial X} + \lambda\int_0^t \frac{\partial f_w}{\partial X}\partial \tau\right]
$$

$$
+ \lambda\left(1 + \frac{V_\infty}{\phi_f}\right)(S_w - S_{wc}) = 0, \tag{11.124}
$$

in which the distance normalized with respect to the reservoir length L is given by

$$
X = x/L. \tag{11.125}
$$

The volumetric flux of the water phase through the fracture system is given by

$$
\mathbf{u}_w = -\frac{k_{rw}}{\mu_w}\mathbf{K}_f \cdot (\nabla p_w + \rho_w g \nabla z). \tag{11.126}
$$

Civan (1994a) defined a scalar fractional water, f_w, according to

$$
\mathbf{u}_w = f_w \mathbf{u}. \tag{11.127}
$$

Thus, equating Eqs. (11.126) and (11.127) for a horizontal, one-dimensional flow and then integrating leads to the following expression for the water phase pressure:

$$p_w = p_{w_{in}} - \frac{u\mu_w}{K_f} \int_0^x \frac{f_w}{k_{rw}} dx. \tag{11.128}$$

For convenience, the subscripts of S_w and f_w are dropped in the following. Function f denotes the fracture medium fractional water function (Marle, 1981) given by

$$f = \varphi + (\Psi/L)\partial S/\partial X, \tag{11.129}$$

in which

$$\varphi = \left[\frac{\mu_o}{k_{r_o}} + \frac{K_f}{u}(\rho_w - \rho_o)g\sin\theta\right]\left[\frac{\mu_w}{k_{r_w}} + \frac{\mu_o}{k_{r_o}}\right]^{-1} \tag{11.130}$$

and

$$\Psi = \frac{K_f}{u}\frac{dp_{cf}}{dS}\left[\frac{\mu_w}{k_{r_w}} + \frac{\mu_o}{k_{r_o}}\right]^{-1}. \tag{11.131}$$

where θ is the inclination angle; μ_w and μ_o are the viscosities, ρ_w and ρ_o are the densities, and k_{r_w} and k_{r_o} are the fracture–media fluid relative permeabilities of the water and oil phases, respectively; K_f is the fracture permeability; and g is the gravitational acceleration. The fracture media fluid capillary pressure was represented by a three-constant hyperbola given by Donaldson et al., (1991):

$$p_{cf} = \frac{A + BS}{1 + CS}, \tag{11.132}$$

where A, B, and C are some empirical constants.

For convenience in numerical solution, Eq. (11.124) is decomposed into two differential equations as

$$\frac{\partial S}{\partial t} + (1 - S_{wc} - S_{or})\frac{v}{L}\left[\frac{\partial f}{\partial X} + \lambda y\right]$$
$$+ \lambda\left(1 + \frac{V_\infty}{\phi_f}\right)(S_w - S_{wc}) = 0 \tag{11.133}$$

and

$$\frac{\partial y}{\partial t} = \frac{\partial f}{\partial X}, \tag{11.134}$$

where y is a dummy variable. The initial and boundary conditions are given, respectively, by

$$S = S_{wc}, y = 0, 0 \le X \le 1, t = 0 \tag{11.135}$$

and

$$f = 1, X = 0, t > 0. \tag{11.136}$$

11.6.2   Quadrature Solution

The quadrature is an algebraic rule by which the numerical value of a linear operation on a function at a given point is approximated as a weighted linear sum of the discrete function values at selected points around that point. The quadrature formula is given by (Civan, 1991, 1994a)

$$L\{f(x)\}_i \cong \sum_{j=1}^{n} w_{ij} f(x_j); i = 1, 2, \dots, n. \tag{11.137}$$

L is a linear operator such as a partial derivative or an integral of any order, or a composite of derivatives, integrals, function values, and constants. Variables x_i; $i = 1, 2, \dots, n$ denote the location of discrete points, each of which are associated with a discrete function value of $f(x_i)$. Variable n is the number of discrete values considered for the quadrature approximation. w_{ij} are the quadrature weights.

To determine the quadrature weights, a convenient function for the local representation of the solution of the problem is considered, generalized in the form of (Civan, 1994a, 1998c)

$$f(x) = \sum_{v=0}^{\infty} a_v B_v(x). \tag{11.138}$$

Variables $B_v(x)$ are some monomials and a_v are the associated weights.

Then, invoking Eq. (11.138), Eq. (11.137) leads to the following moment equations:

$$\sum_{j=1}^{n} B_v(x_j) w_{ij} = L\{B_v(x)\}_i; \quad i = 1, 2, \dots, n \text{ and } v = 0, 1, \dots, n-1. \tag{11.139}$$

For a set of prescribed discrete point values, x_i; $i = 1, 2, \dots, n$, the set of linear equations given by Eq. (11.139) is solved simultaneously for the quadrature weights w_{ij}.

For the application, consider a basic power series representation. Hence, Eq. (11.139) becomes

$$\sum_{j=1}^{n} x_j^v w_{ij} = L\{x^v\}_i; \quad i = 1, 2, \dots, n \text{ and } v = 0, 1, \dots, n-1. \tag{11.140}$$

In the applications, the quadrature weights for first-order partial derivatives are required. Thus, the linear operator is

$$L \equiv \partial/\partial x. \tag{11.141}$$

Applying Eq. (11.141), Eq. (11.140) becomes

$$\sum_{j=1}^{n} x_j^v w_{ij} = \partial/\partial x\{x^v\}_i = vx_i^{v-1}; \quad i = 1, 2, \dots, n \text{ and } v = 0, 1, \dots, n-1. \tag{11.142}$$

Because the coefficients of Eq. (11.142) form a Vandermonde matrix, a unique solution for the quadrature weights w_{ij} is obtained efficiently using the method by Björck and Pereyra (1970). The quadrature weights were obtained for 6, 11, and 21 equally spaced points and were used in the following.

For numerical solution, replace the spatial derivatives in Eqs. (11.133) and (11.134) by the quadrature approximation to obtain the following ordinary differential equations:

$$\frac{dS_i}{dt} = -(1 - S_{wc} - S_{or})\frac{v}{L}\left(\sum_{j=1}^{n} w_{ij}f_j + \lambda y_i\right) - \lambda\left(1 + \frac{V_\infty}{\phi_f}\right)(S_i - S_{wc}) \quad (11.143)$$

and

$$\frac{dy_i}{dt} = \sum_{j=1}^{n} w_{ij}f_j, 1 = 2, 3, \ldots n \text{ and } t > 0. \quad (11.144)$$

Eqs. (11.143) and (11.144) are subject to the initial conditions that

$$S_i = S_{wc}, y_i = 0, i = 1, 2, \ldots n \quad \text{and} \quad t = 0. \quad (11.145)$$

The boundary value at the point $i = 1$, Eq. (11.136), is

$$f_1 = 1, i = 1, t > 0. \quad (11.146)$$

The water phase pressure is calculated by Eq. (11.128) as

$$p_{w_i} = p_{w_{in}} - \frac{u\mu_w}{K_f}\sum_{j=1}^{n} c_{ij}\left(\frac{f}{k_{r_w}}\right)_j : i = 2, 3, \ldots, n, \quad (11.147)$$

where c_{ij}'s denote the integral quadrature weights. Then, the oil phase pressure is calculated by the fracture capillary pressure definition:

$$p_{cf} = p_o - p_w. \quad (11.148)$$

The numerical solution of the simultaneous differential equations was obtained using a variable-step Runge–Kutta–Fehlberg four (five) method (Fehlberg, 1969). First, the solution of Eqs. (11.143) and (11.144) was carried out by approximating the fractional flow of water according to the unit end-point mobility ratio approach by deSwaan (1978):

$$f = S. \quad (11.149)$$

Second, the nonlinear fractional water function given by Eqs. (11.129)–(11.132) was used. By replacing the spatial derivative by the quadrature approximation, Eq. (11.129) yields

$$f_i = \varphi_i + \frac{\Psi_i}{L}\sum_{j=1}^{n} w_{ij}S_j. \quad (11.150)$$

Civan (1994a, 1998c) obtained the numerical solution of Eqs. (11.143) and (11.144) using Eq. (11.146) with Eqs. (11.149) or (11.150) for 6, 11, and 21 equally spaced grid points with the data given in the following: $\phi_f = 0.001$, $k_f = 1.0$ darcy, $L = 1000$ ft, $A = 1000$ ft^2, $\lambda = 0.1$/day, $v = 168.4$ ft/day, $V_\infty = 0.08$, $S_{wc} = 0.0$, $S_{or} = 0.0$, $\mu_w = 1.0$ cP, $\mu_o = 1.0$ or 3.0 cP, $k_{rw} = S$, $k_{ro} = 1 - S$, $p_{cf} = 2(1 - S)/(1 + 5S)$ psi, and $\theta = 0.0$. Figure 11.16 shows a comparison of numerical solutions for $f = s$ and $f = \varphi + \Psi\partial/\partial x$ using 1- and 3-cP oil viscosities. It can be seen that the capillary pressure plays a small role, while the oil viscosity has a large effect on the displacement of oil by water in naturally fractured formations.

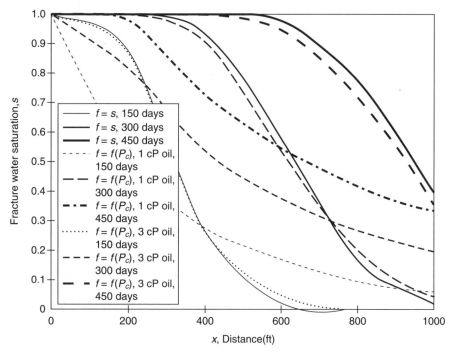

Figure 11.16   Comparison of solutions for $f = s$ and $f = f\,(P_c)$ (after Civan, 1998c; © 1998 SPE, reproduced with permission from the Society of Petroleum Engineers).

The method presented here can be readily extended to multidimensional cases according to Civan (1991, 1994a). However, in contrast to the one-dimensional example presented earlier, the velocity components depend on the multidimensional pressure and permeability distributions. Therefore, a simultaneous solution of the fracture medium fluid pressure and saturation equations, similar to Kazemi et al. (1992), is required.

11.7   FINITE DIFFERENCE NUMERICAL SOLUTION

Civan et al. (1999) investigated the effect of matrix-to-fracture transfer on the oil displacement by water imbibition in naturally fractured porous media in terms of the wettability effects and the governing rate processes. A mathematical model was developed by coupling the two-phase flow in the fracture network and in the porous matrix via an oil–water exchange function that incorporates the rates of transfer of oil from the dead-end pores to the network of pores and then to the network of fractures. The resulting integro-differential equation was solved numerically using the finite difference method. The parametric studies carried out utilizing this model indicated that the rate constants and the matrix wettability play important roles in obtaining an accurate description of the oil recovery during waterflooding in naturally fractured reservoirs.

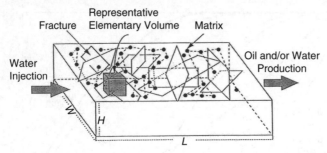

Figure 11.17 Schematic of a network of fractures and pores in a naturally fractured reservoir (after Civan et al., 1999; © 1999 SPE, reproduced with permission from the Society of Petroleum Engineers).

11.7.1 Formulation

Figure 11.17 presents a schematic of a network of fractures and pores in a naturally fractured reservoir. Assuming that the water and oil phases are incompressible, the water injection rate during waterflooding is constant, and the flow is horizontal and one dimensional, the macroscopic equation of continuity for the water phase in the REV of a network of fractures in a naturally fractured reservoir is given as

$$\phi_f \frac{\partial S_w}{\partial t} + u \frac{\partial f_w}{\partial x} + \dot{q}_w = 0. \tag{11.151}$$

$\dot{q}_w$ is the volumetric rate of water lost from the fracture to the porous matrix by imbibition. Neglecting the capillary pressure and gravity effects, the fractional flow of water for horizontal flow in the network of fractures can be expressed as

$$f_w = \frac{1}{1 + \frac{k_{ro}}{\mu_o} \frac{\mu_w}{k_{rw}}}. \tag{11.152}$$

k_{ro} and k_{rw} are the fracture oil and water phase permeabilities, respectively.

Assuming a linear dependency of the relative permeabilities of the fluids in the fracture on the normalized saturations, Eq. (11.152) can be simplified as (Luan, 1995)

$$f_w = \frac{M\bar{S}_w}{1 + (M-1)\bar{S}_w}, \tag{11.153}$$

where M and $\bar{s}_w$ are the end-point mobility ratio and the normalized saturation given, respectively, by

$$M = \frac{k_{rw}^* \mu_o}{k_{ro}^* \mu_w} \tag{11.154}$$

and

$$\bar{S}_w = \frac{S_w - S_{wc}}{1 - S_{wc} - S_{or}}.$$ (11.155)

S_{wc} and S_{or} denote the connate water and residual oil, respectively.

The Gupta and Civan (1994a) model, given by the following three-exponent expression, was considered for the volume of oil transferred into the fracture network:

$$V_f = V_\infty \left(1 - a_1 e^{-\lambda_1 t} - a_2 e^{-\lambda_2 t} - a_3 e^{-\lambda_3 t}\right),$$ (11.156)

where a_1, a_2, and a_3, and λ_1, λ_2, and λ_3 are related by Eqs. (11.72)–(11.75). λ_1, λ_2, and λ_3 can be expressed in a dimensionless form as

$$\lambda_{iD} = \frac{\lambda_i}{\left[\dfrac{\sigma \cos\theta F_s}{\mu_w} \sqrt{\dfrac{k}{\phi_m}}\right]}; i = 1, 2, 3.$$ (11.157)

Here, λ_{iD} is the dimensionless rate constant; σ is the interfacial tension; θ is the contact angle; μ_w is the viscosity of injected water; k is the absolute permeability of matrix; ϕ_m is the porosity of matrix; and F_s is the shape factor. The shape factor F_s, which was proposed by Kazemi et al., (1992), incorporates the variable shape of the matrix blocks as well as the imposed boundary conditions. F_s is calculated by the following equation:

$$F_s = \frac{4}{w_f^2}.$$ (11.158)

Here, w_f is the mean fracture spacing of the system.

By applying Eqs. (11.86), (11.87), and (11.156), the volumetric flow rate of the water lost by imbibition from the fractures to the matrix is obtained as follows:

$$\dot{q}_w = \int_0^t \left[V_\infty \left(\begin{array}{c} a_1\lambda_1 e^{-\lambda_1(t-\tau)} + a_2\lambda_2 e^{-\lambda_2(t-\tau)} \\ + a_3\lambda_3 e^{-\lambda_3(t-\tau)} \end{array}\right)\right] \frac{\partial S_w(\tau)}{\partial \tau} \partial \tau.$$ (11.159)

Therefore, invoking Eqs. (11.155) and (11.159) into Eq. (11.151) leads to the following integro-differential equation for the water saturation in the fracture network:

$$\frac{\partial \bar{S}_w}{\partial t} + v\frac{\partial f_w}{\partial x} + \frac{V_\infty}{\phi_f}\int_0^t \left[\begin{array}{c} a_1\lambda_1 e^{-\lambda_1(t-\tau)} + a_2\lambda_2 e^{-\lambda_2(t-\tau)} \\ + a_3\lambda_3 e^{-\lambda_3(t-\tau)} \end{array}\right] \frac{\partial \bar{S}_w(\tau)}{\partial \tau} \partial \tau = 0,$$ (11.160)

in which

$$v = \frac{u}{\phi_f\left(1 - S_{wc} - S_{or}\right)}.$$ (11.161)

Here, v is the interstitial velocity of the fluid in fracture.

For convenience in the numerical solution, by applying the integration by parts and noting that $\bar{s}_w = 0$ because $\bar{s}_w = s_{wc}$ at $t = 0$, Eq. (11.160) can be rearranged as

$$\frac{\partial \bar{S}_w}{\partial t} + v \frac{\partial f_w}{\partial x} + \frac{V_\infty \bar{S}_w}{\phi_f} \left(a_1 \lambda_1 + a_2 \lambda_2 + a_3 \lambda_3 \right) - \frac{R_\infty}{\phi_f} \left[a_1 \lambda_1^2 \int_0^t e^{-\lambda_1(t-\tau)} \bar{S}_w d\tau \right.$$

$$\left. + a_2 \lambda_2^2 \int_0^t e^{-\lambda_2(t-\tau)} \bar{S}_w d\tau + a_3 \lambda_3^2 \int_0^t e^{-\lambda_3(t-\tau)} \bar{S}_w d\tau \right] = 0. \tag{11.162}$$

Then, Eq. (11.162) can be solved to obtain the distribution of the water saturation in the fractures at different positions and times with the initial and boundary conditions given, respectively, by

$$S_w = S_{wc}, \quad 0 \le x \le L, \quad t = 0 \tag{11.163}$$

and

$$f_w = 1.0, \quad x = 0, \quad t > 0. \tag{11.164}$$

The cumulative oil volume produced at the outlet of the system is given by

$$Q_o = \int_0^t Au f_w dt. \tag{11.165}$$

The initial recoverable oil in the system is given by

$$V_0 = WHL \left[\phi_m \left(1 - S_{wc} - S_{or} \right)_m + \phi_f \left(1 - S_{wc} - S_{or} \right)_f \right]. \tag{11.166}$$

The subscripts m and f denote the matrix and fracture, respectively. Therefore, the oil recovery factor can be calculated by the following equation:

$$R_e = \frac{Q_o}{V_o}. \tag{11.167}$$

11.7.2   Numerical Solutions

In Eq. (11.162), there are three time integral terms that are of the same type but have different values of λ_1, λ_2, and λ_3. Applying the trapezoidal rule, each of the integral terms can be discretized as

$$I^n = \int_0^t e^{-\lambda(t-\tau)} \bar{S}_w d\tau = \frac{\Delta t}{2} \left[e^{-\lambda(t-0)} \bar{S}_{w(t=0)} + 2 \sum_{j=1}^{n-1} e^{-\lambda(t-t_j)} \bar{S}_w^j + e^{-\lambda(t-t)} \bar{S}_w^n \right], \tag{11.168}$$

in which $\bar{S}_{w(t=0)} = 0$ because $S_w = S_{wc}$ at $t = 0$.

The Crank–Nicolson implicit finite difference of Eq. (11.162) (central in time and backward in space) and using Eq. (11.168) leads to the following discretized equation:

$$
\frac{\bar{S}_{wi}^{n+1} - \bar{S}_{wi}^{n}}{\Delta t} + \frac{v}{2}\left(\frac{f_{wi}^{n} - f_{wi-1}^{n} + f_{wi}^{n+1} - f_{wi-1}^{n+1}}{\Delta x} \right) + \frac{V_{\infty}}{2\phi_f}(a_1\lambda_1 + a_2\lambda_2 + a_3\lambda_3)\left(\bar{S}_{wi}^{n} + \bar{S}_{wi}^{n+1} \right)
$$

$$
- \frac{V_{\infty}}{2\phi_f}
\left\{
\begin{array}{l}
\displaystyle\sum_{i=1}^{3} a_i\lambda_i^2 I_i^n + a_1\lambda_1^2 \left\{ \frac{\Delta t}{2}\left[2\displaystyle\sum_{j=1}^{n} e^{-\lambda_1(t_{n+1}-t_j)}\bar{S}_{wi}^{j} + \bar{S}_{wi}^{n+1} \right] \right\} \\[12pt]
+ a_2\lambda_2^2 \left\{ \frac{\Delta t}{2}\left[2\displaystyle\sum_{j=1}^{n} e^{-\lambda_2(t_{n+1}-t_j)}\bar{S}_{wi}^{j} + \bar{S}_{wi}^{n+1} \right] \right\} \\[12pt]
+ a_3\lambda_3^2 \left\{ \frac{\Delta t}{2}\left[2\displaystyle\sum_{j=1}^{n} e^{-\lambda_3(t_{n+1}-t_j)}\bar{S}_{wi}^{j} + \bar{S}_{wi}^{n+1} \right] \right\}
\end{array}
\right\} = 0
$$

$$(11.169)$$

Note that the I^n value is calculated at the end of the previous time step. A Taylor series expansion truncated after the second term yields (Luan, 1995)

$$
f_i^{n+1} \cong f_i^n + \left(\frac{\partial f}{\partial t} \right)_i^n \Delta t, \tag{11.170}
$$

in which, applying Eq. (11.153),

$$
\left(\frac{\partial f_w}{\partial t} \right)_i^n = \frac{M}{[1 + (M-1)\bar{S}_{wi}^n]^2} \frac{\bar{S}_{wi}^{n+1} - \bar{S}_{wi}^{n}}{\Delta t}. \tag{11.171}
$$

Substituting Eq. (11.171) into Eq. (11.169) and solving for S_{wi}^{n+1} results in the following equation used for the explicit numerical solution in the present study:

$$
\bar{S}_{wi}^{n+1} = \left\{ \frac{\bar{S}_{wi}^{n}}{\Delta t} - \frac{v}{2\Delta x}\left[\frac{2f_{wi}^{n} - f_{wi-1}^{n} - f_{wi-1}^{n+1}}{} - \frac{M\bar{S}_{wi}^{n}}{[1+(M-1)\bar{S}_{wi}^n]^2\,\Delta t} \right] \right.
$$

$$
- \frac{R_{\infty}\bar{S}_{wi}^{n}}{2\phi_f}(a_1\lambda_1 + a_2\lambda_2 + a_3\lambda_3)
$$

$$
+ \frac{R_{\infty}}{2\phi_f}\left[\sum_{i=1}^{3} a_i\lambda_i^2 I_i^n + \Delta t \sum_{i=1}^{3}\left(a_i\lambda_i^2 \sum_{j=1}^{n} e^{-\lambda_i(t_{n+1}-t_j)}\bar{S}_{wj} \right) \right] \tag{11.172}
$$

$$
\div \left\{ \frac{1}{\Delta t} + \frac{v}{2\Delta x}\frac{M}{[1+(M-1)\bar{S}_{wi}^n]^2\,\Delta t} \right.
$$

$$
\left. + \frac{R_{\infty}}{2\phi_f}(a_1\lambda_1 + a_2\lambda_2 + a_3\lambda_3) - \frac{R_{\infty}\Delta t}{4\phi_f}(a_1\lambda_1^2 + a_2\lambda_2^2 + a_3\lambda_3^2) \right\}.
$$

Figure 11.18 shows the schematic of the grid system used. Eqs. (11.161)–(11.164) were solved according to the numerical scheme derived previously. The parameters used are listed in the following: S_{wc}, fraction = 0.0; S_{or}, fraction = 0.0;

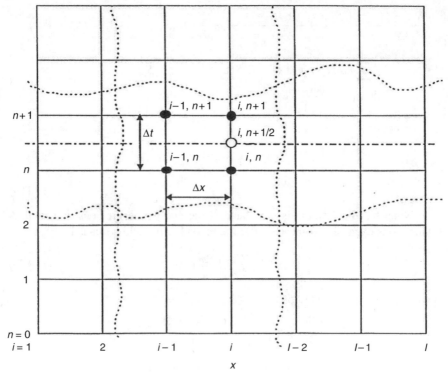

Figure 11.18   Schematic of the grid system used for the numerical solution (after Civan et al., 1999; © 1999 SPE, reproduced with permission from the Society of Petroleum Engineers).

ϕ_f, fraction = 0.001; ϕ_m, fraction = 0.1; K = 1.18E-13 m²; μ_w = 0.0009 Pa·s; R_n, fraction = 0.08; R_d, fraction = 0.02; v = 51.33 m/day; L = 304.8 m; W = 30.48 m; H = 3.05 m; M, dimensionless = 3.0; σ = 0.035 N/m; λ_{D1}, dimensionless = 8.4; λ_{D2}, dimensionless = 0.0807; λ_{D3}, dimensionless = 0.0115; θ_1 = 0°; θ_2 = 30°; θ_3 = 60°; w_{f1} = 30.48 m; w_{f2} = 60.96 m; and w_{f3} = 152.4 m.

The calculations were conducted for five different cases. First, only λ_1 was included in the calculation, which assumed that the flow rate from matrix to fracture is dominantly greater; that is $\lambda_2 \gg \lambda_1$ and $\lambda_2 \gg \lambda_3$. Second, both λ_1 and λ_2 were included and λ_3 was neglected. The third case considered all three parameters, λ_1, λ_2, and λ_3 simultaneously. In the fourth case, the distributions of water saturation at 300 days were calculated for different wettabilities with contact angles of 0°, 30°, and 60° and by considering all three exponent functions. The fifth case studied the effect of the fracture spacing on the distributions of water saturation in fracture. The calculations were carried for a fracture spacing of 30.48 m (100 ft), 60.96 m (200 ft), and 152.4 m (500 ft). Figures 11.19–11.23 show the simulation results for these five cases, respectively. The oil recovery factor was calculated for the aforementioned cases of transfer functions, different wettability, and different fracture spacing. Figures 11.24–11.26 show the effects of these parameters on the oil recovery factor.

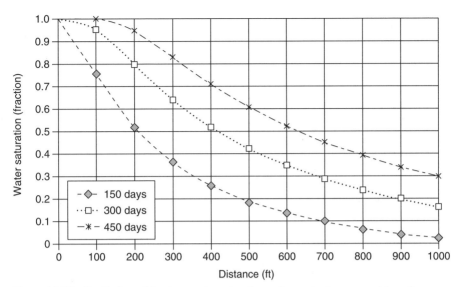

Figure 11.19 Prediction of fracture water saturation using a single-exponent transfer function (after Civan et al., 1999; © 1999 SPE, reproduced with permission from the Society of Petroleum Engineers).

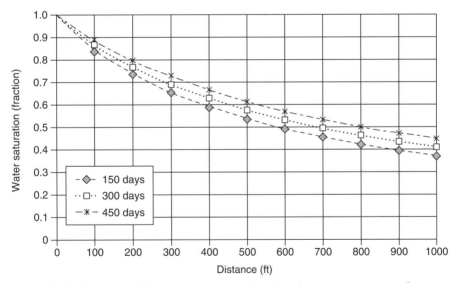

Figure 11.20 Prediction of fracture water saturation using the two-exponent transfer function (after Civan et al., 1999; © 1999 SPE, reproduced with permission from the Society of Petroleum Engineers).

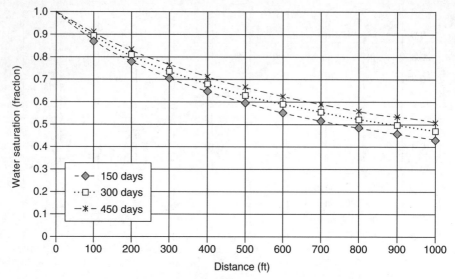

Figure 11.21 Prediction of fracture water saturation using the three-exponent transfer function (after Civan et al., 1999; © 1999 SPE, reproduced with permission from the Society of Petroleum Engineers).

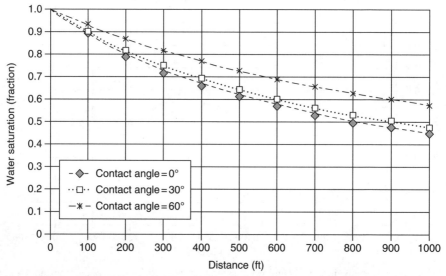

Figure 11.22 Effect of wettability on fracture water saturation at 300 days using the three-exponent transfer function (after Civan et al., 1999; © 1999 SPE, reproduced with permission from the Society of Petroleum Engineers).

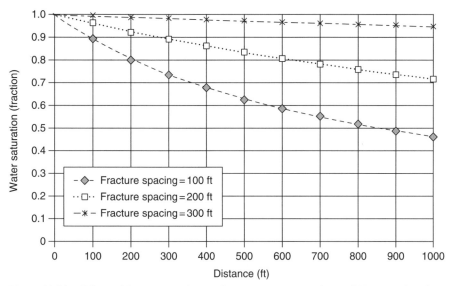

Figure 11.23 Effect of fracture spacing on fracture water saturation at 300 days using the three-exponent transfer function (after Civan et al., 1999; © 1999 SPE, reproduced with permission of the Society of Petroleum Engineers).

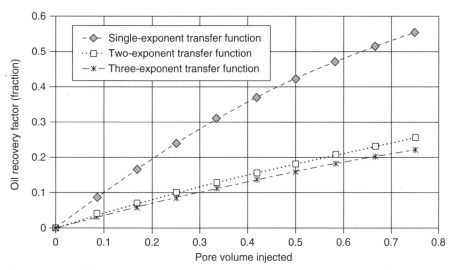

Figure 11.24 Effect of the transfer function on the oil recovery factor (after Civan et al., 1999; © 1999 SPE, reproduced with permission from the Society of Petroleum Engineers).

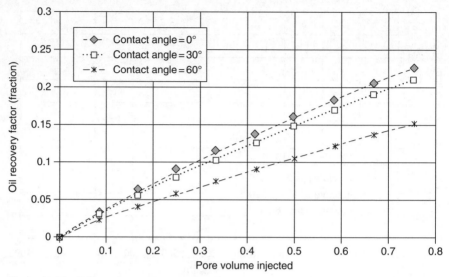

Figure 11.25   Effect of wettability on the oil recovery factor (after Civan et al., 1999; © 1999 SPE, reproduced with permission from the Society of Petroleum Engineers).

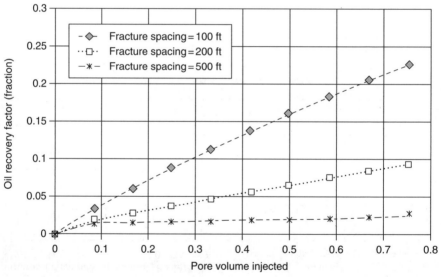

Figure 11.26   Effect of fracture spacing on the oil recovery factor (after Civan et al., 1999; © 1999 SPE, reproduced with permission from the Society of Petroleum Engineers).

From these results, it was observed that λ_1 and λ_2 have significant effects on the distributions of water saturation in fracture systems, but λ_3 has a much smaller effect compared with λ_1 and λ_2. Wettability has significant effects on the distribution of water saturation in the fracture. There is a smaller change of water saturation distribution when the contact angle changes from $0°$ to $30°$, but a bigger change occurs when the contact angle changes from $30°$ to $60°$. Fracture spacing is also important for the distribution of water saturation in fractures. The water saturation in fractures increases with the increase of the fracture spacing. The oil recovery rate appears to increase with a decrease in the number of exponents from three to two and then to one. Therefore, the single-exponent transfer function used by previous researchers may overpredict the recovery rate. The oil recovery rate decreases by increasing contact angles and fracture spacing.

11.8 EXERCISES

1. Determine the shape factor expression to replace Eq. (11.25) when the Cauchy-type boundary condition given by Eq. (11.14) is applied. (Hint: See Moench, 1984.)

2. Determine the shape factor expressions to replace Eqs. (11.33) and (11.34) when the Cauchy-type boundary condition given by Eq. (11.14) is applied. (Hint: See Moench, 1984.)

3. Determine the total volumetric flow across a matrix block using Eq. (11.34) for the shape factor.

4. Derive the expression for the shape factor for an infinite, long, cylindrical-shaped matrix block of radius R using the formulation of Lim and Aziz (1995) given next and verify their solution:

$$\phi_m c_m \frac{\partial p_m}{\partial t} = \frac{1}{\mu} K_m \frac{1}{r} \frac{\partial}{\partial r}\left(r \frac{\partial p_m}{\partial r} \right), 0 \le r \le R, t > 0, \tag{11.173}$$

$$p_m = p_i, 0 \le r \le R, t = 0, \tag{11.174}$$

and

$$p_m = p_f, r = R, t > 0. \tag{11.175}$$

Show that

$$\frac{\bar{p}_m - p_i}{p_f - p_i} = 1 - \frac{4}{R^2} \sum_{n=1}^{\infty} \frac{1}{\alpha_n^2} \exp\left(-\frac{\alpha_n^2 K_m t}{\mu \phi_m c_m} \right), \tag{11.176}$$

where α_n: $n = 1, 2, ...$ are obtained as the roots of the Bessel function of the first kind of order zero, given by $J_o(R\alpha_n) = 0$. Using only the first term in the summation, confirm that the shape factor is given by $\sigma = 18.17/L^2$.

5. Derive the expression for the shape factor for a spherical-shaped matrix block of radius R using the formulation of Lim and Aziz (1995) given next and verify their solution:

$$\phi_m c_m \frac{\partial p_m}{\partial t} = \frac{1}{\mu} K_m \frac{1}{r^2} \frac{\partial}{\partial r}\left(r^2 \frac{\partial p_m}{\partial r} \right), 0 \le r \le R, t > 0, \tag{11.177}$$

$$p_m = p_i, 0 \le r \le R, t = 0, \tag{11.178}$$

and

$$p_m = p_f, r = R, t > 0. \tag{11.179}$$

Show that

$$\frac{\bar{p}_m - p_i}{p_f - p_i} = 1 - \frac{6}{\pi^2} \sum_{n=1}^{\infty} \frac{1}{n^2} \exp\left(-\frac{\pi^2 n^2 K_m t}{\mu \phi_m c_m R^2}\right). \tag{11.180}$$

Using only the first term in the summation, confirm that the shape factor is given by $\sigma = 25.67/L^2$.

6. Estimate how accurately a cylindrical block would approximate a matrix block formed in between two parallel fractures for the determination of the shape factor.

7. Estimate how accurately a spherical block would approximate a rectangular matrix block for the determination of the shape factor.

8. Applying the Laplace transformation, derive an analytical solution for the model of de Swaan (1990) given by Eqs. (11.41)–(11.44).

9. Prepare the plots of the analytical solutions given by deSwaan (1978) and Kazemi et al. (1989) using $q = 168\,\text{ft}^3/\text{day}$; $L = 1000\,\text{ft}$; $\lambda = 0.01$ and $0.1/\text{day}$; $R_\infty = 0.08$; and $\phi_f = 0.001$ for 5, 10, 100, 300, 450, and 600 days as a function of distance. Compare the results.

10. Prepare plots of the short- and long-time saturation profiles at different times for Examples A and B mentioned in this chapter.

11. Consider a parallelepiped-shaped matrix block, which is separated from the surrounding matrix blocks by means of natural fractures. The equation describing the variation of the dimensionless average oil concentration with dimensionless time inside the matrix block is given by the following equation.

   (a) How much dimensionless time does it take for the average oil concentration to decrease to one-half of its initial value in the matrix block?

   (b) How much dimensionless time is required for the recovery factor value to attain a value of 0.75, calculated by the following equation?

$$RF(T) = 1 - \bar{C}(T) \tag{11.181}$$

and

$$\sigma = -\frac{\bar{K}}{L^2} \frac{d\bar{C}/dT}{\bar{C}}, \tag{11.182}$$

where

σ: shape factor $= 0.5$

$\bar{K}$: average permeability $= 100$ md

$\bar{L}$: average dimension of matrix block $= 1.0$ m

$\bar{C}$: dimensionless average concentration of oil in the matrix block

$\bar{C}_o$: initial value of the dimensionless average
    concentration of oil in the matrix block $= 1.0$

T: dimensionless time

12. The following set of equations describes the pressure depression in a matrix block because of a decrease of pressure in the surrounding fracture medium.

(a) Write down the simplified forms of these equations if steady-state condition and one-dimensional flow in the x-direction are considered.

(b) What is the general analytic solution of the one-dimensional model under steady-state conditions obtained without the application of the boundary conditions? Assume a slightly compressible fluid and constant viscosity, and the permeability of the matrix is constant and isotropic.

(c) What is the specific analytic solution of the one-dimensional model under steady-state conditions obtained from the general solution after the application of the boundary conditions?

Consider (Rasmussen and Civan, 2003)

$$\frac{\phi}{\phi}\frac{\partial \psi}{\partial t} = D\nabla \cdot \left[\frac{1}{K}\mathbf{K}\cdot\nabla\psi\right], \tag{11.183}$$

subject to the conditions of solution, given by

$$-(\mathbf{K}\cdot\nabla\psi)\cdot\mathbf{n} \equiv \rho\mathbf{u}\cdot\mathbf{n} = K_s\left[\frac{\psi-\psi_f}{b_s}\right], \tag{11.184}$$

where $x = L/2$, K_s = skin coefficient and

$$-(\mathbf{K}\cdot\nabla\psi)\cdot\mathbf{n} = 0, x = 0. \tag{11.185}$$

Express the pseudopressure function ψ in terms of the fluid pressure p using the following truncated Taylor series:

$$\psi = \int_{p_b}^{p}\frac{\rho}{\mu}dp \cong \psi(p_b) + \frac{\rho}{\mu}(p - p_b). \tag{11.186}$$

13. Does the oil recovery become higher or lower for the waterflooding of a water-wet reservoir when the contact angle is higher?

14. Does the oil recovery become higher or lower for the waterflooding of a water-wet reservoir when the spacing between the natural fractures is larger?

REFERENCES

Aarts, E.H.L. and Korst, J.H.M. 1989. Simulated Annealing and Boltzman Machines. Wiley, New York.

AGA. 1974. Battelle Columbus Laboratories LNG safety program, interim report on Phase II work. American Gas Association Project IS-3-1, Columbus, OH, July 1.

Ahmed, N. and Sunada, D.K. 1969. Nonlinear flow in porous media. Journal of the Hydraulics Division, Proceedings of the ASCE, 95, pp. 1847–1857.

Akai, T.J. 1994. Applied Numerical Merthods for Engineers. John Wiley & Sons, Inc., New York.

Al-Hadhrami, A.K., Elliott, L., and Ingham, D.B. 2003. A new model for viscous dissipation in porous media across a range of permeability values. Transport in Porous Media, 53(1), pp. 117–122.

Al-Hussainy, R. and Ramey, H.J. Jr. 1966. The flow of real gases through porous media. Transactionsof the AIME, 237, p. 637.

Altoe, J.E., Bedrikovetsky, F.P., Siqueira, A.G., Souza, A.L., and Shecaira, F.S. 2006. Correction of basic equations for deep bed filtration with dispersion. Journal of Petroleum Science and Engineering, 51(1–2), pp. 68–84.

Amaefule, J.O., Altunbay, M., Tiab, D., Kersey, D.G., and Keelan, D.K. 1993. Enhanced reservoir description: Using core and log data to identify hydraulic (flow) units and predict permeability in uncored intervals/wells. SPE 26436, Proceedings of the 68th Annual Technical Conference and Exhibition of the SPE held in Houston, TX, October 3–6, pp. 205–220.

Amott, E. 1959. Observations relating to the wettability of porous material. Transactions of the American Institute of Mining and Metallurgical Engineers, 216, pp. 156–162.

Anderson, R.M. and Ford, W.T. 1981. Test Solutions of Porous Media Problems. Journal of Mathematical Analysis and Applications, 79(1), pp. 26–37.

Anderson, D.M. and Tice, A.R. 1972. Predicting Unfrozen Water Contents in Frozen Soils from Surface Area Measurements. Highway Research Board Record, 393, pp. 12–18. Highway Research Board, National Academy of Sciences, Washington, DC.

Anderson, D.M. and Tice, A.R. 1973. The unfrozen interfacial phase in frozen soil water systems. In: Physical Aspects of Soil Water and Salts in Ecosystems, A. Hadas, D. Swartzendruber, P.E. Rijtema, M. Fuchs, and B. Yaron, eds. Springer-Verlag, New York, pp. 107–124.

Anderson, D.M., Tice, A.R., and McKim, H.L. 1973. The unfrozen water and the apparent specific heat capacity of frozen soils. Permafrost: Proceedings of the Second International Conference, National Academy of Science, Washington, DC, pp. 289–295.

Porous Media Transport Phenomena, First Edition. Faruk Civan.
© 2011 John Wiley & Sons, Inc. Published 2011 by John Wiley & Sons, Inc.

Antohe, B.V. and Lage, J.L. 1997. A general two-equation macroscopic turbulence model for incompressible flow in porous media. International Journal of Heat and Mass Transfer, 40(13), pp. 3013–3024.

App, J.F. 2008. Nonisothermal and productivity behavior of high pressure reservoirs. Paper SPE 114705 presented at the SPE Annual Technical Conference and Exhibition, Denver, CO, September 21–24.

App, J.F. 2009. Field cases: Nonisothermal behavior due to Joule-Thomson and transient fluid expansion/compression effects. Paper SPE 124338 presented at the SPE Annual Technical Conference and Exhibition, New Orleans, Louisiana, USA, 4–7 October.

Arastoopour, H. and Semrau, J. 1989. Mathematical analysis of two-phase flow in low-permeability porous media. AIChE Journal, 35(10), pp. 1710–1718.

Archie, G.E. 1942. The electrical resistivity log as an aid in determining some reservoir characteristics. Transactions of the AIME, 146, pp. 54–61.

Aronofsky, J.S., Masse, L., and Natanson, S.G. 1958. A Model for the mechanism of oil recovery from the porous matrix due to water invasion in fractured reservoirs. Transactions of the AIME, 213, pp. 17–19.

Arrhenius, S. 1889. Uber die Reaktionsgeschwindigkeit der inversion von rohrzucker durch saeuren. Zeitschrift fur Physikalische Chemie, 4(2), pp. 226–248.

Auset, M. and Keller, A.A. 2004. Pore-scale processes that control dispersion of colloids in saturated porous media. Water Resources Research, 40, pp. 1–11. W03503. doi: 10.1029/2003WR002800.

Aziz, K. and Settari, A. 1979. Petroleum Reservoir Simulation. Applied Science Publisher Ltd, London.

Bae, J.-S. and Do, D.D. 2005. Permeability of subcritical hydrocarbons in activated carbon. AIChE Journal, 51(2), pp. 487–501.

Bai, M., Bouhroum, A., Civan, F., and Roegiers, J.-C. 1995. Improved model for solute transport in heterogeneous porous media. Journal of Petroleum Science and Engineering, 14, pp. 65–78.

Bai, M. and Civan, F. 1998a. Comments on: Macroscopic modeling of double porosity reservoirs. Journal of of Petroleum Science and Engineering, 20, pp. 87–88.

Bai, M. and Civan, F. 1998b. Reply to: Macroscopic modeling of double porosity reservoirs. Journal of of Petroleum Science and Engineering, 20, pp. 93–94.

Bai, R. and Tien, C. 1999. Particle deposition under unfavorable surface interactions. Journal of Colloid and Interface Science, 218, pp. 488–499.

Barenblatt, G.I., Zheltov, Y.P., and Kochina, I.N. 1960. Basic concepts on the theory of seepage of homogeneous liquids in fissured rocks. Prikladnaya Matematika i Mekhanika, Akad Nauk, 24(5), pp. 852–864.

Barenblatt, G.I., Zheltov, Y.P., and Kochina, I.N. 1960. Basic concepts in the theory of seepage of homogeneous liquids in fissured rocks. Journal of Applied Mathematics and Mechanics (Eng. Trans.), 24, pp. 1286–1303.

Barenblatt, G.I. 1979. Similarity, Self-Similarity, and Intermediate Asymptotics. Consultants Bureau, New York.

Basak, P. 1977. Non-Darcy flow and its implications to seepage problems. Journal of the Irrigation & Drainage Division, American Society of Civil Engineers, 103, pp. 459–473.

Baveye, P. and Sposito, G. 1984. The operational significance of the continuum hypothesis in the theory of water movement through soils and aquifers. Water Resources Research, 20(5), pp. 521–530.

Bear, J. 1972. Dynamics of Fluids in Porous Media. Elsevier North-Holland, Inc., New York.

Bear, J. and Braester, C. 1972. On the flow of two immiscible fluids in fractured porous media. Proceedings of First Symposium of Fundamentals of Transport Phenomena in Porous Media, Elsevier Sci., New York, pp. 177–202.

Bear, J. and Bachmat, Y. 1990. Introduction to Modeling of Transport Phenomena in Porous Media. Kluwer, Norwell, MA. p. 465p.

Bear, J., Zaslavsky, D., and Irmay, S. (eds.) 1968. Physical Principles of Water Percolation and Seepage. UNESCO, Paris.

Beckett, P.M. 1980. Combined natural and forced convection between parallel vertical walls. SIAM Journal on Applied Mathematics, 39, pp. 372–384.

Beckett, P.M. and Friend, I.E. 1984. Combined natural and forced convection between parallel walls: Developing flow at high Rayleigh numbers. International Journal of Heat Mass Transfer, 27, pp. 611–621.

Bejan, A. 1982. Entropy Generation Through Heat and Fluid Flow. John Wiley & Sons, New York.

Bertero, L., Chierici, G.L., Gottardi, G., Mesini, E., and Mormino, G. 1988. Chemical equilibrium models: Their use in simulating the injection of incompatible waters. SPE Reservoir Engineering Journal, 3(1), pp. 288–294.

Bertin, J.J. 1987. Engineering Fluid Mechanics, 2nd ed. Prentice Hall Inc., Englewood Cliffs, NJ, and New York.

Beskok, A. and Karniadakis, G.E. 1999. A model for flows in channels, pipes, and ducts at micro and nano scales. Microscale Thermophysical Engineering, 3, pp. 43–77.

Bett, K.E. and Cappi, J.B. 1965. Effect of pressure on the viscosity of water. Nature, 207, pp. 620–621.

Bhat, S.K. 1998. Modeling permeability evolution in diatomite reservoirs during steam drive. M.S. Thesis, Petroleum Engineering, Stanford University, p. 138p.

Bhat, S.K. and Kovscek, A.R. 1999. Statistical network theory of silica deposition and dissolution in diatomite. In Situ, 23(1), pp. 21–53.

Biot, M.A. 1941. General theory of three dimensional consolidation. Journal of Applied Physics, 12, pp. 155–164.

Bird, G.A. 1983. Definition of mean free path for real gases. Physics of Fluids, 26, p. 3222, 3223.

Bird, G.A. 1994. Molecular Gas Dynamics and the Direct Simulation of Gas Flows. Oxford University Press, Oxford, UK.

Bird, R.B., Stewart, W.E., and Lightfoot, E.N. 1960. Transport Phenomena. John Wiley & Sons, Inc., New York.

Bishnoi, P.R., Natarajan, V., and Kalogerakis, N. 1994. A unified description of the kinetics of hydrate nucleation, growth, and decomposition. In: International Conference on Natural Gas Hydrates, E.D. Sloan, J. Happel and M.A. Hnatow, eds. The New York Academy of Sciences, New York, pp. 311–322.

Björck, A. and Pereyra, V. 1970. Solution of the vandermonde systems of equations. Mathematics of Computation, 24, pp. 893–903.

Blackwell, R.J., Rayne, J.R., and Terry, W.M. 1959. Factors influencing the efficiency of miscible displacement. Transactions of the AIME, 217, pp. 1–8.

Blick, E.F. 1966. Capillary-orifice model of high speed flow through porous media. I&EC, Process Design and Development, 5(1), pp. 90–94.

Blick, E.F. and Civan, F. 1988. Porous-media momentum equation for highly accelerated flow. SPE Reservoir Engineering, 3(3), pp. 1048–1052.

Bolton, E.W. 2002. On the use of effective stress in granular media. American Geophysical Union, Fall Meeting in San Francisco, abstract H71B-0816.

Bravo, M.C. 2007. Effect of transition from slip to free molecular flow on gas transport in porous media. Journal of Applied Physics, 102, p. 074905.

Brooks, R.H. and Corey, A.T. 1966. Properties of porous media affecting fluid flow. Journal of Irrigation and Drainage Division, Proceedings of the American Society of Civil Engineers, 92(IR 2), pp. 61–88.

Brown, G.P., DiNardo, A., Cheng, G.K., and Sherwood, T.K. 1946. The flow of gases in pipes at low pressures. Journal of Applied Physics, 17, pp. 802–813.

Brown, G.O., Hsieh, H.T., and Lucero, D.A. 2000. Evaluation of laboratory dolomite core sample size using representative elementary volume concept. Water Resources Research, 36(6), pp. 1199–1207.

Bruneau, Ch.-H., Fabrie, P., and Rasetarinera, P. 1995. Numerical resolution of in-situ bio-restoration in porous media. In: Mathematical Modeling of Flow Through Porous Media, A.P. Bourgeat, C. Carasso, S. Luckhaus, and A. Mikelic, eds. World Scientific, Singapore, pp. 391–401.

Buckingham, E. 1914. On physically similar systems: Illustrations of the use of dimensional equations. Physical Review, Ser 2, 4(4), p. 345.

Buckley, J.S. (ed.) 2002. Evaluation of reservoir wettability. Journal of Petroleum Science and Engineering, 33(1–3), pp. 1–222.

Buckley, S.E. and Leverett, M.C. 1942. Mechanism of Fluid Displacement in Sands. Transactions of the AIME, 146, pp. 107–116.

Callister, W.D. Jr. 2000. Materials Science and Engineering: An Introduction, 5th ed. John Wiley & Sons, Inc., New York.

Carman, P.C. 1937a. The determination of the specific surface of powder, I. Journal of the Society of Chemical Industries, 57, p. 225.

Carman, P.C. 1937b. Fluid flow through a granular bed. Transactions of the Institution of Chemical Engineers, London, 15, pp. 150–167.

Carman, P.C. 1938. The determination of the specific surfaces of powders. I. Transactions, Journal of the Society of Chemical Industries, 57, pp. 225–234.

Carman, P.C. 1956. Flow of Gases Through Porous Media. Butterworths, London.

Carnahan, C.L. 1990. Coupling of precipitation-dissolution reactions to mass diffusion via porosity changes. In: Chemical Modeling of Aqueous Systems II, Chapter 18, D.C. Melchior and R.L. Basset, eds. ACS Symposium Series 416, American Chemical Society, Washington, DC, pp. 234–242.

Carslaw, H.S. and Jaeger, J.C. 1959. Conduction of Heat in Solids, Oxford University Press, New York.

Carslaw, H.S. and Jaeger, J.C. 1969. Conduction of Heat in Solids, 2nd ed. Clarendon, Oxford.

Chang, F.F. and Civan, F. 1997. Practical model for chemically induced formation damage. Journal of of Petroleum Science and Engineering, 17(1/2), pp. 123–137.

Chang, M.-M. 1993. Deriving the shape factor of a fractured rock matrix. Technical Report NIPER-696 (DE93000170), NIPER, Bartlesville, OK, p. 40 p.

Chang, Y.-I. and Chan, H.-C. 2008. Correlation Equation for Predicting filter coefficient under unfavorable deposition conditions. AIChE Journal, 54(5), pp. 1235–1253.

Chang, M.-M., Bryant, R.S., Stepp, A.K., and Bertus, K.M. 1992. Modeling and Laboratory Investigations of Microbial Oil Recovery Mechanisms in Porous Media, Topical Report

No. NIPER-629, FC22-83FE60149, U.S. Department of Energy, Bartlesville, Oklahoma, p. 27.

Chang, Y.-I., Cheng, W.-Y., and Chan, H.-C. 2009. A proposed correlation equation for predicting filter coefficient under unfavorable deposition conditions. Separation and Purification Technology, 65(3), pp. 248–250.

Chardaire, C., Chavent, G., Liu, J., and Bourbiaux, B. 1989. Simultaneous estimation of relative permeabilities and capillary pressure. Paper SPE 19680 presented at the 1989 Annual Technical Conference and Exhibition, San Antonio, TX, October 8–11.

Chardaire, C., Chavent, G., Jaffre, J., and Liu, J. 1990. Multiscale representation for simultaneous estimation of relative permeabilities and capillary pressure. Paper SPE 20501 presented at the 1990 Annual Technical Conference and Exhibition, New Orleans, LA, September 23–26.

Chardaire-Riviere, C., Forbes, P., Zhang, J.F., Chavent, G., and Lenormand, R. 1992. Improving the centrifuge technique by measuring local saturations. SPE 24882 Paper, Proceedings of the 67th SPE Annual Technical Conference and Exhibition, Washington, DC, October 4–7, pp. 193–208.

Chen, S., Qin, F., Kim, K.-H., and Watson, A.T. 1992. NMR imaging of multiphase flow in porous media. SPE 24760 Paper, Proceedings of the 67th SPE Annual Technical Conference and Exhibition, Washington, DC, October 4–7, pp. 1013–1026.

Cheng, P. and Wang, C.Y. 1996. A multiphase mixture model for multiphase, multicomponent transport in capillary porous media. 2. Numerical simulation of the transport of organic compounds in the subsurface. International Journal of Heat and Mass Transfer, 39(17), pp. 3619–3632.

Chhabra, R.P., Comiti, J., and Machac, I. 2001. Flow of non-newtonian fluids in fixed and fluidized beds. Chemical Engineering Science, 56(1), pp. 1–27.

Chung, M. and Catton, I. 1988. Multi-dimensional two-phase flow in porous media with phase change. In: Heat Transfer-Houston, S.B. Yilmaz, ed. Vol. 84. AIChE Symposium Series, New York, pp. 177–185.

Churchill, S.W. 1997. A new approach to teaching dimensional analysis. Chemical Engineering Education, 30, pp. 158–165.

Civan, F. 1990. A generalized model for formation damage by rock-fluid interactions and particulate processes. Paper SPE 21183, Proceedings of SPE 1990 Latin American Petroleum Engineering Conference, Rio de Janeiro, Brazil, October 14–19.

Civan, F. 1991. Quadrature and cubature methods for numerical solution of integro-differential equations. In: Integral Methods in Science and Engineering, 90, Series in Computational and Physical Processes in Mechanics and Thermal Sciences, A. Haji-Sheikh, C. Corduneanu, J.L. Fry, T. Huang and F.R. Payne, eds. Hemisphere Publ. Co., New York, pp. 282–297.

Civan, F. 1993. Waterflooding of naturally fractured reservoirs: An efficient simulation approach. Paper SPE 25449 presented at the Production Operations Symposium, Oklahoma, OK, March 21–23, pp. 395–407.

Civan, F. 1994a. Solving multivariable mathematical models by the quadrature and cubature methods. Journal of Numerical Methods for Partial Differential Equations, 10, pp. 545–567.

Civan, F. 1994b. Comment on: On scaling immiscible displacements in permeable media. Letter to the Editor, Journal of Petroleum Science Engineering, 11, pp. 351–352.

Civan, F. 1994c. Numerical simulation by the quadrature and cubature methods. Paper SPE 28703, Proceedings of the SPE International Petroleum Conference and Exhibition of Mexico, Veracruz, Mexico, October 10–13, pp. 353–363.

Civan, F. 1994d. A theoretically derived transfer function for oil recovery from fractured reservoirs by waterflooding. Paper SPE/DOE 27745 presented at the Ninth Symposium on Improved Recovery, Tulsa, OK, April 17–20.

Civan, F. 1994e. Rapid and accurate solution of reactor models by the quadrature method. Computers & Chemical Engineering, 18(10), pp. 1005–1009.

Civan, F. 1995. Letter to the Editor, Comment on "Time-space method for multidimensional melting and freezing problems." International Journal of Numerical Methods in Engineering, 38(8), p. 1423.

Civan, F. 1996a. A time-space solution approach for simulation of flow in subsurface reservoirs. Proceedings of the 11th Petroleum Congress of Turkey, UCTEA Chamber of Petroleum Engineers, Ankara, Turkey, April 15–17, pp. 161–167.

Civan, F. 1996b. Convenient formulation for immiscible displacement in porous media. SPE 36701, Proceedings of the 71st SPE Annual Tech Conference and Exhibition, Denver, CO, Oct. 6–9, pp. 223–236.

Civan, F. 1996c. A multi-purpose formation damage model. Paper SPE 31101, Proceedings of the SPE Formation Damage Symposium, Lafayette, LA, pp. 311–326.

Civan, F. 1997. Model for interpretation and correlation of contact angle measurements. Journal of of Colloid and Interface Science, 192, pp. 500–502.

Civan, F. 1998a. Practical model for compressive cake filtration including fine particle invasion. AIChE Journal, 44(11), pp. 2388–2398.

Civan, F. 1998b. Incompressive cake filtration: Mechanism, Parameters, and Modeling. AIChE Journal, 44(11), pp. 2379–2387.

Civan, F. 1998c. Quadrature solution for waterflooding of naturally fractured reservoirs. SPE Reservoir Evaluation & Engineering Journal, 1(2), pp. 141–147.

Civan, F. 2000a. Reservoir Formation Damage- Fundamentals, Modeling, Assessment, and Mitigation. Gulf Publ. Co., Houston, TX, and Butterworth-Heinemann, Woburn, MA.

Civan, F. 2000b. Predictability of porosity and permeability alterations by geochemical and geomechanical rock and fluid interactions. Paper SPE 58746, Proceedings of the SPE International Symposium on Formation Damage, Lafayette, LA, February 23–24, pp. 359–370.

Civan, F. 2000c. Correlation of the pit depth in crystal etching by dissolution. Journal of of Colloid and Interface Science, 222(1), pp. 156–158.

Civan, F. 2000d. Unfrozen water content in freezing and thawing soils- kinetics and correlation. Journal of Cold Regions Engineering, 14(3), pp. 146–156.

Civan, F. 2000e. Leaky-tank reservoir model including the non-darcy effect. Journal of Petroleum Science and Engineering, 28(3), pp. 87–93.

Civan, F. 2001. Scale effect on porosity and permeability- kinetics, model, and correlation. AIChE Journal, 47(2), pp. 271–287.

Civan, F. 2002a. A triple-mechanism fractal model with hydraulic dispersion for gas permeation in tight reservoirs. Paper SPE 74368, SPE Intl. Petroleum Conference and Exhibition, Villahermosa, Mexico, February 10–12.

Civan, F. 2002b. Fractal formulation of the porosity and permeability relationship resulting in a power-law flow units equation- a leaky-tube model. Paper SPE 73785, Proceedings of the SPE International Symposium on Formation Damage, Lafayette, LA, February 23–24.

Civan, F. 2002c. Modeling and analysis of pitting during crystal dissolution. In: Encyclopedia of Surface and Colloid Science, A. Hubbard, ed. Marcel-Dekker Inc., New York, pp. 3463–3465.

Civan, F. 2002d. Relating permeability to pore connectivity using a power-law flow unit equation. Petrophysics, 43(6), pp. 457–476.

Civan, F. 2002e. Implications of alternative macroscopic descriptions illustrated by general balance and continuity equations. Journal of Porous Media, 5(4), pp. 271–282.

Civan, F. 2003. Leaky-tube permeability model for identification, characterization, and calibration of reservoir flow units. Paper SPE 84603, SPE Annual Technical Conference and Exhibition, Denver, CO, October 5–8.

Civan, F. 2004. Temperature dependence of wettability related rock properties correlated by the arrhenius equation. Petrophysics, 45(4), pp. 350–362.

Civan, F. 2005a. Improved permeability equation from the bundle-of-leaky-capillary-tubes model. Paper SPE 94271 presented at the SPE Production Operations Symposium, Oklahoma City, OK, 16–19 April.

Civan, F. 2005b. Applicability of the Vogel-Tammann-Fulcher type asymptotic exponential functions for ice, hydrates, and polystyrene latex. Journal of Colloid and Interface Science, 285, pp. 429–432.

Civan, F. 2006. Including non-equilibrium relaxation in models for rapid multiphase flow in wells. SPE Production & Operations Journal, 21(1), pp. 98–106.

Civan, F. 2007a. Reservoir Formation Damage- Fundamentals, Modeling, Assessment, and Mitigation. Gulf Professional Pub., Elsevier, Burlington, MA.

Civan, F. 2007b. Temperature effect on power for particle detachment from pore wall described by an arrhenius-type equation. Transport in Porous Media, 67(2), pp. 329–334.

Civan, F. 2007c. Brine viscosity correlation with temperature using the vogel-tammann-fulcher equation. Paper SPE-108463, SPE Drilling & Completion Journal, 22(4), pp. 341–355.

Civan, F. 2007d. Critical modification to the vogel-tammann-fulcher equation for temperature effect on the density of water. Industrial & Engineering Chemistry Research, 46(17), pp. 5810–5814.

Civan, F. 2008a. Temperature effect on advection-diffusion transport involving fines migration and deposition in geological porous media. Article No. 452, 2008 IAHR International Groundwater Symposium, Istanbul, Turkey, June 18–20.

Civan, F. 2008b. Use exponential functions to correlate temperature dependence. Chemical Engineering Progress, 104(7), pp. 46–52.

Civan, F. 2008c. Correlation of permeability loss by thermally-induced compaction due to grain expansion. Petrophysics Journal, 49(4), pp. 351–361.

Civan, F. 2008d. Generalized Darcy's law by control volume analysis including capillary and orifice effects. Journal of Canadian Petroleum Technology, 47(10), pp. 1–7.

Civan, F. 2009. Practical finite-analytic method for solving differential equations by compact numerical schemes. Numerical Methods for Partial Differential Equations, pp. 347–379. 25(2).

Civan, F. 2010a. Effective correlation of apparent gas permeability in tight porous media. Transport in Porous Media, 82(2), pp. 375–384.

Civan, F. 2010b. Modeling transport in porous media by control volume analysis. Journal of Porous Media, 13(10), pp. 855–873.

Civan, F. 2010c. Non-isothermal permeability impairment by fines migration and deposition in porous media including dispersive transport. Transport in Porous Media, 85(1), pp. 233–258.

Civan, F. 2010d. A review of approaches for describing gas transfer through extremely tight porous media. The Third ECI International Conference on Porous Media and its Applications in Science, Engineering and Industry, Montecatini Terme, Italy, June 20–25.

Civan, F. 2011. Correlate data effectively. Chemical Engineering Progress, 107(2), pp. 35–44.

Civan, F. and Donaldson, E.C. 1987. Relative Permeability from Unsteady State Displacements: An Analytical Interpretation. Paper SPE 16200, Proceedings of SPE Production Operations Symposium, Oklahoma City, OK, March, pp. 139–155.

Civan, F. and Donaldson, E.C. 1989. Relative permeability from unsteady-state displacements with capillary pressure included. SPE Formation Evaluation, 4(2), pp. 189–193.

Civan, F. and Evans, R.D. 1991. Non-Darcy flow coefficients and relative permeabilities for gas/brine systems. Paper SPE 21516, Proceedings of the 1991 Society of Petroleum Engineers Gas Technology Symposium, Houston, TX, January 23–25, pp. 341–352.

Civan, F. and Evans, R.D. 1993. Relative permeability and capillary pressure data from non-Darcy flow of gas/brine systems in laboratory cores. Paper SPE 26151, Proceedings of the 1993 SPE Gas Technology Symposium, Calgary, Canada, June 28–30, pp. 139–153.

Civan, F. and Evans, R.D. 1995. Discussion of high velocity flow in porous media. SPE Reservoir Engineering, 10(4), pp. 309–310.

Civan, F. and Evans, R.D. 1996. Determination of Non-Darcy flow parameters using a differential formulation of the Forchheimer equation. Paper SPE 35621, Proceedings of the Society of Petroleum Engineers Gas Technology Symposium (April 28–May 1, 1996), Calgary, Alberta, Canada, pp. 415–429.

Civan, F. and Evans, R.D. 1998. Determining the parameters of the forchheimer equation from pressure-squared vs. pseudopressure formulations. SPE Reservoir Evaluation & Engineering, 1(1), pp. 43–46.

Civan, F. and Knapp, R.M. 1991. Correlation method for chemical comminution of coal. AIChE Journal, 37(6), pp. 955–957.

Civan, F. and Nguyen, V. 2005. Modeling particle migration and deposition in porous media by parallel pathways with exchange, Chapter 11, In: Handbook of Porous Media, 2nd ed. K. Vafai, ed. CRC Press, Taylor & Francis, Boca Raton, FL, pp. 457–484.

Civan, F. and Rasmussen, M.L. 2001. Asymptotic analytical solutions for imbibition water floods in fractured reservoirs. Society of Petroleum Engineers Journal, 2, pp. 171–181.

Civan, F. and Rasmussen, M.L. 2003. Analysis and interpretation of gas diffusion in quiescent reservoir, drilling, and completion fluids: Equilibrium vs. non-equilibrium models. Paper SPE 84072, SPE Annual Technical Conference and Exhibition, Denver, CO, October 5–8.

Civan, F. and Rasmussen, M.L. 2005. Determination of parameters for matrix-fracture transfer functions from laboratory data. Paper SPE 94267, the 2005 Production Operations Symposium, Oklahoma City, OK, April 17–19.

Civan, F. and Sliepcevich, C.M. 1984. Efficient numerical solution for enthalpy formulation of conduction heat transfer with phase change. International Journal of Heat Mass Transfer, 27(8), pp. 1428–1430.

Civan, F. and Sliepcevich, C.M. 1985a. Convenient formulations for convection/diffusion transport. Chemistry Engineering Science, 40(10), pp. 973–1974.

Civan, F. and Sliepcevich, C.M. 1985b. Comparison of the thermal regimes for freezing and thawing of moist soils. Water Resources Research, 21(3), pp. 407–410.

Civan, F. and Sliepcevich, C.M. 1987. Limitation in the apparent heat capacity formulation for heat transfer with phase change. Proceedings of the Oklahoma Academy of Science, 67, pp. 83–88.

Civan, F. and Tiab, D. 1989. Second law analysis of petroleum reservoirs for optimized performance. Paper SPE 18855, Proceedings of the SPE Production Operations Symposium, Oklahoma City, OK, March 13–14, pp. 285–294.

Civan, F. and Tiab, D. 1991. Steady and semi-steady state radial flow equations for bounded and partial water-drive oil and gas reservoirs using Darcy, Forchheimer, Brinkman, and capillary-orifice models. SPE 22923, Proceedings of the SPE 66th Annual Tech. Conf. and Exhibition, Dallas, TX, October, pp. 381–386.

Civan, F. and Weers, J.J. 2001. Laboratory and theoretical evaluation of corrosion-inhibiting emulsions. SPE Production and Facilities, 16(4), pp. 260–266.

Civan, F., Wang, W., and Gupta, A. 1999. Effect of wettability and matrix-to-fracture transfer on the waterflooding in fractured reservoirs. Paper SPE 52197, 1999 SPE Mid-Continent Operations Symposium, Oklahoma City, OK, March 28–31.

Civan, F., Alarcon, L.J., and Campbell, S.E. 2004. Laboratory confirmation of new emulsion stability model. Journal of Petroleum Science and Engineering, 43(1–2), pp. 25–34.

Civan, F., Rai, C.S., and Sondergeld, C.H. 2010. Intrinsic shale permeability determined by pressure-pulse measurements using a multiple-mechanism apparent-gas-permeability non-Darcy model. Paper SPE 135087-PP, the SPE Annual Technical Conference and Exhibition held in Florence, Italy, September 19–22.

Civan, F., Rai, C.S., and Sondergeld, C.H. 2011. Shale-gas permeability and diffusivity inferred by improved formulation of relevant retention and transport mechanisms. Transport in Porous Media Journal, 86(3), pp. 925–944.

Coats, K. 1989. Implicit compositional simulation of single-porosity and dual-porosity reservoirs. Paper SPE 18427, Reservoir Simulation Symposium, held in Houston, TX, Feb. 6–8, pp. 239–275.

Collins, E.R. 1961. Flow of Fluids Through Porous Materials. Penn Well Publishing Co., Tulsa, OK.

Comiti, J., Sabiri, N.E., and Montillet, A. 2000. Experimental characterization of flow regimes in various porous media—III: Limit of Darcy's or creeping flow regime in the case of Newtonian and purely viscous non-Newtonian fluids. Chemical Engineering Science, 55, pp. 3057–3061.

Constantz, J. 1991. Comparison of isothermal and isobaric water retention paths in nonswelling porous materials. Water Resources Research, 27(12), pp. 3165–3170.

Couvreur, J.-M., Mertens de Wilmars, A., and Monjoie, A. 2001. Application of a connected tubes model to gas permeability of a silty soil. In: Clay Science for Engineering, K. Adachi and M. Fukue, eds. A.A. Balkema, Rotterdam, pp. 361–367.

Craig, F.F. Jr. 1980. The Reservoir Engineering Aspects of Waterflooding, Third Printing, November 1980, Society of Petroleum Engineers of AIME, New York.

Crank, J. 1957. The Mathematics of Diffusion. Oxford University Press, Amen House, London.

Crowe, C.T., Elger, D.F., and Roberson, J.A. 2001. Engineering Fluid Mechanics. 7th ed. John Wiley & Sons, Inc., New York.

Dake, L.P. 1978. Fundamentals of Reservoir Engineering. Elsevier Scientific Publishing Co, Amsterdam.

Darcy, H. 1856. Les Fontaines Publiques de la Ville de Dijon. Victor Dalmont, Paris.

Das, A.K. 1997. Generalized Darcy's law including source effect. Journal of Canadian Petroleum Technology, 36(6), pp. 57–59.

Das, A.K. 1999. Generalized Darcy's law with source and heterogeneity effects. Journal of Canadian Petroleum Technology, 38(1), pp. 32–38.

Das, D.B., Gauldie, R., and Mirzaei, M. 2007. Dynamic effects for two-phase flow in porous media: Fluid property effects. AIChE Journal, 53, pp. 2505–2520.

Davis, G.B. and Hill, J.M. 1982. Some theoretical aspects of oil recovery from fractured reservoirs. Transactions of the Institution of Chemical Engineers, 60, pp. 352–358.

de Lemos, M.J.S. 2008. Analysis of turbulent flows in fixed and moving permeable media. Acta Geophysica, 56(3), pp. 562–583.

de Nevers, N. 1970. Fluid Mechanics. Addison-Wesley Publ. Co. Inc., Reading, MA.

deSwaan, A. 1978. Theory of waterflooding in fractured reservoirs. SPEJ, Transactions of the AIME, 265, pp. 117–122.

de Swaan, A. 1990. Influence of shape and skin of matrix-rock blocks on pressure transients in fractured reservoirs. SPE Formation Evaluation, 5, pp. 344–352.

Dietrich, J.K. and Scott, J.D. 2007. Modeling thermally induced compaction in diatomite. Society of Petroleum Engineers Journal, 12(1), pp. 130–144.

Do, H.D., Do, D.D., and Prasetyo, L. 2001. Surface Diffusion and adsorption of hydrocarbons in activated carbon. AIChE Journal, 47(11), pp. 2515–2525.

Donaldson, E.C. and Cmaterialer, M.E. 1980a. Characterization of the Crude Oil Polar Compound Extract, DOE/BETC/RI-80/5, NTIS. Springfield, VA.

Donaldson, E.C., Thomas, R.D., and Lorenz, P. 1969. Wettability determination and its effect on recovery efficiency. Society of Petroleum Engineers Journal, 9(1), pp. 13–20.

Donaldson, E.C., Kendall, R.F., Pavelka, E.A., and Cmaterialer, M.E. 1980b. Equipment and procedures for fluid flow and wettability tests of geological materials. Bartlesville Energy Technology Center, Report No. DOE/BETC/IC-79/5, U.S. DOE, May.

Donaldson, E.C., Ewall, N., and Singh, B. 1991. Characteristics of capillary pressure curves. Journal of Petroleum Science and Engineering, 6(3), pp. 249–261.

Douglas, J. Jr, Blair, P.M., and Wagner, R.J. 1958. Calculation of linear waterflood behavior including the effects of capillary pressure. Petroleum Transactions of the AIME, 213, pp. 96–102.

Dranchuk, P.M., Purvis, R.A., and Robinson, D.B. 1974. Computer calculation of natural gas compressibility factors using the standing and Katz correlation. Report, Institute of Petroleum, IP 74-008.

Duguid, J.O. and Lee, C.Y. 1977. Flow in fractured porous media. Water Resources Research, 13(3), pp. 558–566.

Dullien, F.A.L. 1992. Porous media: Fluid Transport and Pore Structure, 2nd ed. Academic Press, San Diego.

Dupuit, J. 1863. Etudes Théoriques et Pratiques sur le Mouvement des Eaux dans les Canaux Découverts et à Travers les Terrains Perméables, 2nd ed. Dunod, Paris, France.

Eglese, R.W. 1990. Simulated annealing: A tool for operational research. European Journal of Operational Research, 46, pp. 271–281.

Elimelech, M., Gregory, J., Jia, X., and Williams, R.A. 1995. Particle Deposition and Aggregation: Measurement, Modeling and Simulation. Butterworth-Heinemann, Oxford, England.

Enwere, M.P. and Archer, J.S. 1992a. NMR imaging for core flood testing. SCA 9218 Paper.

Enwere, M.P. and Archer, J.S. 1992b. NMR imaging for water/oil displacement in cores under viscous-capillary force control. SPE/DOE 24166 Paper, Proceedings of the SPE/DOE Eighth Symposium on Enhanced Oil Recovery, Tulsa, OK, April 22–24.

Ergun, S. 1952. Fluid flow through packed columns. Chemistry Engineering Progress, 48, pp. 89–94.

Ertekin, T. and Watson, R.W. 1991. An experimental and theoretical study to relate uncommon rock-fluid properties to oil recovery, Contract No. AC22-89BC14477. 64, pp. 66–71. EOR-DOE/BC-90/4 Progress Review, U.S. Department of Energy, Bartlesville, OK, May.

Ertekin, T., King, G.R., and Schwerer, F.C. 1986. Dynamic gas slippage: A unique dual-mechanism approach to the flow of gas in tight formations. SPE Formation Evaluation, 1(1), pp. 43–52.

Evans, R.D. and Civan, F. 1994. Characterization of non-Darcy multiphase flow in petroleum bearing formations. Report, U.S. DOE Contract No. DE-AC22-90BC14659, School of Petroleum and Geological Engineering, U. of Oklahoma.

Farmer, C.L. 1989. The mathematical generation of reservoir geology. Paper presented at the IMA/ SPE European Conference on the Mathematics of Oil Recovery, Cambridge, July 25–27.

Fehlberg, E. 1969. Low-order classical runge-kutta formulas with stepsize control and their application to some heat transfer problems. NASA TR R-315, NASA, Huntsville, AL, July.

Firoozabadi, A. and Aziz, K. 1991. Relative permeabilities from centrifuge data. Journal of Canadian Petroleum Technology, 30, p. 5.

Firoozabadi, A., Thomas, L.K., and Todd, B. 1995. High-velocity flow in porous media. SPE Reservoir Engineering, 10(2), pp. 149–152.

Florence, F.A., Rushing, J.A., Newsham, K.E., and Blasingame, T.A. 2007. Improved permeability prediction relations for low permeability sands. Paper SPE 107954, presented at the 2007 SPE Rocky Mountain Oil and Gas Technology Symposium held in Denver, Colorado, April 16–18.

Forchheimer, P. 1901. Wasserbewegung durch Boden. Zeitschrift des Vereines Deutscher Ingenieure, 45, pp. 1782–1788.

Forchheimer, P. 1914. Hydraulik, 1st ed. Teubner, Leipzig and Berlin. Chap. 15.

Fulcher, G.S. 1925. Analysis of recent data of the viscosity of glasses. Journal of the American Ceramic Society, 8, pp. 339–355.

Garrison, J.R., Pearn, W.C., and von Rosenberg, D.U. 1992. The fractal Menger sponge and Sierpinski carpet as models for reservoir rock/pore systems: I. theory and image analysis of Sierpinski carpets. In Situ, 16(4), pp. 351–406.

Garrison, J.R., Pearn, W.C., and von Rosenberg, D.U. 1993. The fractal Menger sponge and Sierpinski carpet as models for reservoir rock/pore systems: II. Image analysis of natural fractal reservoir rocks. In Situ, 17(1), pp. 1–53.

Geertsma, J. 1974. Estimating the coefficient of inertial resistance in fluid flow through porous media. Society of Petroleum Engineers Journal, 14(5), pp. 445–450.

Gelhar, L.W., Welty, C., and Rehfeldt, K.R. 1992. A critical review on field scale dispersion in aquifers. Water Resources Research, 28(7), pp. 1955–1974.

Getachew, D., Minkowycz, W.J., and Lage, J.L. 2000. A modified form of the kappa-epsilon model for turbulent flows of an incompressible fluid in porous media. International Journal of Heat and Mass Transfer, 43(16), pp. 2909–2915.

Gilman, J.R. 1983. Numerical simulation of phase segregation in the primary porosity (matrix blocks) in two-porosity porous media. Paper SPE 12271, Proceedings of the Reservoir Simulation Symposium, San Francisco, CA, November 15–18, pp. 375–384.

Glover, M.C. and Guin, J.A. 1973. Dissolution of a Homogeneous Porous Medium by Surface Reaction. AIChE Journal, 19(6), pp. 1190–1195.

Gobran, B.D., Brigham, W.E., and Ramey, Jr, H.J. 1987. Absolute permeability as a function of confining pressure, pore pressure, and temperature: Paper SPE 10156. SPE Formation Evaluation Journal, 2(1), pp. 77–84.

Goodrich, L.E. 1978. Efficient numerical techniques for one-dimensional thermal problems with phase change. International Journal of Heat Mass Transfer, 21, pp. 615–621.

Gopal, V.N. 1977. Gas Z-factor equations developed for computer. The Oil and Gas Journal, 75(32), pp. 58–60.

Grant, S.A. and Salehzadeh, A. 1996. Calculation of the temperature effects on wetting coefficients of porous solids and their capillary pressure functions. Water Resources Research, 32(2), pp. 261–270.

Gray, W.G. 1975. A derivation of the equations for multi-phase transport. Chemical Engineering Science, 30(2), pp. 229–233.

Gray, W.G. 1982. On the need for consistent manipulation in volume averaging. Chemical Engineering Science, 37(1), pp. 121–122.

Gray, W.G. 1999. Elements of a systematic procedure for the derivation of macroscale conservation equations for multiphase flow in porous media, kinetic and continuum theories of granular and porous media. In: International Centre for Mechanical Sciences (CISM, Udine), Courses and Lectures, Vol. 400, K. Hutter and K. Wilmanski, eds. Springer-Verlag, Wien, New-York, pp. 67–129.

Gray, W.G. 2000. Macroscale equilibrium conditions for two-phase flow in porous media. International Journal of Multiphase Flow, 26(3), pp. 467–501.

Gray, W.G. and Miller, C.T. 2004. Examination of Darcy's law for flow in porous media with variable porosity. Environmental Science and Technology, 38, pp. 5895–5901.

Gray, W.G., Leijnse, A., Kolar, R.L., and Blain, C.A. 1993. Mathematical Tools for Changing Spatial Scales in the Analysis of Physical Systems. CRC Press, Boca Raton, FL.

Green, L. and Duwez, P. 1951. Fluid flow through porous metals. Journal of Applied Mechanics, 18, p. 39.

Greenkorn, R.A. and Cala, M.A. 1986. Scaling dispersion in heterogeneous porous-media. Industrial & Engineering Chemistry Fundamentals, 25(4), pp. 506–510.

Gruesbeck, C. and Collins, R.E. 1982. Entrainment and deposition of fine particles in porous media. Society of Petroleum Engineers Journal, 22(6), pp. 847–856.

Guan, H., Brougham, D., Sorbie, K.S., and Packer, K.J. 2002. Wettability effects in a sandstone reservoir and outcrop cores from NMR relaxation time distributions. Journal of Petroleum Science and Engineering, 34(1–4), pp. 35–54.

Guevara-Jordan, J.M. and Rodriguez-Hernandez, F. 2001. Application of singular value decomposition to determine streamline distribution for sectionally homogeneous reservoirs. Paper SPE 65414, Proc. Of the 2001 SPE International Symposium on Oilfield Chemistry held in Houston, TX, 13–16 February, p. 11p.

Guo, B., Schechter, D.S., and Baker, R.O. 1998. An integrated study of imbibition waterflooding in the naturally fractured spraberry trend area reservoirs. Paper SPE 39801, Proceedings of the 1998 SPE Permian Basin Oil and Gas Recovery Conference, Midland, TX, March 25–27.

Gupta, A. and Civan, F. 1994a. An improved model for laboratory measurement of matrix to fracture transfer function parameters in immiscible displacement. Paper SPE 28929 presented at the 69th Annual Technical Conference and Exhibition, New Orleans, LA, September 25–28.

Gupta, A. and Civan, F. 1994b. Temperature sensitivity of formation damage in petroleum reservoirs. Paper SPE 27368, Proceedings of the 1994 SPE Formation Damage Control Symposium, Lafayette, LA, February 9–10, pp. 301–328.

Hashemi, H.T. and Sliepcevich, C.M. 1967. A numerical method for solving two-dimensional problems of heat conduction with change of phase. Chemistry Engineering Progress in Symposium Series, 63(79), pp. 34–41.

Hassanizadeh, S.M. and Gray, W.G. 1993. Thermodynamic basis of capillary pressure in porous media. Water Resources Research, 29, pp. 3389–3405.

Haughey, D.P. and Beveridge, G.S.G. 1969. Structural properties of packed beds—A review. The Canadian Journal of Chemical Engineering, 47, pp. 130–140.

Hawking, S.W. 1998. A Brief History of Time. Bantam Books, New York.

Hiby, J.W. 1962. Longitudinal and transverse mixing during single phase flow through granular beds. In: Proceedings of the Symposium on the Interaction Between Fluids and Particles, Institute of Chemical Engineers, London, pp. 312–325.

Hirasaki, G.J. 1991. Wettability: Fundamentals and surface forces. SPE Formation Evaluation, 6(2), pp. 217–226.

Hirst, J.P.P., Davis, N., Palmer, A.F., Achache, D., and Riddiford, F.A. 2001. The "tight gas" challenge: Appraisal results from the Devonian of Algeria. Petroleum Geoscience, 7, pp. 13–21.

Holstad, A. 1995. Mathematical modeling of diagenetic processes in sedimentary basins. In: Mathematical Modeling of Flow Through Porous Media, A.P. Bourgeat, C. Carasso, S. Luckhaus and A. Mikelić, eds. World Scientific Publ. Co. Pte. Ltd., New Jersey, pp. 418–428.

Honarpoor, M., Koedritz, L., and Harvey, A.H. 1986. Relative Permeability of Petroleum Reservoirs. CRC Press, Inc., Boca Raton, FL.

Hsu, C.-T. 2005. Dynamic modeling of convective heat transfer in porous media, Chapter 2. In: Handbook of Porous Media, 2nd ed. K. Vafai, ed. CRC Press, Taylor & Francis Group, LLC, Boca Raton, FL, pp. 39–80.

Hubbert, M.K. 1940. The theory of ground-water motion. Journal of Geology, 68(8), pp. 785–944.

Hubbert, M.K. 1956. Darcy's law and the field equations of the flow of underground fluids. AIME Petroleum Transactions, 207, pp. 222–239.

Huyakorn, P.S., Lester, B.H., and Mercer, J.W. 1983. Finite element technique for modeling transport in fractured porous media 1. Single species transport. Water Resources Research, 19(3), pp. 841–854.

Ikoku, C.U. 1984. Natural Gas Production Engineering. John Wiley & Sons, Inc., New York.

Inaba, H. 1983. Heat transfer behavior of frozen soils. Transactions of the ASME Journal of Heat Transfer, 105, pp. 680–683.

Ingham, D.B., Pop, I., and Cheng, P. 1990. Combined free and forced convection in a porous medium between two vertical wails with viscous dissipation. Transport in Porous Media, 5, pp. 381–398.

Ipsen, D.C. 1960. Units, Dimensions and Dimensionless Numbers. McGraw-Hill Book Company, New York.

Ives, K.J. 1967. Deep filters. Filtration and Separation. March/April, pp. 125–135.

Ives, K.J. 1987. Filtration of clay suspensions through sand. Clay Minerals, 22, pp. 49–61.

Iwasaki, T. 1937. Some notes on sand filtration. Journal of American Water Works Association, 29(10), pp. 1591–1602.

Janna, W.S. 1993. Introduction to Fluid Mechanics. PWS Publishing Co., Boston.

Javadpour, F. 2009. Nanopores and apparent permeability of gas flow in mudrocks (shales and siltstone). Journal of Canadian Petroleum Technology, 48(8), pp. 16–21.

Javadpour, F., Fisher, D., and Unsworth, M. 2007. Nanoscale gas flow in shale gas sediments. Journal of Canadian Petroleum Technology, 46(10), pp. 55–61.

Johns, M.L. and Gladden, L.F. 2000. Probing ganglia dissolution and mobilization in a water-saturated porous medium using MRI. Journal of Colloid and Interface Science, 225, pp. 119–127.

Johnson, E.F., Bossler, D.P., and Naumann, V.O. 1959. Calculation of relative permeability from displacement experiments. Transactions of the AIME, 216, pp. 370–372.

Jones, S.C. and Roszelle, W.O. 1978. Graphical techniques for determining relative permeability from displacement experiments. JPT, Transactions of the AIME, 265, pp. 807–817.

Kalaydjian, K.J.-M. 1992. Dynamic capillary pressure curve for water/oil displacement in porous media: Theory vs. experiment. Paper SPE 24813, Proceedings of the 67th SPE Annual Technical Conference and Exhibition, Washington, DC, October 4–7, pp. 491–506.

Kaviany, M. 1991. Principles of Heat Transfer in Porous Media. Springer-Verlag New York Inc., New York.

Kazemi, H., Gilman, J.R., and El-Sharkawy, A.M. 1992. Analytical and numerical solution of oil recovery from fractured reservoirs with empirical transfer functions. SPE Reservoir Engineering Journal, 7(2), pp. 219–227.

Kazemi, H., Merrill, L., Porterfield, K., and Zeman, P. 1976. Numerical simulation of water-oil flow in naturally fractured reservoirs. Society of Petroleum Engineers Journal, 16(6), pp. 317–326.

Kececioglu, I. and Jiang, Y. 1994. Flow through porous media of packed spheres saturated with water. Journal of Fluids Engineering, Transactions of the American Society of Mechanical Engineers, 116, pp. 164–170.

Kemblowski, Z., Dziubinski, M., and Sek, J. 1989. Flow of non-Newtonian fluids through granular media. In: Transport phenomena in polymeric systems, R.A. Mashelkar, A.S. Mujumdar and R. Kamal, eds. Ellis Horwood, Chichester, pp. 117–175.

Kerig, P.D. and Watson, A.T. 1987. A new algorithm for estimating relative permeabilities from displacement experiments. SPE Reservoir Engineering, 2(2), pp. 103–112.

Kim, H.C., Bishnoi, P.R., Heidemann, R.A., and Rizvi, S.S.H. 1987. Kinetics of methane hydrate decomposition. Chemistry Engineering Science, 42(7), pp. 1645–1653.

Kimber, K.D., Ali, S.M.F., and Puttagunta, V.R. 1988. New scaling criteria and their relative merits for steam recovery experiments. Journal of Canadian Petroleum Technology, 27(4), pp. 86–94.

King, G.R. and Ertekin, T. 1991. State-of-the art modeling for unconventional gas recovery. SPE Formation Evaluation Journal, 6(1), pp. 63–72.

Kirkpatric, S., Gelatt, C.D., and Vecchi, M.P. 1983. Optimization by simulated annealing. Science, 220, pp. 671–680.

Kline, J. 1965. Similitude and Approximation Theory. McGrawHill Book Co., New York.

Klinkenberg, L.J. 1941. The permeability of porous media to liquids and gases. API Drilling and Production Practices, Paper No. 41-200, pp. 200–213.

Knudsen, M. 1909. The laws of molecular and viscous flow of gases through tubes. Annals of Physik, 28, pp. 75–177.

Koch, D.L. and Sangani, A.S. 1999. Particle pressure and marginal stability limits for a homogeneous monodisperse gas-fluidized bed: Kinetic theory and numerical simulations. Journal of Fluid Mechanics, 400, p. 229.

Kozeny, J. 1927. Uber Kapillare Leitung des Wasser im Boden. Sitzungsbericht der Akademie der Wissenschaften, Wien, 136, pp. 271–306.

Krantz, W.B. 2000. An alternative method for teaching and implementing dimensional analysis. Chemical Engineering Education, 34(3), pp. 216–221.

Krantz, W.B. and Sczechowski, J.G. 1994. Scaling Initial and Boundary Value Problems: A Tool in Engineering Teaching and Practice. Chemical Engineering Education, 28, pp. 236–241.

Kulkarni, R., Watson, A.T., Nordtvedt, J.-E., and Sylte, A. 1998. Two-phase flow in porous media: Property identification and model validation. AIChE Journal, 44(11), pp. 2337–2350.

Kumar, A. 1977a. Strength of water drive or fluid injection from transient well test data. Journal of Petroleum Technology, 29(11), pp. 1497–1508.

Kumar, A. 1977b. Steady flow equations for wells in partial water-drive reservoirs. Journal of Petroleum Technology, 29(12), pp. 1654–1656.

Ladd, A.J.C. 1990. Hydrodynamic transport coefficients of random dispersions of hard spheres. The Journal of Chemical Physics, 93, p. 3484.

Lake, L.W., Pope, G.A., Carey, G.F., et al. 1984. Isothermal, multiphase, multicomponent fluid-flow in permeable media .1. Description and mathematical formulation. In Situ, 8(1), pp. 1–40.

Langmuir, I. 1916. The constitution and fundamental properties of solids and liquids I: Solids. Journal of the American Chemistry Society, 38, pp. 2221–2295.

Langmuir, I. 1917. The constitution and fundamental properties of solids and liquids II: Liquids. Journal of the American Chemistry Society, 39, pp. 1848–1906.

Le Gallo, Y., Bildstein, O., and Brosse, E. 1998. Coupled reaction-flow modeling of diagenetic changes in reservoir permeability, porosity and mineral compositions. Journal of Hydrology, 209, pp. 366–388.

Lerman, A. 1979. Geochemical Processes, Water and Sediment Environments. John Wiley&Sons, New York.

Leverett, M.C. 1941. Capillary behavior in porous solids. Transactions of the American Institute of Mining and Metallurgical Engineers, 142, pp. 152–169.

Lewis, R.W. and Schrefler, B.A. 1987. The Finite Element Method in the Deformation and Consolidation of Porous Media. John Wiley & Sons, New York.

Li, K. and Horne, R.N. 2001. An experimental and analytical study of steam/water capillary pressure. SPE Reservoir Evaluation & Engineering, 4(6), pp. 477–482.

Li, S., Dong, M., Dai, L., Li, Z., and Pan, X. 2004. Determination of gas permeability of tight reservoir cores without using klinkenberg correlation. Paper SPE 88472-MS, SPE Asia Pacific Oil and Gas Conference and Exhibition, 18–20 October 2004, Perth, Australia.

Lichtner, P.C. 1992. Time-space continuum description of fluid/rock interaction in permeable media. Water Resources Research, 28(12), pp. 3135–3155.

Liepmann, H.W. 1961. Gas kinetics and gas dynamics of orifice flow. Journal of Fluid Mechanics, 10, p. 65.

Lim, K.T. and Aziz, K. 1995. Matrix-fracture transfer shape factors for dual-porosity simulators. Journal of Petroleum Science and Engineering, 13(3/4), pp. 169–178.

Liu, H. and Seaton, H.A. 1994. Determination of the connectivity of porous solids from nitrogen sorption measurements—III. Solids containing large mesopores. Chemical Engineering Science, 49(11), pp. 1869–1878.

Liu, S. and Masliyah, J.H. 1996. Single fluid flow in porous media. Chemical Engineering Communications, 148–150, pp. 653–732.

Liu, S. and Masliyah, J.H. 2005. Dispersion in porous media, Chapter 3. In: Handbook of Porous Media, 2nd ed. K. Vafai, ed. CRC Press, Taylor&Francis Group, LLC, Boca Raton, pp. 81–140.

Liu, X., Civan, F., and Evans, R.D. 1995. Correlation of the non-Darcy flow coefficient. Journal of Canadian Petroleum Technology, 34(10), pp. 50–54.

Liu, X., Ormond, A., Bartko, K., Li, Y., and Ortoleva, P. 1997. A geochemical reaction-transport simulator for matrix acidizing analysis and design. Journal of Petroleum Science and Engineering, 17(1/2), pp. 181–196.

Loeb, L.B. 1934. The kinetic theory of gases, 2nd ed. McGraw-Hill Co. Inc., New York.

Logan, B.E., Jewett, D.G., Arnold, R.G., Bouwer, E.J., and O'Melia, C.R. 1997. Closure of "Clarification of clean-bed filtration models." Journal of Environmental Engineering-ASCE, 123(7), pp. 730–731.

Lominé, F. and Oger, L. 2009. Dispersion of particles by spontaneous interparticle percolation through unconsolidated porous media. Physical Review E, 79(5), pp. 051307.1–051307.12.

Longmuir, G. 2004. Pre-Darcy flow: A missing piece of the improved oil recovery puzzle? Paper SPE 89433, presented at the 2004 SPE/DOE Fourteenth Symposium on Improved Oil Recovery held in Tulsa, Oklahoma, U.S.A., 17–21 April, p. 14p.

Loyalka, S.K. and Hamoodi, S.A. 1990. Poiseuille flow of a rarefied gas in a cylindrical tube: Solution of linearized boltzmann equation. Physical Fluids A, 2(11), pp. 2061–2065.

Luan, Z. 1995. Splitting pseudospectral algorithm for parallel simulation of naturally fractured reservoirs. Paper SPE 30723 presented at the Annual Technical Conference & Exhibition held in Dallas, October 22–25.

Luan, Z. and Kleppe, J. 1992. Discussion of analytical and numerical solution of waterflooding in fractured reservoirs with empirical transfer function. SPE Reservoir Engineering, 7(4), p. 456.

Lukianov, V.S. and Golovko, M.D. 1957. Calculation of the Depth of Freeze in Soils (in Russian). Bull. 23, Union of Transp. Constr., Moscow, (English translation, National Technical Information Service, Springfield, Va., 1957).

Lunardini, V.J. 1981. Heat Transfer in Cold Climates. Van Nostrand Reinhold, New York.

Lunardini, V.H. 1991. Heat Transfer with Freezing and Thawing. Elsevier Science Publishers B.V.

Machac, I. and Dolejs, V. 1981. Flow of generalised Newtonian liquids through fixed beds of non-spherical particles. Chemical Engineering Science, 36, pp. 1679–1686.

Madden, M.P. and Strycker, A.R. 1989. Thermal processes for light oil recovery. 1988 Annual Report, Work Performed Under Cooperative Agreement No. FC22-83FE60149, for the U.S.

Department of Energy, Bartlesville Project Office, Bartlesville, Oklahoma, September, pp. 205–218.

Mahadevan, J., Sharma, M.M., and Yortsos, Y.C. 2007. Evaporative cleanup of water blocks in gas wells, Paper 94215. Society of Petroleum Engineers Journal, 12(2), pp. 209–216.

Makse, H.A., Johnson, D.L., and Schwartz, L.M. 2000. Packing of compressible granular materials. Physical Review Letters, 84(18), pp. 4160–4163.

Manoranjan, V.S. and Stauffer, T.B. 1996. Exact solution for contaminant transport with kinetic langmuir sorption. Water Resources Research, 32(3), pp. 749–752.

Marathe, R.V., Shenoy, U.V., and Khilar, K.C. 1995. Application of singular value decomposition to determine streamline distribution for unit mobility ratio displacements. The Journal of Canadian Petroleum Technology, 34(5), pp. 55–62.

Marle, C.M. 1981. Multiphase Flow in Porous Media. Gulf Publishing Co., Houston.

Marle, C.M. 1982. On macroscopic equations governing multiphase flow with diffusion and chemical reactions in porous media. International Journal of Engineering Science, 20(5), pp. 643–662.

Martin, J.C. and Wegner, R.E. 1979. Numerical solution of multiphase, two-dimensional incompressible flow using stream-tube relationships. Paper SPE 7140-PA, Society of Petroleum Engineers Journal, 19(5), pp. 313–323.

Matejka, M.C., Llanos, E.M., and Civan, F. 2002. Experimental determination of the matrix-to-fracture transfer functions for oil recovery by water imbition. Journal of Petroleum Science and Engineering, 33(4), pp. 253–264.

McCain, W.D. 1990. The Properties of Petroleum Fluids. PennWell, Tulsa.

Metropolis, M., Rosenbluth, M., Rosenbluth, A., Teller, A., and Teller, E. 1953. Equation of state calculations by fast computing machines. Journal of Chemical Physics, 30, pp. 1087–1092.

Metz, C.R. 1976. Physical Chemisty. McGraw-Hill, New York.

Metzner, A.B. 1956. Non-Newtonian technology: Fluid mechanics, mixing and heat transfer. Advances in Chemical Engineering, 1, pp. 79–150.

Miller, C. 1972. Predicting non-Newtonian flow behaviour in ducts of unusual cross-sections. Industrial and Engineering Chemistry, Fundamentals, 11, pp. 524–528.

Moench, A.F. 1984. Double-porosity models for a fissured groundwater reservoir with fracture skin. Water Resources Research, 20(7), pp. 831–846.

Monkos, K.A. 2003. Method of calculations of the parameters in the vogel-tammann-fulcher's equation: An application to the porcine serum albumin aqueous solutions. Current Topics in Biophysics, 27(1–2), pp. 17–21.

Moore, W.J. 1972. Physical Chemistry, Vol. 103, 4th ed. Prentice-Hall, Inc, Englewood Cliffs, New Jersey.

Mowers, T.T. and Budd, D.A. 1996. Quantification of porosity and permeability reduction due to calcite cementation using computer-assisted petrographic image analysis techniques. AAPG Bulletin, 80, pp. 309–322.

Muskat, M. 1937. The Flow of Homogeneous Fluids Through Porous Media. McGraw Hill Book Co., New York.

Nakano, Y. and Brown, J. 1971. Effect of a freezing zone of finite width on the thermal regime of soils. Water Resources Research, 7(5), pp. 1226–1233.

Nakayama, A. and Kuwahara, F. 1999. A macroscopic turbulence model for flow in a porous medium. Journal of Fluids Engineering-Transactions of the ASME, 121(2), pp. 427–433.

Needham, R.B. 1976. Prediction of the permeability of a fragmented oil shale bed during in-situ retorting with hot gas. Paper SPE 6071, SPE Annual Fall Technical Conference and Exhibition, 3–6 October, New Orleans, Louisiana.

Nelson, P.H. 1994. Permeability-porosity relationships in sedimentary rocks. The Log Analyst, 35, pp. 38–62.

Nelson, P.H. 2000. Evolution of permeability-porosity trends in sandstones. SPWLA 41st Annual Logging Symposium, June 4–7, p. 14 p.

Neretnieks, I. 1980. Diffusion in the rock matrix: An important factor in radionuclide retardation? Journal of Geophysical Research, 85(B8), pp. 4379–4397.

Nicholson, D. and Petropoulos, J.H. 1985. Calculation of the surface flow of a dilute gas in model pore from first principles. Part III. Molecular gas flow in single pores and simple model porous media. Journal of Colloid and Interface Science, 106, 538.

Nimmo, J.R. and Miller, E.E. 1986. The temperature dependence of isothermal moisture vs. potential characteristics of soils. Soil Science Society of America Journal, 50(5), pp. 1105–1113.

Norinaga, K., Hayashi, J.-I., Kudo, N., and Chiba, T. 1999. Evaluation of effect of predrying on the porous structure of water-swollen coal based on the freezing property of pore condensed water. Energy and Fuels, 13, pp. 1058–1066.

Novak, V. 1975. Non-isothermal flow of water in unsaturated soils. Journal of Hydrological Sciences (Polish Academy of Sciences), 2(1–2), pp. 37–51.

Ochi, J. and Vernoux, J.-F. 1998. Permeability decrease in sandstone reservoirs by fluid injection- hydrodynamic and chemical effects. Journal of of Hydrology, 208, pp. 237–248.

Oddo, J.E. and Tomson, M.B. 1994. Why scale forms and how to predict it. SPE Production & Facilities, 9(1), pp. 47–54.

Odeh, A.S. 1982. An overview of mathematical-modeling of the behavior of hydrocarbon reservoirs. Siam Review, 24(3), pp. 263–273.

Odeh, A.S. and Dotson, B.J. 1985. A method for reducing the rate effect on oil and water relative permeabilities calculated from dynamic displacement data. Paper SPE 14417-PA. Journal of Petroleum Technology, 37(11), pp. 2051–2058.

Odeh, A.S. and Heinemann, R.F. 1988. Reservoir simulation—The present and the future. In Situ, 12(3), pp. 153–180.

Olivella, S., Gens, A., and Carrera, J. 2000. Water phase change and vapour transport in low permeability undersaturated soils with capillary effects. In: Computational Methods for Flow and Transport in Porous Media, J.M. Crolet, ed. Kluver Academic Publishers, Boston.

Ortoleva, P. 1994. Geochemical Self-Organization. Oxford University Press, New York.

Ouenes, A., Fasanino, G., and Lee, R.L. 1992. Simulated annealing for interpreting gas/water laboratory corefloods. SPE 24870 Paper, Proceedings of the 67th SPE Annual Technical Conference & Exhibition, Washington, DC, October 4–7, Reservoir Engineering, pp. 43–55.

Pandya, V.B., Bhuniya, S., and Khilar, K.C. 1998. Existence of a critical particle concentration in plugging of a packed bed. AIChE Journal, 44(4), pp. 978–981.

Panfilov, M. 2000. Macro scale model of flow through highly heterogeneous porous media. Kluwer Academic Publishers, Boston.

Parker, J.C. 1989. Multi-phase flow and transport in porous media. Review of Geophys, 27, pp. 311–328.

Parvatiyar, M.G. 1998. Entropy generation in ultrafiltration processes. Journal of Membrane Science, 144, pp. 125–132.

Pascal, H. 1986. Rheological effects of non-Newtonian behaviour of displacing fluids on stability of a moving interface in radial oil displacement mechanism in porous media. International Journal of Engineering Science, 24, pp. 1465–1476.

Pascal, H. 1990. Non isothermal flow of non-Newtonian fluids through a porous medium. International Journal of Heat Mass Transfer, 33, pp. 1937–1944.

Pascal, H. and Pascal, F. 1989. On viscoelastic effects in non-Newtonian steady flows through porous media. Transport in Porous Media, 4, pp. 17–35.

Pavone, D.R. 1990. A Darcy's law extension and a new capillary pressure equation for two-phase flow in porous media. SPE 20474 Paper, Proceedings of the 65th SPE Annual Technical Conference & Exhibition, New Orleans, LA, September 23–26, EOR, pp. 111–119.

Peaceman, D.W. 1977. Fundamentals of Numerical Reservoir Simulation. Elsevier Scientific, Amsterdam.

Peaceman, D.W. 1990. Interpretation of wellblock pressures in numerical reservoir simulation. Part 3: Off-center and multiple wells within a wellblock. Paper SPE 16976-PA. SPE Reservoir Engineering Journal, 5(2), pp. 227–232.

Pedras, M.H.J. and Lemos, M.J.S. 2001. Macroscopic turbulence modeling for incompressible flow through undeformable porous media. International Journal of Heat and Mass Transfer, 44(6), pp. 1081–1093.

Penuela, G. and Civan, F. 2001. Two-phase flow in porous media: Property identification and model validation. Letter to the Editor. AIChE Journal, 47(3), pp. 758–759.

Perkins, L.R. and Geankoplis, C.J. 1969. Molecular diffusion in a ternary liquid system with diffusing component dilute. Chemical Engineering Science, 24(7), pp. 1035–1042.

Perrier, E., Rieu, M., Sposito, G., and de Marsily, G. 1996. Models of the water retention curve for soils with a fractal pore size distribution. Water Resources Research, 32(10), pp. 3025–3031.

Peters, E.J., Afzal, N., and Gharbi, R. 1993. On scaling immiscible displacements in permeable media. Journal of Petroleum Science and Engineering, 9(3), pp. 183–205.

Peters, E.J. and Flock, D.L. 1981. The onset of instability during 2-phase immiscible displacement in porous-media. Society of Petroleum Engineers Journal, 21(2), pp. 249–258.

Pong, K.C., Ho, C.M., Liu, J.Q., and Tai, Y.C. 1994. Non-linear pressure distribution in uniform micro-channels. Applications of Microfabrication to Fluid Mechanics, in: ASMEFED, 197, pp. 51–56.

Prada, A. and Civan, F. 1999. Modification of Darcy's law for the threshold pressure gradient. Journal of Petroleum Science and Engineering, 22(4), pp. 237–240.

Prats, M. 1982. Thermal Recovery. Society of Petroleum Engineers of AIME, Dallas, Texas.

Present, R.D. 1958. Kinetic Theory of Gases. McGraw-Hill, New York.

Pruess, K. and Narasimhan, T.N. 1985. A practical method for modeling fluid and heat flow in fractured porous media. Paper SPE 10509-PA, Society of Petroleum Engineers Journal, 25(1), pp. 14–26.

Pruess, K. and Tsang, Y.W. 1989. On relative permeability of rough-walled fractures. Presented at 14th Workshop on Geothermal Reservoir Engineering, Stanford University, January 24–26.

Quintard, M. and Whitaker, S. 2005. Coupled, nonlinear mass transfer and heterogeneous reaction in porous media, Chapter 1. In: Handbook of Porous Media, 2nd ed. K. Vafai, ed. CRC Press, Taylor&Francis Group, LLC, Boca Raton, FL, pp. 3–37.

Radhakrishnan, R., Gubbins, K.E., and Sliwinska-Bartkowiak, M. 2000a. Effect of the fluid-wall interaction on freezing of confined fluids: Towards the development of a global phase diagram. Journal of Chemical Physics, 112, pp. 11048–11057.

Radhakrishnan, R., Gubbins, K.E., Sliwinska-Bartkowiak, M., and Kaneko, K. 2000b. Understanding freezing behavior in pores. In: Adsorption Science and Technology, D.D. Do, ed. World Scientific, Singapore, pp. 234–238.

Radhakrishnan, R., Gubbins, K.E., and Sliwinska-Bartkowiak, M. 2002a. Global phase diagrams for freezing in porous media. Journal of Chemical Physics, 116, pp. 1147–1155.

Radhakrishnan, R., Gubbins, K.E., Sliwinska-Bartkowiak, M., and Kaneko, K. 2002b. Understanding freezing behavior in porous materials. In: Fundamentals of Adsorption 7, K. Kaneko, H. Kanoh and Y. Hanzawa, eds. International Adsorption Society, IK International, Chiba, Japan, p. 341.

Rajani, B.B. 1988. A simple model for describing variation of permeability with porosity for unconsolidated sands. In Situ, 12(3), pp. 209–226.

Ramakrishnan, T.S. and Cappiello, A. 1991. A new technique to measure statical and dynamic properties of a partially saturated porous medium. Chemistry Engineering Science, 46(4), pp. 1157–1163.

Rangarajan, R., Mazid, M.A., Matsuura, T., and Sourirajan, S. 1984. Permeation of pure gases under pressure through asymmetric porous membranes. Membrane characterization and prediction of performance. Ind Engineering Chemistry Process Design and Development, 23, pp. 79–87.

Rasmussen, M.L. and Civan, F. 1998. Analytical solutions for water floods in fractured reservoirs obtained by an asymptotic approximation. Society of Petroleum Engineers Journal, 3(3), pp. 249–252.

Rasmussen, M.L. and Civan, F. 2003. Full, short, and long-time analytical solutions for hindered matrix-fracture transfer models of naturally fractured petroleum reservoirs. Paper SPE 80892, SPE Mid-Continent Operations Symposium, Oklahoma City, OK, 22–25 March.

Revil, A. and Cathles, L.M. III. 1999. Permeability of shaly sands. Water Resources Research, 35(3), pp. 651–662.

Richardson, J.G. 1961. Flow through porous media. In: Handbook at Fluid Dynamics, Section 16, V.L. Streeter, ed. McGraw-Hill, New York, pp. 68–69.

Richmond, P.C. and Watson, A.T. 1990. Estimation of multiphase flow functions from displacement experiment. SPE Reservoir Engineering, 5(1), pp. 121–127.

Ridgway, K. and Tarbuck, K.J. 1967. The random packing of spheres. British Chemical Engineering, 12(3), pp. 384–388.

Robin, M., Rosenberg, E., and Fassi-Fihri, O. 1995. Wettability studies at the pore level: A new approach by use of Cryo-SEM. SPE Formation Evaluation, 10(1), pp. 11–19.

Rojas, G.A., Zhu, T., Dyer, S.B., Thomas, S., and Farouq Ali, S.M. 1991. Scaled model studies of CO_2 floods. SPE Reservoir Engineering, 6(2), pp. 169–178.

Rowlinson, J.S. and Widom, B. 1982. Molecular Theory of Capillarity, International Series of Monographs on Chemistry. Oxford Science Publications, Clarendon Press, Oxford.

Roy, S., Raju, R., Chuang, H.F., Cruden, B.A., and Meyyappan, M. 2003. Modeling gas flow through microchannels and nanopores. Journal of Applied Physics, 93(8), pp. 4870–4879.

Ruth, D. and Ma, H. 1992. On the derivation of the Forchheimer equation by means of the averaging theorem. Transport in Porous Media, 7(3), pp. 255–264.

Ruth, D. and Ma, H. 1997. Physical explanations of non-Darcy effects for fluid flow in porous media. SPE Formation Evaluation, 12(1), pp. 13–18.

Sabiri, N.E. and Comiti, J. 1995. Pressure drop in non-Newtonian purely viscous fluid flow through porous media. Chemical Engineering Science, 50, pp. 1193–1201.

Sabiri, N.E. and Comiti, J. 1997a. Experimental validation of a model allowing pressure gradient determination for non-Newtonian purely viscous fluid-flow through packed beds. Chemical Engineering Science, 52, pp. 3589–3592.

Saito, M.B. and de Lemos, M.J.S. 2010. A macroscopic two-energy equation model for turbulent flow and heat transfer in highly porous media. International Journal of Heat and Mass Transfer, 53(11–12), pp. 2424–2433.

Saito, A., Okawa, S., Suzuki, T., and Maeda, H. 1995. Calculation of permeability of porous media using direct simulation Monte Carlo method (effect of porosity and void distribution on permeability). In: Proceedings of the ASME/JSME Thermal Engineering Joint Conference, Vol. 3, L.S. Fletcher and T. Aihara, eds. The American Society of Mechanical Engineers, New York, pp. 297–304. The Japan Society of Mechanical Engineers, Tokyo.

Salem, H.S. and Chilingarian, G.V. 2000. Influence of porosity and direction of flow on tortuosity in unconsolidated porous media. Energy Sources, 22, pp. 207–213.

Satman, A., Zolotukhin, A.B., and Soliman, M.Y. 1984. Application of the time-dependent overall heat-transfer coefficient concept to heat-transfer problems in porous media. Paper SPE 8909, Society of Petroleum Engineers Journal, 24(1), pp. 107–112.

Schaaf, S.A. and Chambre, P.L. 1961. Flow of Rarefied Gases. Princeton University Press, Princeton, NJ.

Schechter, R.S. and Gidley, J.L. 1969. The change in pore size distribution from surface reactions in porous media. AIChE Journal, 15(3), pp. 339–350.

Schembre, J.M. and Kovscek, A.R. 2005. Mechanism of formation damage at elevated temperature. Journal of Energy Resources Technology-Transactions of the ASME, 127(3), pp. 171–180.

Scholes, O.N., Clayton, S.A., and Hoadley, A.F.A. 2007. Permeability anisotropy due to consolidation of compressible porous media. Transport in Porous Media, 68(3), pp. 365–387.

Schulenberg, T. and Muller, U. 1987. An improved model for two-phase flow through beds of coarse particles. International Journal of Multiphase Flow, 13(1), pp. 87–97.

Selim, M.S. and Sloan, E.D. 1989. Heat and mass transfer during the dissociation of hydrates in porous media. AIChE Journal, 35(6), pp. 1049–1052.

Shapiro, A.A. and Wesselingh, J.A. 2008. Gas transport in tight porous media. Chemical Engineering Journal, 142(1), pp. 14–22.

Sharma, R. 1985. On the application of reversible work to wetting/dewetting of porous media. Colloids and Surfaces, 16(1), pp. 87–91.

Shikhov, V.M. and Yakushin, V.I. 1987. A method for evaluating the flow of fluids in a porous media with sources and sinks. Fluid Mechanics—Soviet Research, 16(3), pp. 108–112.

Shirato, M., Murase, T., Ivitari, E., Tiller, F.M., and Alciatore, A.F. 1987. Filtration in the chemical process industry. In: Filtration: Principle and Practice, 2nd ed. M.J. Matteson and C., Orr, eds. Marcel Dekker, New York.

Shirman, E.I. and Wojtanowicz, A.K. 1996. Analytical modeling of crossflow into wells in stratified reservoirs: Theory and field application. Proceedings of the 7th International

Scientific and Technical Conference, New Methods and Technologies in Petroleum Geology, Drilling, and Reservoir Engineering, Technical University of Mining and Metallurgy, Krakow, Poland, June 20–21, pp. 163–182.

Shook, M., Li, D.C., and Lake, L.W. 1992. Scaling immiscible flow through permeable media by inspectional analysis. In Situ, 16(4), pp. 311–349.

Siddiqui, S., Ertekin, T., and Hicks, P.J. 1993. A comparative analysis of the performance of two-phase relative permeability models in reservoir engineering calculations. Paper SPE 26911 presented at the Eastern Regional Conf. and Exhibition, Pittsburgh, PA, November 2–4.

Singh, M. and Mohanty, K.K. 2000. Permeability of spatially correlated porous media. Chemical Engineering Science, 55, pp. 5393–5403.

Skjaeveland, S.M. and Kleppe, J. 1992. Recent Advances in Improved Oil Recovery Methods for North Sea Sandstone Reservoirs. SPOR Monograph, Norwegian Petroleum Institute.

Slattery, J.C. 1969. Single-phase flow through porous media. AIChE Journal, 15, pp. 866–872.

Slattery, J.C. 1972. Momentum, Energy and Mass Transfer in Continua. McGraw-Hill Book Co., New York.

Slattery, J.C. 1990. Interfacial Transport Phenomena. Springer-Verlag New York Inc., New York.

Smith, G.W. 1985. Geology of the deep Tuscaloosa (Upper Cretaceous) gas trend in Louisiana. GCSSEPM Foundation Fourth Annual Research Conference Proceedings, June.

Somerton, W.H. 1992. Thermal properties and temperature-related behavior of rock/fluid systems, Developments in petroleum science, Vol. 37. Elsevier, Amsterdam, New York.

Stahl, D.E. 1971. Transition range flow through microporous Vycor. Ph.D. thesis, Chemical Engineering Department, The University of Iowa.

Starling, K.E. and Ellington, R.T. 1964. Viscosity correlations for nonpolar dense fluids. AIChE Journal, 10, pp. 11–15.

Steefel, C.I. and Lasaga, A.C. 1990. Evolution of dissolution patterns-permeability change due to coupled flow and reaction. In: Chemical Modeling of Aqueous Systems II, Chapter 16, D.C. Melchior and R.L. Basset, eds. ACS Symposium Series 416, American Chemical Society, Washington, DC, pp. 212–225.

Takatsu, Y. and Masuoka, T. 1998. Turbulent phenomena in flow through porous media. Journal of Porous Media, 1(3), pp. 243–251.

Tammann, G. and Hesse, W. 1926. Die abhängigkeit der viskosität von der temperature bei unterkühlten flüssigkeiten. Zeitscrift für Anorg Allg Chemistry, 156, pp. 245–257.

Teng, H. and Zhao, T.S. 2000. An extension of Darcy's law to non-stokes flow in porous media. Chemical Engineering Science, 55, pp. 2727–2735.

Terzaghi, K. 1923. Die Berechnung der Durchlassigkeitsziffer des Tones aus dem Verlauf der hydrodynamischen Spannungserscheinungen, Sitzungsber. Akad Wiss Wien Math -Naturwiss Kl., Abt, 2A(132), p. 105.

Thomas, G.W. 1982. Principles of Hydrocarbon Reservoir Simulation. International Human Resources Development Corporation, Boston.

Tien, C. 1989. Granular Filtration of Aerosols and Hydrosols. Butterworths, Stoneham, MA.

Tien, C., Bai, R., and Ramarao, B.V. 1997. Analysis of cake growth in cake filtration: Effect of fine particle retention. AIChE Journal, 43(1), pp. 33–44.

Tiller, F.M. and Crump, J.R. 1985. Recent advances in compressible cake filtration theory. In: Mathematical Models and Design Methods in Solid-Liquid Separation, A. Rushton, ed. Martinus Nijhoff, Dordrecht.

Tison, S.A. and Tilford, C.R. 1993. Low density water vapor measurements; the NIST primary standard and instrument response. NIST Internal Report 5241.

Todd, A.C. and Yuan, M.D. 1988. Barium and strontium sulfate solid solution formation in relation to north sea scaling problems. Paper SPE 18200, Proceedings of the Society of Petroleum Engineers 63rd Annual Technical Conference and Exhibition, Houston, TX, October 2–5, pp. 193–198.

Tóth, J. 1995. Determination of the Leverett- and relative permeability functions from waterflood experiments. Kőolaj és Földgáz, 28(128), pp. 65–71.

Tóth, J., Bodi, T., Szucs, P., and Civan, F. 1998. Practical method for analysis of immiscible displacement in laboratory core tests. Transport in Porous Media, 31, pp. 347–363.

Tóth, J., Bodi, T., Szucs, P., and Civan, F. 2001. Direct determination of relative permeability from non-steady-state constant pressure and rate displacements. Paper SPE 67318, SPE Production and Operations Symposium, Oklahoma City, OK, March 24–27.

Tóth, J., Bódi, T., Szücs, P., and Civan, F. 2002. Convenient formulae for determination of relative permeability from unsteady-state fluid displacements in core plugs. Journal of Petroleum Science and Engineering, 36(1–2), pp. 33–44.

Tóth, J., Bódi, T., Szücs, P., and Civan, F. 2006. Near-wellbore field water/oil relative permeability inferred from production with increasing water-cut. Paper SPE 102312, SPE Annual Technical Conference and Exhibition, San Antonio, TX, September 24–27.

Tóth, J., Bódi, T., Szücs, P., and Civan, F. 2010. Determining effective fluid saturation, relative permeability, heterogeneity and displacement efficiency in drainage zones of oil wells producing under waterdrive. Paper 140783-PA, Journal of Canadian Petroleum Technology, 49(11), pp. 69–80.

Tran, T.V., Civan, F., and Robb, I. 2009. Correlating flowing time and condition for perforation plugging by suspended particles. Paper SPE120847-MS, SPE Drilling and Completion Journal, 24(3), pp. 398–403.

Tsypkin, G.G. 1998. Decomposition of gas hydrates in low-temperature reservoirs. Fluid Dynamics, 33(1), pp. 82–90.

Tufenkji, N. and Elimelech, M. 2004. Correlation equation for predicting single-collector efficiency in physicochemical filtration in saturated porous media. Environmental Science & Technology, 38(2), pp. 529–536.

Tufenkji, N. and Elimelech, M. 2005. Response to comment on "Correlation equation for predicting single-collector efficiency in physicochemical filtration in saturated porous media. Environmental Science & Technology, 39(14), pp. 5496–5497.

Tutu, N.K., Ginsberg, T., and Chen, J.C. 1983. Interfacial drag for two-phase flow through high permeability porous beds. In: Interfacial Transport Phenomena, J.C. Chen and S.G. Bankoff, eds. ASME, New York, pp. 37–44.

Ucan, S., Civan, F., and Evans, R.D. 1993. Simulated annealing for relative permeability and capillary pressure from unsteady-state non-darcy displacement. Paper SPE 26670, Proceedings of the 1993 SPE Annual Technical Conference and Exhibition, Houston, TX, October 3–6, pp. 673–687.

Ucan, S., Civan, F., and Evans, R.D. 1997. Uniqueness and simultaneous predictability of relative permeability and capillary pressure by discrete and continuos means. Journal of of Canadian Petroleum Technology, 36(4), pp. 52–61.

Unice, K.M. and Logan, B.E. 2000. Insignificant role of hydrodynamic dispersion on bacterial transport. Journal of Environmental Engineering-ASCE, 126(6), pp. 491–500.

Valliappan, S., Wang, W., and Khalili, N. 1998. Contaminant transport under variable density flow in fractured porous media. International Journal for Numerical and Analytical Methods in Geomechanics, 22, pp. 575–595.

van der Marck, S.C. 1999. Evidence for a nonzero transport threshold in porous media. Water Resources Research, 35(2), pp. 595–599.

Van Driest, E.R. 1946. On dimensional analysis and the pre- sentation of data in fluid flow problems. Journal of Applied Mechanics, 13, p. A34.

Vinegar, H.J. 1986. X-Ray CT and NMR Imaging of Rocks. Technology Today Series, Journal of Petroleum Tehnology, 38(3), pp. 257–259.

Vogel, H. 1921. Das temperature-abhängigketsgesetz der viskosität von flüssigkeiten. Physikalische Zeitschrift, 22, pp. 645–646.

Voller, V. and Cross, M. 1983. An explicit numerical method to track a moving phase change front. International Journal of Heat Mass Transfer, 26, pp. 147–150.

Wa'il Abu-El-Sha'r and Abriola, L.M. 1997. Experimental assessment of gas transport mechanisms in natural porous media- parameter estimation. Water Resources Research, 33(4), pp. 505–516.

Walsh, M.P., Lake, L.W., and Schechter, R.S. 1982. A description of chemical precipitation mechanisms and their role in formation damage during stimulation by hydrofluoric acid. Journal of Petroleum Technology, 34(9), pp. 2097–2112.

Wang, C.Y. 1998. Modeling multiphase flow and transport in porous media. In: Transport Phenomena In Porous Media, D.B. Ingham and I. Pop, eds. Pergamon, Elsevier Science Lmt., Oxford, UK, pp. 383–410. 438p.

Wang, C.Y. and Beckermann, C. 1993. Single- vs. dual-scale volume averaging for heterogeneous multiphase systems. International Journal of Multiphase Flow, 19(2), pp. 397–407.

Wang, C.Y. and Cheng, P. 1996. A multiphase mixture model for multiphase, multicomponent transport in capillary porous media. 1. Model development. International Journal of Heat and Mass Transfer, 39(17), pp. 3607–3618.

Wang, S., Huang, Y., and Civan, F. 2006. Experimental and Theoretical investigation of the zaoyuan field heavy oil flow through porous media. Journal of Petroleum Science and Engineering, 50(2), pp. 83–101.

Warren, J.E. and Root, P.J. 1963. The behavior of naturally fractured reservoirs. Society of Petroleum Engineers Journal, 3(3), pp. 245–255.

Welge, H.J. 1952. A simplified method for computing oil recovery by gas or water drive. Transactions of the AIME, 195, pp. 91–98.

Whitaker, S. 1968. Introduction to Fluid Mechanics. Prentice-Hall, Englewood Cliffs, NJ.

Whitaker, S. 1969. Advances in theory of fluid motion in porous media. Industrial and Engineering Chemistry, 61, pp. 14–28.

Whitaker, S. 1986. Flow in porous media I: A theoretical derivation of Darcy's law. Transport in Porous Media, 1(1), pp. 3–25.

Whitaker, S. 1996. The forchheimer equation: A theoretical development. Transport in Porous Media, 25(1), pp. 27–61.

Whitaker, S. 1999. The Method of Volume Averaging. Kluwer Academic Publishers, Boston.

Wilke, C.R. 1950. Diffusional properties of multicomponent gases. Chemical Engineering Progress, 46(2), pp. 95–104.

Wood, A.S., Ritchie, S.I.M., and Bell, G.E. 1981. An efficient implementation of the enthalpy method. International Journal for Numerical Methods in Engineering, 17, pp. 301–305.

Wyllie, M.R.J. and Gardner, G.H.F. 1958a. The generalized Kozeny-Carman equation, Part 1—Review of existing theories. World Oil, 146, pp. 121–126.

Wyllie, M.R.J. and Gardner, G.H.F. 1958b. The generalized Kozeny-Carman equation, Part 2—A novel approach to problems of fluid flow. World Oil, 146, pp. 210–228.

Yan, J., Plancher, H., and Morrow, N.R. 1997. Wettability changes induced by adsorption of asphaltenes. Paper SPE 37232, Proceedings of the 1997 SPE International Symposium on Oilfield Chemistry, Houston, TX, February 18–21, pp. 213–227.

Yang, D. 1998. Simulation of miscible displacement in porous media by a modified Uzawa's algorithm combined with a characteristic method. Computational Methods in Applied Mechanics and Engineering, 162, pp. 359–368.

Yokoyama, Y. and Lake, L.W. 1981. The effects of capillary pressure on immiscible displacements in stratified porous media. Paper SPE 10109, Presented at the 56th Annual Fall Technical Conference and Exhibition of the Society of Petroleum Engineers of AIME, San Antonio, TX, October 5–7.

Yousif, M.H. and Sloan, E.D. 1991. Experimental investigation of hydrate formation and dissociation in consolidated porous media. SPE Reservoir Engineering, 6(4), pp. 452–458.

Yousif, M.H., Abass, H.H., Selim, M.S., and Sloan, E.D. 1991. Experimental and theoretical investigation of methane gas hydrate dissociation in porous media. SPE Reservoir Engineering, 6, pp. 69–76.

Yu, X., Lei, S., Liangtian, S., and Shilun, L. 1996. A new method for predicting the law of unsteady flow through porous medium on gas condensate well. Paper SPE 35649, presented at the SPE Program Conference, Calgary, Alberta, Canada, April 28–May 1.

Zhang, F., Zhang, R., and Kang, S. 2003. Estimating temperature effects on water flow in variably saturated soils using activation energy. Soil Science Society of America Journal, 67(5), pp. 1327–1333.

Zhou, X., Morrow, N.R., and Ma, S. 2000. Interrelationship of wettability, initial water saturation, ageing time and oil recovery by spontaneous imbibition and waterflooding. Society of Petroleum Engineers Journal, 5(2), pp. 199–207.

Zimmerman, R.W., Chen, G., Hagdu, T., and Bodvarsson, G.S. 1993. A numerical dual-porosity model with semi analytical treatment of fracture/matrix flow. Water Resources Research, 29(7), pp. 2127–2137.

Zolotukhin, A.B. 1979. Analytical definition of the overall heat transfer coefficient. SPE 7964 presented at the SPE California Regional Meeting, Ventura, April 18–20.

INDEX

Porous Media Transport Phenomena, First Edition. Faruk Civan.
© 2011 John Wiley & Sons, Inc. Published 2011 by John Wiley & Sons, Inc.